INSECT ECOLOGY

An Ecosystem Approach

INSECT ECOLOGY

An Ecosystem Approach

TIMOTHY D. SCHOWALTER

Entomology Department
Oregon State University

ACADEMIC PRESS

A Harcourt Science and Technology Company

*San Diego   San Francisco   New York
Boston   London   Sydney   Tokyo*

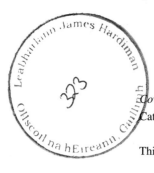

Cover photographs: Butterfly image © 1996 PhotoDisc, Inc.
Caterpillar image © 1996 Digital Stock, Inc.

This book is printed on acid-free paper. ⊗

Academic Press
A Harcourt Science and Technology Company
525 B Street, Suite 1900, San Diego, California 92101-4495, USA
http://www.apnet.com

Academic Press
24-28 Oval Road, London NW1 7DX, UK
http://www.hbuk.co.uk/ap/

Library of Congress Catalog Card Number: 99-65100

International Standard Book Number: 0-12-628975-1

PRINTED IN THE UNITED STATES OF AMERICA
99   00   01   02   03   04   QW   9   8   7   6   5   4   3   2   1

CONTENTS

SECTION I
ECOLOGY OF INDIVIDUAL INSECTS

SECTION V
SYNTHESIS

PREFACE

This book is intended for graduate students and professional entomologists and ecologists who desire a state-of-the-art insect ecology text and reference book that synthesizes feedbacks and interactions between insects and their environment. In particular, I provide background and encouragement for insect ecologists to collaborate with ecosystem ecologists in furthering our understanding of these feedbacks and interactions. The vision for this book grew from the need for such a text for my own insect ecology courses. Work on this book has benefited from feedback from my students over the years.

Insects are recognized as a dominant group of organisms on Earth. Many species conspicuously affect ecosystem structure and function through pollination and seed predation, herbivory, and decomposition, often in conflict with our own management goals. Insect ecology is a theme in all current environmental issues, including biodiversity, forest and rangeland "health," land use practices, climate change, and air and water pollution, especially by pesticides. Many species have demonstrated a remarkable capacity to thrive in human-altered landscapes, whereas other species and their ecological functions are threatened by anthropogenic changes in environmental conditions.

Available texts do not integrate the traditional emphases on insect diversity, life history adaptation, and species interactions with insect roles in ecosystems subject to environmental changes, especially the capacity of some species to function as regulators, potentially stabilizing ecosystem processes in a changing environment. Insects are capable of responding to environmental changes and of engineering further changes in ecosystem structure and function. A major challenge for ecologists and policy-makers is to integrate insect ecology with ecosystem ecology and thereby improve understanding, prediction, and resolution of the consequences of environmental change.

The lack of texts that accomplish this integration may be due, in large part, to the different approaches taken by insect ecologists and ecosystem ecologists to the study of insect ecology. Whereas insect ecologists have taken primarily an evolutionary approach to studying the diversity of insect adaptations and interactions with other organisms, ecosystem ecologists have tended to ignore diversity, often combining diverse assemblages of arthropod species into a single component in models of ecosystem dynamics. Insect ecologists generally have not been involved in multidisciplinary studies of whole ecosystems.

Nevertheless, the evolutionary and ecosystem approaches complement each other. The evolutionary approach provides insight into the selection factors and adaptive pathways that produce observed ecological behaviors and interactions, whereas the ecosystem approach contributes to an understanding of how the community of adapted organisms collectively modifies its environment. Both approaches are necessary to understand and predict the consequences of environmental changes, including anthropogenic changes, for insects and their contributions to ecosystem structure and processes such as primary productivity, biogeochemical cycling, carbon flux, and community dynamics. In fact, evolution can be viewed as feedback between ecosystem conditions and individual adaptations that modify or stabilize ecosystem parameters.

Training of graduate insect ecologists will benefit from this integrated approach. Two trends require that insect ecologists be more broadly trained in both evolutionary and ecosystem approaches. First, environmental issues are increasingly being addressed from a global (rather than local) perspective, with emphasis on integrating processes from various ecosystem types to explain global changes. Second, the changing goals of natural resource management require a shift in emphasis from insect–plant interactions and crop "protection" to integration of ecosystem components and processes that affect sustainability of ecosystem condition.

This book is intended to meet these challenges by using a hierarchical model that focuses on linkages and feedbacks among individual, population, community, and ecosystem properties. This model clarifies the effects of properties at higher levels on the environment perceived at lower levels and the effects of responses at lower levels on properties at higher levels of this hierarchy. Some overlap among sections and chapters is necessary to emphasize linkages among levels. Where possible, overlap is minimized through cross-referencing.

A useful textbook must balance coverage with brevity. My objective has been to organize and deal with topics in a way that clarifies the integration of ecological attributes across multiple levels of ecological resolution. I have particularly emphasized topics that are relevant to development of ecosystem structure and function, and have limited coverage of more evolutionary topics that have received greater emphasis in traditional insect ecology courses, in order to represent advances in ecosystem concepts adequately. My students have guided this effort.

A number of valued colleagues have contributed enormously to my perspectives on insect ecology. I am especially grateful to J. T. Callahan, S. L. Collins, R. N. Coulson, D. A. Crossley, Jr., R. Dame, D. A. Distler, L. R. Fox, J. F. Franklin, F. B. Golley, J. R. Gosz, M. D. Hunter, F. Kozar, M. D. Lowman,

G. L. Lovett, J. C. Moore, E. P. Odum, H. T. Odum, T. R. Seastedt, P. Turchin, R. B. Waide, W. G. Whitford, R. G. Wiegert, and M. R. Willig for sharing ideas and data. I also have benefited from collaboration with colleagues at Oregon State University or associated with Long Term Ecological Research (LTER) projects, International LTER projects in Hungary and Taiwan, the Smithsonian Tropical Research Institute, Wind River Canopy Crane Research Facility, Teakettle Experimental Forest, USDA Forest Service Demonstration of Ecosystem Management Options (DEMO) Project, USDA Western Regional Project on Bark Beetle–Pathogen Interactions, and the National Science Foundation. I particularly thank Drs. Laurel Fox, Tim Seastedt, and Mike Willig for generously reviewing drafts of this text. Furthermore, no text reaches completion without considerable guidance from and assistance by family, editor, and publisher. I especially thank C. Schowalter, C. Crumly, J. Dinsmore, and D. James for encouragement and editorial contributions to this book. I am, of course, solely responsible for the selection and organization of material herein.

Timothy D. Schowalter

Overview

INSECTS ARE THE DOMINANT GROUP OF ORGANISMS ON EARTH, IN terms of both taxonomic diversity (>50% of all described species) and ecological function (Wilson, 1992) (Fig. 1.1). They represent the vast majority of species in terrestrial and freshwater ecosystems and are important components of near-shore marine ecosystems, as well. This diversity of insect species represents an equivalent variety of adaptations to variable environmental conditions. Insects affect other species (including humans) and ecosystem parameters in a variety of ways. The capacity for rapid response to environmental change makes insects useful indicators of change, major engineers and potential regulators of ecosystem conditions, and frequent competitors with human demands for ecosystem resources.

Insects represent important ecological functions (e.g., as herbivores, pollinators, and detritivores). They are food resources for many other organisms, including humans, and they have the capacity to alter rates and directions of energy and matter fluxes in ways that potentially affect global processes. In some ecosystems, insects and other arthropods represent the dominant pathways of energy and matter flow, and their biomass may exceed that of the more conspicuous vertebrates (e.g., Whitford, 1986). Some species are capable of removing virtually all vegetation from a site. They affect, and are affected by, environmental issues as diverse as ecosystem health, air and water quality, frequency and severity of fire and other disturbances, control of exotic species, land use, and climate change. Environmental changes, especially those result-

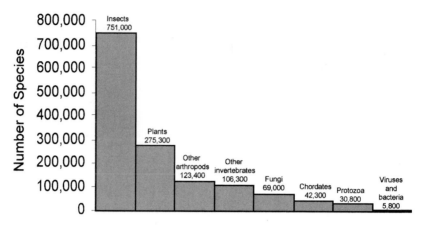

FIG. 1.1   Distribution of described species within major taxonomic groups. Species numbers for insects, bacteria, and fungi likely will increase greatly as these groups become better known. Data from Wilson (1992).

ing from anthropogenic activities, affect abundances of many species in ways that alter ecosystem and, perhaps, global processes.

A primary challenge for insect ecologists is to place insect ecology in an ecosystem context that recognizes the important effects of insects on ecosystem properties, as well as the diversity of their adaptations and responses to environmental conditions. Until relatively recently, insect ecologists have focused on the evolutionary significance of insect life histories and interactions with other species, especially as pollinators, herbivores, and predators (Price, 1997). This focus has yielded much valuable information about the ecology of individual species and species associations and provides the basis for pest management or recovery of threatened and endangered species. However, relatively little attention has been given to the important role of insects as ecosystem engineers, other than to their effects on vegetation (especially commercial crop) dynamics.

Ecosystem ecology has advanced rapidly during the past 50 years. Major strides have been made in understanding how species interactions and environmental conditions affect rates of energy and nutrient fluxes in different types of ecosystems, how these provide free air and water filtering services, and how environmental conditions both affect and reflect community structure. Interpreting the responses of a diverse community to multiple interacting environmental factors in integrated ecosystems requires new approaches, such as multivariate statistical analysis and modeling approaches (e.g., Gutierrez, 1996, Liebhold *et al.*, 1993). Such approaches may involve loss of detail, such as combination of species into phylogenetic or functional groupings. However, an ecosystem approach provides a framework for integrating insect ecology with the changing patterns of ecosystem structure and function and for applying insect ecology to understanding of ecosystem, landscape, and global issues, such as climate change or sustainability of ecosystem resources. Unfortunately, few ecosystem studies have involved insect ecologists and, therefore, have tended to underrepresent insect responses and contributions to ecosystem changes.

I. SCOPE OF INSECT ECOLOGY

Insect ecology is the study of interactions between insects and their environment. Such study must recognize that the environment affects organisms, populations, and communities but also is affected by their activities through a variety of feedback loops (Fig. 1.2). Insect ecology has both basic and applied goals. Basic goals are to understand and model these interactions and feedbacks (e.g., Price, 1997). Applied goals are to evaluate the extent to which insect responses to environmental changes, including those resulting from anthropogenic activities, mitigate or exacerbate ecosystem change (e.g., Croft and Gutierrez, 1991; Kogan, 1998).

Research on insects and associated arthropods (e.g., spiders, mites, centipedes, millipedes, crustaceans) has been critical to development of the fundamental principles of ecology, such as evolution of social organization (Haldane, 1932; Hamilton, 1964; Wilson, 1973), population dynamics (Coulson, 1979; Morris, 1969; Nicholson, 1958; Varley and Gradwell, 1970; Varley *et al.*, 1973; Wellington *et al.*, 1975), competition (Park, 1948, 1954), predator–prey interaction (Nicholson and Bailey, 1935), mutualism (Batra, 1966; Bronstein,

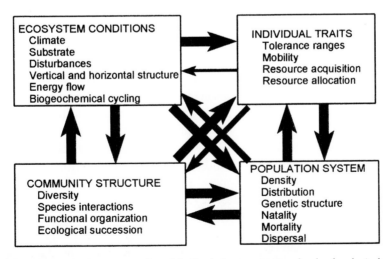

FIG. 1.2 Diagrammatic representation of feedbacks between various levels of ecological organization. Size of arrows is proportional to strength of interaction. Note that individual traits have a declining direct effect on higher organizational levels but are affected strongly by feedback from all higher levels.

1998; Janzen, 1966; Morgan, 1968; Rickson, 1971, 1977), island biogeography (Darlington, 1943; MacArthur and Wilson, 1967; Simberloff, 1969, 1978), metapopulation ecology (Hanski, 1989), and regulation of ecosystem processes, such as primary productivity, nutrient cycling, and succession (Mattson and Addy, 1975; Moore *et al.*, 1988; Schowalter, 1981; Seastedt, 1984). Insects and other arthropods are small and easily manipulated subjects. Their rapid numerical responses to environmental changes facilitate statistical discrimination of responses and make them particularly useful models for experimental study. Insects also have been recognized for their capacity to engineer ecosystem change, making them ecologically and economically important.

Insects fill a variety of important ecological (functional) roles. Many species are key pollinators. Pollinators and plants have adapted a variety of mechanisms for ensuring transfer of pollen, especially in tropical ecosystems where sparse distributions of many plant species require a high degree of pollinator fidelity to ensure pollination among conspecific plants (Feinsinger, 1983). Other species are important agents for dispersal of plant seeds, fungal spores, viruses, or other invertebrates (Moser, 1985; Nault and Ammar, 1989). Herbivorous species are particularly well known as agricultural and forestry "pests," but their ecological roles are far more complex, often stimulating plant growth, affecting nutrient fluxes, or altering the rate and direction of ecological succession (MacMahon, 1981; Maschinski and Whitham, 1989; Schowalter *et al.*, 1986; Schowalter and Lowman, 1999). Insects and associated arthropods are instrumental in processing of organic detritus in terrestrial and aquatic ecosystems and influence soil fertility and water quality (Kitchell *et al.*, 1979; Seastedt and Crossley, 1984). Woody litter decomposition typically is delayed until insects penetrate the bark barrier and inoculate the wood with saprophytic fungi and other microorganisms (Ausmus, 1977; Dowding, 1984; Swift, 1977). In addition, insects are important resources for a variety of fish, amphibians, reptiles, birds, and mammals, as well as other invertebrate predators and parasites.

The significant economic and medical/veterinary importance of many insect species is the reason for distinct entomology programs in various universities and government agencies. Damage to agricultural crops and transmission of human and livestock diseases have stimulated interest in, and support for, the study of factors influencing abundance and effects of these insect species. Much of this research has focused on evolution of life history strategies, interaction with host plant chemistry, and predator–prey interactions as these contribute to our understanding of "pest" population dynamics, especially population regulation by biotic and abiotic factors. However, failure to understand these aspects of insect ecology within an ecosystem context undermines our ability to predict and manage insect populations and ecosystem resources effectively (Kogan, 1998). Suppression efforts may be counterproductive to the extent that insect outbreaks represent ecosystem-level regulation of critical processes.

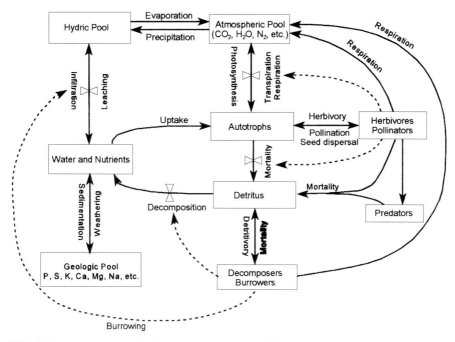

FIG. 1.3 Conceptual model of ecosystem structure and function. Boxes represent storage compartments, lines represent fluxes, and hourglasses represent regulation. Solid lines are direct transfers of energy and matter, and dashed lines are informational or regulatory pathways.

II. ECOSYSTEM ECOLOGY

The ecosystem is a fundamental unit of ecological organization, although its boundaries are not easily defined. An ecosystem generally is considered to represent the integration of a more or less discrete community of organisms and the abiotic conditions at a site (Fig. 1.3). However, research and environmental policy decisions are recognizing the importance of scale in ecosystem studies, i.e., extending research or extrapolating results to landscape, regional, and even global scales (e.g., Holling, 1992; Turner, 1989). Ecosystems are interconnected, just as the species within them are interconnected. Exports from one ecosystem become imports for others. Energy, water, organic matter, and nutrients from terrestrial ecosystems are major sources of these resources for many aquatic ecosystems. Organic matter and nutrients eroded by wind from arid ecosystems are filtered from the airstream by ecosystems downwind. Some ecosystems within a landscape or watershed are the sources of colonists for recently disturbed ecosystems. Therefore, our perspective of the ecosystem needs to incorporate the concept of interactions among ecosystem types (patches) within the landscape or watershed.

Overlapping gradients in abiotic conditions establish the template that limits options for community development, but established communities can modify abiotic conditions to varying degrees. For example, minimum rates of water and nutrient supply are necessary for establishment of grasslands or forests, but once canopy cover and water and nutrient storage capacity in organic material have developed, the ecosystem is relatively buffered from changes in water and nutrient supply (e.g., Odum, 1969). Although ecosystems typically are defined on the basis of the dominant vegetation (e.g., tundra, desert, marsh, grassland, forest) or type of water body (stream, pond, lake), characteristic insect assemblages also differ among ecosystems. For example, wood-boring insects (e.g., ambrosia beetles, wood wasps) are characteristic of communities in wooded ecosystems (shrub and forest ecosystems) but clearly could not survive in ecosystems lacking woody resources.

A. Ecosystem Complexity

Ecosystems are complex systems with **structure**, represented by abiotic resources and a diverse assemblage of component species and their products (such as organic detritus and tunnels), and **function**, represented by fluxes of energy and matter among biotic and abiotic components (Fig. 1.3). This complexity extends to the spatial delineation of an ecosystem. Ecosystems can be identified at micro- and mesocosm scales (e.g., decomposing logs or treehole pools), patch scale (area encompassing a particular community on the landscape), landscape scale (the mosaic of patch types representing different edaphic conditions or successional stages that compose a broader ecosystem type), and regional or biome scale.

Addressing taxonomic, temporal, and spatial complexity has proven a daunting challenge to ecologists, who must decide how much complexity can be ignored safely (Gutierrez, 1996; Polis, 1991a,b). Evolutionary and ecosystem ecologists have taken contrasting approaches to dealing with this complexity in ecological studies. The evolutionary approach emphasizes adaptive aspects of life histories, population dynamics, and species interactions. This approach restricts complexity to interactions among one or a few species and their hosts, competitors, predators, or other biotic and abiotic environmental factors and often ignores the complex feedbacks at the ecosystem level. On the other hand, the ecosystem approach emphasizes rates and directions of energy and matter fluxes. This approach restricts complexity to fluxes among functional groups and often ignores the contributions of individual species. Either approach, by itself, limits our ability to understand feedbacks among individual, population, community, and ecosystem parameters and to predict effects of a changing global environment on these feedbacks.

B. The Hierarchy of Subsystems

Complex systems with feedback mechanisms can be partitioned into component subsystems, which are themselves composed of sub-subsystems. Viewing

TABLE 1.1 Ecological Hierarchy and the Structural and Functional Properties Characterizing Each Level

Ecological level	Structure	Function

Global	Biome distribution
	Atmospheric condition
	Climate
	Sea level
		Gas, water, nutrient exchange
		  between terrestrial and marine
		  systems
		Total NPP

Biome	Landscape pattern Temperature, moisture profile Disturbance regime	Energy and matter fluxes Integrated NPP of ecosystems Migration
Landscape	Disturbance pattern Community distribution Metapopulation structure	Energy and matter fluxes Integrated NPP of ecosystems Colonization and extinction
Ecosystem	Vertical and horizontal structure Disturbance type and frequency Biomass Functional organization	Energy and matter fluxes Succession NPP, herbivory, decomposition, pedogenesis
Community	Diversity Trophic organization	Species interactions Temporal and spatial changes
Population	Density Dispersion Age structure Genetic structure	Natality Mortality Dispersal Gene flow Temporal and spatial changes
Individual	Morphology Physiology Behavior	Resource acquisition Resource allocation Learning

the ecosystem as a nested hierarchy of subsystems (Table 1.1), each with its particular properties and processes (Coulson and Crossley, 1987; Kogan, 1998; O'Neill *et al.*, 1986), facilitates understanding of complexity. Each level of the hierarchy can be studied at an appropriate level of detail and its properties explained by the integration of its subsystems. For example, population responses to changing environmental conditions reflect the net physiological and behavioral responses of individuals that determine their survival and reproduction. Changes in community structure reflect the dynamics of component populations. Fluxes of energy and matter through the ecosystem reflect community organization. Landscape structure reflects ecosystem processes that affect movement of individuals. Hence, the integration of structure and function at each level determines properties at higher levels.

At the same time, the conditions produced at each level establish the context, or template, for responses at lower levels. Population structure resulting from individual survival, dispersal, and reproduction determines future survival, dispersal, and reproduction of individuals. Ecosystem conditions resulting from community interactions affect subsequent behavior of individual organisms, populations, and the community. Recognition of feedbacks from higher levels has led to developing concepts of inclusive fitness (fitness accruing through feedback from benefit to a group of organisms) and ecosystem self-regulation (see Chapter 15). The hypothesis that insects function as cybernetic regulators that stabilize ecosystem properties has been one of the most important and controversial concepts to emerge from insect ecology.

Ecosystem processes represent the integration of processes at the level of component communities. Component communities are subsystems, i.e., more or less discrete assemblages of organisms based on particular resources. For example, the relatively distinct assemblages of herbivores and associated predators and parasites on different plant species (e.g., Schowalter 1994a) constitute component communities within the broader general community of a site. Similarly, the relatively distinct soil faunas associated with fungal, bacterial, or plant root resources represent different component communities (Moore and Hunt, 1988). Component communities are composed of individual species populations with varying strategies for acquiring and allocating resources. Species populations, in turn, are composed of individual organisms with variation in individual physiology and behavior. Ecosystems can be integrated at the landscape or biome levels and biomes integrated at the global (biosphere) level.

Spatial and temporal scales vary across this hierarchy. Whereas individual physiology and behavior operate on small scales of space and time (i.e., limited to the home range and life span of the individual), population dynamics span landscape and decadal scales, and ecosystem processes, such as patterns of resource turnover, recovery from disturbance, or contributions to atmospheric carbon, operate at scales from the patch to the biome and from decades to millennia.

Modeling approaches have greatly facilitated understanding of the complexity and consequences of interactions and linkages within and among these organizational levels of ecosystems. The most significant challenges to ecosystem modelers remain (a) the integration of appropriately detailed submodels at each level to improve prediction of causes and consequences of environmental changes and (b) the evaluation of contributions of various taxa (including particular insects) or functional groups to ecosystem structure and function. In particular, certain species or structures have effects disproportionate to their abundance or biomass. Studies focused on the most abundant or conspicuous species or structures fail to address the substantial contributions of rare or inconspicuous components, such as insects.

C. Regulation

An important aspect of this functional hierarchy is the "emergence" of properties that are not easily predictable by simply adding the contributions of constitutive components. Emergent properties include feedback processes at each level of the hierarchy. For example, individual organisms acquire and allocate energy and biochemical resources, affecting resource availability and population structure in ways that change the environment and determine future options for acquisition and allocation of these resources. Regulation of density and resource use emerges at the population level through negative feedback from resource availability that functions to prevent overexploitation and/or through positive feedback that prevents extinction. Similarly, species populations acquire and transport resources, but regulation of energy flow and biogeochemical cycling emerge at the ecosystem level. Potential regulation of atmospheric and oceanic pools of carbon and nutrients at the global level reflects integration of biogeochemical cycling and energy fluxes among the Earth's ecosystems.

Information flow and feedback processes are the mechanisms of regulation. Although much research has addressed energy and material flow through food webs, relatively little research has quantified the importance of indirect interactions or information flow. Indirect interactions and feedbacks are common features of ecosystems. For example, herbivores feeding above ground alter the availability of resources for root-feeding organisms (Gehring and Whitham, 1991, 1995; Masters *et al.*, 1993); early-season herbivory can affect plant suitability for later-season herbivores (Harrison and Karban, 1986; Hunter, 1987). Information can be transmitted as volatile compounds that advertise the location and physiological condition of prey, the proximity of potential mates, and the population status of predators. Such information exchange is critical to discovery of suitable hosts, attraction of mates, regulation of population density, and defense against predators by many (if not all) insects.

This ecosystem information network among the members of the community, along with resource supply/demand relationships, provides the basis for regulation of ecosystem processes. Levels of herbivory and predation are sensitive to resource availability. If environmental conditions increase resource abundance at any trophic level, communication to, and response by, the next trophic level provides negative feedback that reduces resource abundance. Negative feedback is a primary mechanism for stabilizing population sizes, species interactions, and process rates in ecosystems. Some interactions provide positive feedback, such as cooperation or mutualism. Although positive feedback is potentially destabilizing, it may reduce the probability of population decline to extinction. The apparent ability of many ecosystems to reduce variation in structure and function suggests that ecosystems are self-regulating; i.e., they behave like cybernetic systems (e.g., Odum, 1969; Patten and Odum, 1981). Insects could be viewed as important mechanisms of regulation because their normally small biomass requires relatively little energy or matter to maintain, and their rapid and dramatic population response to environmental changes constitutes an effective and efficient means for reducing deviation in nominal ecosystem structure and function (Schowalter, 1994b, 2000). This developing concept of ecosystem self-regulation has major implications for ecosystem responses to anthropogenic change in environmental conditions and for our approaches to managing insects and ecosystem resources.

III. ENVIRONMENTAL CHANGE AND DISTURBANCE

Environmental changes across temporal and spatial gradients are critical components of an ecosystem approach to insect ecology. Insects are highly responsive to environmental changes, including those resulting from anthropogenic activity (Schowalter, 1985). Many insects have considerable capacity for long-distance dispersal, enabling them to find and colonize isolated resources. Other insects are flightless and vulnerable to environmental change or habitat fragmentation. Because of their small size, short life spans, and high reproductive rates, abundances of many species can change several orders of magnitude on a seasonal or annual time scale, minimizing time lags between environmental changes and population adjustment to new conditions. Such changes are easily detectable and make insects more useful as indicators of environmental changes than are larger or longer-lived organisms. In turn, insect responses to environmental change can affect ecosystem patterns and processes dramatically (Schowalter, 1985). Some phytophagous species are well known for their ability, at high population levels, to reduce host plant density and productivity greatly over large areas. Effects of other species may be more subtle but equally significant from the standpoint of ecosystem structure and function.

Environmental change operates on a continuum of spatial and temporal scales. Although strict definitions of environmental change and disturbance have proven problematic, generally environmental change occurs over a longer term, whereas disturbances are short-term events (White and Pickett, 1985). Chronic changes in temperature or precipitation patterns, such as following the last glaciation, may occur on a scale of 10^3–10^5 years and be barely detectable on human time scales. Long-term changes may be difficult to distinguish from cycles operating over decades or centuries, leading to disagreements over whether measured changes represent a fluctuation or a long-term trend. Acute events, such as fires or storms, are more recognizable as disturbances that have dramatic effects on time scales of seconds to hours. However, the duration at which a severe drought, for example, is considered a climate change rather than a disturbance has not been determined. The combination of climate and geological patterns, disturbances, and environmental changes creates a constantly shifting landscape mosaic of various habitat and resource patches that determine where and how insects and other organisms find suitable conditions.

Insects affect ecosystem structure and function in many ways comparable to physical disturbances and often have been viewed as agents of disturbance (Schowalter and Lowman, 1999). White and Pickett (1985) proposed that disturbance be defined as any relatively discrete event in time that causes measurable change in ecosystem structure or function. This definition clearly incorporates insect outbreaks. Similarly, human activities have become increasingly prominent agents of disturbance and environmental change.

Insect outbreaks are comparable to physical disturbances in terms of severity, frequency and scale. Insects can defoliate or kill all host plants over large areas, up to 10^3–10^6 ha (e.g., Furniss and Carolin, 1977). For example, 39% of a montane forest landscape in Colorado has been affected by insect outbreaks (spruce beetle, *Dendroctonus rufipennis*) since about 1633, compared to 59% by fire and 9% by snow avalanches (Veblen *et al.*, 1994), with an average return interval of 117 years, compared to 202 years for fire. Frequent, especially cyclic, outbreaks of herbivorous insects probably have been important in selection for plant defenses.

However, unlike abiotic disturbances, insect outbreaks are biotic responses to a change in environmental conditions. Outbreaks most commonly reflect anthropogenic redistribution of resources, such as increased density of commercially valuable (often exotic) plant species. Outbreaks typically develop in dense patches of host plants and function to reduce host density, increase vegetation diversity, and increase water and nutrient availability (Schowalter *et al.*, 1986). Therefore, insect outbreaks do not cause random mortality, as do physical disturbances, but are a specific response to changes in ecosystem conditions. Consideration of insects as integral components of potentially self-maintaining ecosystems could improve our management of insects and ecosystem resources, within the context of global change.

Currently, human alteration of Earth's ecosystems is substantial and accelerating (Vitousek *et al.*, 1997). Anthropogenic changes to the global environment affect insects in various ways. Combustion of fossil fuels has elevated atmospheric concentrations of CO_2 (Keeling *et al.*, 1995), methane, nitrous oxides, and sulfur dioxide, leading to increasingly acidic precipitation and prospects of global warming. Some insect species show high mortality as a direct result of atmospheric toxins, whereas other species are affected indirectly by changes in resource conditions induced by atmospheric change (Alstad *et al.*, 1982; Arnone *et al.*, 1995; Heliövaara, 1986; Kinney *et al.*, 1997; Lincoln *et al.*, 1993; Smith, 1981). A thinning ozone layer at higher altitudes and toxic ozone levels at lower altitudes have similar effects (Alstad *et al.*, 1982). However, the anthropogenic changes with the most immediate effects are land use patterns and redistribution of exotic species, including plants, insects, and livestock. These activities are altering and isolating natural communities at an unprecedented rate, leading to outbreaks of insect "pests" in crop monocultures and fragmented ecosystems (Roland, 1993) and potentially threatening species incapable of surviving in increasingly inhospitable landscapes (Samways *et al.*, 1996; Shure and Phillips, 1991). Predicting and mitigating pest outbreaks or species losses depends strongly on our understanding of insect ecology within the context of ecosystem structure and function.

IV. ECOSYSTEM APPROACH TO INSECT ECOLOGY

Insect ecology can be approached using a hierarchical model (Coulson and Crossley, 1987). Ecosystem conditions represent the environment, i.e., the combination of physical conditions, interacting species, and availability of resources, that determine the survival and reproduction by individual insects, but, in turn, insect activities alter vegetation cover, soil properties, community organization, etc. (Fig. 1.2). A hierarchical approach offers a means of integrating evolutionary and ecosystem approaches to studying insect ecology. The evolutionary approach focuses at lower levels of resolution (individual, population, community) and offers explanation (i.e., natural selection) for why individuals and populations act and interact as they do. Such explanation provides context for understanding adaptations and their consequences at the community and ecosystem levels. At the same time, natural selection represents feedback from ecosystem conditions as altered by coevolving organisms. The evolutionary and ecosystem perspectives are most complementary at the community level, where species diversity emphasized by the evolutionary approach is the basis for functional diversity emphasized by the ecosystem approach.

Although the evolutionary approach has provided valuable explanations for how complex interactions have arisen, current environmental issues require an understanding of how insect functional roles affect ecosystem, landscape, and global processes. Insect ecologists have recognized insects as important components of natural communities but have only begun to explore the roles insects play as integral components of ecosystems. Insects affect primary productivity and organic matter turnover in ways that potentially regulate ecological succession, biogeochemical cycling, carbon and energy flux, albedo, and hydrology, perhaps affecting regional and global climate as well (Schowalter and Lowman, 1999). These roles may complement or exacerbate changes associated with human activities. Therefore, the purpose of this book is to provide a text that addresses the fundamental issues of insect ecology as they relate to ecosystem, landscape, and global processes.

V. SCOPE OF THIS BOOK

This book is organized hierarchically, to emphasize feedbacks among individual, population, and community levels and the ecosystems they represent. Three questions have been used to develop this text:

1. How do insects respond to variation in environmental conditions, especially gradients in abiotic factors and resource availability?
2. How do interactions among individuals affect the structure and function of populations and communities?
3. How do insect-induced changes in ecosystem properties affect the gradients in environmental conditions to which individuals respond?

Chapter and topic organization are intended to address these questions by emphasizing key spatial and temporal patterns and processes at each level and their integration among levels. The evaluation of insect functional roles in ecosystems and their responses to global change (Section IV) depends on understanding of species diversity, interactions, and community organization (Section III) that, in turn, depends on understanding of population dynamics and biogeography (Section II), that depends on understanding of individual physiological and behavioral responses to environmental variation (Section I).

Three themes integrate these ecological levels. First, spatial and temporal patterns in environmental variability and disturbance determine survival and reproduction of individuals and patterns of population, community, and ecosystem structure and dynamics. Individual acquisition and allocation of resources, population distribution and colonization and extinction rates, community patterns and successional processes, and ecosystem structure and function depend on environmental conditions. Second, energy and nutrients move through individuals, populations, and communities. The net foraging success and resource use by individuals determine energy and nutrient fluxes at the population level. Trophic interactions among populations determine energy and nutrient fluxes at the community and ecosystem levels. Third, regulatory mechanisms at each level serve to balance resource demands with resource availability (carrying capacity) or to dampen responses to other environmental changes. Regulation results from a balance between negative feedback that reduces population size and positive feedback that increases population size. Regulation of population sizes throughout the community tends to stabilize ecosystem conditions within ranges favorable to most members. The capacity to regulate environmental conditions increases from individual to ecosystem levels. If feedbacks among levels contribute to ecosystem stability, then human influences on ecosystem structure and function could seriously impair this function.

Section I (Chapters 2–4) addresses the physiological and behavioral ecology of insects. Physiology and behavior represent the means by which organisms interact with their environment. Physiology represents "fixed" adaptations to predictable variation in environmental conditions, whereas behavior represents a more flexible means of adjusting to unpredictable variation. Chapter 2 summarizes insect responses to habitat conditions, especially gradients in climate, soil, and chemical conditions. Chapter 3 describes physiological and behavioral mechanisms for acquiring energy and matter resources, and Chapter 4 addresses the allocation of assimilated resources to various metabolic and behavioral pathways. These chapters provide a basis for understanding distribution patterns and movement of energy and matter through populations and communities.

Section II (Chapters 5–7) deals with population ecology. Populations of organisms integrate variation in adaptive strategies and patterns of movement among individuals. This level of organization determines the net effect of organisms of a particular species on ecosystem structure and function. Chapter 5 outlines population systems, including population structure and the processes of reproduction, mortality, and dispersal. Chapter 6 addresses processes and models of population change; Chapter 7 describes biogeography, processes and models of colonization and extinction, and metapopulation dynamics over landscapes. These population parameters determine the role of particular populations in ecological processes through time in various patches across regional landscapes.

Section III (Chapters 8–10) addresses community ecology. Species populations interact with other species in a variety of ways that determine changes in community structure through time and space. Chapter 8 describes species interactions (e.g., competition, predation, symbioses). Chapter 9 addresses measures of diversity and community structure and spatial patterns in community structure. Chapter 10 addresses changes in community structure over time, especially community responses to environmental change. These community characteristics determine spatial and temporal patterns of energy and nutrient storage and flux through ecosystems.

Section IV (Chapters 11–15) focuses on ecosystems and is the major contribution of this text to graduate education in insect ecology. Chapter 11 addresses general aspects of ecosystem structure and function, especially processes of energy and matter storage and flux that determine resource availability. Chapter 12 describes patterns of herbivory and its effects on ecosystem parameters, Chapter 13 describes patterns and effects of pollination and seed predation and dispersal, and Chapter 14 describes patterns and effects of detritivory and burrowing on ecosystem processes. Chapter 15 addresses the developing concept of ecosystem self-regulation and mechanisms, including species diversity and insects, that may contribute to ecosystem stability.

The final chapter (Chapter 16) is a summary and synthesis of major concepts. This chapter also suggests future directions and data necessary to improve understanding of linkages and feedbacks among hierarchical levels in order to solve environmental problems involving aspects of insect ecology at ecosystem, landscape, and global levels. Although the focus of this book clearly is on insects, examples from studies of other organisms are used where appropriate to illustrate concepts.

I

ECOLOGY OF INDIVIDUAL INSECTS

THE INDIVIDUAL ORGANISM IS A FUNDAMENTAL unit of ecology. Organisms interact with their environment and affect ecosystem processes largely through their physio- logical and behavioral responses to environmental variation. Individual success in finding and using necessary habitats and resources determines fitness. Insects have a number of general attributes that have contributed to their ecological success (Romoser and Stoffolano, 1998).

First, small size (an attribute shared with other invertebrates and microorganisms) has permitted exploitation of habitat and food resources at a microscopic scale. Insects can take shelter from adverse conditions in microsites too small for larger organisms, e.g., within the boundary layer of individual leaves. Large numbers of insects can ex- ploit the resources represented by a single leaf, often by partitioning leaf resources. Some species feed on cell contents, others on sap in leaf veins, some on top of the leaf, others on the underside, some internally. At the same time, small size makes in- sects sensitive to changes in temperature and moisture.

Second, the exoskeleton (shared with other arthropods) provides protection against desiccation and predation (necessary for small terrestrial organisms) and in- numerable points of muscle attachment (for flexibility). However, the exoskeleton also limits the size arthropods can attain. The weight of exoskeleton required for a larger body would limit mobility. Larger arthropods occurred prehistorically, before the appearance of more flexible vertebrate predators imposed a risk to slower movement. Larger arthropods occur in aquatic environments, where water helps support their weight.

Third, metamorphosis is a necessary means of growth mandated by the exoskeleton but permits partitioning of habitats and resources among life stages. Immature and adult insects can live in different habitats and feed on different resources, thereby reducing competition. For example, dragonflies and mayflies live in aquatic ecosystems as immatures but in terrestrial ecosystems as adults. Many Lepidoptera feed on foliage as immatures and on nectar as adults. Among the Endopterygota, the quiescent pupal stage facilitates survival during unfavorable environmental conditions. However, insects, as well as other arthropods, are particularly vulnerable to desiccation and predation during ecdysis.

Finally, flight evolved first among insects and conferred a distinct advantage over other organisms. Flight permits rapid long-distance movement that facilitates discovery of new resources, as well as escape from predators or unfavorable conditions. Flight remains a dominant feature of insect ecology.

This section focuses on aspects of physiology and behavior that affect insect interactions with environmental conditions, specifically, their mechanisms for finding, exploiting, and allocating resources. Physiology and behavior are closely integrated. For example, movement, including dispersal, is affected by chemical perception of the environment, fat storage, mechanisms for rapid oxygen supply, etc. Similarly, physiological processes may be affected by insect selection of thermally suitable location, choice of food resources, and other factors. Chemical defenses against predators are based on physiological processes but often are enhanced by behaviors that facilitate expression of chemical defenses. The ways in which organisms affect ecosystem processes, such as energy and nutrient fluxes, are determined by the spatial and temporal patterns of energy and nutrient acquisition and allocation.

Chapter 2 deals with physiological and behavioral responses to habitat conditions and potential responses to changing environmental conditions. Chapter 3 addresses physiological and behavioral mechanisms for finding and exploiting resources. Chapter 4 describes allocation of resources to various metabolic pathways and behaviors that facilitate resource acquisition, such as finding mates, reproducing, and interacting with other organisms. Physiology and behavior interact to determine the conditions under which insects can survive and the means by which they acquire and use available resources.

These ecological attributes affect population ecology (such as biogeography, population responses to environmental change and disturbances, and population genetics; Section II), community attributes (such as use of, or use by, other organisms as resources; Section III), and ecosystem attributes (such as rates and directions of energy and matter flows; Section IV).

Responses to Abiotic Conditions

INSECTS ARE A DOMINANT GROUP OF ORGANISMS IN VIRTUALLY ALL terrestrial, freshwater, and near-coastal marine habitats, including many of the harshest ecosystems on the globe (e.g., deserts, hot springs, and tundra). However, particular species have restricted ranges of occurrence dictated by their tolerance to a variety of environmental factors.

One of the earliest (and still important) objectives of ecologists was explanation of the spatial patterns of species distributions (e.g., Andrewartha and Birch, 1954; Wallace, 1876). The geographical ranges of insect species generally are determined by their tolerances, or the tolerances of their food resources and predators, to variation in abiotic conditions. Insect morphological, physiological, and behavioral adaptations reflect the characteristic physical conditions of the habitats in which they occur. However, variation in physical conditions requires some flexibility in physiological and behavioral traits. All ecosystems experience climatic fluctuation and periodic disturbances that affect the survival of organisms in the community. Furthermore, anthropogenic changes in habitat conditions increase the range of conditions to which organisms must respond.

I. THE PHYSICAL TEMPLATE

A. Biomes

Global patterns of temperature and precipitation, reflecting the interaction between latitude, global atmospheric and oceanic circulation patterns, and topography, establish a regional template of physical conditions that support characteristic communities, biomes (Fig. 2.1) (Finch and Trewartha, 1949). Latitudinal gradients in temperature from Earth's equator to its poles define the tropical, subtropical, temperate, and arctic zones. Precipitation patterns overlay these temperature gradients. Warm, humid air rises in the tropics, drawing air from higher latitudes into this equatorial convergence zone. The rising air cools and condenses moisture, resulting in a band of high precipitation and tropical rainforests centered on the equator. The cooled, dried air flows away from the equatorial zone and warms as it descends in the "horse latitudes," centered around 30° N and S. These latitudes are dominated by arid grassland and desert ecosystems because of high evaporation rates in warm, dry air. Airflow at these latitudes diverges to the equatorial convergence zone and to similar convergence zones at about 60° N and S latitudes. Rising air at 60° N and S latitudes creates bands of relatively high precipitation and low temperature that support boreal forests. These latitudinal gradients in climate restrict the distribution of organisms on the basis of their tolerance ranges for temperature and moisture. No individual species is capable of tolerating the entire range of tropical to arctic temperatures or desert to mesic moisture contents.

Mountain ranges interact with oceanic and atmospheric circulation patterns to modify latitudinal patterns of temperature and precipitation. Mountains force airflow upward, causing cooling, condensation, and precipitation on the windward side. Desiccated air descends on the leeward side where it gains moisture through evaporation. This orographic effect leads to development of mesic environments on the windward side and arid environments on the leeward side of mountain ranges. Mountains are characterized by elevational gradients of temperature, moisture, and atmospheric conditions, i.e., lower elevations tend to be warmer and drier whereas higher elevations are cooler and moister. Concentrations of oxygen and other gases decline with elevation, so that species occurring at higher elevations must be capable of surviving at low oxygen concentrations. The montane gradient is much shorter than the corresponding latitudinal gradient, with the same temperature change occurring in a 1000-m difference in elevation or an 880-km difference in latitude. Hence, the range of habitat conditions that occur over a wide latitudinal gradient occur on a smaller scale in montane areas.

The relatively distinct combinations of temperature and precipitation (MacMahon, 1981) determine the assemblage of species capable of surviving

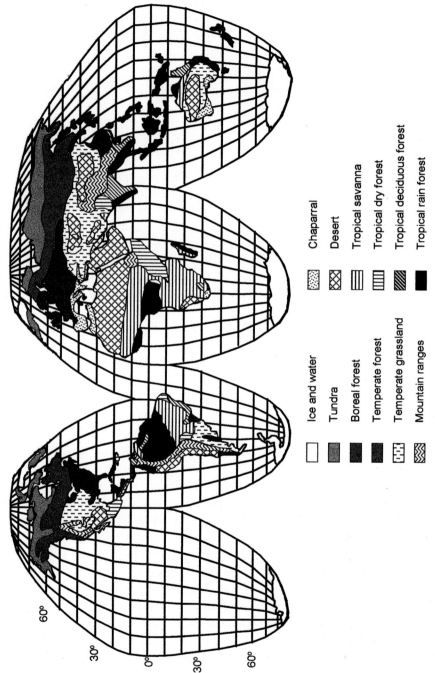

FIG. 2.1 Global distribution of the major terrestrial biomes. The distribution of biomes is affected by latitude, global atmospheric and oceanic circulation patterns, and major mountain ranges. Modified from Finch and Trewartha (1949) *Elements of Geography: Physical and Cultural*, by permission from The McGraw-Hill Companies and adapted from *Fundamentals of Ecology, Third Edition* by Eugene P. Odum, copyright © 1971 by Saunders College Publishing, by permission of the publisher.

Ice and water

Tundra

Boreal forest

Temperate forest

Temperate grassland

Mountain ranges

Chaparral

Desert

Tropical savanna

Tropical dry forest

Tropical deciduous forest

Tropical rain forest

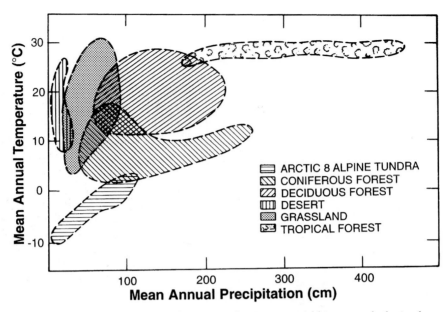

FIG. 2.2 Discrimination of geographic ranges of major terrestrial biomes on the basis of temperature and precipitation. From McMahon (1981) by permission from Springer-Verlag.

and defining the characteristic vegetation type, i.e., tundra, temperate deciduous forest, temperate coniferous forest, tropical rain forest, tropical dry forest, grassland, savanna, chaparral, and desert (Fig. 2.2). Representative terrestrial biomes and their seasonal patterns of temperature and precipitation are shown in Figs. 2.3 and 2.4.

Habitat conditions in terrestrial biomes are influenced further by topographic relief, substrate structure and chemistry, and exposure to wind. For example, topographic relief creates gradients in solar exposure and soil drainage, as well as in temperature and moisture, providing local habitats for unique communities. Local differences in substrate structure and chemistry may limit the ability of many species of plants and animals, characteristic of the surrounding biome, to survive. Some soils (e.g., sandy loams) are more fertile or more conducive to excavation than others; serpentine soils and basalt flows require special adaptations for survival by plants and animals. Insects that live in windy areas, especially alpine tundra and oceanic islands, often are flightless as a result of selection against individuals blown away in flight.

Aquatic biomes are formed by topographic depressions and gradients that create zones of standing or flowing water. Aquatic biomes vary in size, depth, flow rate, and marine influence, i.e., lakes, ponds, streams and rivers, estuaries, and tidal marshes (Fig. 2.5). Lotic habitats often show considerable gradation in temperature and solute concentrations with depth. Because water has high specific heat, water changes temperature slowly relative to air temperature. However, because water is most dense at 4°C, changes in density as tem-

FIG. 2.3 Examples of ecosystem structure in representative terrestrial biomes: (A) tundra (alpine) (western U.S.), (B) desert shrubland (southwestern U.S.), (C) grassland (central U.S.), (D) tropical savanna (note termite mounds in foreground; northern Australia), (E) boreal forest (north-western U.S.), (F) temperate deciduous forest (southeastern U.S.), and (G) tropical rain forest (northern Panama).

FIG. 2.3 (*Continued*)

perature changes result in seasonal stratification of water temperature. Thermal stratification develops in the summer, as the surface of standing bodies of water warms and traps cooler, denser water below the thermocline (the zone of rapid temperature change), and again in the winter, as freezing water rises to the surface, trapping warmer and denser water below the ice. During fall and

FIG. 2.3 (*Continued*)

FIG. 2.3 (*Continued*)

spring, changing surface temperatures result in mixing of water layers and movement of oxygen and nutrients throughout the water column. Hence, deeper zones in aquatic habitats show relatively little variation in temperature, allowing aquatic insects to continue development and activity throughout the year, even in temperate regions.

Habitat conditions in aquatic biomes are influenced further by substrate structure and chemistry, amount and chemistry of regional precipitation, and the characteristics of surrounding terrestrial communities, including conditions upstream. Substrate structure and chemistry determine flow characteristics (including turbulence), pH, and inputs of nutrients from sedimentary sources. Amount and chemistry of regional precipitation determine regularity of water flow and inputs of atmospheric gases and nutrients. Characteristics of surrounding communities determine the degree of exposure to sunlight and the character and condition of allocthonus inputs of organic matter and sediments.

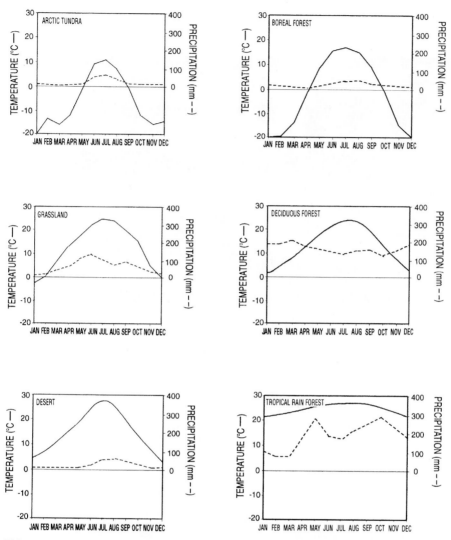

FIG. 2.4 Seasonal variation in temperature and precipitation at sites representing major biomes. Data from Van Cleve and Martin (1991).

B. Environmental Variation

Physical conditions vary seasonally in most biomes (Fig. 2.4). Temperate ecosystems are characterized by obvious seasonality in temperature, with cooler winters and warmer summers and also may show distinct seasonality in precipitation patterns, resulting from seasonal changes in the orientation of Earth's axis relative to the sun. Although tropical ecosystems experience relatively consistent temperatures, precipitation often shows pronounced seasonal variation

FIG. 2.5 Examples of aquatic biomes: (A) stream (Taiwan), (B) beaver pond (western U.S.), (C) cypress swamp (southern U.S.), (D) marsh (southern U.S.), (E) lake (Hungary). Shallower or clearer streams and lakes have greater vegetation cover than do deeper or more turbid streams or lakes.

FIG. 2.5 (*Continued*)

(Fig. 2.4). Aquatic habitats show seasonal variation in water level and circulation patterns related to seasonal patterns of precipitation and evaporation. Seasonal variation in circulation patterns can result in stratification of thermal layers and water chemistry in lotic systems. Intermittent streams and ponds may disappear during dry periods or when evapotranspiration exceeds precipitation.

Physical conditions also vary through time as a result of irregular events. Changes in global circulation patterns can affect biomes globally. For example, the east–west gradient in surface water temperature in the southern Pacific diminishes in some years, altering oceanic and atmospheric currents globally, the El Niño/southern oscillation (ENSO) phenomenon (Rasmussen and Wallace, 1983; Windsor, 1990). The effect of El Niño/southern oscillation varies among regions. Particularly strong El Niño years, e.g., 1982–83 and 1997–98, are characterized by extreme drought conditions in some tropical ecosystems and severe storms and wetter conditions in some higher latitude ecosystems. Sea-

FIG. 2.5 (*Continued*)

sonal patterns of precipitation can be reversed, i.e., drier wet season and wetter dry season. The year following an El Niño year may show a rebound, an opposite but less intense, effect (La Niña). Windsor (1990) found a strong positive correlation between El Niño index and precipitation during the previous year in Panama. Precipitation in Panama typically is lower than normal during El Niño years, in contrast to the greater precipitation accompanying El Niño in Peru and Ecuador (Windsor, 1990).

Solar activity, such as solar flares, may cause irregular departures from typical climatic conditions. Current changes in regional or global climatic conditions also may be the result of deforestation, desertification, fossil fuel combustion, and other anthropogenic factors that affect albedo, global circulation patterns, and atmospheric concentrations of CO_2, other greenhouse gases, and particulates. Characteristic ranges of tolerance to climatic factors determine the seasonal, latitudinal, and elevational distributions of species and potential changes in distributions of species as a result of changing climate.

Terrrestrial and aquatic biomes differ in the type and extent of variation in physical conditions. Terrestrial habitats are sensitive to changes in air temperature, wind speed, relative humidity, and other atmospheric conditions. Aquatic habitats are relatively buffered from sudden changes in air temperature but are sensitive to changes in flow rate, depth, and chemistry, especially changes in pH and concentrations of dissolved gases and nutrients. Vegetation cover insulates the soil surface and reduces albedo, thereby reducing diurnal and seasonal variation in soil and near-surface temperatures. Hence, desert biomes with sparse vegetation cover typically show the widest diurnal and seasonal variation in physical conditions.

Physiological tolerances of organisms, including insects, generally reflect the physical conditions of the biomes in which they occur. Insects associated with the tundra biome tolerate a lower range of temperatures than do insects associated with tropical biomes. The upper threshold temperature for survival of a tundra species might be the lower threshold temperature for survival of a tropical species. Similarly, insects characterizing mesic or aquatic biomes would have less tolerance for desiccation than would insects characterizing xeric biomes.

C. Disturbances

Within biomes, characteristic abiotic and biotic factors interact to influence the pattern of disturbances, relatively discrete events that alter ecosystem conditions, and create a finer-scale landscape mosaic of patches with different disturbance and recovery histories (Willig and Walker, 1999). Disturbances, such as fire, storms, drought, flooding, and anthropogenic change (Fig. 2.6), alter vertical and horizontal gradients in temperature, moisture, and air or water chemistry (White and Pickett, 1985), significantly altering the abiotic and biotic conditions to which organisms are exposed (Agee, 1993; Schowalter, 1985; Schowalter and Lowman, 1999).

Disturbances can be characterized by several criteria that determine their effect on various organisms (see White and Pickett, 1985). Disturbance **type,**

FIG. 2.6 Natural disturbances include (A) fire, especially in grasslands and savannas (north central U.S.), (B) storms (north central U.S.), and (C) floods (northwestern U.S.). Anthropogenic disturbances include (D) forest harvest fragmentation (northwestern U.S.), (E) overgrazing and desertification (right of fence, compared to natural grassland on left; southwestern U.S.), and (F) air pollution (darker band over industrialized valley; western U.S.). These disturbances affect ecosystem components differentially. Adapted species survive whereas non-adapted species may disappear. Overgrazing and desertification photo (E) courtesy of D. C. Lightfoot.

such as fire, drought, flood, or storm, determines which ecosystem components will be most affected. Above-ground versus below-ground species or terrestrial versus aquatic species are affected differently by fire versus flood. **Intensity** is the physical force of the event, whereas **severity** represents the effect on the ecosystem. A fire or storm of given intensity, based on temperature or wind speed, will affect organisms differently in a grassland versus a forest. **Scale** is the area affected by the disturbance and determines the rate at which organisms recolonize the interior portions of the disturbed area. **Frequency** is the mean number of events per time period; **reliability** is an inverse function of variability in the time between successive events (recurrence interval).

Reliability of recurrence, with respect to generation time, of a particular disturbance type is probably the most important characteristic driving directional selection for adaptation to disturbance, e.g., traits that confer tolerance (resistance) to fire or flooding. Effects of disturbances may be most pronounced in ecosystems such as forests and lakes, which have the greatest capacity to modify abiotic conditions and, therefore, have the least exposure and species tolerances to sudden or extreme departures from nominal conditions. Anthropogenic activities often create conditions to which locally adapted species have not been exposed previously. Clearing riparian zones, typically too moist to burn and rarely denuded naturally, for agricultural or urban use drastically alters conditions for aquatic organisms. Converting forests to pasture or crop production often leads to extreme warming and desiccation of exposed soil, compared to changes resulting from natural disturbances, and consequent loss of soil biota. Changes may persist for long periods, even after abandonment of agricultural lands, because local mechanisms are lacking for reversal of such extreme alteration of vegetation, litter, or sediment structure.

Individual insects have specific tolerance ranges to abiotic conditions that dictate their ability to survive local conditions, but may be exposed during some periods to lethal extremes of temperature, water availability or other factors. Variable ecosystem conditions typically select for wider tolerance ranges than do more stable conditions. Although abiotic conditions can affect insects directly (e.g., burning, drowning, particle blocking of spiracles), they also affect insects indirectly through changes in resource quality and availability and in exposure to predation or parasitism. The degree of genetic heterogeneity affects the number of individuals that survive altered conditions. As habitat conditions change, intolerant individuals disappear, leaving a higher frequency of genes for tolerance of the new conditions in the surviving population. Adapted colonists arrive from other areas.

Some species are favored by altered conditions, whereas others disappear. Sap-sucking insects become more abundant, but forest Lepidoptera, detritivores, and predators become less abundant, following canopy-opening disturbances (Schowalter, 1994a, 1995a). However, individual species within these groups may respond quite differently. Among Homoptera, some scale insects increase in numbers and others decline in numbers following canopy disturbance. Schowalter *et al.* (1999) found that species within each resource functional group responded differentially to manipulated change in moisture availability in a desert ecosystem, i.e., some species increased in abundance whereas other species decreased or showed no change. Root bark beetles (e.g., *Hylastes nigrinus*) are attracted to chemicals, emanating from exposed stump surfaces, that advertise suitable conditions for brood development (Fig. 2.7) (Witcosky *et al.*, 1986). Because survival and reproduction of individual insects determine population size, distribution, and effects on community and ecosystem processes, the remainder of this chapter focuses on the physiological and behavioral characteristics that affect individual responses to variable abiotic conditions.

II. SURVIVING VARIABLE ABIOTIC CONDITIONS

Insects are particularly vulnerable to changes in temperature, water availability, and air or water chemistry because of their relatively large ratios of surface area to volume. However, many insects can live within suitable microsites that

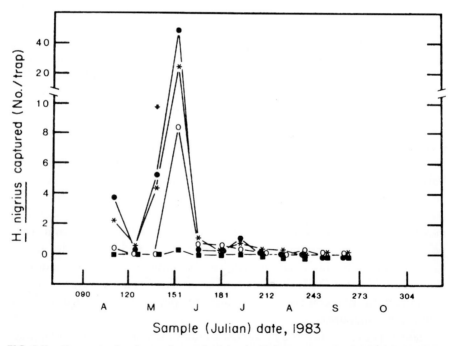

FIG. 2.7 Changes in abundance of a root bark beetle, *Hylastes nigrinus*, in undisturbed, 12-yr.-old plantations (black squares) of Douglas-fir, *Pseudotsuga menziesii*, and plantations thinned in September 1982 (asterisks), January 1983 (black circles), or May 1983 (white circles) in western Oregon. Arrow indicates time of thinning in May 1983. From Witcosky *et al.* (1986), courtesy of the Research Council of Canada.

buffer exposure to environmental changes. Insects in aquatic environments or deep in soil or woody habitats may be relatively protected from large changes in air temperature and relative humidity (e.g., Curry, 1994; Seastedt and Crossley, 1981a). High moisture content of soil can mitigate heat penetration and protect soil fauna.

Most insects are subject to environmental variability that includes periods of lethal or stressful abiotic conditions. Therefore, maintaining optimal body temperature, water content, and chemical processes are challenges for survival in variable environments. Insects possess a remarkable variety of physiological and behavioral mechanisms for surviving in variable environments.

Adaptive physiological responses can mitigate exposure to suboptimal conditions. For example, diapause is a general physiological mechanism for surviving seasonally adverse conditions, typically in a resistant stage such as the pupa of holometabolous insects. Diapause induction and termination are controlled by cues such as degree-day accumulation or photoperiod and may require a minimum duration of freezing temperatures or other factors that maximize synchronization of development with seasonally suitable conditions. Nevertheless, exposed insects often are killed by sudden or unexpected changes in temperature, moisture, or chemical conditions of the habitat.

Behavior represents a more flexible means of responding to environmental variation, compared to physiology, because an animal can respond actively to sensory information to avoid or mitigate lethal conditions. Mobile insects have an advantage over sessile species in avoiding or mitigating exposure to extreme temperatures, water availability, or chemical conditions. Limited mobility often is sufficient within steep environmental gradients. Many small, flightless litter species need move vertically only a few millimeters within the soil profile to avoid higher temperatures and desiccation at the surface following fire or clearcutting (Seastedt and Crossley, 1981a). Some species choose protected habitats prior to entering diapause to reduce their vulnerability to potential disturbances. Miller and Wagner (1984) reported that pandora moth, *Coloradia pandora*, pupae in a ponderosa pine, *Pinus ponderosa*, forest were significantly more abundant on the forest floor in areas with open canopy and sparse litter than in areas with closed canopy and deeper litter. Although other factors also differ between these microhabitats, avoidance of accumulated litter may represent an adaptation to survive frequent ground fires in this ecosystem. In addition, mobile insects often can detect and colonize suitable patches within variable environments.

A. Themoregulation

Insects, as well as other invertebrates, are generally ectothermic, meaning that their body temperatures are determined primarily by ambient temperature. Rates of metabolic activity (hence, energy and carbon flux) generally increase with temperature. Developmental rates and processes also are temperature dependent. However, some species can regulate body temperature to some degree through physiological or behavioral responses to extreme temperatures.

Insect species show characteristic ranges in temperatures suitable for activity. Aquatic ecosystems have relatively consistent temperature, but insects in terrestrial ecosystems often experience considerable temperature fluctuation, even on a daily basis. Most insects can survive at temperatures from well below freezing (as a result of electrolyte concentration in the hemolymph) to 40–50°C (Whitford, 1992), depending on adapted tolerance ranges and acclimation (preconditioning). Some insects occurring at high elevations die at a maximum temperature of 20°C, whereas insects from warm environments often die at higher minimum temperatures. Chironomid larvae living in hot springs survive water temperatures of 49–51°C (Chapman, 1982).

In general, developmental rate of ectotherms increases with temperature. Both terrestrial and aquatic insects respond to the accumulation of thermal units (the sum of degree-days above a threshold temperature) (Baskerville and Emin, 1969; Ward and Stanford, 1982). Degree-day accumulation can be similar under different conditions, e.g., mild winter/cool summer and cold winter/hot summer, or quite different along elevational or latitudinal gradients. Anthropogenic conditions can significantly alter thermal conditions, especially in aquatic habitats. Discharge of heated water, artificial mixing of thermal strata, impoundment, diversion, regulation of water level and flow, and canopy opening in riparian zones through harvest or grazing severely modify the thermal environment for aquatic species and favor heat-tolerant individuals and species over heat-intolerant individuals and species (Ward and Stanford, 1982).

A number of mechanisms permit survival at subfreezing temperatures. Voiding the gut at the onset of cold conditions may prevent food particles from serving as nuclei for ice crystal formation. Hence, nonfeeding stages may have lower supercooling points than do feeding stages (Kim and Kim, 1997). Some diapausing insects can survive temperatures as low as −30°C, either by supercooling of body fluids, such as with high concentrations (up to 25% of fresh weight) of glycerol in the hemolymph or by tolerance to ice formation in body fluids in extracellular compartments (Lundheim and Zachariassen, 1993). Such insects exhibit mechanisms that reduce the rate of cooling at the cell surface, such as partial cellular desiccation, concentrating salts around extracellular ice crystals, or other mechanisms. (Chapman, 1982). Cold tolerance varies with life stage, temperature, and exposure time and can be enhanced by preconditioning to sublethal temperatures (Kim and Kim, 1997).

Many insects can modify body temperature physiologically to some extent (Casey, 1988; Heinrich, 1974, 1979, 1981, 1993). Evaporative cooling may enable insects to survive short-term exposure to higher temperatures when the air is dry. Higher humidity is necessary to survive long-term exposure to high temperatures, because death results from desiccation at low humidity (Chapman, 1982). Humidity has less effect on upper lethal temperature in small insects because of the small volume of water available for evaporative cooling. Prange and Pinshow (1994) reported that both sexes of a sexually dimorphic desert grasshopper, *Poekiloceros bufonius*, depress their internal temperatures through evaporative cooling. However, males lost proportionately more water through evaporation, but retained more water from food, than did the much larger females, indicating that thermoregulation by smaller insects is more constrained by water availability. An Australian montane grasshopper, *Kosciuscola*, can change color from black at night to pale blue during the day (Key and Day, 1954), thereby regulating heat absorption.

Thermoregulation also can be accomplished behaviorally. Heinrich (1974, 1979, 1981, 1993) and Casey (1988) reviewed studies demonstrating that a variety of flying insects are capable of thermoregulation through activities that generate metabolic heat, such as fanning the wings and flexing the abdomen (Fig. 2.8). Flight can elevate body temperature 10–30°C above ambient (Chapman, 1982; Heinrich, 1993). A single bumble bee, *Bombus vosnesenskii*, queen can raise the temperature of the nest as much as 25°C above air temperatures as low as 2°C, even in the absence of insulating materials (Heinrich, 1979).

Insects can sense and often move within temperature gradients to thermally optimal habitats. Light is an important cue that attracts insects to sources of heat or repels them to darker, cooler areas. Insects frequently bask on exposed surfaces to absorb heat during early morning or cool periods and retreat to less exposed sites during warmer periods (Fig. 2.9). Insects use or construct shelters to trap or avoid heat. Some arthropods burrow to depths at which diurnal temperature fluctuation is minimal (Polis *et al.* 1986). Seastedt and Crossley (1981a) reported significant redistribution of soil/litter arthropods from the upper 5 cm of the soil profile to deeper levels following canopy removal and consequent soil surface exposure and warming in a forested ecosystem. Tent caterpillars, *Malacosoma* spp., build silken tents that slow dissipation of metabolic heat and increase colony temperature above ambient (Fig. 2.10) (Fitzgerald, 1995; Heinrich, 1993). Tents of overwintering larvae of the arctiid moth, *Lepesoma argentata*, occur almost exclusively in the exposedupper canopy and significantly more often on the south-facing sides of host conifers in western Washington, U.S.A. (Shaw, 1998).

Some insects regulate body temperature by optimal positioning (Heinrich, 1974, 1993). Web-building spiders adjust their posture to control their exposure to solar radiation (Robinson and Robinson, 1974). Desert beetles, grasshoppers, and scorpions prevent overheating by stilting, i.e., extending their legs and elevating the body above the heated soil surface and by orienting the body to minimize the surface area exposed to the sun (Heinrich, 1993).

B. Water Balance

Maintenance of homeostatic water balance also is a challenge for organisms with high ratios of surface area to volume. The arthropod exoskeleton is an im-

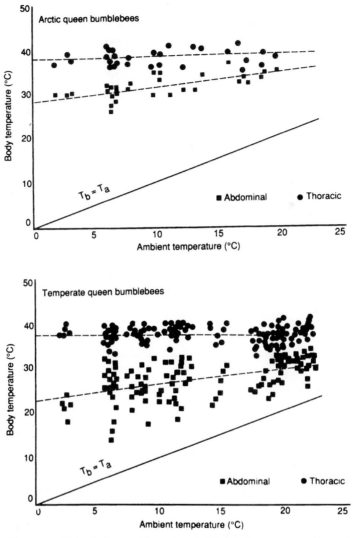

FIG. 2.8 Thermoregulation by insects. Thoracic and abdominal temperatures of high Arctic and temperate queen bumble bees foraging in the field as a function of air temperature; T_b = body temperature, T_a = ambient temperature. Arctic queens forage at significantly higher abdominal temperature than do temperate queens. From Heinrich (1993) by permission from Bernd Heinrich.

portant mechanism for control of water loss. Larger, more heavily sclerotized arthropods are less susceptible to desiccation than are smaller, more delicate species (Alstad *et al.*, 1982; Kharboutli and Mack, 1993). Arthropods in xeric environments typically are larger, have a thicker cuticle, and secrete more waxes to inhibit water loss than do insects in mesic environments (Crawford, 1986; Edney, 1974; Kharboutli and Mack, 1993). Some species in xeric environments

FIG. 2.9 Many insects, such as grasshoppers, raise their body temperatures by basking. Heat absorption is enhanced by dark coloration.

conserve metabolic water (from oxidation of food) or acquire water from condensation on hairs or spines (Chapman, 1982). Many insects, especially those living in xeric environments, also regulate respiratory water loss by controlling spiracular activity under dry conditions (Fielden *et al.*, 1994; Kharboutli and Mack, 1993) or by undergoing anhydrobiosis, a physiological state characterized by an absence of free water and of measurable metabolism (Whitford, 1992). Water conservation is under hormonal control in some species. An antidiuretic hormone is released in desert locusts (*Schistocerca gregaria*) and other species under conditions of water loss (Delphin, 1965).

Insects and other arthropods are most vulnerable to desiccation at times when a new exoskeleton is forming, i.e., during eclosion from eggs, during molts, and during diapause (Crawford, 1978; Willmer *et al.*, 1996). Tisdale and Wagner (1990) found that percentage of sawfly, *Neodiprion fulviceps*, eggs hatched was significantly higher at relative humidities $\geq 50\%$. Yoder *et al.* (1996) found that slow water loss through the integument and respiration by diapausing fly pupae was balanced by passive water vapor absorption from the air at sufficiently high humidities.

Insects in diapause at subfreezing temperatures are subject to freeze-drying. Lundheim and Zachariassen (1993) reported that beetles that tolerate ice formation in extracellular fluids have lower rates of water loss than do insects that have supercooled body fluids, perhaps because the hemolymph in frozen bee-

FIG. 2.10   Tent caterpillars, *Malacosoma* spp., and other tent-constructing Lepidoptera, reduce airflow and variation in temperatures within their tents.

tles is in vapor pressure equilibrium with surrounding ice whereas the hemolymph in supercooled insects has vapor pressure higher than the environment.

On the other hand, some insects must contend with excess water intake. Insects that ingest liquid food immediately excrete large amounts of water to concentrate dissolved nutrients. Elimination of excess water (and carbohydrates) in sap-feeding Homoptera is accomplished in the midgut by rapid diffusion across a steep moisture gradient created by a filter loop (Chapman, 1982). Honeydew excreted by phloem-feeding Homoptera (Fig. 2.11) is an important resource for ants, hummingbirds, predaceous Hymenoptera, and sooty molds (Dixon, 1985; Edwards, 1982; Elliott *et al.*, 1987; Huxley and Cutler, 1991). The abundant water excreted by xylem-feeding spittlebugs is used to create the frothy mass that hides the insect. Excretion in some species, such as the blood-feeding *Rhodnius* (Heteroptera) is controlled by a diuretic hormone (Maddrell, 1962).

Water balance also can be maintained behaviorally, to some extent, by retreating to cooler or moister areas to prevent desiccation. Burrowing provides access to more mesic subterranean environments (Polis *et al.*, 1986). The small size of most insects makes them vulnerable to desiccation, but also permits habitation within the relatively humid boundary layer around plant surfaces or at the soil surface.

Termites construct their colonies to optimize temperature and moisture

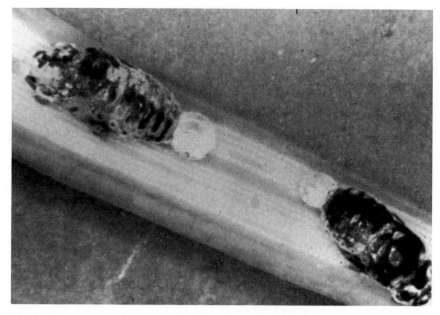

FIG. 2.11   Sap-feeding Homoptera (*Adelges cooleyi* on Douglas-fir) egest excess water and carbohydrates as honeydew.

conditions. Metabolic heat generated in the core of the nest rises by convection into large upper cavities and diffuses to the sides of the nest where air is cooled and gaseous exchange occurs through the thin walls. Cooled air sinks into lower passages (Lüscher, 1961). The interior chambers of termite colonies typically have high relative humidities.

C. Air and Water Chemistry

Air and water chemistry affect insect physiology. Oxygen supply is, of course, critical to survival but may be limited under certain conditions. Airborne or dissolved chemicals can affect respiration and development. Soil or water pH can affect exoskeleton function and other physiological processes. Changes in concentrations of various chemicals, especially those affected by industrial activities, affect many organisms, including insects.

Oxygen supply can limit activity and survival of aquatic species and some terrestrial species living in enclosed habitats. Less oxygen can remain dissolved in warm water than in cold water. Stagnant water can undergo oxygen depletion as a result of algal and bacterial respiration. Insect species living in oxygen-poor environments typically have more efficient oxygen delivery systems, such as increased tracheal supply, gills, or breathing tubes that extend to an air supply (Chapman, 1982). For example, the hemolymph of some aquatic chironomid larvae and endoparasitic fly larvae is unique among insects in containing a hemoglobin that has a higher affinity for oxygen than does mammalian hemoglobin (Chapman, 1982; Pinder and Morley, 1995). Wood-boring species tolerate low oxygen concentrations deep in decomposing wood.

Increased atmospheric CO_2 appears to have little direct effect on insects or other arthropods. However, relatively few insect species have been studied with respect to CO_2 enrichment. Increased atmospheric CO_2 can significantly affect the quality of plant material for some herbivore (Arnone et al., 1995; Fajer et al., 1989; Kinney et al., 1997; Lincoln et al., 1993; Roth and Lindroth, 1994) and decomposer (Grime et al., 1996; Hirschel et al., 1997) species, although plant response to CO_2 enrichment depends on a variety of environmental factors (e.g., Lawton, 1995; Watt et al., 1995; see Chapter 3). For insects in general, elevated CO_2 causes slower growth rates, higher consumption rates, longer development times, and higher mortality rates (Watt et al., 1995).

Airborne and dissolved materials can include volatile emissions or secretions from plant, animal, and industrial origin. Fluorides, sulfur compounds, nitrogen oxides, and ozone affect many insect species directly, although the physiological mechanisms of toxicity are not well known (Alstad et al., 1982; Heliövaara, 1986; Heliövaara and Väisänen, 1986, 1993; Pinder and Morley, 1995). Disruption of epicuticular or spiracular tissues by these reactive chemicals may be involved. Dust and ash kill many insects, apparently because they absorb and abrade the thin epicuticular wax-lipid film that is the principal barrier to water loss. Insects then die of desiccation (Alstad et al., 1982). Brown (1995) concluded that there is little evidence for direct effects of realistic concentrations of these major air pollutants on terrestrial herbivores but considerable evidence that many herbivorous species respond to changes in the quality of plant resources or abundance of predators resulting from exposure to these pollutants.

Soil and water pH affects a variety of chemical reactions, including enzymatic activity. Changes in pH resulting from acidification (such as from volcanic or anthropogenic activity) affect osmotic exchange, gill and spiracular surfaces, and digestive processes. Changes in pH often are correlated with other chemical changes, such as increased N or S, and effects of pH change may be difficult to separate from other factors. Pinder and Morley (1995) reported that many chironomid species are relatively tolerant of alkaline water, but few are tolerant of pH < 6.3. Acid deposition and loss of pH buffering capacity likely will affect survival and reproduction of aquatic and soil/litter arthropods (Curry, 1994; Pinder and Morley, 1995).

D. Other Abiotic Factors

Many aquatic insects are sensitive to water level and flow rate. These factors can fluctuate dramatically, especially in seasonal habitats such as desert playas, intermittent streams, and perched pools in treeholes and bromeliads (phytotelmata). Water level affects both temperature and water quality, temperature because smaller volumes absorb or lose heat more quickly than do larger volumes, water quality because various solutes become more concentrated as water evaporates. Insects and other aquatic arthropods show life history adaptations to seasonal patterns of water availability or quality, often undergoing physiological diapause as water resources disappear. Flow rate affects temperature and oxygenation, with cooler temperature and higher oxygen content at higher flow rates, but high flow rates can physically dislodge and remove exposed insects. Adler and McCreadie (1997) and McCreadie and Colbo (1993) reported that sibling species of black flies, *Simulium*, select different stream microhabitats on the basis of their adaptations to water velocity (Fig. 2.12).

III. FACTORS AFFECTING DISPERSAL BEHAVIOR

Insects have a considerable capacity to escape adverse conditions and to find optimal conditions within temperature, moisture and chemical gradients. Dispersal is the process of individuals leaving an area to find and colonize other habitats. This is an important adaptive behavior that minimizes the risk that the entire population will be destroyed by disturbance or resource depletion, maximizes the chance that some individuals will find and exploit new resources, and maximizes genetic heterogeneity (Schowalter, 1985; Wellington, 1980; see Chapter 5).

Nevertheless, dispersal entails considerable risk and requires considerable energy expenditure (Rankin and Burchsted, 1992). Torres (1988) documented cases of exotic insects being introduced into Puerto Rico by hurricane winds, including a swarm of desert locusts blown across the Atlantic Ocean from Africa. Many insects (and other organisms) fail to find or reach suitable habitats.

Flight capacity contributes enormously to insect ability to disperse. Adult aquatic insects can disperse from an intermittent pond or stream before the water disappears and search for other bodies of water. Dispersal may be particularly important for distributing populations and minimizing risk in ecosystems characterized by frequent disturbances. The consequences of poor dispersal ability can be illustrated by recovery of insect populations from the eruption of Mount St. Helens in western Washington in May, 1980. Ashfall from the volcanic eruption severely reduced abundances of insects and spiders in the fallout zone to the east of Mount St. Helens (Oman, 1987). Female leafhoppers in the genus *Errhomus* are brachypterous and incapable of flight. Other life stages, except for adult males, also lack means of long-distance dispersal. Although many more mobile insects and spiders reached predisturbance abundances by 1981, *Errhomus* populations did not. A number of factors affect the probability of successful dispersal, i.e., arrival at suitable habitats, including life history strategy, crowding, nutritional status, habitat and resource conditions, and the mechanism of dispersal.

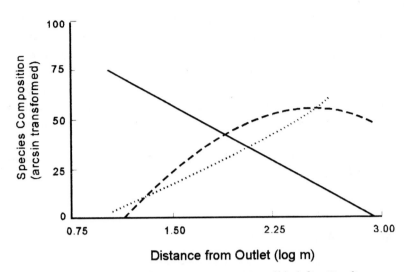

FIG. 2.12   Relationship between sibling species composition of black flies (*Simulium truncatum*, solid line, *S. verecundum* AA, dashed line, and *S. venustum* CC2, dotted line) and distance from lake outlets on the Avalon Peninsula, Newfoundland, in early June. Least squares regression equations were significant at $P < 0.01$; adjusted R^2 values were 92, 85, and 68% for the three species, respectively. From Adler and McCreadie (1997) by permission from the Entomological Society of America.

A. Life History Strategy

The degree of adaptation to disturbance affects the predisposition of individuals to disperse. Species characterizing relatively stable, infrequently disturbed habitats tend to disperse slowly, i.e., produce few offspring and move short distances (see Chapter 5). Infrequent disturbance and consistent resource availability provides little or no selection for greater dispersal ability. Many forest species (especially Lepidoptera and Coleoptera) are flightless or, at least, poor fliers. By contrast, species such as aphids that characterize temporary, frequently disturbed habitats produce large numbers of individuals and a high proportion of dispersers. Such traits are important adaptations for species exploiting temporary, unstable conditions (Janzen, 1977).

B. Crowding

Crowding affects insect inclination to disperse and, in some cases, may stimulate morphological or physiological transformations that facilitate dispersal. Survival and fecundity are often density dependent, i.e., inversely related to population density. Therefore, dispersing individuals may achieve higher fitnesses than do nondispersing individuals at high population densities (Price, 1997). For example, some bark beetle species oviposit their full complement of eggs in one tree under low-density conditions but only a portion of their eggs in one tree under high-density conditions, leaving that tree and depositing remaining eggs in other trees (Wagner *et al.*, 1981). If all eggs were laid in the first tree under crowded conditions, the large number of offspring could deplete resources before completing development.

Crowding has been shown to stimulate feeding and developmental rates. Under crowded conditions, some insects spend more time eating and less time resting (Chapman, 1982). Crowding may increase the incidence of cannibalism in many species (Fox, 1975a, b), encouraging dispersal. In addition, crowding can induce morphological changes that promote dispersal. Uncrowded desert locusts tend to repel one another and feed quietly on clumps of vegetation, whereas crowded locusts are more active, attract one another and march en masse, spending little time on vegetation (Matthews and Matthews, 1978). Accompanying physiological changes in color, wing length, and ability to feed on a wider variety of plants facilitate migration and the chances of finding suitable resources.

C. Nutritional Status

Nutritional status affects the endurance of dispersing insects. Populations of many insects show considerable variation in fat storage and vigor as a result of variation in food quality and quantity and maternal partitioning of nutrient resources to progeny (Wagner *et al.*, 1981; Wellington, 1980). Many species exhibit obligatory flight distances that are determined by the amount of energy and nutrient reserves: dispersing individuals respond to external stimuli only after depleting these reserves. Hence, less vigorous individuals tend to colonize more proximal habitats, whereas more vigorous individuals fly greater distances and colonize more remote habitats. Because crowding and nutritional status have opposite effects on dispersal, the per capita accumulation of adequate energy reserves and the number of dispersing individuals tend to peak when resource quality and quantity are still sufficient to promote insect development and vigor.

D. Habitat and Resource Conditions

The likelihood that an insect will find a suitable patch depends strongly on patch size and proximity to insect population sources. The probability of survival declines with distance, as a result of depletion of metabolic resources and protracted exposure to various mortality factors (Pope *et al.*, 1980). Hence, more insects reach closer resources or sites. Sartwell and Stevens (1975) and Schowalter *et al.* (1981b) reported that, under nonoutbreak conditions, probability of bark beetle colonization of living pine trees declined with distance from currently attacked trees. Trees more than 6 m from currently colonized trees had negligible probability of colonization. Under outbreak conditions, the effect of distance disappears (Schowalter *et al.*, 1981b). Similarly, He and Alfaro (1997) reported that, under nonoutbreak conditions, colonization of white spruce by the white pine weevil, *Pissodes strobi*, depended on host condition and distance from trees colonized the previous year, but during outbreaks most trees were sufficiently near occupied trees to be colonized.

Larger or more conspicuous habitats or resources are more likely to be perceived by dispersing insects and to be intercepted by a given direction of flight (see Chapter 6). For example, Courtney (1985, 1986) reported that the pierid butterfly, *Anthocharis cardamines*, preferentially oviposited on the most conspicuous (in terms of flower size) host species that were less suitable for larval development than less conspicuous hosts. This behavior by the adults represented a trade-off between the prohibitive search time required to find the most suitable hosts and the reduced larval survival on the most conspicuous hosts. Larger habitat patches also intersect a longer arc centered on a given starting point. Insects dispersing in any direction have a higher probability of contacting larger patches than they do smaller patches.

E. Mechanism of Dispersal

The probability that suitable resources can be found and colonized depends on the mode of dispersal. Three general mechanisms can be identified: random, phoretic, and directed.

Random dispersal direction and path is typical of most small insects with little capacity to detect or orient toward environmental cues. Such insects are at the mercy of physical barriers or wind or water currents, and their direction and path of movement are determined by obstacles and patterns of air or water movement. For example, first instar nymphs of a *Pemphigus* aphid that lives on the roots of sea aster growing in salt marshes climb the sea asters and are set adrift on the rising tide. Sea breezes enhance movement, and successful nymphs are deposited at low tide on new mud banks where they seek new hosts (Kennedy, 1975). Aquatic insect larvae often are carried downstream during floods. Hatching gypsy moth, *Lymantria dispar*, and tussock moth larvae (Lymantriidae), scale insect crawlers, and spiders (as well as other arthropods,) disperse by launching themselves into the airstream. Lymantriid and scale insect adults have poor (if any) flight capacity. The wind-aided dispersal by larval Lepidoptera and spiders is facilitated by extrusion of silk strands, a practice known as "ballooning." Western spruce budworm, *Choristoneura occidentalis*, adults aggregate in mating swarms above the forest canopy and are carried by wind currents to new areas (Wellington, 1980). The probability that these insects will arrive at suitable resources depends on the number of dispersing insects and the predictability of wind or water movement in the direction of new resources. Most individuals fail to colonize suitable sites, and many become part of the aerial or aquatic plankton that eventually "falls out" and are deposited in remote, unsuitable locations. For example, Edwards and Sugg (1990) documented fallout deposition of many aerially dispersed insect species on mountains in western Washington.

FIG. 2.13   Phoretic mesostigmatid mites in coxal cavity of a scarab beetle.

Phoretic dispersal is a special case in which a flightless insect or other arthropod hitches a ride on another animal (Fig. 2.13). Phoresy is particularly common among wingless Hymenoptera and mites. For example, scelionid wasps ride on the backs of female grasshoppers, benefitting from both transport and the eventual opportunity to oviposit on the grasshopper's eggs. Wingless Malaphaga attach themselves to hippoboscid flies that parasitize the same bird hosts. Many species of mites attach themselves to dispersing adult insects that feed on the same dung or wood resources (Krantz and Mellott, 1972; Stephen et al., 1993). The success of phoresy (as with wind or water-aided dispersal) depends on the predictability of host dispersal. However, in the case of phoresy, success is enhanced by the association of both the hitchhiker and its mobile (and perhaps cue-directed) host with the same resource.

Directed dispersal provides the highest probability of successful colonization and is observed in larger, stronger fliers capable of orienting within concentration gradients of chemicals that indicate suitable or unsuitable resources (see Chapter 3). Many wood-boring insects, such as wood wasps (Siricidae) and beetles (especially Buprestidae) are attracted to sources of smoke or volatile tree chemicals emitted from burned or injured trees over distances of up to several kilometers (Gara et al., 1984; Mitchell and Martin, 1980; Raffa et al., 1993; Wickman, 1964). Attraction to suitable hosts often is significantly enhanced by mixing with pheromones emitted by early colonists (see Chapters 3 and 4).

Migration is an active mass movement of individuals that functions to displace entire populations. Migration always involves females but not always males. Examples of migratory behavior in insects include locusts, monarch butterflies, *Danaus plexippus*, and ladybird beetles. Locust (*Schistocerca gregaria* and *Locusta migratoria*) migration depends, at least in part, on wind patterns. Locust swarms remain compact, not because of directed flight, but because randomly oriented locusts reaching the swarm edge reorient toward the body of the swarm. Swarms are displaced downwind into equatorial areas where converging air masses rise, leading to precipitation and vegetation growth favorable to the locusts (Matthews and Matthews, 1978). In this way, migration displaces the swarm from an area of crowding and insufficient food to an area with more abundant food resources. Monarch butterfly and ladybird beetle migration occurs seasonally and displaces large numbers to and from overwintering sites, Mexico for monarch butterflies and sheltered sites for ladybird beetles.

IV. RESPONSES TO ANTHROPOGENIC CHANGES

Insect responses to environmental changes caused by anthropogenic activity remain largely unknown. A number of studies have documented insect responses to elevated temperature and to increased atmospheric or aqueous concentrations of CO_2 or various pollutants, including pesticides (e.g., Alstad et al., 1982; Arnone et al., 1995; Heliövaara and Väisänen, 1986, 1993; Kinney et al., 1997; Lincoln et al., 1993; Marks and Lincoln, 1996). Some studies have addressed effects of ecosystem fragmentation on abiotic variables that affect survival or movement of various insects (e.g., Chen et al., 1995; Franklin et al., 1992; Roland, 1993; Rubenstein, 1992).

However, humans are changing environmental conditions in many ways simultaneously, through fossil fuel combustion, industrial effluents, water impoundment and diversion, and land use practices. Although global atmospheric concentrations of CO_2 and other greenhouse gases are clearly increasing, their effect on global climate remains inconclusive (e.g., Keeling *et al.*, 1995). Acidic precipitation has greatly reduced the pH of many aquatic ecosystems in northern temperate countries, with more dramatic effects. Nitrogen subsidies resulting from increased atmospheric NO_x may provide a short-term fertilization effect in N-limited ecosystems, until pH buffering capacity of the soil is depleted. Deforestation, desertification, and other changes in regional landscapes are altering habitat suitability for organisms around the globe. Few studies have measured insect responses to multiple changes in ecosystem conditions. Chapin *et al.* (1987) addressed plant responses to multiple stressors and concluded that multiple factors can have additive or synergistic effects. Some insects will disappear as their host plants decline. Others may facilitate host plant decline by exploiting stressed and poorly defended hosts (see Chapter 3). Clearly, studies are needed on the effects of multiple natural and anthropogenic changes on insects in order to improve prediction of insect responses.

V. SUMMARY

Insects are affected by abiotic conditions that reflect latitudinal gradients in temperature and moisture, as modified by circulation patterns and mountain ranges. At the global scale, latitudinal patterns of temperature and precipitation produce bands of tropical rainforests along the equatorial convergence zone (where warming air rises and condenses moisture), deserts centered at 30°N and S latitudes (where cooled, dried air descends), and moist boreal forests centered at 60°N and S latitudes (where converging air masses rise and condense moisture). Mountains affect the movement of air masses across continents, forcing air to rise and condense on the windward side and dried air to descend on the leeward side. The combination of mountain ranges and latitudinal gradients in climatic conditions creates a template of regional ecosystem types known as biomes, characterized by distinctive vegetation (e.g., tundra, desert, grassland, forest). Aquatic biomes are distinguished by size, depth, flow rate and marine influence (e.g., ponds, lakes, streams, rivers, estuaries).

The inherent problems of maintaining body heat and water content and avoiding adverse chemical conditions by small, ectothermic organisms has led to an astounding variety of physiological and behavioral mechanisms by which insects adjust to and interact with environmental conditions. Mechanisms for tolerating or mitigating effects of variation in abiotic factors determine the seasonal, latitudinal, and elevational distributions of insect species.

Many insects have a largely unappreciated physiological capacity to cope with the extreme temperatures and relative humidities found in the harshest ecosystems on the planet. However, even insects in more favorable environments must cope with variation in abiotic conditions through diapause, color change, evaporative cooling, supercooling, voiding of the gut, control of respiratory water loss, etc. Many species exhibit at least limited homeostatic ability, i.e., ability to regulate internal temperature and water content.

Behavior represents the active means by which animals respond to their environment. Insects are sensitive to a variety of environmental cues, and most insects are able to modify their behavior in response to environmental gradients or changes. Insects, especially those that can fly, can move within gradients of temperature, moisture, chemicals, or other abiotic factors to escape adverse conditions. Many species are able to regulate body heat or water content by rapid muscle contraction, elevating the body above hot surfaces, seeking shade, or burrowing. Social insects appear to be particularly flexible in the use of colony activity and nest contruction to facilitate thermoregulation.

Many insects are capable of flying long distances but dispersal entails considerable risk, and many individuals do not reach suitable habitats. The probability that an insect will discover a suitable patch is a function of the inclination to disperse (as affected by life history strategy and crowding), endurance (determined by nutritional condition), patch size, distance, and the mechanism of dispersal (whether random, phoretic, or oriented toward specific habitat cues).

Environmental changes resulting from anthropogenic activities are occurring at an unprecedented rate. The effects of these changes on insects are difficult to predict because few studies have addressed the effects of multiple interacting changes on insects.

Resource Acquisition

ALL ORGANISMS ARE EXAMPLES OF NEGATIVE ENTROPY. THEY REQUIRE energy to synthesize the organic molecules that are the basis for life processes, growth, and reproduction. Hence, the acquisition and concentration of energy and matter are necessary goals of all organisms and largely determine individual fitness.

Insects, like other animals, are heterotrophic, i.e., they must acquire their energy and material resources from other organisms (see Chapter 11). As a group, insects exploit a wide range of resources, including plant, animal, and detrital material, but individual organisms must find and acquire appropriate resources to support growth, maintenance, and reproduction.

The organic resources used by insects vary widely in suitability (nutritional value), acceptability (preference ranking, given choices and trade-offs), and availability, depending on environmental conditions. Physiological and behavioral mechanisms for evaluating and acquiring resources and their efficiencies under different developmental and environmental conditions are the focus of this chapter.

I. RESOURCE SUITABILITY

Resource quality is the net energy and nutrient value of chemical resources after accounting for an individual's ability (and energetic or nutrient cost) to digest the resource. The energy and nutrient value of organic molecules is a product of the number, type, and bonding energy of constituent atoms. However, organic resources are not equally digestible into usable components. Some resources provide little nutritional value for the expense required to acquire and digest them, and others cannot be digested by common enzymes. Many organic molecules are essentially unavailable, or even toxic, to a majority of organisms. Vascular plant tissues are composed largely of lignin and cellulose, digestible only by certain microorganisms. Nitrogen is particularly limiting to animals that feed on wood or dead plant material. Some organic molecules are cleaved into toxic components by commonly occurring digestive enzymes (see following).

A. Resource Requirements

Insects feed on a wide variety of plant, animal, and dead organic matter. Dietary requirements for all insects include carbohydrates, amino acids, cholesterol, B vitamins, and inorganic nutrients, such as P, K, Ca, Na, etc. Insects lack the ability to produce their own cellulases to digest cellulose. The nutritional value of plant material often is limited further by deficiency in certain requirements, such as low content of N (Mattson, 1980), Na (Seastedt and Crossley, 1981b; Smedley and Eisner, 1995), or linoleic acid (Fraenkel and Blewett, 1946). High lignin content toughens foliage and other tissues and limits feeding by many herbivores. Toxins or feeding deterrents in the food resource increase the cost in terms of search time, energy, and nutrients needed to exploit their nutritional value.

For particular arthropods, several factors influence food requirements. The most important of these are the size and maturity of the arthropod and the quality of food resources. Larger organisms require more food and consume more oxygen per unit time than do smaller organisms, although smaller organisms consume more food and oxygen per unit biomass. Insects require more food and often are able to digest a wider variety of resources as they mature. Increased feeding may be particularly pronounced in holometabolous species which must store sufficient resources during larval feeding for pupal diapause and, often, for dispersal and reproduction by nonfeeding adult stages.

Some insects may require long periods (several years to decades) of larval feeding in order to concentrate sufficient nutrients (especially N and P) from nutritionally poor resources to complete development. Many arthropods that feed on nutrient-poor detrital resources have obligate associations with other organisms that provide or increase access to limiting nutrients. Microbes can be internal or external associates. For example, termites host mutualistic gut bacteria or protozoa that catabolize cellulose, fix nitrogen, and concentrate or synthesize other nutrients and vitamins needed by the insect. Termites and some other detritivores feed on feces (coprophagy) after sufficient incubation time for microbial digestion and enhancement of nutritive quality of egested material. If coprophagy is prevented, these organisms often compensate by increasing their rate of consumption of detritus (McBrayer, 1975).

B. Variation in Food Quality

Food quality varies widely among resource types. Plant material has relatively low nutritional quality, because N typically occurs at low concentrations and most carbohydrates are in the form of indigestible cellulose and lignin. Woody tissues are particularly low in labile resources readily available to insects or other animals. Detrital resources may be impoverished in important nutrients as a result of weathering, leaching, or plant resorption prior to shedding senescent tissues.

The nutritional value of plant resources frequently changes seasonally and ontogenically. Filip *et al.* (1995) reported that the foliage of many tropical trees has higher nitrogen and water content early in the wet season than late in the wet season. Lawrence *et al.* (1997) caged several cohorts of western spruce budworm, *Choristoneura occidentalis*, larvae on white spruce at different phenological stages of the host. Cohorts that began feeding 3–4 weeks before budbreak and completed larval development prior to the end of shoot elongation developed significantly faster and showed significantly greater survival rate and adult mass than did cohorts caged later (Fig. 3.1). These results indicate that the phenological window of opportunity for this insect was sharply defined by the period of shoot elongation, during which foliar nitrogen, phosphorus, potassium, copper, sugars, and water were higher than in mature needles.

Food resources often are defended in ways that limit their utilization by consumers. Physical defenses include spines, toughened exterior layers, and other barriers. Spines and hairs can inhibit penetration by small insects or interfere with ingestion by larger organisms. These structures often are associated with glands that augment the defense by delivering toxins. Some plants entrap phytophagous insects in adhesives and may obtain nutrients from insects trapped in this way (Simons, 1981). Toughened exteriors include lignified epidermis of foliage and bark of woody plants and heavily armored exoskeletons of arthropods. Bark is a particularly effective barrier to penetration by most organisms (Ausmus, 1977), but lignin also reduces ability of many insects to utilize the toughened foliage (e.g., Scriber and Slansky, 1981). The viscous oleoresin (pitch) produced by conifers and some hardwoods can push insects out of their entrances into plant tissues (Fig. 3.2).

Many plant and animal species gain protection through interactions with other organisms, especially ants. A number of plant species provide habitable structures (domatia) suitable for colonies of ants or predaceous mites (e.g., Huxley and Cutler, 1991). Cecropia trees in the tropics are one of the best-

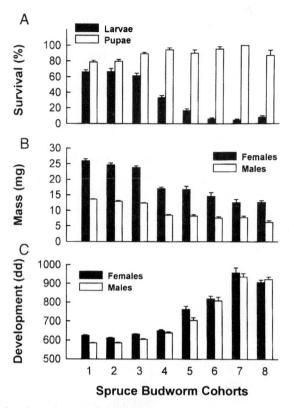

FIG. 3.1 Larval and pupal survival, adult dry mass, and development time from 2nd instar through adult for eight cohorts of spruce budworm caged on white spruce in 1985. The first six cohorts were started at weekly intervals beginning on Julian date 113 (April 23) for cohort 1. Cohort 7 started on Julian date 176 (June 25), and cohort 8 on Julian date 204 (July 23). Each cohort remained on the tree through completion of larval development, 6–7 weeks. Budbreak occurred during Julian dates 118–136 and shoot elongation during Julian dates 118–170. From Lawrence *et al.* (1997) by permission from the Entomological Society of Canada.

known plants protected by aggressive ants, *Azteca* spp., housed in its hollow stems (Rickson, 1977). Bullthorn acacia is defended against herbivores by colonies of aggressive ants housed in its thorns (Janzen, 1966). Many species of plants produce extrafloral nectaries that attract ants for protection. A variety of insect and other species gain protection through adaptation that permit them to mimic ants (Blum, 1980, 1981) or to live in ant or termite colonies.

Both plants and insects produce a remarkable range of compounds that have been the source of important pharmaceuticals, as well as effective defenses. Biochemical interactions between herbivores and their host plants and between predators and their prey have been one of the most stimulating areas of ecological and evolutionary research since the 1970s. This topic cannot be treated adequately in this chapter, but major points affecting ecological processes are summarized following. Readers desiring additional information are referred to Bernays (1989), Bernays and Chapman (1994), Brown and Trigo (1995), Coley and Barone (1996), Edwards (1989), Harborne (1994), Hedin (1983), Rosenthal and Berenbaum (1991, 1992), and Rosenthal and Janzen (1979).

C. Plant Chemical Defenses

Plant chemical defenses generally are classified as nonnitrogenous and nitrogenous. Ecologically, this distinction reflects the availability of C versus N for allocation to defense at the expense of maintenance, growth, and reproduction. Each of these categories is represented by innumerable compounds, often differing only in the structure and composition of attached radicals.

1. Nonnitrogenous Defenses

Nonnitrogenous defenses include phenolics, terpenoids, pyrethroids, photooxidants, aflatoxins, and insect hormone analogs (Figs. 3.2–3.5). Phenolics, or flavenoids, are distributed widely among terrestrial plants and are likely among the oldest of plant secondary (i.e., nonmetabolic) compounds. Although phenolics are perhaps best known as defenses against herbivores and plant pathogens, they also protect plants from damage by UV radiation, provide support for vascular plants (lignins), are the pigments that determine flower color for angiosperms, and play a role in nutrient acquisition by affecting soil chemistry. Phenolics include the hydrolyzable tannins, derivatives of simple phenolic acids, and condensed tannins, polymers of higher molecular weight hydroxyflavenol units (Fig. 3.3). Polymerized tannins are highly resistant to decomposition, eventually composing the humic materials that largely determine soil properties. Tannins are distasteful, typically bitter and astringent, and act as feeding deterrents for many herbivores. When ingested, tannins chelate N-bearing molecules to form indigestible complexes (Feeny, 1969). Insects incapable of catabolizing tannins or preventing chelation suffer gut damage and inability to assimilate nitrogen from their food. Some flavenoids, such as rotenone, are directly toxic to insects and other animals.

Terpenoids are widely represented among plant groups. These compounds are synthesized by linking isoprene subunits. The lower molecular weight monoterpenes and sesquiterpenes are highly volatile compounds that function as floral scents that attract pollinators and other plant scents that herbivores use to recognize their hosts. Terpenoids with higher molecular weights include plant resins, cardiac glycosides, and saponins (Figs. 3.2 and 3.3). Terpenoids typically are distasteful or toxic to herbivores. In addition, they are primary resin components of pitch, produced by many plants to seal wounds. Pitch flow in response to injury by insect feeding can physically push the insect away, deter further feeding, and kill associated microorganisms (Nebeker *et al.*, 1993).

Becerra (1994) reported that the tropical succulent shrub, *Bursera schlechtendalii*, stores terpenes under pressure in a network of canals in its leaves and

FIG. 3.2 The wound response of conifers constitutes a physical–chemical defense against invasion by insects, such as this bark beetle, and pathogens. The oleoresin, or pitch, flowing from severed resin ducts hinders penetration of the bark.

stems. When these canals are broken, the terpenes are squirted up to 150 cm, bathing the herbivore and drenching the leaf surface. A specialized herbivore, the chrysomelid, *Blepharida* sp., partially avoids this defense by severing leaf veins before feeding, but nevertheless suffers high mortality and may spend more time cutting veins than feeding, thereby suffering reduced growth.

Cardiac glycosides are terpenoids best known as the compounds in milkweeds (Euphorbiaceae) sequestered by monarch butterflies to deter predation by birds. Ingestion of these compounds by vertebrates either induces vomiting or results in cardiac arrest.

Photooxidants, such as the quinones (Fig. 3.3) and furanocoumarins, increase epidermal sensitivity to solar radiation. Assimilation of these compounds can result in severe sunburn, necrosis of the skin, and other epidermal damage upon exposure to sunlight. Feeding on furanocoumarin-producing plants in daylight can cause 100% mortality to insects whereas feeding in the dark causes only 60% mortality. Insect herbivores can circumvent this defense by becoming leaf rollers or nocturnal feeders (Harborne, 1994) or by sequestering antioxidants (Blum, 1992).

Insect development and reproduction are governed primarily by two hormones, molting hormone (ecdysone) and juvenile hormone (Fig. 3.4). The relative concentrations of these two hormones dictate the timing of ecdysis and the subsequent stage of development. A large number of phytoecdysones have

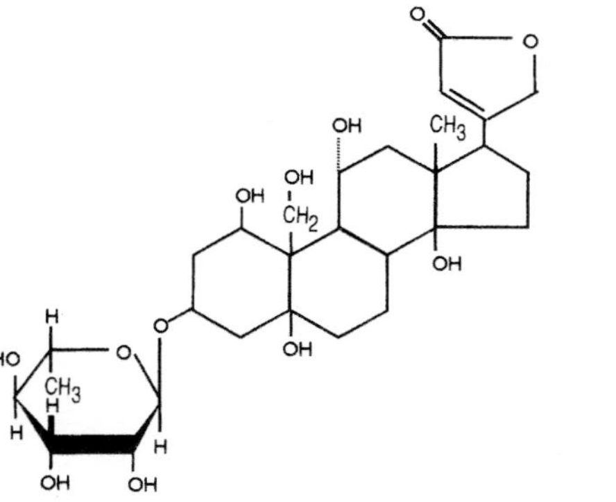

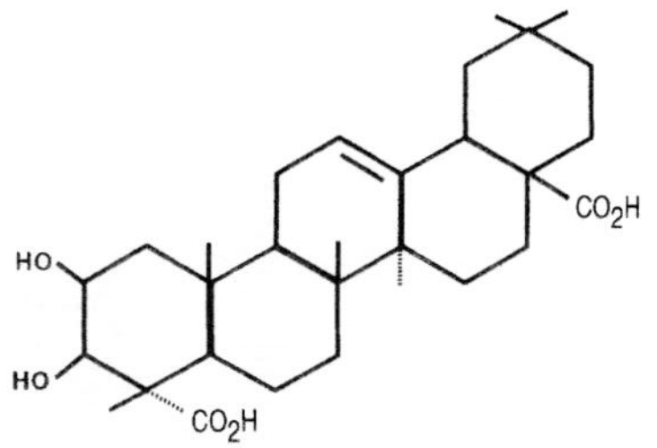

Terpenoid saponin, medicagenic acid, from *Medicago sativa*

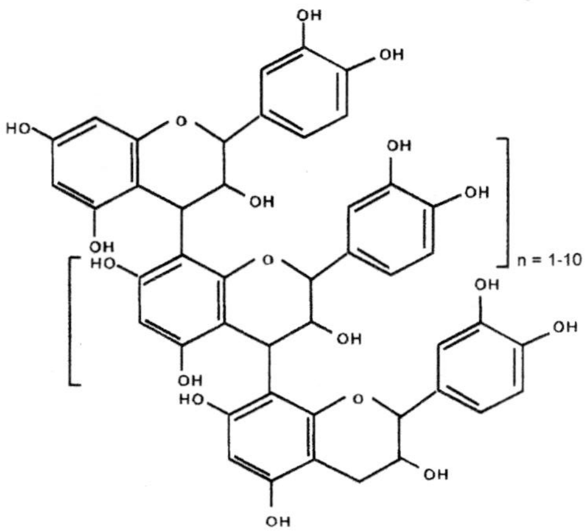

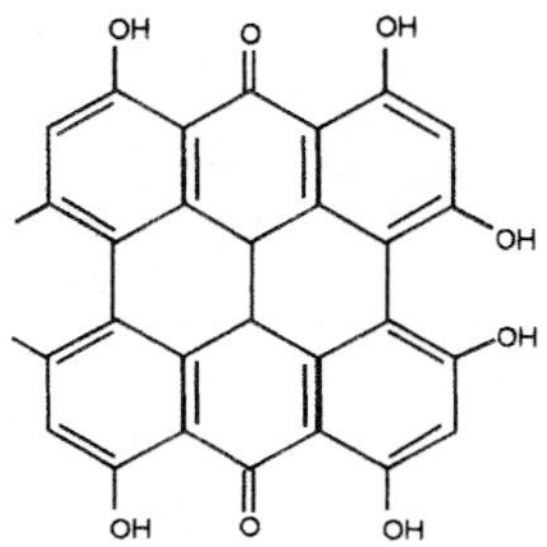

Quinone, hypericin, from *Hypericum perforatum*

Terpenoid cardiac glycoside, ouabain, from *Acokanthera ouabaio*

Flavonoid tannin, procyanidin, from *Quercus* spp.

FIG. 3.3 Examples of nonnitrogenous defenses of plants. From Harborne (1994).

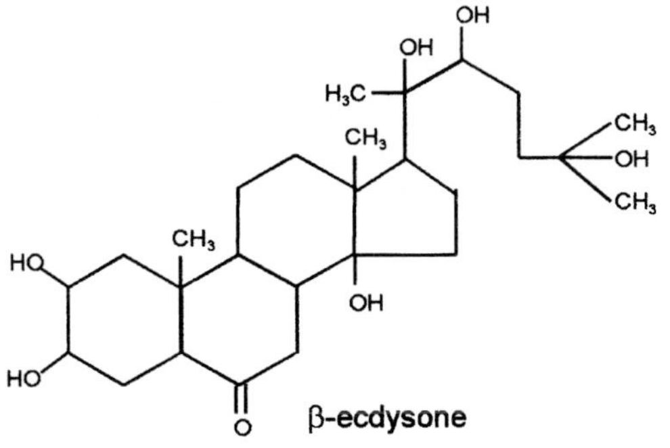

β-ecdysone

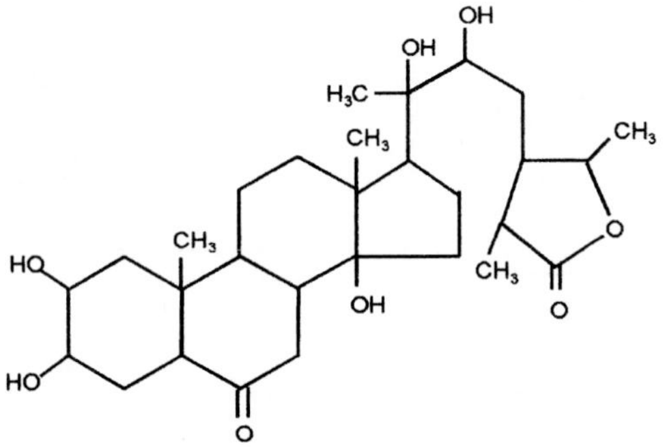

Cycasterone from *Cycas*

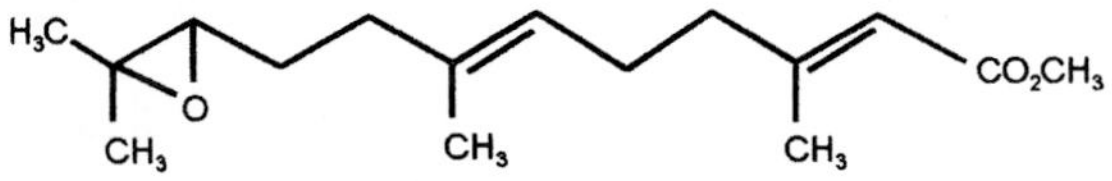

Juvenile hormone III

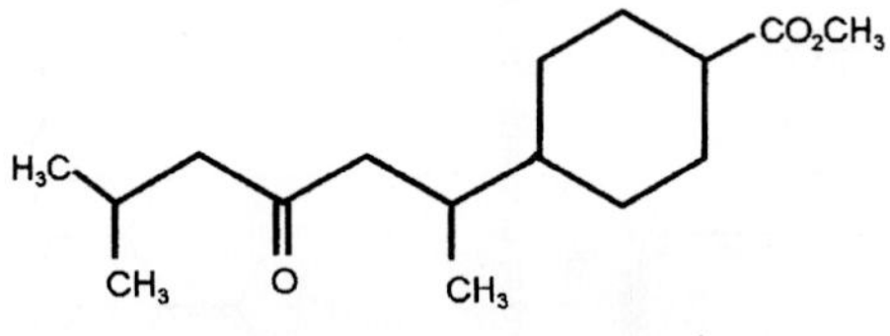

Juvabione from *Abies balsamea*

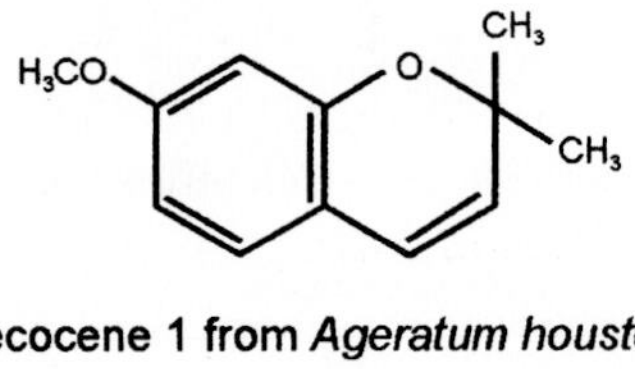

Precocene 1 from *Ageratum houstoniatum*

FIG. 3.4 Insect developmental hormones and examples of their analogues in plants. From Harborne (1994).

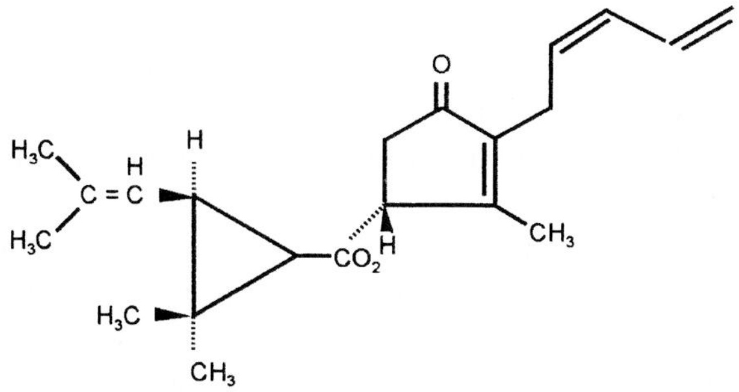

Pyrethrin I, from *Chrysanthemum cinearifolium*

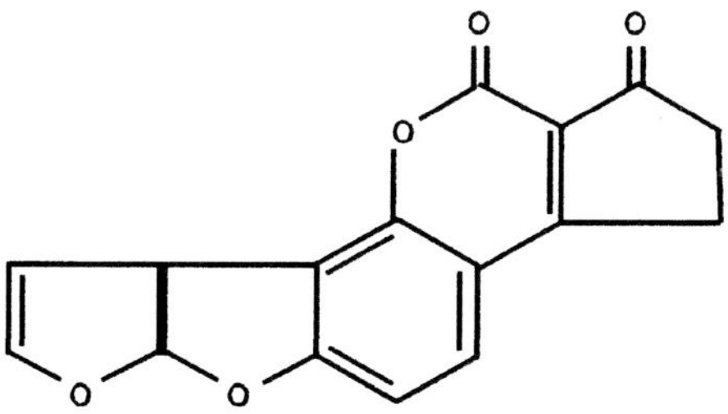

Aflatoxin B, from *Aspergillus flavus*

FIG. 3.5   Examples of pyrethroid and aflatoxin defenses. From Harborne (1994).

been identified, primarily from ferns and gymnosperms. Some of the phytoecdysones are as much as 20 times more active than the ecdysones produced by insects (Harborne, 1994). These compounds resist inactivation by insects. Plants also produce some juvenile hormone analogues (primarily juvabione) and compounds that interfere with juvenile hormone activity (primarily precocene). The antijuvenile hormones typically cause precocious development. Plant-derived hormone analogues are highly disruptive to insect development, typically preventing maturation or producing imperfect and sterile adults (Harborne, 1994).

Many other categories of nonnitrogenous compounds also have been identified. Pyrethroids (Fig. 3.5) are an important group of contact toxins, i.e., absorbed through the exoskeleton. Aflatoxins (Fig. 3.5) produced by fungi are highly toxic to vertebrates and, perhaps, to invertebrates (Carroll, 1988; Harborne, 1994). Higher plants may augment their own defensive chemistry through mutualistic associations with endophytic or mycorrhizal fungi (Carroll, 1988; Clay, 1990; Clay *et al.*, 1993).

2. Nitrogenous Defenses

Nitrogenous defenses include nonprotein amino acids, cyanogenic glucosides, glucosinolates, and alkaloids (Fig. 3.6). These compounds are highly toxic due to their interference with numerous animal physiological processes.

Nonprotein amino acids are analogues of essential amino acids (Fig. 3.6). Their substitution for essential amino acids in proteins results in improper configuration, loss of enzyme function, and inability to maintain other functions critical to survival. Some nonprotein amino acids are toxic for other reasons, such as interference with tyrosinase (an enzyme critical to hardening of the insect cuticle) by 3,4-dihyrophenylalanine (L-DOPA). Over 300 nonprotein amino acids are known, primarily from seeds of legumes (Harborne, 1994).

Cyanogenic glycosides are distributed widely among plant families (Fig. 3.6). These compounds are inert in the glycoside form in plant cells. Plants also produce specific enzymes to control hydrolysis of the glycoside. When crushed plant cells enter the herbivore gut, the glycoside is hydrolyzed into glucose and a cyanohydrin, which spontaneously decomposes into a ketone or aldehyde, and hydrogen cyanide. Hydrogen cyanide is toxic to most organisms because of its inhibition of cytochromes in the electron transport system.

Alkaloids include over 5000 known structures from about 20% of higher plant families (Harborne, 1994). Molecules range in size from the relatively simple coniine of poison hemlock (Fig. 3.6) to multicyclic compounds, such as solanine. Familiar examples include atropine, caffeine, nicotine, belladonna, digitalis, and strychnine. They are highly toxic and teratogenic, even at relatively low concentrations, because of their interference with major physiological processes, especially cardiovascular and nervous system functions.

D. Arthropod Defenses

Arthropods also employ various defenses against predators and parasites. Physical defenses include hardened exoskeleton, spines, claws, and mandibles. Chemical defenses are nearly as varied as plant defenses. Hence, predaceous species also must be capable of evaluating and exploiting defended prey resources. The compounds used by arthropods, including predaceous species, generally belong to the same categories of compounds already described for plants.

Many insect herbivores sequester plant defenses for their own use (Blum, 1981, 1992). The relatively inert exoskeleton provides an ideal site for storage of toxic compounds. Toxins can be stored in scales on the wings of Lepidoptera, e.g., cardiac glycoside storage in the wings of monarch butterflies. Some insects make more than such passive use of their sequestered defenses. Sawfly larvae store the resinous defenses from host conifer foliage in diverticular pouches in the foregut and regurgitate the fluid to repel predators (Codella and Raffa, 1993).

Many arthropods synthesize their own toxins (Meinwald and Eisner,

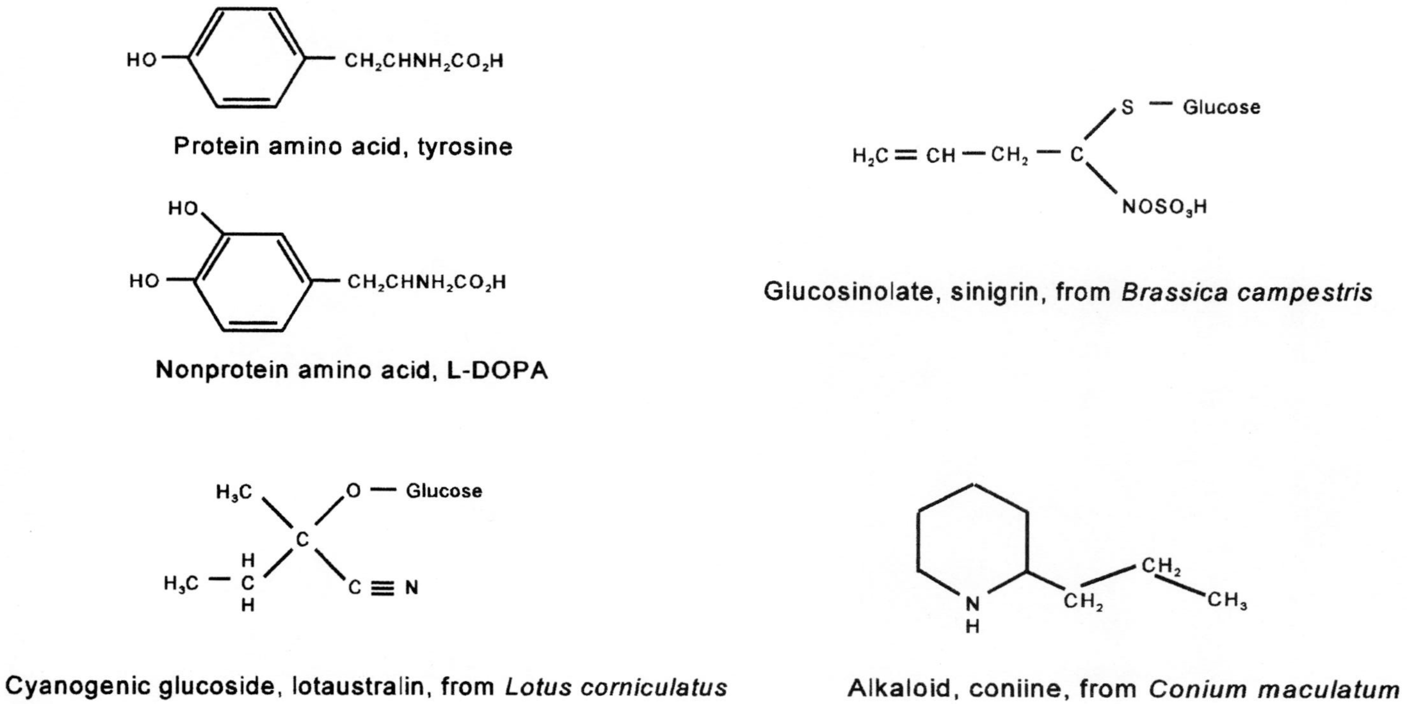
HO
CH₂CHNH₂CO₂H
Protein amino acid, tyrosine
HO
HO
CH₂CHNH₂CO₂H
Nonprotein amino acid, L-DOPA
H₂C = CH — CH₂ — C
S — Glucose
NOSO₃H
Glucosinolate, sinigrin, from Brassica campestris
H₃C
O — Glucose
C
H₃C — C
H
H
C ≡ N
Cyanogenic glucoside, lotaustralin, from Lotus corniculatus
N
H
CH₂
CH₂
CH₃
Alkaloid, coniine, from Conium maculatum

FIG. 3.6 Examples of nitrogenous defenses of plants. From Harborne (1994).

FIG. 3.7   Defensive froth of an adult lubber grasshopper, *Romalea guttata*. This secretion includes repellent chemicals sequestered from host plants. From Blum (1992) by permission from the
Entomological Society of America.

1995). Irritant froths or sprays are produced by a number of arthropods. Several grasshopper and beetle species exude irritating or repellent froths when disturbed (Fig. 3.7). Examples include the terpenoid, cantharidin, synthesized by
blister beetles (Meloidae) and the alkaloid, coccinelline, synthesized by lady-

bird beetles (Coccinellidae) (Meinwald and Eisner, 1995). Both compounds are unique to insects. These compounds occur in the hemolymph and are exuded by reflex bleeding from leg joints. They deter both invertebrate and vertebrate predators. Whiptail scorpions spray acetic acid from their "tails," and the millipede, *Harpaphe*, sprays cyanide (Meinwald and Eisner, 1995). The bombardier beetle, *Brachynus*, sprays a hot (100°C) cloud of benzoquinone produced by mixing, at the time of discharge, a phenolic substrate (hydroquinone), peroxide, and an enzyme catalase (Harborne, 1994).

Several arthropod groups produce venoms, primarily peptides, that include phospholipases, histamines, proteases, and esterases, for defense as well as predation (Meinwald and Eisner, 1995; Schmidt, 1982). Both neurotoxic and hemolytic venoms are represented among insects. Phospholipases are particularly well known because of their high toxicity and their strong antigen activity capable of inducing life-threatening allergy. Larvae of several families of Lepidoptera, especially the Saturniidae and Limacodidae (Fig. 3.8), deliver venoms passively through urticating spines, although defensive flailing behavior by many species increases the likelihood of striking an attacker. A number of Heteroptera, Diptera, Neuroptera, and Coleoptera produce orally derived venoms that facilitate prey capture as well as defense (Schmidt, 1982). Venoms are particularly well known among the Hymenoptera and consist of a variety of enzymes, biogenic amines (such as histamine and dopamine), epinephrine, norepinephrine, and acetylcholine. This combination produces severe pain and affects cardiovascular, central nervous, and endocrine systems in vertebrates (Schmidt, 1982). Some venoms include nonpeptide components. For example, venom of the imported fire ant, *Solenopsis invicta*, contains piperidine alkaloids, with hemolytic, insecticidal, and antibiotic effects.

Arthropods also defend themselves against internal parasites and pathogens. Major mechanisms include ingested antibiotics (Blum, 1992; Tallamy *et al.*, 1998), gut modifications that prevent growth or penetration by pathogens, and cellular immunity against parasites and pathogens in the hemocoel (Tanada and Kaya, 1993). For example, Lepidoptera susceptible to *Bacillus thuringiensis* typically have high gut pH and large quantities of reducing substances and proteolytic enzymes, conditions that facilitate dissolution of the crystal protein and subsequent production of the delta-endotoxin. By contrast, resistant species have a lower gut pH and lower quantities of reducing substances and proteolytic enzymes (Tanada and Kaya, 1993).

Cellular immunity is based on cell recognition of "self" and "nonself," and includes endocytosis and cellular encapsulation. Endocytosis is the process of infolding of the plasma membrane and enclosure of foreign substances within a phagocyte, without penetration of the plasma membrane. This process removes viruses, bacteria, fungi, protozoans, and other foreign particules from the hemolymph, although some of these pathogens then can infect the phagocytes. Cellular encapsulation occurs when the foreign particle is too large to be engulfed by phagocytes. Aggregation and adhesion by hemocytes forms a dense

FIG. 3.8 Physical and chemical defensives of a limacodid (Lepidoptera) larva, *Isa textula*. The urticating spines can inflict severe pain on attackers.

covering around the particle. Surface recognition may be involved because parasitoid larvae normally protected (by viral associates) from encapsulation are encapsulated when wounded or their surfaces are altered (Tanada and Kaya, 1993). Hemocytes normally encapsulate hyphae of the fungus, *Entomophtho-*

ra egressa, but do not adhere to hyphal bodies that have surface proteins protecting them from attachment of hemocytes (Tanada and Kaya, 1993).

E. Factors Affecting Expression of Defenses

Some plant groups are characterized by particular defenses. For example, ferns and gymnosperms rely primarily on terpenoids and insect hormone analogues, whereas angiosperms more commonly produce alkaloids, phenolics, and many other types of compounds. However, most plants apparently produce compounds representing a variety of chemical classes (Harborne, 1994; Newman, 1990). Each plant species can be characterized by a unique "chemical fingerprint" conferred by these chemicals. Production of alkaloids and other physiologically active nitrogenous defenses depends on the availability of nitrogen (Harborne, 1994). However, at least four species of spruce and seven species of pines are known to produce piperidine alkaloids (Stermitz *et al.*, 1994), despite high C:N ratios. Feeding by phytophagous insects can be reduced substantially by the presence of plant defensive compounds, but insects also identify potential hosts by their chemical fingerprints.

Defensive compounds may be energetically expensive to produce, and their production competes with production of other necessary compounds and tissues (Chapin *et al.*, 1987; Herms and Mattson, 1992). Some, such as the complex phenolics and terpenoids, are highly resistant to degradation and cannot be catabolized to retrieve constituent energy or nutrients for other needs. Others, such as alkaloids and nonprotein amino acids, can be catabolized and the nitrogen, in particular, retrieved for other uses, but such catabolism involves metabolic costs that reduce net gain in energy or nutrient budgets. Given the energy requirements and competition among metabolic pathways for limiting nutrients, production of defensive compounds should be sensitive to risk of herbivory or predation and to environmental conditions (Chapin *et al.*, 1987; Coley, 1986; Coley, *et al.*, 1985; Herms and Mattson, 1992; Hunter and Schultz, 1995; Karban and Niiho, 1995).

Organisms are subjected to a variety of selective factors in the environment. Intense herbivory is only one factor that affects plant fitness. Plant genotype also is selected by climatic and soil conditions, various abiotic disturbances, etc. Factors that select intensively and consistently among generations are most likely to result in directional adaptation. The variety of biochemical defenses against herbivores testifies to the significance of herbivory in the past. Nevertheless, at least some biochemical defenses have multiple functions (e.g., phenolics as UV filters, pigments, and structural components, as well as defense), implying that their selection was enhanced by meeting multiple plant needs. Similarly, insect survival is affected by climate, disturbances, and condition of host(s) as well as a variety of predators. Short generation time confers a capacity to adapt quickly to strong selective factors, such as consistent and widespread exposure to particular plant defenses.

Plants balance the trade-off between the expense of defense and the risk of severe herbivory (Coley, 1986; Coley *et al.*, 1985). Plants are capable of producing **constitutive defenses** that are present in plant tissues at any given time and determine the chemical fingerprint of the plant and **inducible defenses** that are produced in response to injury (e.g., Haukioja, 1990; Nebeker *et al.*, 1993). Constitutive defenses might be expected to consist primarily of relatively less specific, but generally effective, compounds, whereas inducible defenses might be expected to consist of more specific compounds produced in response to particular types of injury. Hence, pitch, consisting of relatively low molecular weight terpenoids is a constitutive defense of many conifers that seals wounds and prevents penetration of the bark by insects (Fig. 3.2). Successful penetration of this barrier by bark beetles induces production of more complex phenolics that cause necrosis and lesion formation in the phloem and cambium tissues surrounding the wound and kill the beetles and associated microorganisms (Nebeker *et al.*, 1993).

Tissues vary in their concentration of defensive compounds, depending on risk of herbivory and value to the plant (Dirzo, 1984; Feeny, 1970; McKey, 1979; Scriber and Slansky, 1981). Foliage tissues, which are the source of photosynthates and have a high risk of herbivory, typically have high concentrations of defensive compounds. Similarly, defensive compounds in shoots are concentrated in bark tissues, perhaps reducing risk to subcortical tissues, which have relatively low concentrations of defensive compounds (e.g., Schowalter *et al.*, 1992). Defensive compounds generally accumulate during leaf maturation, reflecting the increasing value of leaves as sources, rather than sinks, of photosynthates. This results in a concentration of herbivore activity during periods of leaf emergence (Coley and Aide, 1991; Feeny, 1970; Hunter and Schultz, 1995; Jackson *et al.*, 1999; Lowman, 1985, 1992; McKey, 1979).

Defensive strategies also change as plants mature (Dirzo, 1984). A visible example is the reduced production of thorns by black locust, *Robinia pseudoacacia*, during the transition from sapling to tree. Seasonal growth patterns also affect plant defense. Lorio (1993) reported that production of resin ducts by loblolly pine, *Pinus taeda*, is restricted to latewood formed during summer. The rate of earlywood formation in the spring determines the likelihood that southern pine beetles, *Dendroctonus frontalis*, colonizing trees in spring will sever resin ducts and induce pitch flow. Hence, tree susceptibility to colonization by this insect increases with stem growth rate.

Healthy plants growing under optimal environmental conditions are capable of meeting the full array of metabolic needs and may provide greater nutritional value to insects capable of countering plant defenses. However, unhealthy plants or plants growing under adverse environmental conditions (such as water or nutrient limitation) favor some metabolic pathways over others (Herms and Mattson, 1992; Lorio, 1993; Mattson and Haack, 1987; Tuomi *et al.*, 1984; Waring and Pitman, 1983). In particular, maintenance and replacement of photosynthetic (foliage), reproductive, and support (root) tissues represent

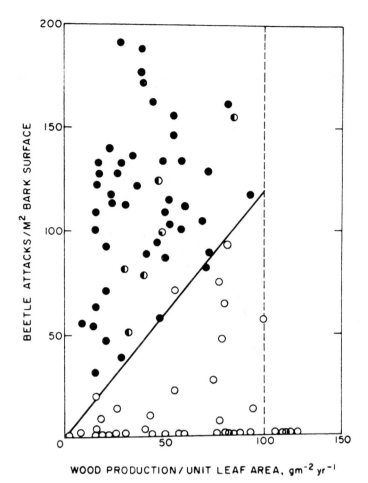

FIG. 3.9 The density of mountain pine beetle (*Dendroctonus ponderosae*) attacks necessary to kill lodgepole pine increases with increasing host vigor, measured as growth efficiency. The blackened portion of circles represents the degree of tree mortality. The solid line indicates the attack level predicted to kill trees of a specified growth efficiency (index of radial growth); the dotted line indicates the threshold above which beetle attacks are unlikely to cause mortality. From Waring and Pitman (1983) by permission from Blackwell Wissenschafts Verlag GmbH.

higher metabolic priorities than does production of defensive compounds under conditions that threaten survival. Therefore, stressed plants often sacrifice production of defenses in order to maximize allocation of limited resources to maintenance pathways and thereby become relatively more vulnerable to herbivores (Fig. 3.9).

Spatial and temporal variability in plant defensive capability creates variation in food suitability for herbivores (Brower *et al.*, 1968). In turn, herbivore employment of plant defenses affects their vulnerability to predators (Brower *et al.*, 1968; Malcolm, 1992; Stamp *et al.*, 1997; Traugott and Stamp, 1996). Herbivore feeding strategies represent a trade-off between maximizing food quality and minimizing vulnerability to predators (e.g., Schultz, 1983, see following).

The frequent association of insect outbreaks with stressed plants, including plants stressed by atmospheric pollutants (e.g., Brown, 1995; Heliövaara, 1986; Heliövaara and Väisänen, 1986, 1993; Smith, 1981), led White (1969, 1976, 1984) to propose the **plant stress hypothesis**, i.e., that stressed plants are more suitable hosts for herbivores. However, experimental studies have indicated that some herbivore species prefer more vigorous plants (Waring and Price, 1990), leading Price (1991) to propose the alternative **plant vigor hypothesis**. Reviews by Koricheva et al. (1998) and Waring and Cobb (1992) revealed that responses to plant condition varies widely among herbivore species. Schowalter et al. (1999) manipulated water supply to creosote bushes, Larrea tridentata, in New Mexico and found positive, negative, nonlinear, and nonsignificant responses to moisture availability among the assemblage of herbivore species on this single plant species.

Regardless of the direction of response, water and nutrient subsidy or limitation clearly affect herbivore–plant interactions (Coley et al., 1985; Hunter and Schultz, 1995; Mattson and Haack, 1987). Therefore, resource acquisition is moderated, at least in part, by ecosystem processes that affect the availability of water and nutrients.

Some plant species respond to increased atmospheric concentrations of CO_2 by allocating more carbon to defenses, especially if other critical nutrients, such as water or nitrogen, remain limiting (Arnone et al., 1995; Chapin et al., 1987; Grime et al., 1996; Kinney et al., 1997; Roth and Lindroth, 1994). However, plant responses to CO_2 enrichment vary considerably among species and as a result of environmental conditions such as light, water, and nutrient availability (Bazzaz, 1990; Dudt and Shure, 1994; Edwards, 1989; Niesenbaum, 1992), with equally varied responses among herbivore species (e.g., Salt et al., 1996; Watt et al., 1995). Such complexity of factors interacting with atmospheric CO_2 precludes general prediction of effects of increased atmospheric CO_2 on insect–plant interactions (Bazzaz, 1990; Watt et al., 1995).

F. Mechanisms for Exploiting Variable Resources

In a classic paper that stimulated much subsequent research on factors affecting herbivory, Hairston et al. (1960) argued that herbivore populations are not limited by food supply because vegetation is normally abundant and herbivores, when numerous, are able to deplete plant resources. We now know, as described previously, that plant resources are not equally suitable or acceptable and that herbivore populations often are limited by availability of suitable food. Herbivore populations are regulated by a combination of factors, as discussed in Chapter 6. The various defenses already outlined are responsible for the feeding preferences observed among herbivores and predators. However, feeding preferences reflect a variety of mechanisms for detoxifying, avoiding, or circumventing host defenses.

Insects can produce a variety of catalytic enzymes to detoxify plant or prey defenses. These enzymes typically are microsomal monooxygenases, glutathione S-transferases, and carboxylesterases (Hung *et al.*, 1990) that fragment defensive compounds into inert molecules. Microsomal monooxygenases are a general-purpose detoxification system in most herbivores and have higher activity in generalist species, compared to specialist species or sap-sucking species that contact only water-soluble materials (Hung *et al.*, 1990). More specific digestive enzymes also are produced by some species. The compounds produced through detoxification pathways may be used to meet the insect's nutritional needs (Bernays and Woodhead, 1982), as in the case of the sawfly, *Gilpinia hercyniae,* that detoxifies and uses the phenolics from its conifer host (Schöpf *et al.*, 1982).

Gut pH is a factor affecting the chelation of nitrogenous compounds by tannins. Some insect species are adapted to digest food at high gut pH to inhibit chelation. The insect thus is relatively unaffected by high tannin contents of its food. Examples include the gypsy moth, *Lymantria dispar*, feeding on oak, *Quercus* spp., and chrysomelid beetles, *Paropsis atomaria*, feeding on *Eucalyptus* spp. (Feeny, 1969; Fox and Macauley, 1977).

Several physiological or behavioral mechanisms are employed to avoid or circumvent host defensive chemicals. Life history phenology of many species is synchronized with periods of most favorable host nutritional chemistry (Feeny, 1970; Varley and Gradwell, 1970). Diapause can be an important mechanism for surviving periods of adverse host conditions as well as adverse climatic conditions. In fact, diapause during certain seasons may reflect seasonal patterns of resource availability more than abiotic conditions. For example, many tropical herbivores become dormant during the dry season when their host plants cease production of foliage or fruit and become active again when production of foliage and fruit resumes in the wet season. Diapause can be prolonged in cases of unpredictable availability of food resources, as in the case of insects feeding on seeds of trees that produce seed crops irregularly. Turgeon *et al.* (1994) reported that 70 species of Diptera, Lepidoptera, and Hymenoptera that feed on conifer cones or seeds can remain in diapause for as long as seven years. In other words, the adult insect population emerging to reproduce in any given year can represent cohorts that developed during each of the previous seven years.

Sequestration and excretion are somewhat related means of avoiding the effects of host toxins. In both cases, toxic compounds are transported quickly to specialized storage tissues (the exoskeleton or protected pouches) or to the Malphigian tubules for isolation from metabolically active tissues. Sequestered toxins become part of the insect's own defensive strategy (Blum, 1981, 1992).

Some herbivores sever the petiole or major leaf veins to inhibit translocation of induced defenses during feeding (Becerra, 1994). Sawflies sever the resin canals of their conifer hosts or feed gregariously to consume foliage before defenses can be induced (McCullough and Wagner, 1993). Species feeding on plants with photooxidant defenses often feed at night or inside rolled leaves to avoid sunlight. The carabid beetle, *Promecognathus*, a specialist predator on *Harpaphe* and other polydesmid millipedes, avoids the defensive secretions of its prey by quickly biting through the ventral nerve cord at the neck, inducing paralysis. Nevertheless, these defenses increase handling time and risk of injury and mortality for the consumer (Becerra, 1994).

Many predaceous insects use their venoms primarily for subduing prey and secondarily for defense. Venoms produced by predaceous Heteroptera, Diptera, Neuroptera, Coleoptera, and Hymenoptera function either to paralyze or kill prey (Schmidt, 1982), thereby minimizing injury to the predator during prey capture.

II. RESOURCE ACCEPTABILITY

The variety of resources and their physical and biochemical properties, including defensive mechanisms, is too great in any ecosystem for any species to exploit all possible resources. The particular physiological and behavioral adaptations of insects to obtain sufficient nutrients and avoid toxic or undigestable materials determine their feeding preferences, i.e., which resources they can exploit. Insects specialized to exploit particular physical and chemical conditions often lose their ability to exploit other resources. Even species feeding on a wide variety of resource types (e.g., host species) have limited ability to exploit all potential resources. For example, the gypsy moth feeds on a variety of plant species (representing many plant families) that generally contain phenolics but not terpenoids or alkaloids (Miller and Hanson, 1989).

Particular compounds can be effective defenses against nonadapted herbivores and, at the same time, be phagostimulants for adapted herbivores. For example, Tallamy *et al.* (1997) reported that cucurbitacins (bitter triterpenes characterizing the Cucurbitaceae) deter feeding and oviposition by nonadapted mandibulate insect herbivores but stimulate feeding by haustellate insect herbivores.

Malcolm (1992) identified three types of consumers with respect to a chemically defended prey species. Excluded predators cannot feed on the chemically defended prey, whereas included predators can feed on the chemically defended prey with no ill effect. Peripheral predators experience growth loss, etc., when fed chemically defended prey, due to the effects of the defensive chemicals on predator physiology or on the nutritional quality of the prey. The effectiveness of peripheral predators on prey differing in chemical defense may be a key to understanding the ecology and evolution of predator/prey interactions. Choice of resource(s) generally depends on three integrated factors: resource suitability, susceptibility, and acceptability.

Resource suitability represents the nutritional value of the resource and reflects both the nutrients available to the insect and the concentration of defenses that must be detoxified. The nutrients represent benefits, and the defenses represent costs, of the resource. Just as production of defensive compounds is expensive for the host in terms of energy and resources, production of detoxification enzymes or development of avoidance mechanisms are expensive in terms of energy, resources, time searching, and exposure to predators. Some of the nutrients in the food must be allocated to production of detoxification enzymes or to energy expended in searching for more suitable food (see Chapter 4). However, defenses can have beneficial side effects for the consumer. Hunter and Schultz (1993) found that phenolic defenses in oak leaves reduce susceptibility of gypsy moth larvae to nuclear polyhedrosis virus.

Resource susceptibility represents the physiological condition of the host, whether for herbivores or predators. Injury or adverse environmental conditions can stress organisms and impair their ability to defend themselves. Initially, stress may prevent expression of induced defenses, an added cost, or may prevent production of constitutive defenses but allow induction of defenses as needed. Nutrient limitation may prevent production of nitrogenous defenses but increase production of nonnitrogenous defenses. Reduced production of biochemical defenses reduces the cost of detoxification or avoidance to the predator. Hence, specialist species can allocate more energy and resources to growth and reproduction, and generalists may be able to expand their host range as biochemical barriers are removed.

Resource acceptability represents the willingness of the insect to feed, given the probability of finding more suitable resources or in view of other trade-offs. Most insects have relatively limited time and energy resources to spend searching for food. Hence, marginally suitable resources may become sufficiently profitable when the probability of finding more suitable resources is low, such as in diverse communities composed primarily of nonhosts. Courtney (1985, 1986) reported that oviposition by a pierid butterfly, *Anthocharis cardamines*, among several potential host plant species was inversely related to the suitability of those plant species for larval development and survival (Fig. 3.10). The more suitable host plant species were relatively rare and inconspicuous compared to the less suitable host species. Hence, butterfly fitness was maximized by laying eggs on the most conspicuous plants, thereby ensuring reproduction, rather than by risking death during continued searching for more suitable hosts. Nevertheless, insects forced to feed on less suitable resources show reduced growth and survival rates (Bozer *et al.*, 1996; Courtney 1985, 1986).

Insects face an evolutionary choice between maximizing the range of resources exploited (generalists) or maximizing the efficiency of exploiting a particular resource (specialists). Generalists maximize the range of resources exploited through generalized detoxification or avoidance mechanisms, such as broad-spectrum microsomal monooxygenases, but sacrifice efficiency in exploiting any particular resource because some unique biochemicals will reduce digestion or survival (Bowers and Puttick, 1988). However, generalists may in-

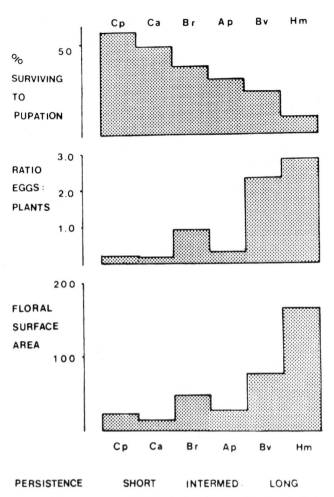

FIG. 3.10 Trade-off between plant suitability for larval survival (top) and efficiency of oviposition site selection by adult pierids, *Anthocharis cardamines*, as indicated by the ratio of eggs per host species (middle) and plant apparency, i.e., floral surface area and longevity (bottom). Searching females preferentially oviposit on the most conspicuous plants, although these are not the most suitable food plants for their larvae. Cp, *Cardamine pratensis*; Ca, *C. amara*; Br, *Brassica rapa*; Ap, *Allaria petiolate*; Br, *Barbarea vulgaris*, and Hm, *Hesperis matronalis*. From Courtney (1985) by permission from Oikos.

crease their effectiveness by exploiting stressed hosts that have sacrificed production of defenses. Specialists maximize the efficiency of exploiting a particular host through more specific detoxification or avoidance strategies, minimizing the effect of more of the host's constitutive and induced defenses, but sacrifice ability to feed on other species with different defenses (Bowers and Puttick 1988).

Specialists and generalists contribute to host population dynamics and to community structure and function in different ways, as described in Chapters

6 and 8. Generalists typically exploit more abundant host species and may reduce competition among hosts but do not effectively target rapidly increasing
populations. By contrast, specialists focus on particular host species and control host population growth more effectively but must be able to discover
sparsely distributed hosts.

Insects initially select particular host species. They then select their particular host tissues based on nutritional value. Nutritional quality of foliage is
higher than that of root tissue. Nutritional value varies between bark, sapwood,
and heartwood tissues (Hodges *et al.*, 1968; Schowalter *et al.*, 1998). In fact,
exploitation of sapwood requires mutualistic interaction with fungi or bacteria
or other adaptations to acquire sufficient nutrients from a resource that is largely indigestible cellulose (see Chapter 8). Nutritional quality can vary even within tissues. For example, insects may target particular portions of leaves, based
on balance among amino acids (Haglund, 1980; Parsons and de la Cruz, 1980),
and particular heights on tree boles, based on balances among amino acids and
carbohydrates (Hodges *et al.*, 1968).

Many insects feed on different resources at different stages of development.
Most Lepidoptera feed on plant foliage, stems, or roots as larvae, but many feed
on nectar as adults. Some cerambycid beetles feed on wood as larvae but feed
on pollen or nectar as adults. A number of Diptera and Hymenoptera have
predaceous or parasitic larvae but feed on pollen or nectar resources as adults.
Many aphids alternate generations between two host plant species (Dixon,
1985). Clearly, these changes in food resources require changes in digestive abilities between life stages. Furthermore, population survival requires the presence
of all necessary resources at an appropriate landscape scale.

The primacy of resource exploitation for development and survival places
strong selective pressure on insects to adapt to changing host quality. This has
led to the so-called "evolutionary arms race," in which herbivory selects for
new plant defenses and the new plant defenses select for insect countermea-
sures. This process has driven reciprocal speciation in both plants and insects
(Becerra, 1997). The long exposure of insects to a wide variety of host (especially plant) toxins has led to flexible detoxification and other mechanisms to
circumvent those defenses, especially among generalists.

Generalists are especially likely to adapt rapidly to new crop varieties, including transgenic crops, and to insecticides (Hung *et al.*, 1990), especially
when these are used consistently over wide areas. For example, over 500 insect
species targeted by crop protection strategies are now resistant to a variety of
insecticides. Even though transgenic (Bt toxin-expressing) crops, such as corn,
cotton, and potato, have been available for a relatively short time, their widespread planting has maximized exposure of herbivorous insects to the Bt toxin
and led to development of resistance to this new pest management tool. Such
directional selection is contrary to the principles of integrated pest management
(IPM) that include using a variety of suppression tactics to retard selection for
resistance.

III. RESOURCE AVAILABILITY

The abundance and distribution of acceptable resources determine their availability to organisms. Resources are most available when distributed evenly at nonlimiting concentrations or densities. Organisms living under such conditions need not move widely to locate new resources and tend to be relatively sedentary. Microorganisms suspended in a concentrated solution of organic molecules (such as in eutrophic aquatic ecosystems or in decomposing detritus) and filter feeders and scale insects that capture resources from flowing solutions of resources enjoy relatively nonlimiting resources for many generations.

However, necessary resources usually are less concentrated or are unevenly distributed at the scale of use by most terrestrial organisms. This requires that organisms search for locations where limiting resources are most concentrated and find new concentrations as current resources become depleted. Although active searching is facilitated by locomotory ability, plant roots are capable of growing in the direction of more concentrated resources. Insects and other animals employ various physiological and behavioral mechanisms to detect, orient toward, and move to concentrations of food.

A. Foraging Strategies

Insect behavior can be viewed from the standpoint of efficiency of resource acquisition and allocation (see also Chapter 4). Foraging should focus on resources that provide the best return and minimum risk for the effort expended. Hence, bumble bees, *Bombus* spp., forage on low-energy resources only at high temperatures when the insects do not require large amounts of energy to maintain sufficiently high body temperature for flight (Bell, 1990; Heinrich, 1979, 1993). Other host-seeking insects tend to focus their searching where the probability of host discovery is highest, i.e., where hosts are concentrated (Bell, 1990; Kareiva, 1983).

Foraging theory focuses on optimization of diet, risk, and foraging efficiency (Kamil *et al.*, 1987; Schultz, 1983; Stephens and Krebs, 1986; Townsend and Hughes, 1981). Profitable resources provide a gain to the consumer, but nonnutritive or toxic resources represent a cost in terms of time, energy, or nutrient resources expended in detoxification or continued search. Continued search also increases exposure to predators or other mortality agents. Sublethal doses of defensive chemicals reduce nutritional value of the resource, so should be avoided when resources are abundant but may be eaten when more profitable resources are unavailable. Consumers should maximize foraging efficiency by focusing on patches with high profitability (and ignore low profitability patches) until their profitability declines below average for the matrix. Orientation toward cues indicating suitable resources improves the efficiency of food acquisition. Furthermore, learning confers an ability to improve resource acquisition as a result of experience.

Most insects not living in their food resource (as do gall-formers, miners, and wood borers) must search for suitable food at some spatial scale. Even within a particular plant, nutritional quality may vary considerably among individual leaves, e.g., between sun and shade leaves, young and old leaves, (Schultz, 1983; Whitham, 1983). Foraging strategy represents a trade-off between costs (in terms of reduced growth and survival) of searching, costs of feeding on less suitable food, and costs of exposure to predators.

Schultz (1983) developed a trade-off surface to illustrate four foraging strategies for arboreal caterpillars (Fig. 3.11). Foraging can be optimized by searching for more nutritive food and risking attention of predators, accepting less nutritive food, or defending against predation. Natural selection can favor a reduction in cost along any of the three axes, within constraints of the other two costs. Feeding selectively on the most suitable food during the day incurs a moderate cost of movement (energy expenditure at higher temperature) and high risk of predation. Similar selective feeding at night reduces predation risk but increases the energetic cost of movement, which now restricts the time spent feeding, especially at high latitudes. Crypsis reduces the risk of predation for day-feeders, but the cost of movement (in terms of attraction of predators) is substantial and limits ability to search for the most suitable food. Conversely, aposomatic (warning) coloration can reduce the risk of predation and allow

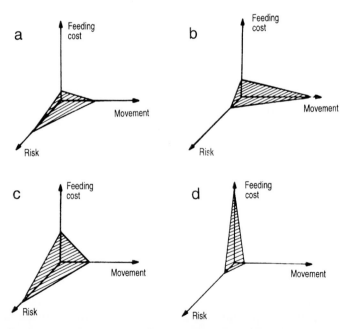

FIG. 3.11 Trade-off planes of selected caterpillar foraging strategies. Costs of feeding (i.e., metabolic costs of digestion, reduced growth, etc.), movement (metabolic costs of reduced growth), and risks (e.g., probability of capture or reduced growth due to time spent hiding) increase in the direction of the arrows: (a) selective diurnal feeder, (b) selective nocturnal aposomatic feeder, (c) diurnal cryptic feeder, and (d) food mimic. From Schultz (1983).

greater freedom of movement, but the energy expenditure of movement and cost of biochemical sequestration must be considered. Species that mimic, and live on or in, their food can avoid both movement and predation costs but have no choice in their food after initial colonization. Hence, feeding costs may be quite high.

Predators also face trade-offs among hunting, ambush, or intermediate strategies. Hunting requires greater energy expenditure searching for prey, but a high return, depending on ability to detect prey from a distance. Detection can be increased by orienting toward prey odors. Accordingly, many predaceous species are attracted to mating pheromones of their prey (Stephen *et al.*, 1993). Ambushers either sit and wait or employ traps to capture prey. As examples, dragonfly larvae hide in the substrate of aquatic habitats and grasp prey coming within reach, antlion larvae excavate conical depressions in loose sandy soil that prevents escape of ants and other insects that wander into the pit, and webspinning spiders construct sticky orb or tangled webs that trap flying or crawling insects. Movement costs are minimal for these species, but prey encounter is uncertain. Frequency of prey encounter can be increased by selecting sites along prey foraging trails, etc.

B. Orientation

Some insects may search randomly, eventually (at some risk) discovering suitable resources (Dixon, 1985; Raffa *et al.*, 1993). However, most insects respond to various cues that indicate the suitability of potential resources. The cues to which searching insects respond may differ among stages in the search process. For example, gross cues, indicative of certain habitats, might initially guide insects to a potentially suitable location. They then respond to cues that indicate suitable patches of resources and, finally, home in on cues characteristic of the necessary resources (Bell, 1990; Mustaparta, 1984). Experience at the habitat scale can affect search at finer scales. Insects search longer in patches where suitable resources have been detected than in patches without suitable resources, resulting in gradual increase in population density on hosts (Bell, 1990; Risch, 1980, 1981; Root, 1973; Turchin, 1988). Orientation toward cues involves the following steps.

1. Information Processing

Several types of information are processed by searching insects. Some cues are nondirectional but alert insects to the presence of resources or initiate search behavior. A nondirectional cue may alter the threshold for response to other cues (cross-channel potentiation) or initiate behaviors that provide more precise information (Bell, 1990). For example, flying bark beetles typically initiate search for their host trees only after exhausting their fat reserves. Emerging adults of parasitic wasps gather information about their host from odors emanating from host frass or food plant material associated with the emergence site

(Godfray, 1994). Wasps emerging in the absence of these cues may be unable to identify potential hosts.

Directional information provides the stimulus to orient in the direction of the perceived resource. For example, detection of attractive chemicals without airflow initiates nondirectional local search, whereas addition of airflow stimulates orientation upwind (Bell, 1990). Accuracy of orientation increases with signal intensity. Signal intensity decreases with distance and density of the source (Stanton, 1983). Concentration of attractive odors remains higher at greater distances from patches of high host density compared to patches of low host density (Fig. 3.12). Insects move upwind in circuitous fashion at low vapor concentration, but movement becomes increasingly directed as vapor concentration increases upwind (Cardé, 1996). Insects integrate visual, chemical, and acoustic signals to find their resources, switching from less precise to more precise signals as these become available (Bell, 1990).

2. Responses to Cues

Visual cues include host silhouettes and radiant energy. Some species find arboreal resources by orienting toward light. Aphids are attracted to young, succulent foliage and to older senescent foliage by longer-wavelength yellow, but this cue is not a good indicator of host species (Dixon, 1985). Aphids, *Pemphigus betae*, migrating in autumn may discriminate among susceptible and resistant poplar trees on the basis of prolonged leaf retention by more susceptible hosts (Moran and Whitham, 1990). Many bark beetles are attracted to dark-colored silhouettes of tree boles and can be attracted to other cylindrical objects or prevented from landing on tree boles painted white (Strom *et al.*, 1999). Some parasitic wasps detect their wood-boring hosts by means of infrared receptors on their antennae (Matthews and Matthews, 1978).

The importance of flower color and color patterns to attraction of pollinators has been a topic of considerable research (Chittka and Menzel, 1992; Heinrich, 1979; Wickler, 1968). Reds and blues are more easily detected in open or well-lighted ecosystems, hence, are more common in tropical and grassland ecosystems, whereas white is more readily detected under low-light conditions, such as in the understories of temperate forests. Ultraviolet designs, detectable by insects, provide important cues to insect pollinators. Insects can detect ultraviolet "runways" or "nectar guides" that guide the insect to the nectaries (Eisner *et al.*, 1969; Heinrich, 1979; Matthews and Matthews, 1978). Some floral designs in the orchid genus *Ophrys* resemble female bees or wasps and produce odors similar to the mating pheromones of these insects. Male bees or wasps are attracted and unwittingly pollinate these flowers while attempting to copulate (Wickler, 1968).

Many plant chemicals are highly volatile and are the basis for floral and other plant odors. Prominent among these are the monoterpenes, such as verbenone in verbena flowers and α-pinene in conifers, and aromatic compounds, such as vanillin (Harborne, 1994). These represent attractive signals to polli-

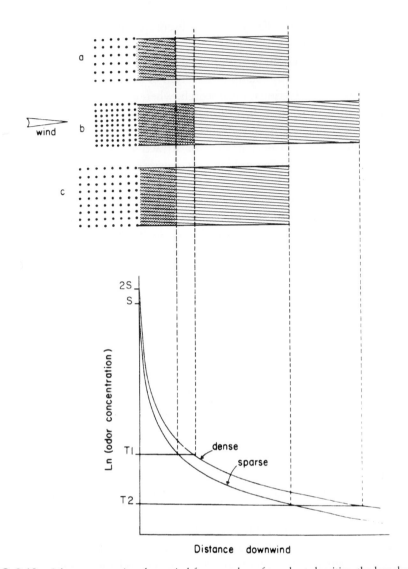

FIG. 3.12 Odor concentration downwind from patches of two host densities: the low-density odor curve represents patches a and c, whereas the high-density curve represents patch b. The curves reflect an ideal situation in which diffusion is overshadowed by convection due to wind. In still air, odor concentration cannot be changed by altering host-plant density. Attractive areas shown as rectangles are actually irregular in shape. Attractive zones for low-sensitivity herbivores (threshold T1) are stippled; those for high-sensitivity herbivores (T2) are shaded. From Stanton (1983).

nators and to herbivores adapted to feeding on a particular plant. Some odors repel some insects. For example, verbenone and 4-allylanisole, in the resin of various trees, repel some bark beetle species (Hayes *et al.*, 1994). Verbenone is present in the bark of certain conifers (including western redcedar, *Thuja plicata*, and Pacific silver fir, *Abies amabilis*) and likely influences orientation by bark beetles among tree species in diverse forests (Schowalter *et al.*, 1992).

Attractiveness of volatile biochemicals to insects is species specific (Mustaparta, 1984). A chemical that is attractive to one species may be nonattractive or even repellent to other, even related, species. For example, among sympatric *Ips* species in California, *I. pini* and *I. paraconfusus* both are attracted to ipsdienol, but *I. paraconfusus* also incorporates ipsenol and cis-verbenol in its pheromone blend, whereas *I. latidens* is attracted to ipsenol and cis-verbenol but not in the presence of ipsdienol (Raffa *et al.*, 1993). Plants that depend on dipteran pollinators often produce odors that resemble those of carrion or feces to attract these insects. Peakall *et al.* (1987) reported that male ants are attracted chemically to an Australian orchid which they pollinate during pseudocopulation, suggesting that the chemical stimulus is similar to the mating pheromone produced by the queen ant. Nonattractive or repellent odors from nonresources can mix with attractive odors in the airstream of more diverse ecosystems and disrupt orientation (Fig. 3.13). Sex pheromones (see Chapter 4) often are more attractive when mixed with host volatiles (e.g., Raffa *et al.*, 1993), indicating prior discovery and evaluation of suitable hosts.

Once a chemical attractant is detected in the air or water current, the insect begins a circuitous search pattern that involves continually turning in the direction of increasing odor concentration (Cardé, 1996). However, insects following the odor trail, or plume, are far from assured of reaching the source. Odor plumes often are disrupted by turbulence, resulting from habitat hetero-

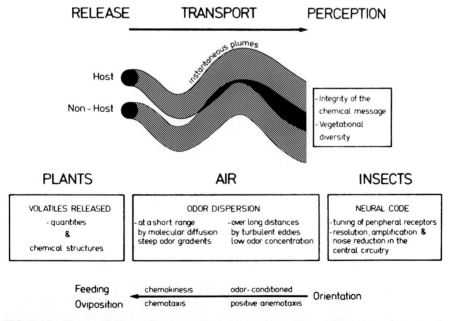

FIG. 3.13 Elements of host odor perception in insects. From Visser (1986), *Annual Review of Entomology*. With permission from the *Annual Review of Entomology*, Vol. 31, 121–144, © 1986 by Annual Reviews.

geneity, such as substrate or canopy irregularities (Mafra-Neto and Cardé, 1995; Murlis *et al.*, 1992). For example, Fares *et al.* (1980) found that openings in forest canopies created sites of soil warming and convective eddies that dissipated chemical plumes.

If an insect successfully arrives at the source of attractive cues, it engages in close-range gustatory, olfactory, or sound reception (Dixon, 1985; Raffa *et al.*, 1993; Städler, 1984). Contact chemoreceptive sensilla generally have a single pore at the tip, with unbranched dendrites from 2–10 receptor cells (Städler, 1984) and are located on antennae, mouthparts, or feet (Dixon, 1985; Städler, 1984). These sensors provide information about nutritive value and defensive chemistry of the resource (Raffa *et al.*, 1993). Certain plant chemicals act as phagostimulants or as deterrents. For example, cucurbitacins (the bitter triterpenes common to Cucurbitaceae) deter feeding and oviposition by nonadapted mandibulate insects but are phagostimulants for diabroticine chrysomelid beetles (Tallamy and Halaweish, 1993). Predators also may avoid prey containing toxic or deterrent chemicals (Stamp *et al.*, 1997; Stephens and Krebs, 1986). Many parasitic wasps avoid hosts marked by wasps that oviposited previously in that host (Godfray, 1994). Because hosts may support a limited number of parasitoid offspring, often no more than one, avoidance of previously parasitized hosts reduces competition among larvae within a host.

Acoustic signals include the sounds produced by cavitating plant cells and by potential mates. Water-stressed plants often produce audible signals from the collapse of cell walls as turgor pressure falls (Mattson and Haack, 1987). Cavitation thus is a valuable cue to stressed, and potentially more suitable, plants. Attraction to this signal may partly explain the association of bark beetles with water-stressed trees (Mattson and Haack, 1987; Raffa *et al.*, 1993).

3. Attraction of Conspecific Insects

Insects also may signal the presence of suitable resources to conspecific insects. Such cooperation increases opportunities for acquisition of larger prey and improves mating success (see Chapter 4).

Acoustic signals (stridulation) from potential mates, especially if combined with attractive host cues, advertise discovery and evaluation of suitable resources. Stridulation contributes to optimal spacing and resource exploitation by colonizing bark beetles (Raffa *et al.*, 1993; Rudinsky and Ryker, 1977).

Attractive and repellent chemicals produced by insects (pheromones) also advertise the location of suitable resources and potential mates (Raffa *et al.*, 1993; Rudinsky and Ryker, 1977). Most insects produce pheromones, but those of Lepidoptera, bark beetles, and social insects have been studied most widely. Social insects produce trail pheromones that are deposited along foraging trails to guide other members of a colony to food resources and back to the colony. These pheromones are used extensively among Isoptera and social Hymenoptera and also by tent caterpillars (Fitzgerald, 1995). A variety of chemical structures are used to mark trails (Fig. 3.14). A plant-derived monoterpene,

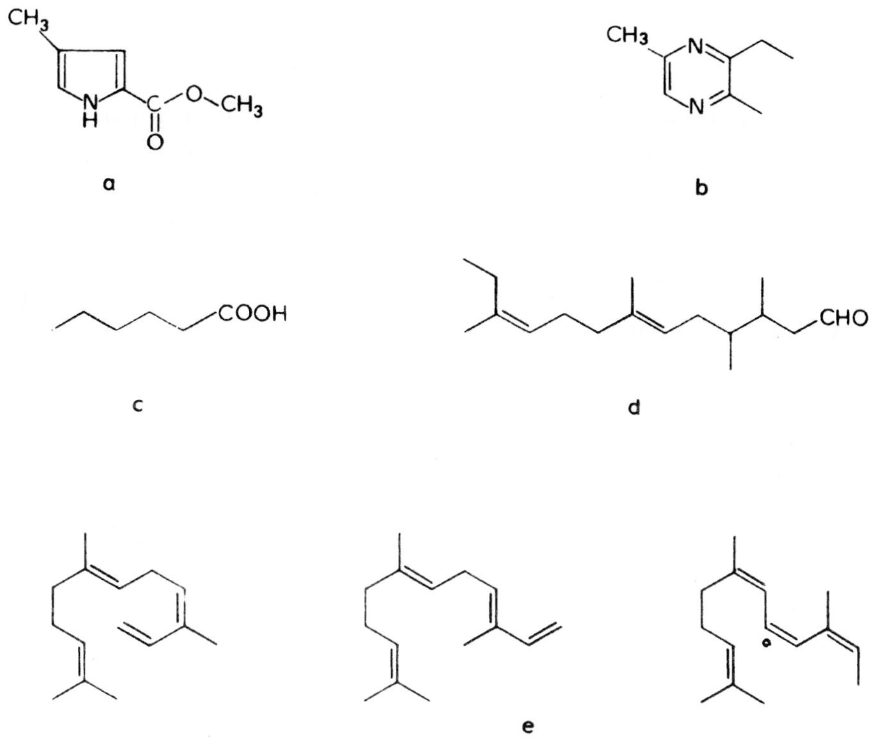

FIG. 3.14 Trail pheromones of myrmicine ants. (a) *Atta texana* and *A. cephalotes*, (b) *A. sexdens rubropilosa* and *Myrmica* spp., (c) *Lasius fuliginosus*, (d) *Monomorium pharaonis*, (e) *Solenopsis invicta*. From Bradshaw and Howse (1984) by permission from Chapman and Hall.

geraniol, is obtained from flower scents, concentrated, and used by honey bees, *Apis mellifera*, to mark trails (Harborne, 1994). The "bee dance" is used to direct other colony members to suitable resources. Trail markers can be highly effective. The trail marker produced by the leaf-cutting ant, *Atta texana*, is detectable by ants at concentrations of 3.48×10^8 molecules cm^{-1}, indicating that 0.33 mg of the pheromone would be sufficient to mark a detectable trail around the world (Harborne, 1994).

C. Learning

Insects can become more efficient at acquiring suitable resources over time as a result of learning. Learning is difficult to demonstrate, because improved performance with experience often may result from maturation of neuromuscular systems rather than from learning (Papaj and Prokopy, 1989). Although an unambiguous definition of learning has eluded ethologists, a simple definition involves any repeatable and gradual improvement in behavior due to experience

(Papaj and Prokopy, 1989; Shettleworth, 1984). From an ecological viewpoint, learning increases the flexibility of responses to unpredictable variation in resource availability.

Learning by insects has been appreciated less widely than has learning by vertebrates, but a number of studies over the past half century have demonstrated learning by various insect groups (cf. Schneirla, 1953; Gould and Towne, 1988; Lewis, 1986). Schneirla was among the first to report that ants can improve their ability to find food in a maze. However, the ants learned more slowly and applied experience less efficiently to new situations than did rats. Learning is best developed in the social and parasitic Hymenoptera and in some other predaceous insects. Nonetheless, learning also has been demonstrated in phytophagous species representing six orders (Chapman and Bernays, 1989; Papaj and Prokopy, 1989). Several types of learning by insects have been identified: habituation, imprinting, associative learning, observational learning, and even cognition.

Habituation is the loss of responsiveness to an unimportant stimulus as a result of continued exposure. Habituation may be the mechanism that induces parasitoids to emigrate from patches that are depleted of unparasitized hosts (Papaj and Prokopy, 1989). Although host odors are still present, a wasp is no longer responsive to these odors. Habituation to deterrent chemicals in the host plant may be a mechanism underlying eventual acceptance of less suitable host plants by some insects (Papaj and Prokopy, 1989).

Imprinting is the acceptance of a particular stimulus in a situation in which the organism has an innate tendency to respond. Parasitic wasps may imprint on host or plant stimuli at the site of adult emergence. Odors from host frass or the host's food plant present at the emergence site offer important information used by the emerging wasp during subsequent foraging (Lewis and Tumlinson, 1988). A number of studies have demonstrated that if the parasitoid is removed from its cocoon or reared on artificial diet, it may be unable to learn the odor of its host or its host's food plant, and hence, be unable to locate hosts (Godfray, 1994).

Associative learning is the linking of one stimulus with another, based on a discerned relationship between the stimuli. Most commonly, the presence of food is associated with cues consistently associated with food. Godfray (1994) summarized a number of examples of associative learning among parasitic Hymenoptera. Information gathered during searching contributes to increased efficiency of host discovery (Lewis and Tumlinson, 1988). Various studies have demonstrated that once parasitic wasps locate a host, they preferentially search similar microhabitats (Godfray, 1994). However, exposure to new hosts or hosts in novel habitats can lead to increased responsiveness to the new cues.

Classical conditioning involves substitution of one stimulus for another. Laboratory studies have demonstrated classical conditioning in parasitic wasps. These insects respond to empty food trays after learning to associate food trays with hosts or respond to novel odors after learning to associate them with provision of hosts (Godfray, 1994).

Operant conditioning, or trial-and-error learning, is associative learning in which an animal learns to associate its behavior with reward or punishment and then tends to repeat or avoid that behavior accordingly. Association of ingested food with post-ingestion malaise often results in subsequent avoidance of that food (Chapman and Bernays 1989, Papaj and Prokopy 1989). For example, laboratory experiments by Stamp (1992) and Traugott and Stamp (1996) demonstrated that predatory wasps initially attack caterpillars that sequester plant defenses, but after a few days will reject unpalatable prey. Honey bees trained to approach a particular flower from different directions at different times of day will subsequently approach other flowers from the direction appropriate to the time of day at which rewards were provided during training (Fig. 3.15).

Observational learning occurs when animals gather information and mod-

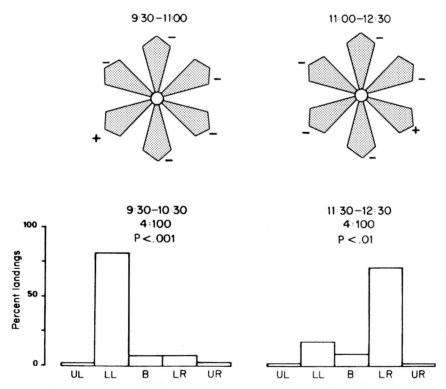

FIG. 3.15 Honey bees can remember how to approach specific flowers in relation to time of day. Bees trained to land at different positions (+) of an artificial flower at different times in the morning subsequently preferred to land on the petal on which they were trained during the same part of the morning. From Gould and Towne (1988).

ify their behavior in response to observation of other individuals. Observational learning is epitomized by social bees, which communicate the location of rich floral resources to other members of the colony through the "bee dance" (Gould and Towne, 1988). Movements of this dance, oriented with reference to the sun, inform other foragers of the direction and distance to a food source.

Cognition, characterized by awareness and judgment, can be demonstrated by application of information gathered during previous experiences to performance in novel situations. This basic form of thinking is widely associated with higher vertebrates. However, Gould (1986) demonstrated that honey bees are capable of constructing cognitive maps of their foraging area. Bees were trained to forage at either of two widely separated sites, then captured at the hive and transported in the dark to an unfamiliar site, the same distance from the hive but in a different direction, within a complex foraging area (open areas interspersed with forest). If released bees were disoriented or could not accommodate a sudden change in landmarks, they should fly in random directions. If they have only route-specific landmark memory and were familiar with a foraging route to their release point, they should be able to return to the hive and, from there, fly to their intended destination (the site to which they had been trained). Only if bees are capable of constructing true cognitive maps should they be able to fly from the release point directly to their intended destination. Gould (1986) found that all bees flew directly to their intended destinations. Therefore, honey bees appear to be capable of constructing cognitive maps of their home ranges and may remember photograph-like images, referenced to the time of day and to landmarks and line angles to floral resources (Gould, 1985, 1986; Gould and Towne, 1988). Such advanced learning greatly facilitates the efficiency with which resources can be acquired.

IV. SUMMARY

Insects, as do all organisms, must acquire energy and material resources to synthesize the organic molecules necessary for the life processes of maintenance, growth, and reproduction. Dietary requirements reflect the size and life stage of the insect and the quality of food resources. Insects exhibit a variety of physiological and behavioral strategies for finding, evaluating and exploiting potential resources.

Defensive chemistry of plants and insects affects their quality as food and is a basis for host choice by herbivorous and entomophagous insects, respectively. Nutritional value of resources varies among host species, among tissues of a single organism, and even within tissues of a particular type. Production of defensive chemicals is expensive in terms of energy and nutrient resources and may be sacrificed during unfavorable periods (such as during water or nutrient shortages or following disturbances) to meet more immediate metabolic needs. Such hosts become more vulnerable to predation. Insect adaptations to detoxify or otherwise circumvent host defenses determines host choice and range of host species exploited. Generalists exploit a relatively broad range of host species but exploit each host species rather inefficiently, whereas specialists are more efficient in exploiting a single or a few related hosts that produce similar chemical defenses.

Chemicals also communicate the availability of food and provide powerful cues that influence insect foraging behavior. Insects are capable of detecting food resources over considerable distances. Perception of chemical cues that indicate availability of hosts is influenced by concentration gradients in air or water, environmental factors that affect downwind or downstream dispersion of the chemical, and sensitivity to particular odors. Orientation to food resources over shorter distances is affected by visual cues (such as color or pattern) and acoustic cues (such as stridulation). Once an insect finds a potential resource, it engages in tasting or other sampling behaviors that permit evaluation of resource acceptability.

Efficiency of resource acquisition may improve over time as a result of learning. Although much of insect behavior may be innate, learning has been documented for many insects. Ability to learn among insects ranges from simple habituation to continuous unimportant stimuli, to widespread associative learning among both phytophagous and predaceous species, to observational learning, and even cognitive ability. Learning represents the most flexible means of responding to environmental variation and allows many insects to adjust to changing environments during short lifetimes.

Resource Allocation

INSECTS ALLOCATE ACQUIRED RESOURCES IN VARIOUS WAYS, DEPEND-ing on the energy and nutrient requirements of their physiological and behav-ioral processes. In addition to basic metabolism, foraging, growth, and repro-duction, individual organisms also allocate resources to pathways that influence their interactions with other organisms.

Interestingly, much of the early data on energy and nutrient allocation by insects was a by-product of studies during 1950–1970 on anticipated effects of nuclear war on radioisotope movement through ecosystems (e.g., Crossley and Howden, 1961; Crossley and Witkamp, 1964). Research also addressed effects of radioactive fallout on organisms that affect human health and food supply. Radiation effects on insects and other arthropods were perceived to be of spe-cial concern because of the recognized importance of these organisms to human health and crop production. Radioactive isotopes, such as ^{31}P, ^{137}Cs (assimi-lated and allocated as is K), and ^{85}Sr (assimilated and allocated as is Ca), be-came useful tools for tracking the assimilation and allocation of nutrients through organisms, food webs, and ecosystems.

I. RESOURCE BUDGET

The energy or nutrient budget of an individual can be expressed by the equation

$$I = P + R + E,$$

where I = consumption, P = production, R = respiration, E = egestion, and $I - E = P + R$ = assimilation. Energy is required to fuel metabolism, so only part of the assimilated energy is available for growth and reproduction (Fig. 4.1). The remainder is lost through respiration. Insects and other ectotherms require little energy to maintain thermal homeostasis. Hence, arthropods generally respire only 60–90% of assimilated energy, compared to > 97% for endotherms (Fitzgerald, 1995; Golley, 1968; Phillipson, 1981; Schowalter, et al., 1977; Wiegert and Petersen, 1983).

Arthropods vary considerably in their requirements for, and assimilation of, energy and nutrients. Reichle et al. (1969) and Gist and Crossley (1975) reported significant variation in cation concentrations among forest floor arthropods, and Schowalter and Crossley (1983) reported significant variation in cation concentrations among forest canopy arthropods. Caterpillars and sawfly larvae accumulated the highest concentrations of K and Mg, spiders accumulated the highest concentrations of Na among arboreal arthropods (Schowalter and Crossley, 1983), and millipedes accumulated the highest concentrations of Ca among litter arthropods (Reichle et al., 1969; Gist and Crossley, 1975).

Assimilation efficiency also varies among developmental stages. Schowalter et al. (1977) found that assimilation efficiency (A/I) of the range caterpillar, Hemileuca oliviae, declined significantly from 69% for first instars to 41% for the prepupal stage (Table 4.1). Respiration by pupae was quite low, amounting to only a few percent of larval production. This species does not feed as an adult, so resources acquired by larvae must be sufficient for adult dispersal and reproduction.

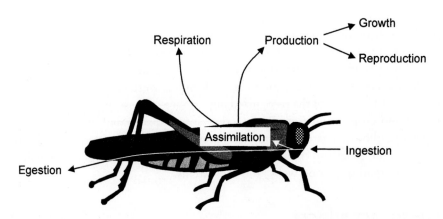

FIG. 4.1 Model of energy and nutrient allocation by insects and other animals. Ingested food is only partially assimilable, depending on digestive efficiency. Unassimilated food is egested. Assimilated food used for maintenance is lost as carbon and heat energy; the remainder is used for growth and reproduction.

TABLE 4.1 Assimilation Efficiency, A/I, Gross Production Efficiency, P/I, and Net Production Efficiency, P/A, for Larval Stages of the Saturniid Moth, *Hemileuca oliviae*

Instar	1	2	3	4	5	6	7	Total
A/I	0.69	0.64	0.60	0.55	0.48	0.43	0.41	0.54
P/I	0.41	0.26	0.28	0.22	0.25	0.26	0.20	0.23
P/A	0.59	0.43	0.47	0.42	0.56	0.63	0.53	0.52

Means underscored by the same line are not significantly different ($p > .05$). Reproduced from Schowalter *et al.* (1977) by permission from Springer-Verlag.

II. ALLOCATION OF ASSIMILATED RESOURCES

Assimilated resources are allocated to various metabolic pathways. The relative amounts of resources used in these pathways depend on stage of development, quality of food resources, physiological condition, and metabolic demands for physiological processes (such as digestion and thermoregulation), activities (such as foraging and mating), and interactions with other organisms (including competitors, predators, and mutualists). For example, many immature insects are relatively inactive and expend energy primarily for feeding and defense, whereas adults expend additional energy and nutrient resources for dispersal and reproduction. Major demands for energy and nutrient resources include foraging activity, mating and reproduction, and competitive and defensive behavior.

A. Resource Acquisition

Foraging activity is necessary for resource acquisition. Movement in search of food requires energy expenditure. Energy requirements vary among foraging strategies, depending on distances covered and the efficiency of orientation toward resource cues. Hunters expend more energy to find resources than do ambushers. The defensive capabilities of the food resource also require different levels of energy and nutrient investment. As described in Chapter 3, defended prey require production of detoxification enzymes or expenditure of energy during capture. Alternatively, energy must be expended for continued search if the resource cannot be acquired successfully.

Larger animals travel more efficiently than do smaller animals. Hence, larger animals often cover larger areas in search of resources. Flight is more efficient than walking, and efficiency increases with flight speed (Heinrich, 1979), enabling flying insects to cover large areas with relatively small energy reserves.

TABLE 4.2   Components of the Benefit-to-Cost (B/C) Ratio for Individual
Paraponera clavata and *Pogonomyrmex occidentalis* Foragers

	Paraponera		*Pogonomyrmex*
	Nectar forager	Prey forager	
Energy cost per m ($J\,m^{-1}$)	0.042	0.042	0.007
Foraging trip distance (m)	125	125	12
Energy expenditure per trip (J)	5.3	5.3	0.09
Average reward per trip (J)	20.8	356	100
B/C	3.9	67	1111

Reprinted from Fewell *et al.* (1996) by permission from Springer-Verlag.

Dispersal activity is an extension of foraging activity and also constitutes an energy drain. Most insects are short lived, as well as energy-limited, and so maximize fitness by accepting less suitable, but available, resources in lieu of continued search for superior resources (Courtney, 1985, 1986; Kogan, 1975).

The actual energy costs of foraging have been measured rarely. Fewell *et al.* (1996) compared the ratios of benefit to cost for a canopy-foraging tropical ant, *Paraponera clavata*, and an arid-grassland seed-harvesting ant, *Pogonomyrmex occidentalis*. They found that the ratio ranged from 3.9 for nectar foraging *P. clavata* and 67 for predaceous *P. clavata* to > 1000 for granivorous *P. occidentalis* (Table 4.2). Differences were due to the quality and amount of the resource, the distance traveled, and the individual cost of transport. In general, the smaller *P. occidentalis* had a higher ratio of benefit to cost because of the higher energy return of seeds, shorter average foraging distances, and lower energy cost m^{-1} traveled. The results indicate that *P. clavata* colonies have similar daily rates of energy intake and expenditure, potentially limiting colony growth, whereas *P. occidentalis* colonies have a much higher daily intake rate, compared to expenditure, reducing the likelihood of short-term energy limitation.

Insects produce a variety of biochemicals to exploit food resources. Trail pheromones provide an odor trail that guides other members of a colony to food resources and back to the colony (Fig. 3.14). Insects that feed on chemically defended food resources often produce more or less specific enzymes to detoxify these defenses (see Chapter 3). On the one hand, production of detoxification enzymes (typically, complex, energetically expensive molecules) reduces the net energy and nutritional value of food. On the other hand, these enzymes permit exploitation of a resource and derivation of nutritional value otherwise unavailable to the insect. Some insects not only detoxify host defenses but digest the products for use in their own metabolism and growth (e.g., Schöpf *et al.*, 1982).

Many insects gain protected access to food (and habitat) resources through symbiotic interactions, i.e., living on or in food resources (see Chapter 8). Phytophagous species frequently spend much of their developmental period on host resources. A variety of myrmecophilous or termitophilous species are tolerated, or even share food with their hosts, as a result of morphological (size, shape, and coloration), physiological (chemical communication), or behavioral (imitation of ant behavior, trophallaxis) adaptations (Wickler, 1968). Resemblance to ants also may confer protection from other predators (see following).

B. Mating Activity

Mate attraction and courtship behavior often are highly elaborated and ritualized and can be energetically costly. Nevertheless, such behaviors that distinguish species, especially sibling species, ensure appropriate mating and reproductive success and contribute to individual fitness through improved survival of offspring of sexual, as opposed to asexual, reproduction.

1. Attraction

Chemical, visual, and acoustic signaling are used to attract potential mates. Attraction of mates can be accomplished by either sex in Coleoptera, but only females of Lepidoptera release sex pheromones and only males of Orthoptera stridulate.

Sex pheromones greatly improve the efficiency with which insects find potential mates over long distances in heterogeneous environments (Law and Regnier, 1971; Mustaparta, 1984; Cardé, 1996). The particular blend of compounds and their enantiomers, as well as the time of calling, varies considerably among species. These mechanisms represent the first step in maintaining reproductive isolation. For example, among tortricids in eastern North America, *Archips mortuanus* uses a 90:10 blend of (Z)-11- and (E)-11-tetradecenyl acetate, *A. argyrospilus* uses a 60:40 blend, and *A. cervasivoranus* uses a 30:70 blend. A related species, *Argyrotaenia velutinana*, also uses a 90:10 blend but is repelled by (Z)-9-tetradecenyl acetate that is incorporated by *A. mortuanus* (Cardé and Baker, 1984). Among three species of saturniids in South Carolina, *Callosamia promethea* is active from about 10:00–16:00, *C. securifera* from about 16:00–19:00, and *C. angulifera* from 19:00–24:00 (Cardé and Baker, 1984). Representative bark beetle pheromones are shown in Fig. 4.2.

Sex pheromones may be released passively, as in the feces of bark beetles (Raffa *et al.*, 1993), or actively through extrusion of scent glands and active "calling" (Cardé and Baker, 1984). The attracted sex locates the signaler by following the concentration gradient (Fig. 4.3). Early studies suggested that the odor from a point source diffuses in a cone-shaped plume that expands downwind, the shape of the plume depending on airspeed and vegetation structure (e.g., Matthews and Matthews 1978). However, more recent work (Cardé, 1996; Mafra-Neto and Cardé, 1995; Murlis *et al.*, 1992; Roelofs, 1995) indicates that this plume is neither straight nor homogeneous over long distances but is influenced by turbulence in the airstream that forms pockets of higher

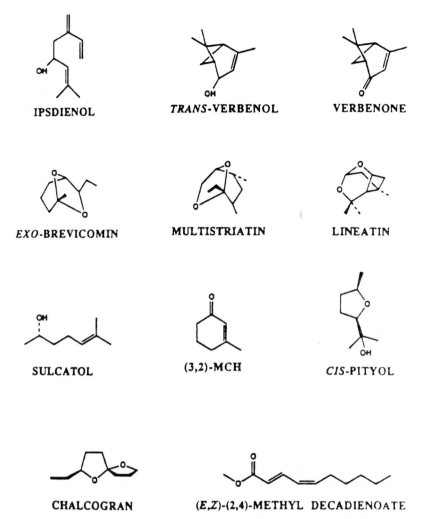

FIG. 4.2 Representative pheromones produced by bark beetles. Pheromones directly converted from plant compounds include ipsdienol (from myrcene), trans-verbenol, and verbenone (from α-pinene). The other pheromones shown are presumed to be synthesized by the beetles. From Raffa *et al.* (1993).

concentration or absence of the vapor (Fig. 4.4). An insect downwind would detect the plume as odor bursts rather than as a constant stream. Heterogeneity in vapor concentration is augmented by pulsed emission by many insects.

Pulses in emission and reception may facilitate orientation, because the antennal receptors require intermittent stimulation to sustain upwind flight (Roelofs, 1995). However, Cardé (1996) noted that the heterogeneous nature of the pheromone plume may make direct upwind orientation difficult over long distances. Pockets of little or no odor may cause the attracted insect to lose the odor trail. Detection can be inhibited further by openings in the vegetation canopy that create warmer convection zones or "chimneys" that carry the

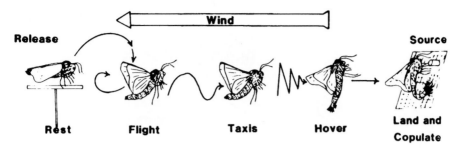

FIG. 4.3 Typical responses of male noctuid moths to the sex pheromone released by female moths. From Tumlinson and Teal (1987).

pheromone through the canopy (Fares *et al.*, 1980). Attracted insects may increase their chances of finding the plume again by casting, i.e., sweeping back and forth in an arcing pattern until the plume is contacted again (Cardé, 1996). Given the small size of most insects and limited quantities of pheromones for release, mates must be able to respond to very low concentrations. Release of less than 1 μg s^{-1} by female *Lymantria dispar* or *Bombyx mori* can attract males, which respond at molecular concentrations as low as 100 molecules ml^{-1} of air (Harborne, 1994).

Visual signaling is exemplified by the fireflies (e.g., Lloyd, 1983). In this group of insects, different species distinguish one another by variation in the rhythm of flashing and by the perceived "shape" of flashes produced by distinctive movements while flashing. Other insects, including the glow worms and several midges, also attract mates by producing luminescent signals.

Acoustic signaling is produced by stridulation, particularly in the Orthoptera, Heteroptera, and Coleoptera, or by muscular vibration of a membrane, common in the Homoptera. Resulting sounds can be quite loud and detectable over considerable distances. For example, the acoustic signals of mole crickets, *Gryllotalpa vinae*, as amplified by the double horn configuration of the cricket's burrow, are detectable by humans up to 600 meters away (Matthews and Matthews, 1978).

During stridulation, one body part, the file (consisting of a series of teeth or pegs), is rubbed over an opposing body part, the scraper. Generally, these structures occur on the wings and legs (Chapman, 1982), but in some Hymenoptera, sound is also produced by the friction between abdominal segments as the abdomen is extended and retracted. The frictional sound produced can be modulated by various types of resonating systems. Frequency and pattern of sound pulses are species specific.

Sound produced by vibrating membranes (tymbals) is accomplished by contracting the tymbal muscle to produce one sound pulse and relaxing the muscle to produce another sound pulse. Muscle contraction is so rapid (170–480 contractions per second) that the sound appears to be continuous (Matthews and Matthews, 1978). The intensity of the sound is modified by air

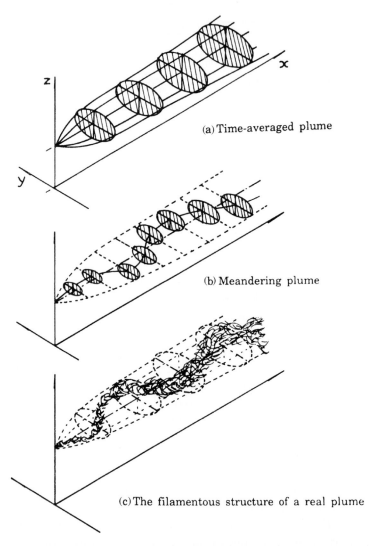

FIG. 4.4 Models of pheromone diffusion from a point source. The time-averaged Gaussian plume model (a) depicts symmetrical expansion of a plume from the point of emission. The meandering plume model (b) depicts concentration in each disk distributed normally around a meandering center line. The most recent work has demonstrated that pheromone plumes have a highly filamentous structure (c). From Murlis *et al.* (1992) by permission from the *Annual Review of Entomology*, Vol. 37, © 1992 by Annual Reviews.

sacs operated like a bellows and by opening and closing opercula that cover the sound organs (Chapman, 1982).

2. Courtship Behavior

Courtship often involves an elaborate, highly ritualized sequence of stimulus and response actions that must be completed before copulation occurs (Fig. 4.5). This provides an important mechanism that identifies species and sex,

FEMALE BEHAVIOR   MALE BEHAVIOR

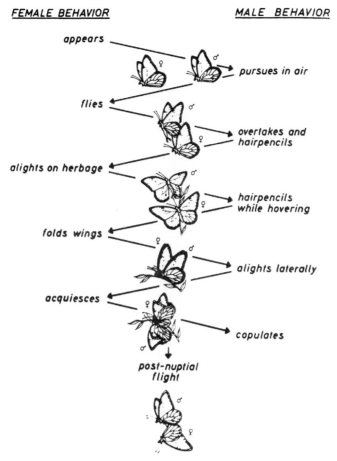

appears ——————→ pursues in air

flies ←——————

overtakes and
hairpencils

alights on herbage ←——————

hairpencils
while hovering

folds wings ←——————

alights laterally

acquiesces ←——————

copulates

post-nuptial
flight

FIG. 4.5 Courtship stimulus–response sequence of the Queen butterfly, *Danaus gilippus berenice*, from top to bottom, with male behavior on the right and female behavior on the left. From Brower *et al.* (1965) by permission of the Wildlife Conservation Society.

thereby enhancing reproductive isolation. Color patterns, odors, and tactile stimuli are important aspects of courtship. For many species, ultraviolet patterns are revealed, close-range pheromones are emitted, or legs or mouthparts stroke the mate as necessary stimuli (Brower *et al.*, 1965; Matthews and Matthews, 1978).

Another important function of courtship displays in predatory insects is appeasement, or inhibition of predatory responses, especially of females. Nuptial feeding occurs in several insect groups, particularly the Mecoptera, empidid flies, and some Hymenoptera and Heteroptera (Fig. 4.6). The male provides a food gift (such as a prey item, nectar, seed, or glandular product) that serves at least two functions (Matthews and Matthews, 1978; Price, 1997; Thornhill, 1976). Males with food may be more conspicuous to females, and feeding the

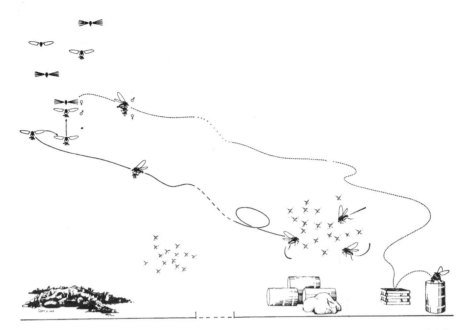

FIG. 4.6 Example of lekking and appeasement behavior in the courtship of an empidid fly, *Rhamphomyia nigripes*. Males capture a small insect, such as a mosquito or midge, then fly to a mating swarm (lek) which attracts females. Females select their mates and obtain the food offering. The pair then leaves the swarm and completes copulation on nearby vegetation. From Downes (1970) by permission from the Entomological Society of Canada.

female prior to oviposition may increase fecundity and fitness. Nuptial feeding has become ritualistic in some insects. Rather than prey, some flies simply offer a silk packet.

Males of some flies, euglossine bees, Asian fireflies, and some dragonflies gather in groups, called leks, to attract and court females (Fig. 4.6). Such aggregations allow females to compare and chose among potential mates and facilitate mate selection.

C. Reproductive and Social Behavior

Insects, like other organisms, invest much of their assimilated energy and nutrient resources in the production of offspring. Reproductive behavior includes varying degrees of parental investment in offspring that determines the survival of eggs and juveniles. Selection of suitable sites for oviposition affects the exposure of eggs to abiotic conditions suitable for hatching. The choice of oviposition site also affects the exposure of hatching immatures to predators and parasites and their proximity to suitable food resources. Nesting behavior, brood care, and sociality represent stages in a gradient of parental investment in survival of offspring.

1. Oviposition Behavior

Insects deposit their eggs in a variety of ways. Most commonly, the female is solely responsible for selection of oviposition site(s). The behaviors leading to oviposition are as complex as those leading to mating, because successful oviposition contributes to individual fitness and is under strong selective pressure.

A diversity of stimuli affects choice of oviposition sites by female insects. Mosquitoes are attracted to water by the presence of vegetation and reflected light but lay eggs only if salt content, pH, or other factors sensed through tarsal sensillae are suitable (Matthews and Matthews, 1978). Grasshoppers assess the texture, salinity, and moisture of soil selected for oviposition.

Many phytophagous insects assess host suitability for development of offspring. This assessment may be on the basis of host chemistry or existing feeding pressure. Ovipositing insects tend to avoid host materials with deleterious levels of secondary chemicals. They also may avoid ovipositing on resources that are already occupied by eggs or competitors. For example, female bean weevils, *Callosobruchus maculatus*, assess each potential host bean by comparison to the previous bean and lay an egg only if the present bean is larger or has fewer eggs. The resulting pattern of oviposition nearly doubles larval survival compared to random oviposition (Mitchell, 1975). Many parasitic wasps mark hosts in which eggs have been deposited and avoid ovipositing in marked hosts, thereby minimizing larval competition within a host (Godfray, 1994). Parasitic wasps can minimize hyperparasitism by not ovipositing in more than one host in an aggregation. This reduces the risk that all of its offspring are found and parasitized (Bell, 1990). Cannibalistic species, such as *Heliconius* butterflies, may avoid laying eggs near each other to minimize cannibalism and predation.

Selection also determines whether insects lay all their eggs during one period (semelparity) or produce eggs over more protracted periods (iteroparity). Most insects with short life cycles (e.g., < 1 year) typically have relatively short adult life spans and lay all their eggs in a relatively brief period. Insects with longer life spans, especially social insects, reproduce continually for many years.

Some insects influence host suitability for their offspring. For example, female sawflies typically sever the resin ducts at the base of a conifer needle prior to laying eggs in slits cut distally to the severed ducts. This behavior prevents or reduces egg mortality resulting from resin flow into the oviposition slits (McCullough and Wagner, 1993). Parasitic Hymenoptera often inject mutualistic viruses into the host along with their eggs. The virus inhibits cellular encapsulation of the egg or larva by the host (Tanada and Kaya, 1993).

In some cases, choices of oviposition sites by adults clearly conflict with suitability of resources for offspring. Kogan (1975) and Courtney (1985, 1986) reported that some species preferentially oviposit on the most conspicuous host species that are relatively unsuitable for larval development (Fig. 3.10). However, this behavior represents a trade-off between the prohibitive search time required to find the most suitable hosts and the reduced larval survival on the more easily discovered hosts.

2. Nesting and Brood Care

Although brood care is best known among the social insects, many other insects exhibit maternal care of offspring and even maternal tailoring of habitat conditions to enhance survival of offspring. Primitive social behavior appears as parental involvement extends further through the development of the offspring.

Several environmental factors are necessary for evolution of parental care (Wilson, 1975). A stable environment favors larger, longer-lived species that reproduce at intervals rather than all at once. Establishment in new, physically stressful environments may select for protection of offspring, at least during vulnerable periods. Intense predation may favor species that guard their young to improve their chances of reaching breeding age. Finally, selection may favor species that invest in their young which, in turn, help the parent find, exploit, or guard food resources. Cooperative brood care, involving reciprocal communication, among many adults is the basis of social organization (Wilson, 1975).

A variety of insect species from several orders exhibit protection of eggs by a parent (Matthews and Matthews, 1978). In most cases, the female remains near her eggs and guards them against predators. However, in some species of giant water bugs (*Belostoma* and *Abedus*), the eggs are laid on the back of the male, which carries them until they hatch. Among dung beetles (Scarabaeidae), adults of some species limit their investment in offspring to providing protected dung balls in which eggs are laid, whereas females in the genus *Copris* remain with the young until they reach adulthood. Extended maternal care, including provision of food for offspring, is seen in crickets, cockroaches, some Homoptera and nonsocial Hymenoptera.

A number of arthropod species are characterized by aggregations of individuals. Groups can benefit their members in a number of ways. Large groups often are able to modify environmental conditions, such as through retention of body heat or moisture. Aggregations also increase the availability of potential mates (Matthews and Matthews, 1978) and minimize exposure of individuals to plant toxins (McCullough and Wagner, 1993; Nebeker *et al.*, 1993) and to predators (Fitzgerald, 1995). Aggregated, cooperative feeding on plants, such as by sawflies and bark beetles, can remove plant tissues or kill the plant before induced defenses become effective (McCullough and Wagner, 1993; Nebeker *et al.*, 1993). Groups limit predator ability to avoid detection and to separate an individual to attack from within a fluid group. Predators are more vulnerable to injury by surrounding individuals, compared to attacking isolated individuals.

Cooperative behavior is evident within groups of some spiders and communal herbivores, such as tent-building caterpillars and gregarious sawflies. Dozens of individuals of the spider, *Mallos gregalis*, cooperate in construction of a communal web and in subduing prey (Matthews and Matthews, 1978). Tent-building caterpillars cooperatively construct their web, which affords protection from predators and may facilitate feeding and retention of heat and moisture (Fitzgerald, 1995). Similarly, gregarious sawflies cooperatively defend against predators and distribute plant resin among many individuals, thereby limiting the effectiveness of the resin defense (McCullough and Wagner, 1993).

Primitive social behavior is exhibited by the woodroach, *Cryptocercus punctulatus*, by passalid beetles, and by many Hymenoptera. In these species, the young remain with the parents in a family nest for long periods of time, are fed by the parents, and assist in nest maintenance (Matthews and Matthews, 1978). However, these insects do not exhibit coordinated behavior or division of labor among distinct castes.

The complex eusociality characterizing termites and the social Hymenoptera has attracted considerable attention (e.g., Matthews and Matthews, 1978; Wilson, 1975). Eusociality is characterized by multiple adult generations and highly integrated cooperative behavior, with efficient division of labor, among all castes (Matthews and Matthews, 1978; Michener, 1969). Members of these insect societies cooperate in food location and acquisition, feeding of immatures, and defense of the nest. This cooperation is maintained through complex pheromonal communication, including trail and alarm pheromones (Hölldobler, 1995; see Chapter 3), and reciprocal exchange of regurgitated liquid foods (trophallaxis) between colony members. Trophallaxis facilitates recognition of nest mates by maintaining a colony-specific odor, ensures exchange of important nutritional resources, and (in the case of termites) of microbial symbionts that digest cellulose, and may be critical to colony survival during periods of food limitation (Matthews and Matthews, 1978). Trophallaxis distributes material rapidly throughout a colony. Wilson and Eisner (1957) fed honey mixed with radioactive iodide to a single worker ant and within one day detected some tracer in every colony member, including the two queens.

Development of altruistic behaviors such as social cooperation can be explained largely as a consequence of kin selection and reciprocal cooperation (Axelrod and Hamilton, 1981; Haldane, 1932; Hamilton, 1964; Trivers, 1971; Wilson, 1973; Wynne-Edwards, 1963, 1965; see also Chapter 15). Self-sacrifice that increases reproduction by closely related individuals increases inclusive fitness, i.e., the individual's own fitness plus the fitness accruing to the individual through its contribution to reproduction of relatives. In the case of the eusocial insects, relatedness among siblings is greater than that between parent and offspring, making cooperation among colony members highly adaptive. The epitome of "altruism" among insects may be the development of the barbed sting in the worker honey bee, *Apis mellifera*, that ensures its death in defense of the colony (Haldane, 1932; Hamilton, 1964).

D. Competitive, Defensive, and Mutualistic Behavior

Insects, like all animals, interact with other species in a variety of ways—as competitors, predators, prey, and mutualists. Interactions among species will be discussed in greater detail in Chapter 8. These interactions require varying degrees of energy and/or nutrient expenditure. Contests among individuals for resources occasionally involve combat. Subduing prey and defending against predators also involve strenuous activity. Mutualism requires reciprocal exchange of resources or services. Obviously, these activities affect the energy and nutrient budgets of individual organisms.

1. Competitive Behavior

Competition is prevalent among individuals using the same limiting resources at the same site. Energy expended, or injury suffered, defending resources or searching for uncontested resources affect fitness. Competition often is mediated by mechanisms that determine a dominance hierarchy. Establishment of dominant and subordinate status among individuals limits the need for physical combat to determine access to resources and ensures sufficient resources for dominant individuals.

Visual determination of dominance status is relatively rare among insects, largely because of their small size, the complexity of the environment which restricts visual range, and the limitations of fixed-focus compound eyes for long-distance vision (Matthews and Matthews, 1978). Dragonflies have well-developed eyes and exhibit ritualized aggressive displays that maintain spacing among individuals. For example, male *Plathemis lydia* have abdomens that are bright silvery-white above. Intrusion of a male into another male's territory initiates a sequence of pursuit and retreat, covering a distance of 8–16 m. The two dragonflies alternate roles and directions, with the abdomens raised during pursuit and lowered during retreat, until the intruder moves to another site (Corbet, 1962).

Mediation of competition by pheromones has been documented for several groups of insects. Adult flour beetles, *Tribolium*, switch from aggregated distribution at low densities to random distribution at intermediate densities, to uniform distribution at high densities. This spacing is mediated by secretion of quinones, repellent above a certain concentration, from thoracic and abdominal glands (Matthews and Matthews, 1978). Larvae of the flour moth, *Anagasta kunniella*, secrete compounds from the mandibular glands that increase dispersal propensity, lengthen generation time, and lower the fecundity of females that were crowded as larvae (Matthews and Matthews, 1978). Bark beetles employ repellent pheromones, as well as acoustic signals, to maintain minimum distances between individuals boring through the bark of colonized trees (Raffa *et al.*, 1993; Rudinsky and Ryker, 1977). Ant colonies also maintain spacing through marking of foraging trails with chemical signals (see previous and also Chapter 3).

Acoustic signals are used by many Orthoptera and some Coleoptera to deter competitors. Bark beetles stridulate to deter other colonizing beetles from the vicinity of their gallery entrances (Rudinsky and Ryker, 1977). Subsequently, excavating adults and larvae respond to the sounds of approaching excavators by mining in a different direction, thus preventing intersection of galleries. Some male crickets and grasshoppers produce a distinctive rivalry song when approaching each other (Matthews and Matthews, 1978; Schowalter and Whitford, 1979). The winner (continued occupant) typically is the male that produces more of this aggressive stridulation.

When resources are relatively patchy, males may increase their access to females by marking and defending territories that contain resources attractive to females. Territorial behavior is less adaptive (i.e., costs of defending resources exceeds benefits) when resources are highly concentrated and competition is severe or when resources are uniformly distributed and female distribution is less predictable (Baker, 1972; Schowalter and Whitford, 1979).

Marking territorial boundaries takes a variety of forms among animal taxa. Male birds mark territories by calling from perches along the perimeter. Male deer rub scent glands and scrape trees with their antlers to advertise their territory. Social insects, including ants, bees, and termites, mark nest sites and foraging areas with trail pheromones that advertise their presence. These trail markers can be perceived by other insects at minute concentrations (see Chapter 3). Many orthopterans and some beetles advertise their territories by stridulating.

However, many insects advertise their presence simply to maintain spacing and do not actively defend territories. Similarly, males of many species, including insects, fight over receptive females. Wilson (1975) considered defense of occupied areas to be the defining criterion for territoriality. Territorial defense is best known among vertebrates, but a variety of insects representing at least eight orders defend territories against competitors (Matthews and Matthews, 1978; Price, 1997; Schowalter and Whitford, 1979). Because territorial defense represents an energetic cost, an animal must gain more of the resource by defending it against competitors than by searching for new resources. Nonaggressive males often "cheat" by nonadvertisement and quiet interception of resources or of females attracted to the territory of the advertising male (Schowalter and Whitford, 1979).

The type of territory differs among insect taxa but usually is associated with competition for food or mates (Matthews and Matthews, 1978; Price, 1997). Male crickets defend the area around their dens and mate with females attracted to their stridulation. Male eastern woodroaches, *Cryptocercus punctulatus*, defend mating chambers in rotten wood (Ritter, 1964). Some insects that form leks defend small territories within the lek. Presumably, more females are attracted to this concentration of males, increasing mating success, than to isolated males (Price, 1997). Such mating territories apparently are not related to food or oviposition sites but may maximize attraction of females.

Two grasshopper species, *Ligurotettix coquilletti* and *Bootettix argentatus*, that feed on creosote bush in the deserts of the southwestern U.S. are perhaps the only territorial acridoids (Otte and Joern, 1975; Schowalter and Whitford, 1979). These grasshoppers defend individual creosote bushes. The larger bushes are more likely to harbor females, and opportunities for mating are increased by defending larger shrubs, especially at low grasshopper population densities. Schowalter and Whitford (1979) reported that male movement from small shrubs was greater than movement from larger shrubs, and contests for larger shrubs occurred more frequently. However, fewer males defended territories at high population densities, apparently because interception of females by non-stridulating males and more frequent combat decreased mating success of territorial defenders.

Males of the speckled wood butterfly, *Pararge aegeria* (Satyridae), defend sunspots on the forest floor, apparently because females are attracted to resources that occur in sunspots (Price, 1997). Only 60% of the males held such territories, but these encountered many more females than did the nonterritorial males that searched for mates in the forest canopy. Defense of an oviposition site may be advantageous where sperm competition cannot be avoided by anatomical or physiological means, such as with mating plugs that prevent subsequent mating. Another butterfly, *Inachis io*, defends territories at the approach to oviposition sites, perhaps because of selective pressure from strong competition at the oviposition sites (Baker, 1972). Other insects, especially the social Hymenoptera, defend nests, foraging trails, or food (Price, 1997).

The benefits of defending food resources or mates must be weighed against the costs of fighting, in terms of time, energy, and risk of injury. Territorial insects may abandon territorial defense at high population densities when time spent fighting detracts from feeding or mating success (Schowalter and Whitford, 1979).

2. Defensive Behavior

Most insects are capable of physically defending themselves. Mandibulate species frequently bite, and haustelate species may stab with their stylets. Many species eject or inject toxic or urticating chemicals, as described in Chapter 3 (Figs. 3.7 and 3.8). Many insects armed with urticating spines or setae increase the effectiveness of this defense by thrashing body movements that increase contact of the spines or setae with an attacker. Many caterpillars and sawfly larvae rear up and strike like a snake when attacked (Fig. 4.7).

As we have seen, insects produce a variety of defensive compounds that can deter or injure predators. Many of these compounds are energetically expensive to produce, and defensive compounds often are toxic to the producer as well as to predators, requiring special mechanisms for storage or delivery. Nevertheless, their production sufficiently improves the probability of survival and reproduction to represent a net benefit to the producer.

Defense conferred by protective coloration reduces the energy costs of ac-

FIG. 4.7 Defensive posture of an unidentified lepidopteran caterpillar in Panama. This posture may increase resemblance to snakes or permit ejection of sequestered plant compounds.

tive defense but may require greater efficiency in foraging or other activities that attract attention (Schultz, 1983). Insects that rely on resemblance to their background (crypsis) must minimize movement to avoid detection (Fig. 4.8). Such insects may restrict necessary movement to nighttime or acquire their food with minimal movement. For example, many Homoptera that are cryptically colored or that resemble thorns or debris are largely sedentary while siphoning plant fluids. Many aquatic insects resemble benthic debris and remain motionless as they filter suspended matter.

Disruptive and deceptive coloration involve color patterns that break up the body form, distract predators from vital body parts, or resemble other predators. For example, many insects have distinct bars of color or other patterns that disrupt the outline of the body and inhibit their identification as prey by passing predators. Insect behavior enhances the effect of these color patterns. The underwing moths (Noctuidae) are noted for their brightly colored hind wings that are hidden at rest by the cryptically colored front wings. When threatened, the moth suddenly exposes the hind wings and has an opportunity to escape its startled attacker. The giant silkworm moths (Saturniidae) and eyed elater, *Alans oculatus* (Coleoptera: Elateridae), have conspicuous eyespots that make these insects look like birds (especially owls) or reptiles. The eyespots of moths typically are hidden on the hind wings during rest and can be exposed suddenly to startle would-be predators. The margin of the front wings in some

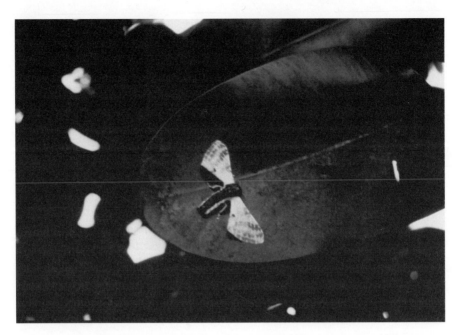

FIG. 4.8   Moth protected from casual observers by its resemblance to patterns of light transmitted through holes in and between leaves.

saturniids are shaped and colored to resemble the heads of snakes (Fig. 4.9) (Grant and Miller, 1995). Sudden wing movement during escape may enhance the appearance of a striking snake.

Mimicry is resemblance to another, usually venomous or unpalatable, species and typically involves conspicuous, or aposematic, coloration. Mimicry can take two forms, **Batesian** and **Müllerian**. Batesian mimicry is resemblance of a palatable or innocuous species to a threatening species, whereas Müllerian mimicry is resemblance among threatening species. Both are exemplified by insects. A variety of insects (representing several orders) and other arthropods (especially spiders) benefit from resemblance to stinging Hymenoptera. For example, clearwing (Sessidae) and some sphingid moths (Lepidoptera), several cerambycid beetles (Coleoptera), and many asilid and syrphid flies (Diptera) resemble ants, bees, or wasps (Fig. 4.10). Müllerian mimicry is exemplified by species of Hymenoptera and heliconiid butterflies (Lepidoptera) that sting or are unpalatable and resemble each other (e.g., Brower, 1996; Sheppard *et al.*, 1985).

Mimicry systems can be complex, including a number of palatable and unpalatable species and variation in palatability among populations depending on food source. For example, the resemblance of the viceroy, *Limenitis archippus* (Nymphalidae), butterfly to the monarch, *Danaus plexippus*, (Daneidae), butterfly is generally considered to be an example of Batesian mimicry. However,

FIG. 4.9   Image of a snake's head on the wing margins of *Attacus atlas*. From Grant and Miller (1995) by permission from the Entomological Society of America.

monarch butterflies show a spectrum of palatability over their geographic range, depending on the quality of their milkweed and other hosts (Brower *et al.*, 1968). Furthermore, populations of the viceroy and monarch in Florida are equally distasteful (Ritland and Brower, 1991). Therefore, this mimicry system may be Batesian in some locations and Müllerian in others. Conspicuous color patterns and widespread movement of the co-models/mimics maximizes exposure to predators and reinforces predator avoidance, providing overall protection against predation.

Some insects alert other members of the population to the presence of predators. Alarm pheromones are widespread among insects. These compounds typically are relatively simple hydrocarbons, but more complex terpenoids occur among ants. The venom glands of stinging Hymenoptera frequently include alarm pheromones. Alarm pheromones function either to scatter members of a group when threatened by a predator or to concentrate attack on the predator, especially among the social insects. A diverse group of ground-dwelling arthropods produce compounds that mimic ant alarm pheromones. These function to scatter attacking ants, allowing the producer to escape (Blum, 1980). Alarm pheromones released with the venom are used by stinging Hymenoptera to mark a predator. This marker serves to attract and concentrate attack by other members of the colony.

FIG. 4.10 Batesian mimicry by two insects. The predaceous asilid fly on the left and its prey, a cerambycid beetle, both display the black and yellow coloration typical of stinging Hymenoptera.

3. Mutualistic Behavior

Organisms also divert resources to production of rewards or inducements that maintain mutualistic interactions. Various pollinators and predators exploit resources allocated by plants to production of nectar, domatia, root exudates, etc. and thereby contribute substantially to plant fitness. Similarly, bark beetles secrete lipids into mycangia for nourishment of mutualistic microorganisms that improve suitability of woody substrates for the beetles. Obviously, the benefit gained from this association must outweigh these energetic and nutritional costs (see Chapter 8).

III. EFFICIENCY OF RESOURCE USE

Fitness accrues to organisms to the extent that they survive and produce more offspring than do their competitors. Hence, the efficiency with which assimilated resources are allocated to growth and reproduction determines fitness. However, except for sessile organisms, much of the assimilated energy and material must be allocated to activities pursuant to food acquisition, dispersal, mating, and defense. The amount of assimilated resources allocated to these activities reduces relative growth efficiency (Schultz, 1983; Zera and Denno, 1997). Clearly, the diversion of resources from growth and reproduction to these other pathways must represent a net benefit to the insect.

A. Factors Affecting Efficiency

Efficiency is affected by a number of constraints on energy and resource allocation. Clearly, selection should favor physiological and behavioral adaptations that improve overall efficiency. However, adaptive strategies reflect the net current result of many factors that have variable and interactive effects on survival and reproduction. Hence, individual responses to current conditions vary in efficiency. Whereas physiological, and many behavioral, responses are innate (genetically based, hence, relatively inflexible), the capacity to learn can improve efficiency greatly by reducing the time and resources expended in responding to environmental variation.

Hairston *et al.* (1960) stimulated research on the constraints of food quality on efficiency of herbivore use of resources by assuming that all plant material is equally suitable for herbivores. Just as plant chemistry can reduce herbivore efficiency, various animal defenses increase the resource expenditure necessary for predators to capture and assimilate prey. In addition to factors affecting the efficiency of resource acquisition, several factors affect the efficiency of resource allocation, including food quality, physiological condition, and learning.

1. Food Quality

Food quality affects the amount of food required to obtain sufficient nutrition and the energy and nutrients required for detoxification and digestion (see Chapter 3). Insects feeding on hosts with lower levels of defensive compounds invest fewer energy and nutrient resources in detoxification enzymes or continued searching behavior than do insects feeding on better defended hosts. Herbivores process much undigestable plant material, especially cellulose, whereas predators process animal material that generally is more similar to their own tissues. Accordingly, we might expect higher assimilation efficiencies for predators than for herbivores (Turner, 1970). Although undigestible and toxic compounds in plant tissues reduce assimilation efficiency for herbivores (Scriber and Slansky, 1981), toxins sequestered or produced by prey likely also reduce assimilation efficiency of predators. However, few studies have addressed the effect of toxic prey on assimilation efficiency of predators (Dyer, 1995; Stamp *et al.*, 1997; Stephens and Krebs, 1986).

Insects may ingest relatively more food to obtain sufficient nutrients or energy to offset the costs of detoxification or avoidance of plant defensive chemicals. Among herbivores, species that feed on mature tree leaves have relative growth rates that are generally half the values for species that feed on forbs, because tree leaves are a poor food resources compared to forbs (Scriber and Slansky, 1981). Although specialists might be expected to feed more efficiently on their hosts than do generalists, Futuyma and Wasserman (1980) reported that a specialist (the eastern tent caterpillar, *Malacosoma americana*) had no greater assimilation or growth efficiencies than did a generalist (the forest tent caterpillar, *M. disstria*). Some wood-boring insects may require long periods (several years to decades) of larval feeding to concentrate nutrients (especially N and P) sufficient to complete development.

2. Size and Physiological Condition

Body size is a major factor affecting efficiency of energy use. Larger organisms have greater energy requirements than do smaller organisms. However, smaller organisms with larger surface area/volume ratios are more vulnerable to heat loss than are larger organisms. Accordingly, maintenance energy expenditure per unit body mass decreases with increasing body size (Phillipson, 1981). In addition, larger organisms tend to use energy more efficiently during movement and resource acquisition, have a competitive advantage in cases of direct aggression, and have greater immunity from predators (Ernsting and van der Werf, 1988; Heinrich, 1979; Phillipson, 1981; Streams, 1994), reducing relative energy expenditures for these activities.

Physiological condition, including the general vigor of the insect as affected by parasites, also influences food requirements and assimilation efficiency. For example, hunger may induce increased effort to gain resources that would be ignored by less desperate individuals (Ernsting and van der Werf, 1988; Holling, 1965; Iwasaki, 1990, 1991; Richter, 1990; Streams, 1994). Slansky (1978) reported that *Pieris rapae* (Lepidoptera) parasitized by *Apanteles glomeratus* (Hymenoptera) increased food consumption, growth rate, and nitrogen assimilation efficiency. Schowalter and Crossley (1982) found that cockroaches, *Gromphadorhina portentosa*, with associated mites, *Gromphadorholaelaps schaeferi*, had a significantly more rapid egestion rate than did cockroaches with mites excluded, although assimilation efficiency did not differ significantly between mite-infested and mite-free cockroaches (Fig. 4.11).

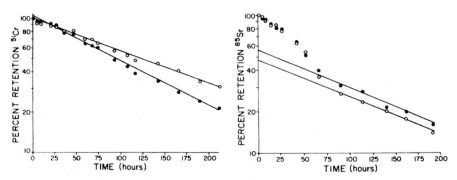

FIG. 4.11 Bioelimination of ^{51}Cr (left) and ^{85}Sr (right) by the cockroach, *Gromphadorhina portentosa* with (solid circles) and without (open circles) the associated mite, *Gromphadorholaelaps schaeferi*. ^{51}Cr has no biological function and its elimination represents egestion; ^{85}Sr is an analogue of Ca and its elimination represents both egestion (regression lines similar to those for ^{51}Cr) and excretion of assimilated isotope (rapid initial loss). This insect appears to assimilate and begin excreting nutrients before gut passage of unassimilated nutrients is complete. From Schowalter and Crossley (1982) by permission from the Entomological Society of America.

3. Learning

Learning is a powerful tool for improving efficiency of resource use (see Chapter 3). Learning reduces the effort wasted in unsuccessful trials (Fig. 3.15). Learning to distinguish appropriate from inappropriate prey (e.g., search image), to respond to cues associated with earlier success, and to improve capture technique greatly facilitates energy and nutrient acquisition. Honey bees represent the epitome of resource utilization efficiency among insects through their ability to communicate foraging success and location of nectar resources to nestmates (Gould and Towne, 1988; Heinrich, 1979).

B. Trade-offs

Allocation efficiency often is optimized by adaptations that generally tailor insect morphology, life histories, or behavior to prevailing environmental conditions or resource availability. For example, synchronization of life histories with periods of suitable climatic conditions and food availability reduces the energy required for thermoregulation or search activity. Bumble bee anatomy optimizes heat retention during foraging in cool temperate and arctic habitats (Heinrich, 1979). Davison (1987) compared the energetics of two harvester ant species, *Chelaner rothsteini* and *C. whitei*, in Australia and found that the smaller *C. rothsteini* had lower assimilation efficiency, but higher production efficiency (largely in production of offspring) than did the larger *C. whitei*. *Chelaner rothsteini* discontinued activity during the winter, perhaps to avoid excessive metabolic heat loss, whereas *C. whitei* remained active all year.

Selection should favor individuals and species that acquire and allocate resources most efficiently. Males that defend territories when the time or energy spent on this activity interferes with mating and reproduction are less likely to contribute to the genetic composition of the next generation than are males that sacrifice territorial defense for mating opportunities under such conditions (Schowalter and Whitford, 1979).

However, as discussed above, female pierid butterflies oviposited preferentially on more conspicuous hosts, a more energetically efficient search strategy for the adult, but these hosts were less suitable for larval development than were less conspicuous hosts (Fig. 3.10) (Courtney, 1985, 1986). Similarly, females of the noctuid moth, *Autographa precationis*, preferentially oviposit on soybeans rather than dandelions, perhaps because the shape of dandelions is a less effective oviposition stimulus, although larvae show a marked feeding preference for dandelions (Kogan, 1975). Hence, conflicts of interest among pathways or life stages often reduce overall efficiency. Resource allocation by insects, as well as other organisms, reflects trade-offs among alternative strategies, even between life stages.

Heinrich (1979) evaluated the trade-offs among various allocation strategies seen among bees. Some bee species begin producing queens and drones (offspring) concurrent with colony development (i.e., production of combs and workers), whereas other bee species achieve large colony sizes before producing queens and drones. The first strategy yields immediate but small returns because of the competing activities of workers, and the second strategy yields no immediate returns but eventually yields much larger returns. In addition, workers must weigh the cost of foraging from particular flowers against the expected nectar returns, especially at low temperatures when nectar return must be at least sufficient to maintain the high thoracic temperatures necessary for continued foraging. Because different flowers provide different amounts of nectar, bees tend to forage at flowers with high yields over a range of temperatures but visit flowers with small nectar rewards only at high temperatures. Similarly, bees must weigh the benefits of foraging at various distances from the colony. Bees will fly several kilometers, given adequate floral rewards, but respond quickly to indications of declining nectar availability, e.g., leave an inflorescence or patch after encountering empty flowers.

Heterogeneous habitats force many herbivores and predators to expend energy searching for scattered resources. Many individuals will be unable to maintain energy or nutrient balance under such conditions. By contrast, abundant suitable resources reduce costs of searching for or detoxifying resources and facilitate maintenance of energy and nutrient budgets. Frequent encounters with predators, especially when combined with low availability of food resources, may restrict the time an individual can spend foraging and increase the expenditure of energy to avoid predators, lowering net energy acquisition and potentially leading to energy balance inadequate for survival.

Species survival represents the net result of various traits that often conflict. Environmental changes, especially rapid changes occurring as a result of anthropogenic activities, will change the balance among these trade-offs, affecting the net result in various ways. Warmer global temperatures may improve energy balance for some arctic species but increase respiration loss or time spent seeking shade for other species. Ecosystem fragmentation will require greater energy expenditure for foraging and dispersal, thereby impeding movement of intolerant species over inhospitable landscapes. Some species will benefit from changes that improve overall performance (e.g., survival and reproduction), whereas other species will decline or disappear.

IV. SUMMARY

Acquired resources are allocated to various pathways. First, they either are assimilated or egested. Assimilated resources either are allocated to production or are expended, e.g., through respiration. Consumption and allocation of resources are influenced by insect size, maturity, food quality, and parasitism. Fitness accrues to the extent that assimilated resources are used for growth and reproduction. However, insect allocation patterns represent trade-offs among competing requirements of growth, reproduction, and activities necessary for food acquisition, mating, reproduction, and interactions with other organisms.

Species persist to the extent that the benefits of these behaviors outweigh the costs, i.e., survival and reproduction are increased by the investment of energy in particular behavior and associated biochemicals. Foraging and reproductive behaviors should provide the best return for the time and effort spent searching. Reproductive behavior should maximize survival of offspring. Among insects, selection of appropriate oviposition sites determines egg development and survival. Brood care is well represented among insects, with examples ranging from protection of young, to provision of food resources, to development of complex social systems for brood care and colony maintenance. However, efficiency of adult behaviors may be in conflict with efficiency of juvenile behaviors. For example, adults may oviposit on the most easily found hosts, whereas survival of immatures may depend on discovery of more suitable food hosts.

Competition and defense against predators often involve considerable expenditure of resources. In many species, males engage in various forms of combat to decide which males mate successfully. Territorial behavior is characterized by both the marking of territorial boundaries and the defense of the territory against intruders. Defense of territories may maximize access to food or mates at low population densities but becomes less advantageous and may be abandoned at high population densities. Insects defend themselves against predators physically and chemically. Behavior often enhances the effectiveness of protective coloration or toxins. For example, cryptically colored insects typically avoid movement during times when predators are active, whereas other insects may suddenly expose eyespots or brightly colored body parts to startle an attacker. Some insects imitate snakes or other predators through color patterns or movements. Such strategies minimize the energetic cost of physical defense but require greater efficiency in foraging or reproductive movements to avoid detection.

The efficiency of foraging, reproductive, competitive, and defensive behavior may be increased by use of visual, chemical, or acoustic signals that communicate information to recipients. Insects can improve foraging and mating efficiency by orienting toward chemical cues produced by suitable resources or potential mates. Discovery of a potential mate initiates a courtship ritual that improves fitness by ensuring species recognition and receptivity. Competition for food or mates can be minimized by signals that deter other individuals.

Environmental changes will affect the efficiency of resource acquisition and allocation strategies. For example, global warming will improve energy balance for some species (e.g., early season or high latitude pollinators) but increase respiration costs beyond ability to acquire energy and nutrients for others. Ultimately, insect strategies for acquiring and allocating energy and nutrient resources affect community interactions, energy flow, and nutrient cycling processes.

II

POPULATION ECOLOGY

A POPULATION IS A GROUP OF INTERBREEDING MEMBERS of a species. A number of more or less discrete subpopulations may be distributed over the geographic range for a species population. Movement of individuals among these "demes" (composing a "metapopulation") and newly available resources compensates for local extinctions resulting from disturbances or biotic interactions (Hanski and Gilpin, 1997). Populations are characterized by structural attributes, such as density, dispersion pattern, and age, sex, and genetic composition (Chapter 5) that change through time (Chapter 6) and space (Chapter 7) as a result of responses to changing environmental conditions.

Population structure and dynamics of insects have been the subject of much ecological research. This is the level of ecological organization that is the focus of evolutionary ecology, ecological genetics, biogeography, development of sampling methods, pest management, and recovery of endangered species. These disciplines all have contributed enormously to our understanding of population-level phenomena.

Abundance of many insects can change orders of magnitude on very short time scales because of their small size and rapid reproductive rates. Such rapid and dramatic change in abundance in response to often subtle environmental changes facilitates statistical evaluation of population response to environmental factors and makes insects useful indicators of environmental change. The reproductive capacity of many insects enables them to colonize new habitats and exploit favorable conditions or new resources quickly. However, their small size, short life span, and dependence on chemical communication to find mates at low densities limit persistence of small or local populations during periods of adverse conditions, frequently leading to local extinction.

Population dynamics reflect the net effects of differences among individuals in their physiological and behavioral interactions with the environment. Changes in individual success in finding and exploiting resources, mating and reproducing, and avoiding mortality agents determine numbers of individuals, their spatial distribution, and genetic composition at any point in time. Population structure is a component of the environment for the members of the population and provides information that affects individual physiology and behavior, hence, fitness (see Section I). For example, population density affects competition for food and oviposition sites (as well as other resources), propensity of individuals to disperse, and the proximity of potential mates.

Population structure and dynamics also affect community structure and ecosystem processes (Sections III and IV). Each population constitutes a part of the environment for other populations in the community. Changes in abundance of any one species population affect population(s) on which it feeds and population(s) which prey on or compete with it. Changes in size of any population also affect the importance of its ecological functions. A decline in pollinator abundance will reduce fertilization and seed production of host plants, thereby affecting aspects of nutrient uptake and primary productivity. An increase in phytophage abundance can increase canopy "porosity," increasing light penetration, and increasing fluxes of energy, water, and nutrients to the soil. A decline in predator abundance will release prey populations from regulation and contribute to increased exploitation of the prey's resources. A decline in detritivore abundance can reduce decomposition rate and lead to bottlenecks in biogeochemical cycling that affect nutrient availability.

Population structure across landscapes also influences source–sink relationships that determine population viability and ability to recolonize patches following disturbances. For example, the size and distribution of demes determines their ability to maintain gene flow or to diverge into separate species. Distribution of demes also determines the source(s) and initial genetic composition of colonists arriving at a new habitat patch. These population attributes are critical to protection or restoration of rare or endangered species. Isolation of demes as a result of habitat fragmentation can reduce their ability to reestablish local demes and lead to permanent changes in community structure and ecosystem processes across landscapes.

Population Systems

THE VARIABLES THAT DETERMINE THE ABUNDANCE AND DISTRIBUTION of a population, in time and space, constitute a population system (Berryman, 1981). The basic elements of this system are the individual members of the population, variables describing population size and structure, processes that affect population size and structure, and the environment. These elements of the population system largely determine the capacity of the population to increase in size and maintain itself within a shifting landscape mosaic of habitable patches. This chapter summarizes these population variables and processes, their integration in life history strategies, and their contribution to change in population size and distribution.

I. POPULATION STRUCTURE

Population structure is described by several variables that describe the spatial distribution of individuals and their age, sex, and genetic composition. These variables provide the means for measuring changes in population size or composition through time and space. Population variables reflect life history and the physiological and behavioral attributes that dictate habitat preferences, home ranges, oviposition patterns, and affinity for other members of the population.

A. Density

Population density is the number of individuals per unit geographic area, e.g., number per square meter, hectare, or square kilometer. This variable affects a number of other population variables. For example, mean density can be used to assess population viability or probability of colonizing habitat patches based on average distances between current location and potential mates or new resources. Density also affects population dispersion pattern (see following). A related measure, population intensity, is a commonly used measure of insect population structure. Intensity is the number of individuals per habitat unit: number per leaf, per meter branch length, per square meter leaf area or bark surface, per kilogram foliage or wood, etc. Mean intensity indicates the degree of resource exploitation, competition for space, food, or mates, and magnitude of effect on ecosystem processes. Intensity measures often can be converted to density measures if the density of habitat units is known (Southwood, 1978).

Densities and intensities of insect populations can vary widely. Bark beetles, for example, often appear to be absent from a landscape (very low density) but, with sufficient examination, can be found at high intensities on widely scattered injured or diseased trees or in the dead tops of trees (Schowalter, 1985). Under favorable conditions of climate and host abundance and condition, populations of these beetles can reach sizes of up to 10^5 individuals per tree over areas as large as 10^7 ha (Coulson, 1979; Furniss and Carolin, 1977; Schowalter *et al.*, 1981b). Schell and Lockwood (1997) reported that grasshopper population densities can increase an order of magnitude over areas of several thousand hectares within one year.

B. Dispersion

Dispersion is the spatial pattern in distribution of individuals. Dispersion is an important characteristic of populations that affects spatial patterns of resource use and population effect on community and ecosystem attributes. Dispersion pattern can be regular, random, or aggregated.

A regular (uniform) dispersion pattern is seen when individuals space themselves at regular intervals within the habitat. This dispersion pattern is typical of species that contest resource use, especially territorial species. For example, bark beetles attacking a tree show a regular dispersion pattern (Fig. 5.1). Such spacing reduces competition for resources. Variability in mean density is low. From a sampling perspective, the occurrence of one individual in a sample unit reduces the probability that other individuals will occur in the same sample unit; sample densities tend to be normally distributed. Hence, regularly dispersed populations are most easily monitored because a relatively small number of samples will provide the same estimates of mean and variance in population density as will a larger number of samples.

In a randomly dispersed population, individuals neither space themselves apart nor are attracted to each other. The occurrence of one individual in a sample unit has no effect on the probability that other individuals will occur in the same sample unit (Fig. 5.1); sample densities show a skewed (Poisson) distribution.

Aggregated (or clumped) dispersion results from grouping behavior or restriction to particular habitat patches. Aggregation is typical of species that occur in herds, flocks, schools, etc. (Fig. 5.1), for enhancement of resource exploitation or protection from predators (see Chapter 3). Gregarious sawfly larvae and tent caterpillars are examples of aggregated dispersion resulting from the tendency of individuals to form groups (Fig. 2.10). Filter feeding aquatic insects tend to be aggregated in riffles or other zones of higher flow rate within the stream continuum (e.g., Fig. 2.12), whereas predators that hide in benthic detritus, such as dragonfly larvae or water scorpions, are aggregated in pools, due to their habitat preferences. For sampling purposes, the occurrence of an individual in a sample unit increases the probability that additional individuals occur in that sample unit; sample densities are distributed as a negative binomial function. Populations with this dispersion pattern require the greatest number of samples and attention to experimental design. A large number of samples is necessary to minimize the obviously high variance in numbers of individuals among sample units and to ensure adequate representation of aggregations. Attention to experimental design can facilitate adequate representation with smaller sample sizes if factors influencing the probability of aggregations in different habitat units are known.

Dispersion pattern can change during insect development, during change in population density, or across spatial scales. For example, larval stages of tent caterpillars and gregarious sawflies are aggregated at the plant branch level, but adults are randomly dispersed at this scale (Fitzgerald, 1995; McCullough and Wagner, 1993). Many host-specific insects are aggregated on particular hosts in diverse communities but are more regularly or randomly dispersed in more homogeneous communities dominated by hosts. Some insects, such as the western box elder bug, *Boisea rubrolineatus*, and ladybird beetle, *Hippodamia convergens*, aggregate for overwintering purposes and redisperse in the spring (Schowalter, 1986). Aphids are randomly dispersed at low population densities but become more aggregated as scattered colonies increase in size (Dixon, 1985). Bark beetles show a regular dispersion pattern on a tree bole, due to spacing behavior, but are aggregated on injured or diseased trees (Coulson, 1979).

C. Metapopulation Structure

The irregular distribution of many populations across landscapes creates a pattern of relatively distinct (often isolated) local demes (aggregations) that com-

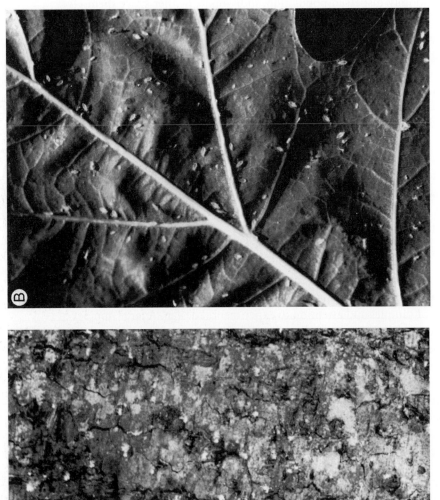

FIG. 5.1 Dispersion patterns: (A) regular dispersion of ambrosia beetle entrances (marked by the small piles of white wood fragments) through bark on a fallen Douglas-fir tree, (B) random dispersion of aphids on an oak leaf, (C) aggregated dispersion of overwintering ladybird beetles on a small shrub in a forest clearing.

FIG. 5.1 (*Continued*)

pose the greater metapopulation (Hanski and Gilpin, 1997). Insect species characterizing discrete habitat types often are dispersed as relatively distinct local demes as a result of environmental gradients or disturbances that affect the distribition of habitat types across the landscape. Obvious examples include insects associated with lotic or high elevation ecosystems. Populations of insects associated with ponds or lakes show a dispersion pattern reflecting dispersion of their habitat units. Demes of lotic species are more isolated in desert ecosystems than in mesic ecosystems. Populations of western spruce budworm, *Choristoneura occidentalis*, and fir engraver beetle, *Scolytus ventralis*, historically occurred in western North America in relatively isolated high elevation and riparian fir forests separated by more xeric patches of pine forest (Wickman, 1992).

Population distribution and degree of isolation among local demes affect gene structure and viability of the metapopulation. If local demes become too

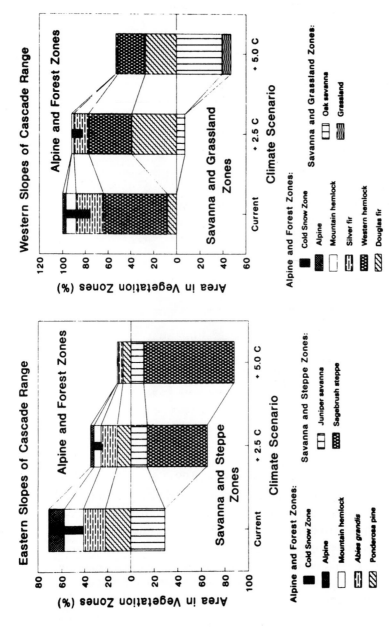

FIG. 5.2 Changes in the percent area in major vegetation zones on the eastern (left) and western (right) slopes of the Cascade Range in Oregon as a result of temperature increases of 2.5°C and 5°C. Major changes are predicted in elevational boundaries and total area occupied by vegetation zones under these global climate change scenarios. Vegetation zones occupying higher elevations will decrease in area or disappear as a result of the smaller conical surface at higher elevations. Other species associated with vegetation zones also will become more or less abundant. Reproduced from Franklin *et al.* (1992) in *Global Warming and Biological Diversity* (R. L. Peters and T. E. Lovejoy, Eds.), by permission of Yale University Press.

isolated, they become inbred and lose their ability to recolonize habitable patches following local extinction (Hedrick and Gilpin, 1997). As human activities increasingly fragment natural ecosystems, local demes become isolated more rapidly than greater dispersal ability can evolve, and species extinction becomes more likely. These effects of fragmentation could be exacerbated by climate change. For example, a warming climate will push high elevation ecosystems into smaller areas on mountaintops, and some mountaintop ecosystems will disappear (Fig. 5.2) (Franklin *et al.*, 1992). Rubenstein (1992) showed that individual tolerances to temperature changes could affect range changes by insects under warming climate scenarios. A species with a linear response to temperature could extend its range to higher latitudes (provided that expansion is not limited by habitat fragmentation) without reducing its current habitat. Conversely, a species with a dome-shaped response to temperature could extend into higher latitudes but would be forced to retreat from lower latitudes that become too warm. If the pathway for range adjustment for this species were blocked by unsuitable habitat, it would face extinction. Metapopulation dynamics are discussed in more detail in Chapter 7.

D. Age Structure

Age structure reflects the proportions of individuals at different life stages. This variable is an important indicator of population status. Growing populations generally have larger proportions of individuals in younger age classes, whereas declining populations typically have smaller proportions of individuals in these age classes. Stable populations typically have relatively more individuals in reproductive age classes. However, populations with larger proportions of individuals in younger age classes also may reflect low survivorship in these age classes, whereas populations with smaller proportions of individuals in younger age classes may reflect high survivorship (see following).

For most insect species, life spans are short (usually <1 year) and revolve around seasonal patterns of temperature and rainfall. Oviposition typically is timed to ensure that feeding stages coincide with the most favorable seasons and that diapausing stages occur during unfavorable seasons, e.g., during winter in temperate regions and during drought in tropical and arid regions. Adults typically die after reproducing. Although there are many exceptions, most temperate species have discrete, annual generations, whereas tropical species are more likely to have overlapping generations.

E. Sex Ratio

The proportion of females indicates the reproductive potential of a population. Sex ratio also reflects a number of life history traits, such as the importance of sexual reproduction, mating system, and ability to exploit harsh or ephemeral habitats (Pianka, 1974).

A 50:50 sex ratio generally indicates equally important roles of males and females, given that selection would minimize the less productive sex. Sex ratio approaches 50:50 in species where males select resources or protect or feed females. This sex ratio maximizes availability of males to females and, hence, maximizes genetic heterogeneity. High genetic heterogeneity is particularly important for population survival in heterogeneous environments. However, when the sexes are equally abundant, only half of the population is capable of producing offspring. By contrast, a parthenogenetic population (with no males) has little or no genetic heterogeneity, but the entire population is capable of producing offspring. Parthenogenetic individuals can disperse and colonize new resources without the additional challenge of finding mates, and successful colonists can generate large population sizes rapidly, ensuring exploitation of suitable resources and large numbers of dispersants in the next generation.

Sex ratio can be affected by environmental factors. For example, haploid males of many insect species are more sensitive to environmental variation, and greater mortality to haploid males may speed adaptation to changing conditions by quickly eliminating deleterious genes (Edmunds and Alstad, 1985).

F. Genetic Composition

All populations show variation in genetic composition (frequencies of various alleles) among individuals and through time. The degree of genetic variability and the frequencies of various alleles depend on a number of factors, including mutation rate, environmental heterogeneity, and population size and mobility (Hedrick and Gilpin, 1997; Mopper, 1996; Mopper and Strauss, 1998). Genetic variation may be partitioned among isolated demes or affected by patterns of habitat use (Hirai *et al.*, 1994). Genetic structure, in turn, affects various other population parameters, including population viability (Hedrick and Gilpin, 1997).

The genetic variation of the founders of a population is relatively low, simply because of the small number of colonists and the limited proportion of the gene pool that they represent. Colonists from a population with low genetic variability may found a population with even lower genetic variability (Hedrick and Gilpin, 1997). Therefore, the size and genetic variability of the source populations, as well as the number of colonists, determine genetic variability in founding populations. Genetic variability remains low during population growth unless augmented by new colonists. This is especially true for parthenogenetic species, such as aphids, for which an entire population could represent clones derived from a founding female. Differential dispersal ability among genotypes affects heterozygosity of colonists. Florence *et al.* (1982) reported that the frequencies of four alleles for an esterase (esB) converged in southern pine beetles, *Dendroctonus frontalis*, collected along a 150 m transect extending from an active infestation in east Texas. As a result, heterozygosity increased significantly with distance, approaching the theoretical maximum of 0.75 for a gene locus with four alleles. These data suggested a system that compensates for loss of genetic variability due to inbreeding by small founding populations and maximizes genetic variability in new populations coping with different selection regimes (Florence *et al.*, 1982). Nevertheless, dispersal among local populations is critical to maintaining genetic variability (Hedrick and Gilpin, 1997). If isolation restricts dispersal and infusion of new genetic material into local demes, inbreeding may reduce population ability to adapt to changing conditions, and recolonization following local extinction will be more difficult.

Polymorphism occurs commonly among insects and may underlie their rapid adaptation to environmental change or other selective pressures, such as predation (Brower, 1996; Sheppard *et al.*, 1985). Among the best known examples of population response to environmental change is the industrial melanism that developed in the peppered moth, *Biston betularia*, in England following the industrial revolution (Kettlewell, 1956). Selective predation by insectivorous birds was the key to the rapid shift in dominance from the white form, cryptic on light surfaces provided by lichens on tree bark, to the black form, which is more cryptic on trees blackened by industrial effluents. Birds preying on the more conspicuous morph maintained low frequencies of the black form in pre-industrial England, but later they greatly reduced frequencies of the white form. Other examples of polymorphism also appear to be maintained by selective predation. In some cases, predators focusing on inferior Müllerian mimics of multiple sympatric models may select for morphs or demes that mimic different models (e.g., Brower, 1996; Sheppard *et al.*, 1985).

Genetic polymorphism can develop in populations that use multiple habitat units or resources (Mopper, 1996; Mopper and Strauss, 1998; Via, 1990). Sturgeon and Mitton (1986) compared allelic frequencies among mountain pine beetles, *Dendroctonus ponderosae*, collected from three host trees [ponderosa (*Pinus ponderosa*), lodgepole (*P. contorta*), and limber pines (*P. flexilis*)] at each of five sites in Colorado. Significant variation occurred in morphological traits and allelic frequencies at five polymorphic enzyme loci among the five populations and among the three host species, suggesting that the host species is an important contributor to genetic structure of polyphagous insect populations.

Via (1991a) compared the fitnesses (longevity, fecundity, and capacity for population increase) of pea aphid, *Acyrthosiphon pisum*, clones from two host plants (alfalfa and red clover) on their source host or the alternate host. She reported that aphid clones had higher fitnesses on their source host, compared to the host on which they were transplanted, indicating local adaptation to factors associated with host conditions. Furthermore, significant negative correlations for fitness between source host and alternate host indicated increasing divergence between aphid genotypes associated with different hosts. In a subsequent study, Via (1991b) evaluated the relative importances of genetics and experience on aphid longevity and fecundity on source and alternate hosts. She maintained replicate lineages of the two clones (from alfalfa versus clover) on both host plants for three generations, then tested performance of each lineage on both hosts. If genetics is the more important factor affecting aphid performance on source and alternate host, then aphids should have highest fitness on the host to which they were adapted, regardless of subsequent rearing on the alternate host. On the other hand, if experience is the more important factor, then aphids should have highest fitness on the host from which they were reared. Via found that three generations of experience on the alternate host did not significantly improve fitness on that host. Rather, fitness was higher on the plant from which the clone was derived originally, supporting the hypothesis that genetics is the more important factor. These data indicate that continued genetic divergence of the two subpopulations is likely, given that individuals moving between alternate hosts cannot improve their performance through time as a result of experience.

Insect populations can adapt to environmental change more rapidly than can longer lived, more slowly reproducing, organisms (Mopper, 1996; Mopper and Strauss, 1998). Heterogeneous environmental conditions tend to mitigate directional selection: any strong directional selection by any environmental factor during one generation can be modified in subsequent generations by a different prevailing factor. However, changes in genetic composition occur quickly in insects when environmental change does impose directional selective pressure, such as in the change from preindustrial to postindustrial morphotypes in the polymorphic peppered moth (Kettlewell, 1956).

Biological factors that determine mating success also affect gene frequencies, perhaps in concert with environmental conditions. In a laboratory experiment with sex-linked mutant genes in *Drosophila melanogaster* (Peterson and Merrell, 1983), mutant and wild male phenotypes exhibited about the same viability, but mutant males showed a significant mating disadvantage, leading to rapid elimination (i.e., within a few generations) of the mutant allele. In addition, whereas the wild male phenotype tended to show a rare male advantage in mating, i.e., a higher proportion of males mating at low relative abundance, mutant males showed a rare male disadvantage, i.e., a lower proportion of males mating at low relative abundance, increasing their rate of elimination. These data indicate that gene frequencies in insect populations can change quickly in response to sexual selection, as well as to environmental changes.

The shift from pesticide-susceptible to pesticide-resistant genotypes may be particularly instructive. Selective pressure imposed by insecticides caused rapid development of insecticide-resistant populations for many species. Resistance development was facilitated by the widespread occurrence in insects, especially herbivores, of genes that encode for enzymes that detoxify plant defenses, since ingested insecticides also are susceptible to detoxification by these enzymes. Although avoidance of directional selection for resistance to any single tactic is a major objective of integrated pest management (IPM), pest management in practice still involves widespread use of the most effective tactic. Following the appearance of transgenic insect-resistant crop species in the late 1980s, genetically engineered, Bt toxin-producing crop varieties have replaced nontransgenic varieties over large areas, despite warnings that such crops should be only part of a multiple-tactic approach to pest management (Alstadt and Andow, 1995; Tabashnik, 1994; Tabashnik *et al.*, 1996). At least 16 species of Lepidoptera, Coleoptera, and Diptera already have developed resistance to the Bt gene as a result of strong selection (Tabashnik, 1994). For example, a single gene in the diamondback moth, *Plutella xylostella*, confers resistance to four Bt toxins (Tabashnik *et al.*, 1997) and >5000-fold resistance can be achieved in a few generations (Tabashnik *et al.*, 1996). Resistance can be reversed when exposure to Bt toxin is eliminated for several generations (Tabashnik *et al.*, 1994), but some strains can maintain resistance in the absence of Bt for more than 20 generations (Tabashnik *et al.*, 1996)

II. POPULATION PROCESSES

The population variables described above change as a result of variable reproduction, movement, and death of individuals. These individual contributions to population change are integrated as three population processes: natality (birth rate), mortality (death rate), and dispersal (rate of movement of individuals in or out of the population). For example, density can increase as a result of increased birth rate and/or immigration; frequencies of various alleles change as a result of differential reproduction, survival, and dispersal. The rate of change in these processes determines the rate of population change, described in the next chapter. Therefore, these processes are fundamental to understanding population responses to changing environmental conditions.

A. Natality

Natality is the population birth rate, i.e., the per capita production of new individuals per unit time. Realized natality is a variable that approaches potential natality, the maximum reproductive capacity of the population, only under ideal environmental conditions. Natality is affected by factors that influence production of eggs (fecundity) or production of viable offspring (fertility) by individual insects. For example, resource quality can affect the numbers of eggs produced by female insects (Chapman, 1982). Ohgushi (1995) reported that females of the herbivorous ladybird beetle, *Henosepilachna niponica*, feeding on the thistle, *Cirsium kagamontanum*, resorbed eggs in the ovary when leaf damage became high. Female blood-feeding mosquitoes often require a blood meal before first or subsequent oviposition can occur (Chapman, 1982); the ceratopogonid, *Culicoides barbosai*, produces eggs in proportion to the size of the blood meal (Linley, 1966). Hence, poor quality or insufficient food resources can reduce natality. Inadequate numbers of males can reduce fertility in sparse populations. Similarly, availability of suitable oviposition sites also affects natality.

Natality typically is higher at intermediate population densities than at low or high densities. At low densities, difficulties in attracting mates may limit mating or may limit necessary cooperation among individuals, as in the case of bark beetles that must aggregate in order to overcome host tree defenses prior to oviposition (Berryman, 1981). At high densities, competition for food, mates, and oviposition sites reduces fecundity and fertility (e.g., Southwood 1975, 1977). The influence of environmental conditions can be evaluated by comparing realized natality to potential natality, e.g., estimated under laboratory conditions.

Differences among individual fitnesses are integrated in natality. Differential reproduction among genotypes in the population determines the frequency of various alleles in the filial generation. As discussed above, gene frequencies can change dramatically within a relatively short time, given strong selection and the short generation times and high reproductive capacity of insects.

B. Mortality

Mortality is the population death rate, i.e., the per capita number of individuals dying per unit time. As with natality, we can distinguish a potential longevity or life span, resulting only from physiological senescence, from the realized longevity, resulting from the action of mortality factors. Hence, mortality can be viewed both as reducing the number of individuals in the population and reducing survival. Both have importance consequences for population dynamics.

Organisms are vulnerable to a variety of mortality agents, including unsuitable habitat conditions (e.g., extreme temperature or water conditions), toxic or unavailable food resources, competition, predation (including cannibalism), parasitism, and disease (see Chapters 2–4). These factors are a focus of studies to enhance pest management efforts. Death can result from insufficient energy or nutrient acquisition to permit detoxification of, or continued search for, suitable resources. Life stages are affected differentially by these various mortality agents (e.g., Fox, 1975b; Varley et al., 1973). For example, immature insects are particularly vulnerable to desiccation during molts, whereas flying insects are more vulnerable to predation by birds or bats. Many predators and parasites selectively attack certain life stages. Among parasitic Hymenoptera, for example, species attacking the same host have different preferences for host egg, larval, or pupal stages. Predation also can be greater on hosts feeding on particular plant species compared to other plant species, based on differential toxin sequestration or predator attraction to plant volatiles (Stamp, 1992; Traugott and Stamp, 1996; Turlings et al., 1990).

In general, mortality due to predation tends to peak at intermediate population densities, when density is sufficient for a high rate of encounter with predators and parasites, but prior to predator satiation (Fig. 5.3) (Southwood, 1975, 1977; see Chapter 8). Mortality due to competition and cannibalism increases at higher population densities (Fig. 5.3) (Fox, 1975a, b; Southwood, 1975, 1977). Competition may cause mortality through starvation, cannibal-

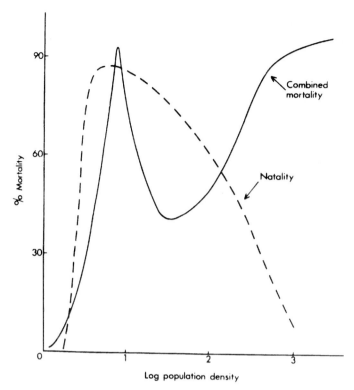

FIG. 5.3 Relationship between population density, natality, and mortality caused by predators and parasites (peaking at lower population density) and interspecific competition (peaking a higher population density). From Southwood (1975), with permission.

ism, increased disease among stressed individuals, displacement of individuals from optimal habitats, and increased exposure and vulnerability to predation as a result of displacement or delayed development.

Survival rate represents the number of individuals still living in relation to time. These individuals continue to feed and reproduce, thereby contributing most to population size as well as to genetic and ecological processes. Hence, survival rate is an important measure in studies of populations.

Survivorship curves reflect patterns of mortality and can be used to compare the effect of mortality in different populations. Lotka (1925) pioneered the comparison of survivorship curves among populations, by plotting the log of number or percent of living individuals against time. Pearl (1928) later identified three types of survivorship curves, based on the log of individual survival through time (Fig. 5.4). Type 1 curves represent species, including most large mammals, but also starved *Drosophila* (Price, 1997), in which mortality is concentrated near the end of the maximum lifespan. Type 2 curves represent species in which the probability of death is relatively constant with age, leading to a linear decline in survivorship. Many birds and reptiles approach the Type 2

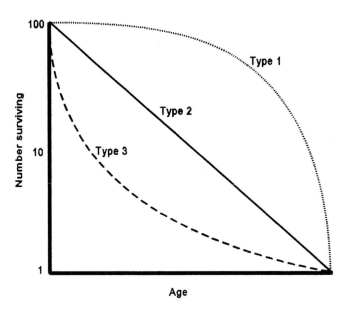

FIG. 5.4   Three generalized types of survivorship curves. Type 1 represents species with high survival rates maintained through the potential life span. Type 2 represents species with relatively constant survivorship with age. Type 3 represents species with low survival rates during early stages but relatively high survival of individuals reaching more advanced ages.

curve. Type 3 curves are seen for most insects, as well as for many other invertebrates and fish, which have high rates of mortality during early life stages but relatively low mortality during later life stages (Begon and Mortimer, 1981; Pianka, 1974). Species representing Type 3 survivorship must have very high rates of natality to ensure that some offspring reach reproductive age, compared to Type 1 species which have a high probability of reaching reproductive age.

The form of the survivorship curve can change during population growth. Mason and Luck (1978) showed that survivorship curves for the Douglas-fir tussock moth, *Orgyia pseudotsugata*, changed with population growth from stable to increasing, then decreasing. Survivorship decreased more steeply during population decline and decreased less steeply during population growth, compared to stable populations.

As described for natality, mortality integrates the differential survival among various genotypes, the basis for evolution. Survivors live longer and have greater capacity to reproduce. Hence, selective mortality can alter gene frequencies rapidly in insect populations.

C. Dispersal

Dispersal is movement of individuals from a normally favorable area to other areas that may or may not be favorable. Dispersal is contrasted with **spread,** the local movement of individuals within a favorable area, and **migration**, the directed movement from one favorable area to another favorable area (Clark *et al.*, 1967). As discussed in Chapter 2, dispersal maximizes the probability that habitat or food resources created by environmental changes or disturbances are colonized before the source population depletes or destroys its resources. However, dispersal also contributes to infusion of new genetic material into populations. This contribution to genetic heterogeneity enhances population capacity to adapt to changing conditions.

Dispersing individuals typically emigrate from a source population and eventually immigrate into another population or found a new deme in a vacant habitat patch. Immigration adds new individuals to the population, whereas emigration reduces the number of individuals in the population. This movement and its effect on population variables is influenced by a number of factors (see also Chapter 2).

In general, the number of dispersants that successfully immigrate or found new demes is the product of the number of dispersants and the individual probability of successful dispersal (Price, 1997). The number of dispersing individuals is governed by life history strategy and crowding, and the individual probability of successful dispersal is governed by dispersal mechanism, individual capacity for long distance dispersal, the distance between population source and destination (i.e., either a new population or a new resource to colonize), and habitat heterogeneity, as will be described (see also Chapter 2).

Species characterizing ephemeral habitats or resources have adapted by showing a greater tendency to disperse than do species characterizing more stable habitats or resources. For example, species found in vernal pools or desert playas tend to produce large numbers of dispersing offspring before water level begins to decline. This ensures that other suitable ponds are colonized and buffers the population against local extinctions. Some dispersal-adapted species produce a specialized morph for dispersal. For example, the dispersal form of most aphids and many scale insects is winged, whereas the feeding form typically is wingless and sedentary. Migratory locusts develop into a specialized long-winged morph for migration, distinct from the shorter-winged nondispersing morph. Some mites have dispersal stages specialized for attachment to phoretic hosts, e.g., ventral suckers in the hypopus of astigmatid mites and anal pedicel in uropodid mites (Krantz, 1978). Crowding increases competition for resources and may interfere with foraging or mating activity, thereby encouraging individuals to seek less crowded conditions.

Dispersal mechanism determines the likelihood that individuals will reach a habitable patch. Individuals that disperse randomly have a low probability of reaching a habitable destination, whereas individuals that can orient toward cues indicating suitable resources have a higher probability of reaching a habitable destination (Schowalter, 1985).

The capacity of individuals for long-distance dispersal is determined by their nutritional status. Individuals feeding on adequate resources can store suf-

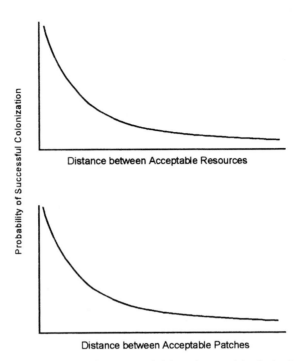

FIG. 5.5 Generalized relationship between probability of successful colonization of acceptable hosts or patches and distance from population sources. Modified from Schowalter and Means (1989) by permission of Timber Press.

ficient energy and nutrients to live longer and travel further than can individuals feeding on marginal or inadequate resources. Although dispersal should increase as population density increases, increased competition for food may limit individual energy reserves and endurance at high densities. Hence, dispersal may peak before density reaches a level that interferes with dispersal capacity (Leonard, 1970; Schowalter, 1985). As the distance between the source population and suitable destinations increases, the proportion of dispersing individuals capable of reaching that destination decreases (Fig. 5.5). Similarly, as habitat heterogeneity increases, the probability of encountering inhospitable patches increases, reducing dispersal success (Schowalter, 1996).

Dispersing individuals become vulnerable to new mortality factors. Whereas nondispersing individuals are relatively buffered against temperature extremes and protected from predators through selection of microsites, dispersing individuals are exposed to ambient temperature and humidity, high winds, and predators. Exposure to higher temperatures increases metabolic rate and depletes energy reserves more quickly, reducing the time and distance an insect can travel (Pope *et al.*, 1980). Actively moving insects also are more conspicuous and more likely to attract the attention of predators (Schultz, 1983).

The mating status of dispersing individuals determines their value as founders when they colonize new resources. Clearly, if unmated individuals must find a mate in order to reproduce after finding a habitable patch, their value as founders is negligible. For some species, mating occurs prior to dispersal of fertilized females (Mitchell, 1970). In species capable of parthenogenetic reproduction, fertilization is not required for dispersal and successful founding of populations. Some species ensure breeding at the site of colonization, such as through long-distance attraction via pheromones, e.g., by bark beetles (Raffa *et al.*, 1993), or through males accompanying females on phoretic hosts, e.g., some mesostigmatid mites (Springett, 1968) or mating swarms, e.g., eastern spruce budworm, *Choristoneura fumiferana* (Greenbank, 1957).

The contribution of dispersing individuals to genetic heterogeneity in a population depends on a number of factors. The genetic heterogeneity of the source population determines the gene pool from which dispersants come. Dispersing individuals represent a proportion of the total gene pool for the population. More heterogeneous demes have greater contributions to the genetic heterogeneity of target or founded demes than do less heterogeneous demes (Fig. 5.6) (Hedrick and Gilpin, 1997). The number or proportion of individuals that disperse affects their genetic heterogeneity. If certain genotypes are more likely to disperse, then the frequencies of these genotypes in the source population may decline, unless balanced by immigration. Distances between demes influence the degree of gene exchange through dispersal. Local demes will be influenced more by the genotypes of dispersants from neighboring demes than from more distant demes. Gene flow may be precluded for sufficiently fragmented populations. This is an increasing concern for demes restricted to isolated refugia. Populations consisting of small, isolated demes may be incapable of sufficient interaction to sustain viability.

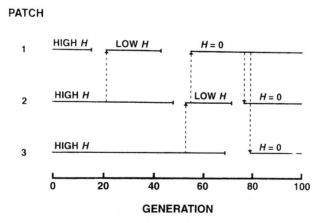

FIG. 5.6 Simulated population heterozygosity (H) over time in three habitat patches. Extinction is indicated by short vertical bars on the right end of horizontal lines; recolonization is indicated by arrows. From Hedrick and Gilpin (1998).

III. LIFE HISTORY CHARACTERISTICS

Life history adaptation to environmental conditions typically involves complementary selection of natality and dispersal strategies. General life history strategies appear to be related to habitat stability.

MacArthur and Wilson (1967) distinguished two life history strategies related to habitat stability and importance of colonization and rapid population establishment. The r-strategy generally characterizes "weedy" species adapted to colonize and dominate new or ephemeral habitats quickly (Janzen, 1977). These species are opportunists that quickly colonize new resources but are poor competitors and cannot persist when competition increases in stable habitats. By contrast, the K strategy is characterized by low rates of natality and dispersal but high investment of resources in storage and individual offspring to ensure their survival. These species are adapted to persist under stable conditions, where competition is intense, but reproduce and disperse too slowly to be good colonizers. Specific characteristics of the two strategies (Table 5.1) have been the subject of debate (Boyce, 1984). For example, small size with smaller resource requirements might be favored by K selection (Boyce, 1984), although larger organisms typically show more efficient resource use. Nevertheless, this model has been useful for understanding selection of life history attributes (Boyce, 1984).

Insects generally are considered to exemplify the r-strategy because of their relatively short life spans, Type 3 survivorship, and rapid reproductive and dispersal rates. However, among insects, a wide range of r–K strategies has been identified. For example, low order streams (characterized by narrow constrained channels and steep topographic gradients) experience wider variation

TABLE 5.1   Life History Characteristics of Species Exemplifying the r- and K-strategies

	Ecological strategy	
Attribute	r (opportunistic)	K (equilibrium)
Homeostatic ability	Limited	Extensive
Development time	Short	Long
Life span	Short	Long
Mortality rate	High	Low
Reproductive mode	Often asexual	Sexual
Age at first brood	Early	Late
Offspring/brood	Many	Few
Broods/lifetime	Usually one	Often several
Size of offspring	Small	Large
Parental care	None	Extensive
Dispersal ability	High	Limited
Numbers dispersing	Many	Few
Dispersal mode	Random	Oriented

in water flow and substrate movement, compared to higher order streams (characterized by broader floodplains and shallower topographic gradients). Insects associated with lower order streams tend to be more r-selected than are insects associated with slower water and greater accumulation of detritus (Reice, 1985). Reice (1985) experimentally disturbed benthic invertebrate communities in a low order stream in the eastern U.S. by tumbling patches of cobbles 0, 1, or 2 times in a 6-week period. Most insect and other invertebrate taxa decreased in abundance with increasing disturbance. Two invertebrate taxa increased in abundance following a single disturbance, but no taxa increased in abundance with increasing disturbance. However, all populations rebounded quickly following disturbance, suggesting that these taxa were adapted to this disturbance.

Similarly, ephemeral terrestrial habitats are dominated by species with higher natality and dispersal rates, e.g., aphids and Collembola, compared to more stable habitats, dominated by Lepidoptera, Coleoptera, and oribatid mites (Schowalter, 1985; Seastedt, 1984). Many species associated with relatively stable habitats are flightless and often wingless, evidence of weak selection for escape and colonization of new habitats. Such species may be at risk if environmental change increases the frequency of disturbance.

Grime (1977) modified the r–K model by distinguishing three primary life history strategies in plants, based on their relative tolerances of disturbance, competition, and stress. Clearly, these three factors are interrelated, since disturbance can affect competition and stress and stress can increase vulnerability to disturbance. Nevertheless, this model has proven useful for distinguishing the following strategies, characterizing harsh versus frequently disturbed and infrequently disturbed habitats.

The **ruderal** strategy generally corresponds to the r-selected strategy and characterizes unstable habitats; the **competitive** strategy generally corresponds to the K strategy and characterizes relatively stable habitats. The **stress-adapted** strategy characterizes species adapted to persist in harsh environments. These species typically are adapted to conserve resources and minimize exposure to extreme conditions, e.g., traits that prevent freezing or desiccation (see Chapter 2). Insects showing the stress-adapted strategy include those adapted to tolerate freezing in arctic ecosystems or minimize water loss in desert ecosystems (see Chapter 2).

Fielding and Brusven (1995) explored correlations between plant community correspondence to Grime's (1977) strategies and the species traits (abundance, habitat breadth, phenology, and diet breadth) of the associated grasshopper assemblages. They found that the three grasshopper species associated with the ruderal plant community had significantly wider habitat and diet breadths (generalists) and had higher densities than did grasshoppers associated with the competitive or stress-adapted plant communities (Fig. 5.7). Grasshopper assemblages also could be distinguished between the competitive and stress-adapted plant communities, but these differences were only margin-

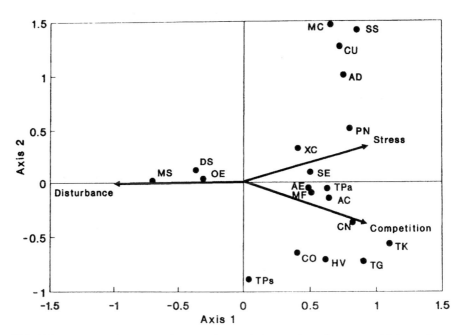

FIG. 5.7 Constrained correspondence analysis ordination of grasshopper species in southern Idaho, using Grime's (1977) classification of life history strategies based on disturbance, competition, and stress variables (arrows). Grasshoppers are denoted by the initials of their genus and species. The length of arrows is proportional to the influence of each variable on grasshopper species composition. Eigenvalues for axes 1 and 2 are 0.369 and 0.089, respectively. From Fielding and Brusven (1995) by permission from the Entomological Society of America.

ally significant. Nevertheless, their study suggested that insects can be classified according to Grime's (1977) model, based on their life history adaptations to disturbance, competition, or stress.

IV. PARAMETER ESTIMATION

Whereas population structure can be measured by sampling the population, estimates of natality, mortality, and dispersal require measurement of changes through time in overall rates of birth, death, and movement. A number of methods used to estimate these population processes (Southwood, 1978) are described briefly here.

Fecundity can be estimated by measuring the numbers of eggs in dissected females or recording the numbers of eggs laid by females caged under natural conditions. Fertility can be measured if the viability of eggs can be assessed. Natality then can be estimated from data for a large number of females. Mortality can be measured by subtracting population estimates for successive life stages, by recovering and counting dead or unhealthy individuals, or by dissection or immunoassays to identify parasitized individuals. Dispersal capacity can be measured in the laboratory using flight chambers to record duration of tethered flight. Natality, mortality, and dispersal can also be estimated from sequential recapture of marked individuals. However, these techniques require a number of assumptions about the constancy of natality, mortality, and dispersal and their net effects on population structure of the sample and do not measure these factors directly.

Deevy (1947) was the first ecologist to apply the methods of actuaries, for determining life expectancy at a given age, to development of survival and reproduction budgets for animals. Life table analysis is the most reliable method to account for survival and reproduction of a population (Begon and Mortimer, 1981; Price, 1997; Southwood, 1978). The advantage of this technique over others is the accounting of survival and reproduction in a way that allows for verification and comparison. For example, a change in cohort numbers at a stage when dispersal cannot occur could signal an error that requires correction or causal factors that merit examination.

Two types of life tables have been widely used by ecologists. The age-specific life table is based on the fates of individuals in a real cohort, a group of individuals born in the same time interval, whereas a time-specific life table is based on the fate of individuals in an imaginary cohort derived from the age structure of a stable population with overlapping generation, at a point in time. Because most insects have discrete generations and unstable populations, the age-specific life table is more applicable than the time-specific life table.

Life tables permit accounting for the survival and reproduction of members of a cohort (Table 5.2). For simplicity, the starting size of the cohort generally is corrected to a convenient number, generally 1 or 1000 females. Females are the focus of life table budgets because of their reproductive potential. Data from many cohorts representing different birth times, population densities, and environmental conditions should be analyzed and compared to gain a broader view of natality and mortality over a wide range of conditions.

Life tables partition the life cycle into discrete time intervals or life stages (Table 5.2). The age of females at the beginning of each period is designated by x, the proportion of females surviving at the beginning of the period, the age-specific survivorship, by l_x, and the number of daughters produced by each female surviving at age x, or age-specific reproductive rate, by m_x. Age-specific survivorship and reproduction can be compared between life stages to reveal patterns of mortality and reproduction. The products of per capita production and proportion of females surviving for each stage ($l_x \cdot m_x$) can be added to yield the net production by the cohort, or the net replacement rate (R_0). Net replacement rate indicates population trend. A stable population would have $R_0 = 1$, an increasing population would have $R_0 > 1$, and a decreasing population would have $R_0 < 1$. These measurements can be used to describe population dynamics, as discussed in the next chapter.

The intensive monitoring necessary to account for survival and reproduction permits identification of factors affecting survival and reproduction. Mor-

TABLE 5.2   Examples of Life Tables

x	l_x	m_x	$l_x m_x$
0	1.0	0	0
1	0.5	0	0
2	0.2	6	1.2
3	0.1	0	0
4	0	0	0
			1.2
0	1.0	0	0
1	0.5	0	0
2	0.2	0	0
3	0.1	12	1.2
4	0	0	0
			1.2
0	1.0	0	0
1	0.5	0	0
2	0.2	0	0
3	0.1	6	0.6
4	0	0	0
			0.6

Note: In these examples, the same or different cohort replacement rates are obtained by the way in which per capita production of offspring is distributed among life stages. x, life stage; l_x, proportion surviving at x; m_x, per capita production at x; and $l_x m_x$, net production at x. The sum of $l_x m_x$ is the replacement rate, R_0.

tality factors, as well as numbers of immigrants and emigrants are conveniently identified and evaluated. Survivorship between cohorts can be modeled as a line with a slope of $-k$. This slope variable can be partitioned among factors affecting survivorship, i.e., $-k_1$, $-k_2$, $-k_3$, ... $-k_i$. Such K-factor analysis has been used to assess the relative contributions of various factors to survival or mortality (e.g., Curry, 1994; Price, 1997; Varley *et al.*, 1973). Factors having the greatest effect on survival and reproduction are designated **key factors** and may be useful in population management. For example, key mortality agents can be augmented for control of pest populations or mitigated for recovery of endangered species.

Measurement of insect movement and dispersal is necessary for a number of objectives (Turchin, 1998). Mortality must be distinguished from disappearance due to emigration for life table analysis. Movement affects the probability of contact among organisms, determining their interactions. Spatial redistribution of organisms determines population structure, colonization, and metapopulation dynamics (see also Chapter 7).

Several methods for measuring and modeling animal movement have been summarized by Turchin (1998). All are labor intensive, especially for insects. Mark–recapture methods can indicate displacement of individuals but do not indicate the path, which requires direct observation. Direct observation has limited value for rapidly moving individuals, although marking individuals in various ways can enhance detection at greater distances. Radio or microwave reflectors or transmitters can be used with a receiver that records the location of an individual continuously or at intervals. However, marking and electronic signaling methods can affect the behavior of the tagged individual.

V. SUMMARY

Population systems can be described in terms of structural variables and processes that produce changes in structure. These variables indicate population status and capacity for change in response to environmental heterogeneity.

Structural variables include density, dispersion pattern of individuals and demes, age structure, sex ratio, and genetic composition. Density is the number of individuals per unit area. Dispersion is a measure of how populations are distributed in space. Regular dispersion occurs when organisms are spaced evenly among habitat or sampling units. Aggregated dispersion occurs when individuals are found in groups, for mating, mutual defense or resource exploitation, or because of the distribution of resources. Random dispersion occurs when the locations of organisms are independent of the locations of others. Metapopulation structure describes the distribution and interaction among relatively distinct subpopulations, or demes, occurring among habitable patches over a landscape. The degree of isolation of demes influences gene flow among demes and ability to colonize or recolonize vacant patches. Age structure represents the proportion of individuals in each age class and may indicate survivorship patterns or direction of change in population size. Sex ratio is the proportion of males in the population and indicates the importance of sexual reproduction, mating system, and capacity for reproduction. Genetic composition is described by the frequencies of various alleles in the population and reflects population capacity to adapt to environmental change. Some insect populations have been shown to change gene frequencies within relatively short times in response to strong directional selection, due to short generation times and high reproductive rates. This capacity for rapid change in gene frequencies makes insects especially capable of adapting to anthropogenic changes in environmental conditions.

Processes that produce change in population structure include natality, mortality, and dispersal. Natality is birth rate and represents the integration of individual fecundity and fertility. Natality is affected by abundance and nutritional quality of food resources, abundance and suitability of oviposition sites, availability of males, and population density. Mortality is death rate and reflects the influence of various mortality agents, including extreme weather conditions, food quality, competition, and predation. Generally, predation has a greater effect at low to moderate densities, whereas competition has a greater effect at high densities. Survivorship curves indicate three types of survivorship, based on whether mortality is consistent or concentrated near the beginning or end of the life span. Dispersal is the movement of individuals from a source deme to other demes or to vacant patches. Individuals colonizing vacant patches have a considerable influence on the genetic composition and development of the deme.

Life history strategies reflect the integration of natality, mortality, and dispersal strategies selected by habitat stability. Two life history classifications have been widely used. Both reflect the importance of disturbance and environmental stress on evolution of complementary strategies for reproduction and dispersal in harsh, stable, or unstable habitats.

Whereas population structure can be described readily by sampling the population, measurement of population processes is more difficult and requires accounting for the fate of individuals. Life table analysis is the most reliable method to account for age-specific survival and reproduction by members of a cohort. The net production of offspring by the cohort is designated the replacement rate and indicates population trend. Changes in these variables and processes are the basis for population dynamics. Regulatory factors and models of population change in time and space are described in the next two chapters.

Population Dynamics

POPULATIONS OF INSECTS CAN CHANGE DRAMATICALLY IN SIZE OVER relatively short periods of time as a result of changes in natality, mortality, immigration, and emigration. Under favorable environmental conditions, some species have the capacity to increase population size by orders of magnitude in a few years, given their short generation times and high reproductive rates. Under adverse conditions, populations can virtually disappear for long time periods. This capacity for significant and measurable change in population size makes insects potentially useful indicators of environmental change, often serious "pests" affecting human activities, and important engineers of ecosystem properties which may also affect global conditions. The role of insects as pests has provided the motivation for an enormous amount of research to identify factors affecting insect population dynamics, to develop models to predict population change and, more recently, to evaluate effects of insect populations on ecosystem properties. Consequently, methods and models for describing population change are most developed for economically important insects.

Predicting the effects of global change has become a major goal of research on population dynamics. Insect populations respond to changes in habitat conditions and resource quality (Heliövaara and Väisänen, 1993; Lincoln *et al.*, 1993; Chapter 2). Their responses to current environmental changes help us to anticipate responses to future environmental changes. Disturbances, in particular, influence population systems abruptly in a variety of ways, as discussed in previous chapters, but these effects are integrated by changes in natality, mortality, and dispersal rates. Factors that affect population size, such as resource availability and predation, also are affected by disturbance. As a result, the normal controls on population growth are alleviated by disturbances, for some insect species. Models of population change generally do not incorporate effects of disturbance. This chapter addresses temporal patterns of abundance, factors causing or regulating population fluctuation, and models of population dynamics.

I. POPULATION FLUCTUATION

Insect populations can fluctuate dramatically over time. If environmental conditions change in a way that favors insect population growth, the population will increase until regulatory factors reduce and finally stop population growth rate. Some populations can vary as much as 10^5-fold (Mason, 1996; Mason and Luck, 1978; Royama, 1984; Schell and Lockwood, 1997), but most populations vary less than this (Berryman, 1981; Strong et al., 1984). The amplitude and frequency of population fluctuations can be used to describe three general patterns. Stable populations fluctuate relatively little over time, whereas irruptive and cyclic populations show wide fluctuations.

Irruptive populations sporadically increase to peak numbers followed by a decline. Certain combinations of life history traits may be conducive to irruptive fluctuation. Larsson et al. (1993) and Nothnagle and Schultz (1987) reported that comparison of irruptive and nonirruptive species of sawflies and Lepidoptera from European and North American forests indicated differences in attributes between these two groups. Irruptive species generally are controlled by only one or a few factors, whereas populations of nonirruptive species are controlled by many factors. In addition, irruptive Lepidoptera and sawfly species tend to be gregarious and have a single generation per year, and are sensitive to changes in quality or availability of their particular resources, whereas nonirruptive species do not share this combination of traits.

Cyclic populations oscillate at regular intervals. Cyclic patterns of population fluctuation have generated the greatest interest among ecologists. Cyclic patterns can be seen over different time scales and may reflect a variety of interacting factors.

Strongly seasonal cycles of abundance can be seen for multivoltine species such as aphids. Aphid population size is correlated with periods of active nutrient translocation by host plants (Dixon, 1985). Hence, populations of most species peak in the spring when nutrients are being translocated to new growth, and populations of many species (especially those feeding on deciduous hosts) peak again in the fall when nutrients are being resorbed from senescing foliage. This pattern can be altered by disturbance. Schowalter and Crossley (1988) re-

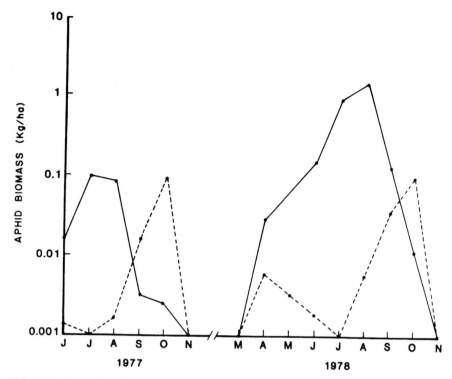

FIG. 6.1 Seasonal trends in aphid biomass in an undisturbed (dotted line) and an early successional (solid line) mixed-hardwood forest in North Carolina. The early successional forest was clearcut in 1976–1977. Peak abundances in spring and fall on the undisturbed watershed reflect nutrient translocation during periods of foliage growth and senescence; continued aphid population growth during the summer on the disturbed watershed reflects the continued production of foliage by regenerating plants. From Schowalter (1985).

ported that sustained growth of early successional vegetation following clearcutting of a deciduous forest supported continuous growth of aphid populations during the summer (Fig. 6.1).

Longer term cycles are apparent for many species. Several forest Lepidoptera exhibit cycles with periods of approximately 10 years or 20–30 years (Berryman, 1981; Mason and Luck, 1978; Price, 1997; Royama, 1992). For example, spruce budworm, *Choristoneura fumiferana*, populations peak at approximately 25- to 30-year intervals in eastern North America (Fig. 6.2). In many cases, population cycles are synchronized over large areas, suggesting the influence of a common widespread trigger such as climate, sunspot, lunar, or ozone cycles (Clark, 1979; Price, 1997; Royama, 1984, 1992). Alternatively, Moran (1953) suggested, and Royama (1992) demonstrated (using models), that synchronized cycles could result from correlations among controlling factors. Hence, the cause of synchrony can be independent of the cause of the cyclic pattern of fluctuation. Generally, peak abundances are maintained only for a few (2–3) years, followed by relatively precipitous declines (Fig. 6.2).

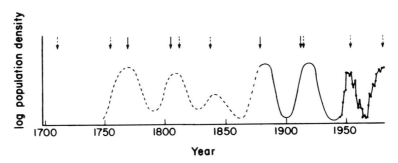

FIG. 6.2 Spruce budworm population cycles in New Brunswick and Quebec over the past 200 years, from sampling data since 1945, from historical records between 1978 and 1945, and from radial growth-ring analysis of surviving trees prior to 1878. Arrows indicate the years of first evidence of reduced ring growth. Data since 1945 fit the log scale, but the amplitude of cycles prior to 1945 are arbitrary. From Royama (1984) by permission from the Ecological Society of America.

Explanations for cyclic population dynamics include climatic cycles and changes in insect gene frequencies or behavior, food quality, or susceptibility to disease that occur during large changes in insect abundance (Myers, 1988). Climatic cycles may trigger insect population cycles directly through changes in mortality or indirectly through changes in host condition or susceptibility to pathogens. Changes in gene frequencies or behavior may permit rapid population growth during a period of reduced selection. In particular, reduced selection under conditions favorable for rapid population growth may permit increased frequencies of deleterious alleles that become targets of intense negative selection when conditions become less favorable. Depletion of food resources during an outbreak may impose a time lag for recovery of depleted resources to levels capable of sustaining renewed population growth (e.g., Clark, 1979). Epizootics of entomopathogens may occur only above threshold densities. Sparse populations near their extinction threshold (see following) may require several years to recover sufficient numbers for rapid population growth. Berryman (1996), Royama (1992), and Turchin (1990) have demonstrated the importance of delayed effects (time lags) of factors (especially predation or parasitism) controlling population growth to the generation of cyclic pattern.

Changes in population size can be described by four distinct phases (Mason and Luck, 1978). The **endemic phase** is the low population level maintained between outbreaks. The beginning of an outbreak cycle is triggered by a disturbance or other environmental change that allows the population to increase in size above its **release threshold**. This threshold represents a population size at which reproductive momentum results in escape of at least a portion of the population from normal regulatory factors, e.g., predation. Despite the importance of this threshold to population outbreaks, few studies have established its size for any insect species. Schowalter *et al.* (1981b) reported that local outbreaks of southern pine beetle, *Dendroctonus frontalis*, occurred when demes reached a critical size of about 100,000 beetles by early June. Above the release threshold, survival is relatively high and population continues to grow uncontrolled during the **release phase**. During this period, emigration peaks and the population spreads to other suitable habitat patches (see Chapter 7). Resources eventually become limiting, as a result of depletion by the growing population, and predators and pathogens respond to increased prey/host density and stress. Population growth slows and abundance reaches a **peak**. Continued predation and pathogen epizootics accelerate population **decline**. Intraspecific competition and predation rates also decline as the population reenters the endemic phase.

Outbreaks of insect populations have become more frequent and intense in crop systems or natural monocultures where food resources are relatively unlimited or where manipulation of disturbance frequency has created favorable conditions (e.g., Kareiva, 1983; Wickman, 1992). However, populations of many species fluctuate at amplitudes that are insufficient to cause economic damage and, therefore, do not attract attention. Some of these species may experience more conspicuous outbreaks under changing environmental conditions, e.g., introduction into new habitats or large-scale conversion of natural ecosystems to managed ecosystems.

II. FACTORS AFFECTING POPULATION SIZE

Populations are affected by a variety of intrinsic and extrinsic factors. Intrinsic factors include intraspecific competition, cannibalism, and territoriality. Extrinsic factors include abiotic conditions and other species. Populations showing wide amplitude of fluctuation may have weak intrinsic ability to regulate population growth, e.g., through depressed natality in response to crowding. Rather, such populations may be regulated by available food supply, predation, or other extrinsic factors. These factors can influence population size in two primary ways. If the proportion of organisms affected by a factor is constant for any population density or the effect of the factor does not depend on population density, the factor is considered to have a density-independent effect. Conversely, if the proportion of organisms affected varies with density or the effect of the factor depends on population density, then the factor is considered to have a density-dependent effect (Begon and Mortimer, 1981; Berryman, 1981; Clark *et al.*, 1967; Price, 1997).

The distinction between density independence and density dependence is often confused, for various reasons. First, many factors may act in both density-independent and density-dependent manners, depending on circumstances. For example, climatic factors or disturbances often are thought to affect populations in a density-independent manner, because the same proportion of exposed individuals typically is affected at any population density. However, for species able to seek shelter from unfavorable conditions, the proportion of individuals exposed (and, therefore, the effect of the climatic factor or disturbance) may be related to population density. Furthermore, a particular factor may have a density-independent effect over one range of population densities and a density-dependent effect over another range of densities. A plant defense may have a density-independent effect until herbivore densities reach a level that triggers induced defenses. Generally, population size is modified by abiotic factors, such as climate and disturbance, but is maintained near an equilibrium level by density-dependent biotic factors.

A. Density-Independent Factors

Insect populations are highly sensitive to changes in abiotic conditions, such as temperature, and water availability, which affect insect growth and survival (see Chapter 2). Changes in population size of some insects have been related directly to changes in climate or to disturbances (e.g., Greenbank, 1963; Kozár, 1991; Porter and Redak, 1996; Reice, 1985; Schowalter, 1985). In some cases, climate fluctuation or disturbance affects resource values for insects. For example, loss of riparian habitat as a result of agricultural practices in western North America may have led to extinction of the historically important Rocky Mountain grasshopper, *Melanoplus spretus* (Lockwood and DeBrey, 1990).

Many environmental changes occur relatively slowly and cause gradual changes in insect populations as a result of subtle shifts in genetic structure and individual fitness. Other environmental changes occur more abruptly and may trigger rapid change in population size because of sudden changes in natality, mortality, or dispersal.

Disturbances are particularly important triggers for inducing population change because of their acute disruption of community or population structure and of resource, substrate, and other ecosystem conditions. The disruption of population structure can alter community structure and cause changes in physical, chemical, and biological conditions of the ecosystem. Disturbances can promote or truncate population growth, depending on species tolerances to particular disturbance or postdisturbance conditions.

Some species are more tolerant of particular disturbances, based on adaptation to regular recurrence. For example, plants in fire-prone ecosystems tend to have attributes that protect meristematic tissues, whereas those in frequently flooded ecosystems can tolerate root anaerobiosis. Generally, insects do not have specific adaptations to survive disturbance, given their short generation times relative to disturbance intervals, and unprotected populations may be greatly reduced. Species that do show some disturbance-adapted traits, such as orientation to smoke plumes or avoidance of litter accumulations in fire-prone ecosystems (Miller and Wagner, 1984; Schowalter, 1985), generally have longer (2–5 year) generation times that would increase the frequency of generations experiencing a disturbance. Most species are affected by postdisturbance conditions. Disturbances affect insect populations both directly and indirectly.

Disturbances can create lethal conditions for many insects. For example, fire can burn exposed insects (Porter and Redak, 1996; Shaw *et al.*, 1987) or raise temperatures to lethal levels in unburned microsites. Tumbling cobbles in flooding streams can crush benthic insects (Reice, 1985). Flooding of terrestrial habitats can create anaerobic soil conditions. Drought can raise air and soil temperatures and cause desiccation (Mattson and Haack, 1987). Populations of many species can suffer severe mortality as a result of these factors, and rare species may be eliminated (Shaw *et al.*, 1987; Schowalter, 1994a). Willig and Camilo (1991) reported the virtual disappearance of two species of walkingsticks, *Lamponius portoricensis* and *Agamemnon iphimedeia*, from tabonuco forests in Puerto Rico following Hurricane Hugo. Drought can reduce water levels in aquatic ecosystems, reducing or eliminating habitat for some aquatic insects. In contrast, storms may redistribute insects picked up by high winds. Torres (1988) reviewed cases of large numbers of insects being transported into new areas by hurricane winds, including swarms of African desert locusts, *Schistocerca gregaria*, deposited on Caribbean islands.

Mortality depends on disturbance intensity and scale and species adaptation. Miller and Wagner (1984) reported that the pandora moth, *Coloradia pandora*, in ponderosa pine forests in western North America preferentially pupates on soil with sparse litter cover, under open canopy, where it is more likely to survive frequent understory fires. This habit would not protect pupae during more severe fires. Small-scale disturbances affect a smaller proportion of the population than do larger-scale disturbances. Large-scale disturbances, such as volcanic eruptions or hurricanes, could drastically reduce populations over much of the species range, making such populations vulnerable to extinction. The potential for disturbances to eliminate small populations or critical local demes of fragmented metapopulations has become a serious obstacle to restoration of endangered (or other) species (Foley, 1997).

Disturbances indirectly affect insect populations by altering the postdisturbance environment. Disturbance affects abundance or physiological condition of hosts and abundances or activity of other associated organisms (Mattson and Haack, 1987; Paine and Baker, 1993; Schowalter, 1996). Selective mortality to disturbance-intolerant plant species reduces the availability of a resource for associated herbivores. Similarly, long disturbance-free intervals can lead to eventual replacement of ruderal plant species and their associated insects. Changes in canopy cover or plant density alter vertical and horizontal gradients in light, temperature, and moisture that influence habitat suitability for insect species and can alter vapor diffusion patterns that influence chemoorientation by insects (Cardé, 1996; Mattson and Haack, 1987).

Disturbances injure or stress surviving hosts or cause changes in nutrient allocation patterns. The grasshopper, *Melanoplus differentialis*, prefers the wilted foliage of sunflower to turgid foliage (Lewis, 1979). Fire or storms can wound surviving plants and increase their susceptibility to herbivorous insects. Lightning-struck (Fig. 6.3) or wind-thrown trees are particular targets for many bark beetles and provide refuges for these insects at low population levels (Flamm *et al.*, 1993; Paine and Baker, 1993). Drought stress can cause audible

FIG. 6.3   Lighting strike impairs tree defense systems. Injured, diseased, or stressed trees are typical targets of bark beetle colonization.

cell wall cavitation that may attract insects adapted to exploit water-stressed hosts (Mattson and Haack, 1987). If drought or other disturbances stress large numbers of plants surrounding these refuges, these small populations can reach

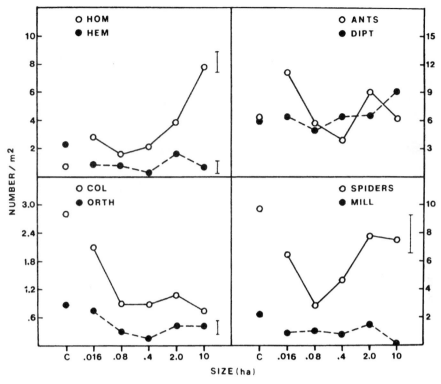

FIG. 6.4 Densities of arthropod groups during the first growing season in uncut forest (C) and clearcut patches ranging in size from 0.016 ha to 10 ha. For groups showing significant differences between patch sizes, vertical bars indicate the least significant difference ($P < .05$). HOM, Homoptera; HEM, Hemiptera; COL, Coleoptera; ORTH, Orthoptera; DIPT, Diptera; and MILL, millipedes. From Shure and Phillips (1991) by permission from Springer-Verlag.

epidemic sizes quickly (Mattson and Haack, 1987). In some cases, crowding of plants, as a result of planting or long disturbance-free intervals, can exacerbate drought stress. Stressed plants may alter their production of particular amino acids or suppress production of defensive chemicals to meet more immediate metabolic needs, thereby affecting their suitability for particular herbivores (Haglund, 1980; Lorio, 1993; Waring and Pitman, 1983). High densities of stressed plants frequently trigger outbreaks of herbivorous species (Mattson and Haack, 1987; Schowalter, 1996).

Changes in abundances of competitors, predators, and pathogens also affect postdisturbance insect populations. For example, phytopathogenic fungi establishing in, and spreading from, stumps following fire, wind-throw, or harvest can stress infected survivors and increase their susceptibility to bark beetles and other wood-boring insects (Paine and Baker, 1993). Drought or solar exposure resulting from disturbance can reduce the abundance or virulence of entomopathogenic fungi, bacteria, or viruses (Mattson and Haack, 1987;

Roland and Kaupp, 1995). Disturbance and/or fragmentation reduce the abundances and activity of some predators and parasites (Kruess and Tscharntke, 1994; Roland and Taylor, 1997; Schowalter and Lowman, 1999) and may induce or support outbreaks of defoliators (Roland, 1993). Alternatively, fragmentation can interrupt spread of some insect populations by creating inhospitable barriers (Schowalter et al., 1981b).

Population responses to these indirect effects again vary, depending on scale (see Chapter 7). Few natural experiments have addressed the effects of scale. Clearly, a larger scale should affect environmental conditions and populations within the disturbed area more than would a smaller scale. Shure and Phillips (1991) compared arthropod abundances in clear-cuts of different sizes in the southeastern U.S. (Fig. 6.4). They suggested that the greater differences in arthropod densities in larger clear-cuts reflect the steepness of environmental gradients from the clear-cut into the surrounding forest. The surrounding forest has a greater effect on environmental conditions within a small canopy opening than within a larger opening.

The capacity for insect populations to respond quickly to abrupt changes in environmental conditions (disturbances) indicates their capacity to respond to more gradual environmental changes. Insect outbreaks have become particularly frequent and severe in landscapes that have been significantly altered by human activity (Hadley and Veblen, 1993; Huettl and Mueller-Dombois, 1993; Wickman, 1992). In particular, anthropogenic suppression of fire, channelization and clearing of riparian areas, and conversion of natural, diverse vegetation to rapidly growing, commercially valuable crop species on a regional scale have resulted in more severe disturbances and dense monocultures of susceptible species that support widespread outbreaks of adapted insects (e.g., Schowalter and Lowman, 1999).

Insect populations also are likely to respond to changing global temperature, precipitation patterns, atmospheric and water pollution, and atmospheric concentrations of CO_2 and other trace gases (e.g., Alstad et al., 1982; Franklin et al., 1992; Heliövaara, 1986; Heliövaara and Väisänen, 1993; Hughes and Bazzaz, 1997; Lincoln et al., 1993; Marks and Lincoln, 1996). Grasshopper populations are favored by warm, dry conditions (Capinera, 1987) predicted by climate change models to increase in many regions. Interaction among multiple factors changing simultaneously may affect insects differently than predicted from responses to individual factors (e.g., Marks and Lincoln, 1996; Franklin et al., 1992).

The similarity in insect population responses to natural versus anthropogenic changes in the environment depends on the degree to which anthropogenic changes create conditions similar to those created by natural changes. For example, natural disturbances typically remove less biomass from a site than do harvest or livestock grazing. This difference likely affects insects that depend on postdisturbance biomass, such as large woody debris, either as a food resource or refuge from exposure to altered temperature and moisture (Seastedt and Crossley, 1981a). Anthropogenic disturbances (because of ownership or management boundaries) leave straighter and more distinct boundaries between disturbed and undisturbed patches, affecting the character of edges and the steepness of environmental gradients into undisturbed patches (Chen *et al.*, 1995; Roland and Kaupp, 1995). Similarly, the scale, frequency, and intensity of prescribed fires may differ from natural fire regimes. For example, in northern Australia, natural ignition would come from lightning during storm events at the onset of monsoon rains, whereas prescribed fires often are set during drier periods to maximize fuel reduction (Braithwaite and Estbergs, 1985). Consequently, prescribed fires burn hotter, are more homogeneous in their severity, and cover larger areas than do lower intensity, more patchy fires burning during cooler, moister periods.

Few studies have evaluated the responses of insect populations to changes in multiple factors. For example, habitat fragmentation, climate change, acid precipitation, and introduction of exotic species may influence insect populations interactively in many areas.

B. Density-Dependent Factors

Primary density-dependent factors include intra- and interspecific competition for limited resources and predation. The relative importance of these factors has been the topic of much debate. Malthus (1789) wrote the first theoretical treatise describing the increasing struggle for limited resources by growing populations. Effects of intraspecific competition on natality, mortality, and dispersal have been demonstrated widely (see Chapter 5). As competition for finite resources becomes intense, fewer individuals obtain sufficient resources to survive, reproduce, or disperse. Similarly, a rich literature on predator–prey interactions generally, and biocontrol agents in particular, has shown the important density-dependent effects of predators, parasitoids, parasites, and pathogens on prey populations (e.g., Carpenter *et al.*, 1985; Marquis and Whelan, 1994; Parry *et al.*, 1997; Price, 1997; Tinbergen, 1960; van den Bosch *et al.*, 1982; Van Driesche and Bellows, 1996). Predation rates typically increase as prey abundance increases, up to a point at which predators become satiated. Predators respond both behaviorally and numerically to changes in prey density (see Chapter 8). Predators can be attracted to an area of high prey abundance, a behavioral response, and increase production of offspring as food supply increases, a numerical response.

Cooperative interactions among individuals lead to inverse density-dependence. Mating success (and thus natality) increases as density increases. Some insects show increased ability to exploit resources as density increases. Examples include bark beetles that must aggregate in order to kill trees, a necessary prelude to successful reproduction (Berryman, 1997; Coulson, 1979), and social insects that increase thermoregulation and recruitment of nestmates to harvest suitable resources as colony size increases (Heinrich, 1979; Matthews and Matthews, 1978).

Factors affecting population size can operate over a range of time delays. For example, fire affects numbers immediately (no time lag) by killing exposed individuals, whereas predation requires some period of time (time lag) for predators to aggregate in an area of dense prey and to produce offspring. Hence, increased prey density is followed by increased predator density only after some time lag. Similarly, as prey abundance decreases, predators disperse or cease reproduction but only after a time lag.

C. Regulatory Mechanisms

When population size exceeds the number of individuals that can be supported by existing resources, competition and other factors reduce population size until it reaches levels in balance with resource supply. This equilibrium population size, that can be sustained indefinitely by resource availability, is termed the **carrying capacity** of the environment and is designated as **K**. Carrying capacity is not constant but depends on factors that affect both the abundance and suitability of necessary resources, including the intensity of competition with other species that also use those particular resources.

Density-independent factors modify population size, but only density-dependent factors can regulate population size, in the sense of stabilizing abundance near carrying capacity. Regulation requires environmental feedback, such as through density-dependent mechanisms that reduce population growth at high densities but allow population growth at low densities (Isaev and Khlebopros, 1979). Nicholson (1933, 1954a,b, 1958) first proposed that density-dependent biotic interactions are the primary factors determining population size. Andrewartha and Birch (1954) challenged this view, suggesting that density-dependent processes generally are of minor importance in determining abundance. This debate was resolved with recognition that regulation of population size requires density-dependent processes, but abundance is determined by all factors that affect the population (Begon and Mortimer, 1981; Isaev and Khlebopros, 1979). However, debate continues over the relative importances of competition and predation, the so-called "bottom-up" (or resource concentration/limitation) and "top-down" (or "trophic cascade") hypotheses, for regulating population sizes (see also Chapter 8).

Bottom-up regulation is accomplished through the dependence of populations on resource supply. Suitable food is most often invoked as the limiting resource, but suitable shelter and oviposition sites also may be limiting. As populations grow, these resources become the objects of intense competition, reducing natality and increasing mortality and dispersal (Chapter 5), and eventually reducing population growth. As population size declines, resources become relatively more available and support population growth. Hence, a population should tend to fluctuate around the size (carrying capacity) that can be sustained by resource supply.

Top-down regulation is accomplished through the response of predators and parasites to increasing host population size. As prey abundance increases, predators and parasites encounter more prey. Predators respond functionally to increased abundance of a prey species by learning to acquire prey more efficiently and respond numerically by increasing population size as food supply increases. Increased intensity of predation reduces prey numbers. Reduced prey availability limits food supply for predators and reduces the intensity of predation. Hence, a prey population should fluctuate around the size determined by intensity of predation.

A number of experiments have demonstrated the dependence of insect population growth on resource availability, especially the abundance of suitable food resources (e.g., M. Brown *et al.*, 1987; Cappuccino, 1992; Harrison, 1994; Lunderstädt, 1981; Ohgushi and Sawada, 1985; Polis and Strong, 1996; Price, 1997; Schowalter and Turchin, 1993; Schultz, 1988; Scriber and Slansky, 1981; Varley and Gradwell, 1968). For example, Schowalter and Turchin (1993) demonstrated that growth of southern pine beetle populations, measured as number of host trees killed, was significant only under conditions of high host density and low nonhost density (Fig. 6.5). However, some populations appear not to be food limited (Wise, 1975). Many exotic herbivores are generalists that are regulated poorly in the absence of coevolved predators.

Population regulation by predators has been supported by experiments demonstrating population growth following predator removal (Carpenter and Kitchell, 1987, 1988; Dial and Roughgarden, 1995; Marquis and Whelan, 1994; Oksanen, 1983). Manipulations in multiple trophic level systems have shown that a manipulated increase at one predator trophic level causes reduced abundance of the next lower trophic level and increased abundance at the second trophic level down (Carpenter and Kitchell, 1987, 1988). However, in many cases, predators appear simply to respond to prey abundance without regulating prey populations (Parry *et al.*, 1997), and the effect of predation and parasitism often is delayed, and hence, less obvious than the effects of resource supply.

Regulation by lateral factors does not involve other trophic levels. Interference competition, territoriality, cannibalism, and density-dependent dispersal have been considered to be lateral factors that may have a primary regulatory role (Harrison and Cappuccino, 1995). For example, Fox (1975a) reviewed studies indicating that cannibalism is a predictable part of the life history of some species, acting as a population control mechanism that rapidly decreases the number of competitors, regardless of food supply. In the backswimmer, *Notonecta hoffmanni*, cannibalism of young nymphs by older nymphs occurred even when alternative prey were abundant (Fox, 1975b). In other species, any exposed or unprotected individuals are attacked (Fox, 1975a). However, competition clearly is affected by resource supply.

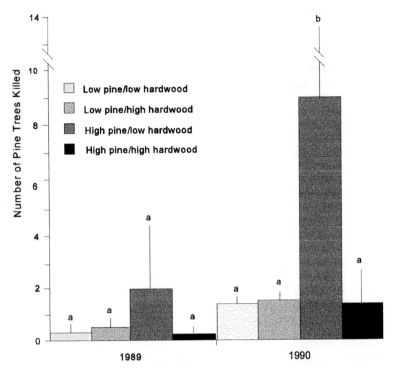

FIG. 6.5 Effect of host (pine) and nonhost (hardwood) densities on population growth of the southern pine beetle, measured as pine mortality in 1989 (Mississippi) and 1990 (Louisiana). Low pine = 11–14 m^2 ha^{-1} basal area; high pine = 23–29 m^2 ha^{-1} basal area; low hardwood = 0–4 m^2 ha^{-1} basal area; high hardwood = 9–14 m^2 ha^{-1} basal area. Vertical lines indicate standard error of the mean. Bars under the same letter did not differ at an experiment-wise error rate of $P < .05$ for data combined for the two years. Data from Schowalter and Turchin (1993).

All populations probably are regulated simultaneously by bottom-up, top-down, and lateral factors. Some resources are more limiting than others for all species, but changing environmental conditions can affect the abundance or suitability of particular resources and directly or indirectly affect higher trophic levels (Hunter and Price, 1992; Polis and Strong, 1996). For example, environmental changes that stress vegetation can increase the suitability of a food plant without changing its abundance. Under such circumstances, the disruption of bottom-up regulation results in increased prey availability and, perhaps, suitability (Stamp, 1992; Traugott and Stamp, 1996) for predators and parasites, resulting in increased abundance at that trophic level. Density-dependent competition and dispersal, as well as increased predation, eventually cause population decline to levels at which these regulatory factors become less operative.

Harrison and Cappuccino (1995) compiled data from 60 studies in which bottom-up, top-down, or lateral density-dependent regulatory mechanisms were evaluated for populations of invertebrates, herbivorous insects, and vertebrates. They reported that bottom-up regulation was apparent in 89% of the studies overall, compared to observation of top-down regulation in 39% and lateral regulation in 79% of the studies.

Top-down regulation was observed more frequently than bottom-up regulation only for the category that included fish, amphibians, and reptiles. Bottom-up regulation may predominate in (primarily terrestrial) systems where resource suitability is more limiting than is resource availability, i.e., resources are defended in some way (especially through incorporation of carbohydrates into indigestible lignin and cellulose). Top-down regulation may predominate in (primarily aquatic) systems where resources are relatively undefended, and production can compensate for consumption (Strong, 1992; see also Chapter 11).

Whereas density dependence acts in a regulatory (stabilizing) manner through negative feedback, i.e., acting to slow or stop continued growth, inverse density dependence has been thought to act in a destabilizing manner. Allee (1931) first proposed that positive feedback creates unstable thresholds, i.e., an extinction threshold below which a population inevitably declines to extinction, and a release threshold above which the population grows uncontrollably until resource depletion or epizootics decimate it (Begon and Mortimer, 1981; Berryman, 1996, 1997; Isaev and Khlebopros, 1979). Between these thresholds, density dependent factors should maintain stable populations near K, a property known as the Allee effect. However, positive feedback likely enhances population persistence at low densities and is counteracted, in most species, by the effects of crowding and resource depletion at higher densities

Clearly, conditions that bring populations near release or extinction thresholds are of particular interest to ecologists, as well as to resource managers. Bazykin *et al.* (1997), Berryman *et al.* (1987), and Turchin (1990) demonstrated the importance of time lags to the effectiveness of regulatory factors. They demonstrated that time lags weaken negative feedback and reduce the rigidity of population regulation. Hence, populations that are controlled primarily by factors that operate through delayed negative feedback should exhibit greater amplitude of population fluctuation, whereas populations that are controlled by factors with more immediate negative feedback should be more stable. Myers (1988) and Mason (1996) concluded that delayed effects of density-dependent factors can generate outbreak cycles with an interval of about 10 years. For irruptive and cyclic populations, decline to near or below local extinction thresholds may affect the time necessary for population recovery between outbreaks.

III. MODELS OF POPULATION CHANGE

Models are representations of complex phenomena and are used to understand and predict changes in those phenomena. Population dynamics of various organisms, especially insects, are of particular concern as population changes affect human health, production of ecosystem commodities, and the quality of terrestrial and aquatic ecosystems. Hence, development of models to improve our ability to understand and predict changes in population abundances has a rich history.

Models take many forms. The simplest are conceptual models that clarify relationships between cause and effect. For example, box-and-arrow diagrams can be used to show which system components interact with each other (e.g., Fig. 1.3). More complex statistical models represent those relationships in quantitative terms, e.g., regression models that depict the relationship between population size and environmental factors (e.g., Figs. 5.3 and 5.4). Advances in computational technology have led to development of biophysical models that can integrate large datasets to predict responses of insect populations to a variety of interacting environmental variables. The most sophisticated models are the computerized decision support systems that integrate a user interface with component submodels that can be linked in various ways, based on user-provided key words, to provide output that addresses specific questions (e.g., Shaw and Eav, 1993).

A. Exponential and Geometric Models

The simplest model of population growth describes change in numbers as the initial population size times the per capita rate of increase (Fig. 6.6) (Berryman, 1997; Price, 1997). This model integrates per capita natality, mortality, immi-

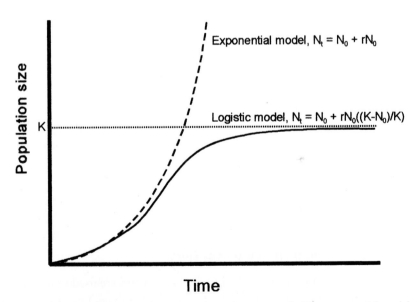

Exponential model, $N_t = N_0 + rN_0$

Logistic model, $N_t = N_0 + rN_0((K-N_0)/K)$

FIG. 6.6 Exponential and logistic models of population growth. The exponential model describes an infinitely increasing population, whereas the logistic model describes a population reaching an asymptote at the carrying capacity of the environment (K).

gration, and emigration per unit time as the instantaneous or **intrinsic rate of increase**, designated **r**

$$r = (N + I) - (M + E),$$ (6.1)

where N = natality, I = immigration, M = mortality, and E = emigration, all instantaneous rates.

Where cohort life table data, rather than time-specific natality, mortality, and dispersal, have been collected, r can be estimated as

$$r = \frac{\log_e R_0}{T},$$ (6.2)

where R_0 is replacement rate and T is generation time.

The rate of change for populations with overlapping generations is a function of the intrinsic (per capita) rate of increase and the current population size. The resulting model for exponential population growth is

$$N_{t+1} = N_t + rN_t,$$ (6.3)

where N_t is the population size at time t, and N_0 is the initial population size. This equation also can be written as

$$N_t = N_0 e^{rt}.$$ (6.4)

For insect species with nonoverlapping cohorts (generations), the replacement rate, R_0, represents the per capita rate of increase from one generation to the next. This parameter can be used in place of r for such insects. The resulting expression for geometric population growth is

$$N_t = R_0^t N_0,$$ (6.5)

where N_t is the population size after t generations.

Equations 6.3–6.5 describe density-independent population growth (Fig. 6.6). However, as discussed above, density-dependent competition, predation and other factors interact to limit population growth.

B. Logistic Model

A mathematical model to account for density-dependent regulation of population growth was developed by Verhulst in 1838 and again, independently, by Pearl and Reed (1920). This logistic model (Fig. 6.6) often is called the Pearl–Verhulst equation (Berryman, 1981; Price, 1997). The logistic equation is

$$N_{t+1} = N_t + rN_t \frac{(K - N_t)}{K},$$ (6.6)

where K is the carrying capacity of the environment. This model describes a sigmoid (S-shaped) curve (Fig. 6.6) that reaches equilibrium at K. If $N < K$, then the population will increase up to $N = K$. If the ecosystem is disturbed in a way that $N > K$, then the population will decline to $N = K$.

C. Complex Models

General models such as the Pearl–Verhulst model usually do not predict the dynamics of real systems accurately. For example, the use of the logistic growth model is limited by several assumptions. First, individuals are assumed to be equal in their reproductive potential. Clearly, immature insects and males do not produce offspring, and females vary in their productivity, depending on nutrition, access to oviposition sites, etc. Second, population adjustment to changing density is assumed to be instantaneous, and effects of density-dependent factors are assumed to be a linear function of density. These assumptions ignore time lags, which may control dynamics of some populations and obscure the importance of density dependence (Turchin, 1990). Finally, r and K are assumed to be constant. In fact, factors (including K) that affect natality, mortality, and dispersal affect r. Changing environmental conditions, including depletion by dense populations, affect K. Therefore, population size fluctuates with an amplitude that reflects variation in both K and the life history strategy of particular insect species. Species with the r strategy (high reproductive rates and low competitive ability) tend to undergo boom-and-bust cycles because of their tendency to overshoot K, deplete resources, and decline rapidly, often approaching their extinction threshold, whereas species with the K strategy (low reproductive rates and high competitive ability) tend to approach K more slowly and maintain relatively stable population sizes near K (Boyce, 1984). Modeling of real populations of interest, then, requires development of more complex models with additional parameters that correct these shortcomings, some of which are described below.

Nonlinear density dependent processes and delayed feedback can be addressed by allowing r to vary as

$$r = r_{max} - sN_{t-T},\qquad(6.7)$$

where r_{max} is the maximum per capita rate of increase, s represents the strength of interaction between individuals in the population, and T is the time delay in the feedback response (Berryman, 1981). The sign and magnitude of s also can vary, depending on the relative dominance of competitive and cooperative interactions

$$s = s_p - s_m N_t,\qquad(6.8)$$

where s_p is the maximum benefit from cooperative interactions and s_m is the competitive effect, assuming that s is a linear function of population density at time t (Berryman, 1981). The extinction threshold, E, can be incorporated by adding a term forcing population change to be negative below this threshold

$$N_{t+1} = N_t + rN_t(K - N_t)/K)((N_t - E)/E).\qquad(6.9)$$

Similarly, the effect of factors influencing natality, mortality, and dispersal can be incorporated into the model to improve representation of r.

The effect of other species interacting with a population was addressed first by Lotka (1925) and Volterra (1926). The Lotka–Volterra equation for the effect of a species competing for the same resources includes a term that reflects the degree to which the competing species reduces carrying capacity

$$N_{1(t+1)} = N_{1t} + r_1 N_{1t}(K_1 - N_{1t} - \alpha N_{2t})/K_1, \qquad (6.10)$$

where N_1 and N_2 are populations of two competing species, and α is a competition coefficient that measures the per capita inhibitive effect of species 2 on species 1.

Similarly, the effects of a predator on a prey population can be incorporated into the logistic model (Lotka, 1925; Volterra, 1926) as

$$N_{1(t+1)} = N_{1t} + r_1 N_{1t} - pN_{1t}N_{2t}, \qquad (6.11)$$

where N_1 is prey population density, N_2 is predator population density, and p is a predation constant. This equation assumes random movement of prey and predator, prey capture and consumption for each encounter with a predator, and no self-limiting density effects for either population (Pianka, 1974; Price, 1997).

Pianka (1974) suggested that competition among prey could be incorporated by modifying the Lotka–Volterra competition equation as

$$N_{1(t+1)} = N_{1t} + r_1 N_{1t} - r_1 N_{1t}^2/K_1 - r_1 N_{1t}\alpha_{12}N_{2t}/K_1, \qquad (6.12)$$

where α_{12} is the per capita effect of the predator on the prey population. The prey population is density-limited as carrying capacity is approached.

May (1981) and Dean (1983) modified the logistic model to include effects of mutualists on population growth. Species interaction models are discussed more fully in Chapter 8.

Gutierrez (1996) and Royama (1992) discuss additional population modeling approaches, including incorporation of age and mass structure and population refuges from predation. Clearly, the increasing complexity of these models, as more parameters are included, requires computerization for prediction of population trends.

D. Computerized Models

Computerized simulation models have been developed to project abundances of insect populations affecting crop and forest resources (e.g., Gutierrez, 1996; Royama, 1992; Rykiel et al., 1984). The models developed for several important forest and range insects are arguably the most sophisticated population dynamics models developed to date because they incorporate long time frames, effects of a variety of interacting factors (including climate, soils, host plant variables, competition, and predation) on insect populations and effects of population change on ecosystem structure and processes. Often, the population dynamics model is integrated with plant growth models, impact models that address effects of population change on ecological, social, and economic variables, and management models that address effects of manipulated resource availability and insect mortality on the insect population (Colbert and Campbell, 1978; Leuschner, 1980). As more information becomes available on population responses to various factors or effects on ecosystem processes, the model can be updated, increasing its representation of population dynamics and the accuracy of predictions.

Effects of various factors can be modeled as deterministic (fixed values), stochastic (values based on probability functions), or chaotic (random values) variables (e.g., Croft and Gutierrez, 1991; Hassell *et al.*, 1991). If natality, mortality, and survival are highly correlated with temperature, these rates would be modeled as a deterministic function of temperature. On the other hand, these processes might be described best by probability functions and modeled stochastically (Fargo *et al.*, 1982; Matis *et al.*, 1994). Advances in chaos theory are contributing to development of population models that more accurately represent the erratic behavior of many insect populations (Cavalieri and Koçak, 1994; Constantino *et al.*, 1997; Hassell *et al.*, 1991; Logan and Allen, 1992). Chaos theory addresses the unpredictable ways in which initial conditions of a system can affect subsequent system behavior. In other words, population trend at any instant is the result of the unique combination of population and environmental conditions at that instant. For example, changes in gene frequencies and behavior of individuals over time affect the way in which populations respond to environmental conditions. Cavalieri and Koçak (1994) found that small changes in weather-related parameters of a European corn borer, *Ostrinia nubilalis*, population dynamics model (increased mortality of pathogen-infected individuals or decreased natality of uninfected individuals) can cause a regular population cycle to become erratic, and that when this chaotic state is reached, the population can reach higher abundances than it does during stable cycles.

E. Model Evaluation

The utility of models often is limited by a number of problems. The effects of multiple interacting factors typically must be modeled as the direct effects of individual factors, in the absence of multifactorial experiments to assess interactive effects. Effects of host condition often are particularly difficult to quantify for modeling purposes because factors affecting host biochemistry remain poorly understood for most species. Moreover, models must be initialized with adequate data on current population parameters and environmental conditions. Finally, most models are constructed from data representing relatively short time periods.

Most models accurately represent the observed dynamics of the populations from which the model was developed (e.g., Varley *et al.*, 1973), but confidence in their utility for prediction of future population trends under a broad range of environmental conditions depends on proper validation of the model.

Validation requires comparison of predicted and observed population dynamics using independent data, i.e., data not used to develop the model. Such comparison using data representing a range of environmental conditions can indicate the generality of the model and contribute to refinement of parameters subject to environmental influence until the model predicts changes with a reasonable degree of accuracy (Hain, 1980).

Departure of predicted results from observed results can indicate several possible weaknesses in the model. First, important factors may be underrepresented in the model. For example, unmeasured changes in plant biochemistry during drought periods could significantly affect insect population dynamics. Second, model structure may be flawed. Major factors affecting populations may not be appropriately integrated in the model. Finally, the quality of data necessary to initialize the model may be inadequate. Initial values for r, N_0, or other variables must be provided or derived from historic data within the model. Clearly, inadequate data or departure of particular circumstances from tabular data will reduce the utility of model output.

Few studies have examined the consequences of using different types of data for model initialization. The importance of data quality for model initialization can be illustrated by evaluating the effect of several input options on predicted population dynamics of the southern pine beetle, *Dendroctonus frontalis*. The TAMBEETLE population dynamics model is a mechanistic model that integrates submodels for colonization, oviposition, and larval development with variable stand density and microclimatic functions to predict population growth and tree mortality (Fargo *et al.*, 1982; Turnbow *et al.*, 1982). Nine variables describing tree (diameter, infested height and stage of colonization of colonized trees), insect (density of each life stage at multiple heights), and environmental (landform, tree size class distribution and spatial distribution, and daily temperature and precipitation) variables are required for model initialization. Several input options were developed to satisfy these requirements. These options range in complexity from correlative information based on aerial survey or inventory records to detailed information about distribution of beetle life stages and tree characteristics that requires intensive sampling. In the absence of direct data, default values are derived from tabulated data based on intensive population monitoring studies.

Schowalter *et al.* (1982) compared tree mortality predicted by TAMBEETLE using four input options: all data needed for initialization (including life stage and intensity of beetles in trees), environmental data and diameter and height of each colonized tree only, environmental data and infested surface area of each colonized tree only, and environmental data and number of colonized trees only. Predicted tree mortality when all data were provided was twice the predicted mortality when only environmental and tree data were provided and most closely resembled observed beetle population trends and tree mortality.

Insect population dynamics models typically are developed to address "pest" effects on commodity values. Few population dynamics models explicitly incorporate effects of population change on ecosystem processes. In fact, for most insect populations, effects on ecosystem productivity, species composition, hydrology, nutrient cycling, soil structure and fertility, etc., have not been documented. However, a growing number of studies are addressing the effects of insect herbivore or detritivore abundance on primary productivity, hydrology, nutrient cycling, and/or diversity and abundances of other organisms (Klock and Wickman, 1978; Leuschner, 1980; Schowalter and Sabin, 1991; Schowalter et al., 1991; Seastedt, 1984, 1985; Seastedt and Crossley, 1984; Seastedt et al., 1988; see also Chapters 12–14). For example, Colbert and Campbell (1978) documented the structure of the integrated Douglas-fir tussock moth, *Orgyia pseudotsugata*, model and the effects of simulated changes in moth density (population dynamics submodel) on density, growth rate, and timber production by tree species (stand prognosis model). Leuschner (1980) described development of equations for evaluating direct effects of southern pine beetle population dynamics on timber, grazing and recreational values, hydrology, understory vegetation, wildlife, and likelihood of fire. Effects of southern pine beetle on these economic values and ecosystem attributes were modeled as functions of the extent of pine tree mortality resulting from changes in beetle abundance. However, for both the Douglas-fir tussock moth and southern pine beetle models, the effects of population dynamics on noneconomic variables are based on limited data.

Modeling of insect population dynamics requires data from continuous monitoring of population size over long time periods, especially for cyclic and irruptive species, in order to evaluate the effect of changing environmental conditions on population size. However, relatively few insect populations, including pest species, have been monitored for longer than a few decades and most have been monitored only during outbreaks (e.g., Curry, 1994; Turchin, 1990). Historic records of outbreak frequency during the past 100–200 years exist for a few species, (e.g., Fitzgerald, 1995; Greenbank, 1963; Turchin, 1990; White, 1969) and, in some cases, outbreak occurrence over long time periods can be inferred from dendrochronological data in old forests (e.g., Royama, 1992; Swetnam and Lynch, 1989; Veblen et al., 1994). However, such data do not provide sufficient detail on concurrent trends in population size and environmental conditions for most modeling purposes. Data on changes in population densities cover only a few decades for most species (e.g., Berryman, 1981; Mason, 1996; Price, 1997; Rácz and Bernath, 1993; Varley et al., 1973; Waloff and Thompson, 1980). For populations that irrupt infrequently, validation often must be delayed until future outbreaks occur.

Despite limitations, population dynamics models are a valuable tool for synthesizing a vast and complex body of information, for identifying critical gaps in our understanding of factors affecting populations, and for predicting or simulating responses to environmental changes. Therefore, they represent our state-of-the-art understanding of population dynamics, can be used to focus future research on key questions, and can contribute to improved efficiency of management or manipulation of important processes. Population dynamics models are the most rigorous tools available for projecting survival or recovery of endangered species and outbreaks of potential pests and their effects on ecosystem resources.

IV. SUMMARY

Populations of insects can fluctuate dramatically through time, with varying effect on community and ecosystem patterns and processes, as well as on the degree of crowding among members of the population. The amplitude and frequency of fluctuations distinguish irruptive populations, cyclic populations, and stable populations. Cyclic populations have stimulated the greatest interest among ecologists. The various hypotheses to explain cyclic patterns of population fluctuation all include density-dependent regulation with a time lag that generates regular oscillations.

Disturbances are particularly important to population dynamics, triggering outbreaks of some species, locally exterminating others. Disturbances can affect insect populations directly by killing intolerant individuals or indirectly by affecting abundance and suitability of resources or abundance and activity of predators, parasites, and pathogens. The extent to which anthropogenic changes in environmental conditions affect insect populations depends on the similarity in conditions produced by natural versus anthropogenic changes.

Population growth can be regulated (stabilized) to a large extent by density dependent factors whose probability of effect on individuals increases as density increases and declines as density declines. Primary density dependent factors are intraspecific competition and predation. Increasing competition for food (and other) resources as density increases leads to reduced natality and increased mortality and dispersal, eventually reducing density. Similarly, predation increases as prey density increases. Although the relative importance of these two factors has been debated extensively, both clearly are critical to population regulation. Regulation by bottom-up factors (resource limitation) may be relatively more important in systems where resources are defended or vary significantly in quality, whereas regulation by top-down factors (predation) may be more important where resources are relatively abundant and show little variation in quality. Inverse density dependence results from cooperation among individuals and represents a potentially destabilizing property of populations. However, this positive feedback may prevent population decline below an extinction threshold. Populations declining below their extinction threshold may be doomed to local extinction, whereas populations increasing above a critical number of individuals (release threshold) continue to increase during an outbreak period. These thresholds represent the minimum and maximum population sizes for species targeted for special management.

Development of population dynamics models has been particularly important for forecasting changes in insect abundance and effects on crop, range, and forest resources. General models include the logistic (Verhulst–Pearl) equation that incorporates initial population size, per capita natality, mortality, and dispersal (instantaneous rate of population change), and environmental carrying capacity. The logistic equation describes a sigmoid curve that reaches an asymptote at carrying capacity. This general model can be modified for particular species by adding parameters to account for nonlinear density dependent factors, time lags, cooperation, extinction, competition, predation, etc. Models are necessarily simplifications of real systems and may poorly represent effects of multiple interacting factors and chaotic processes. Few models have been adequately validated and fewer have evaluated the effects of input quality on accuracy of predictions. Few population models have been developed to predict effects of insect population dynamics for ecosystem processes other than commodity production. Nevertheless, models represent powerful tools for synthesizing information, identifying priorities for future research, and simulating population responses to future environmental conditions.

Biogeography

GEOGRAPHIC RANGES OF OCCURRENCE GENERALLY REFLECT THE TOLerances of individual organisms to geographic gradients in physical conditions (see Chapter 2). However, most species do not occupy the entire area of potentially suitable environmental conditions. Discontinuity in geographical range reflects a number of factors, particularly geographic barriers. By contrast, suitable habitats often have been colonized over large distances from population sources, as a result of dispersal processes, often aided by anthropogenic movement. Factors determining the geographic distribution of organisms have been a particular subject of investigation for the past several centuries (e.g., Andrewartha and Birch, 1954; Price, 1997), spurred in large part by European and American exploration, and floral and faunal collections, in continental interiors during the 1800s.

The spatial distribution of populations changes with population size. Growing populations expand over a larger area as individuals in the high density core disperse to the fringe of the population or colonize new patches. Declining populations shrink into refuges that maintain isolated demes of a metapopulation. Spatial distribution of populations is influenced to a considerable extent by anthropogenic activities that determine landscape structure and introduce (intentionally or unintentionally) commercial and "pest" species to new regions. However, the use of insects as biological indicators of ecosystem condition depends on the degree of specialization of particular species (Rykken *et al.*, 1997). Changes in the presence and abundance of particular species affects various ecosystem properties, encouraging efforts to predict changes in distributions of insect populations.

I. GEOGRAPHIC DISTRIBUTION

Geographic distribution of species populations can be described over a range of scales. At the largest scale, some species have population distributions that span large areas of the globe, including multiple continents. At smaller scales, individual species may occur in a suitable portion of a biome or in suitable patches scattered across a biome or landscape. At the same time, species often are absent from suitable habitats. The geographical distribution of individual species can change as a result of changing conditions and/or dispersal.

A. Global Patterns

Global patterns of distribution reflect latitudinal gradients in temperature and moisture, as well as natural barriers to dispersal. Wallace (1876) identified six relatively distinct faunal assemblages that largely coincide with major continental boundaries but also reflect the history of continental movement, as discussed below. Wallace's **biogeographic realms** (Fig. 7.1) remain a useful template for describing species distributions on a global scale. Many taxa occupy large areas within a particular biogeographic realm, e.g., the unique Australian flora and fauna. Others, because of the narrow gap between the Palearctic and

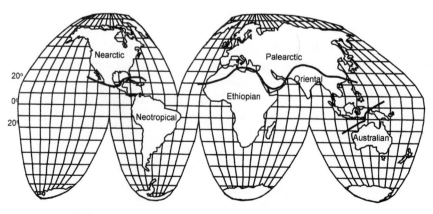

FIG. 7.1   Biogeographic realms identified by Wallace (1876).

Nearctic realms, are able to cross this barrier and exhibit a Holarctic distribution pattern. Of course, many species occupy much smaller geographic ranges, limited by topographic barriers or other factors.

Some distribution patterns, especially of fossil species, are noticeably disjunct. Hooker (1847, 1853, 1860) was among the first to note the similarity of floras found among lands bordering the southern oceans, including Antarctica, Australia, and Tasmania, New Zealand, Tierra del Fuego, the Falklands, and other islands. Many genera, and even some species, of plants were shared among these widely separated lands, suggesting a common origin.

Later in the 1800s, evidence of stratigraphic congruence of various plant and animal groups among the southern continents supported a hypothetical separation of northern and southern supercontinents. Wegener (1924) was the first to outline a hypothetical geologic history of drift for all the continents, concentrated during Cenozoic time. Wegener's **continental drift hypothesis** was criticized because this history appeared to be incompatible with nonmarine paleontology. However, a growing body of geologic and biological evidence, including stratigraphic congruence, rift valleys, uplift and subsidence zones, and distributions of both extinct and extant flora and fauna, eventually were unified into the **theory of plate tectonics**.

According to this theory, a single landmass (Pangaea) split about 200 million years ago and separated into northern (Laurasia) and southern (Gondwanaland) supercontinents that moved apart as a result of volcanic upwelling in the rift zone. About 135 million years ago, India separated from Gondwanaland, moved northward, and eventually collided with Asia to form the Himalayas. Africa and South America separated about 65 million years ago, prior to the adaptive radiation of angiosperms and mammalian herbivores. South America eventually rejoined North America at the Isthmus of Panama, permitting the placental mammals that evolved in North America to invade and displace the marsupials (other than the generalized opossum) that had continued to dominate South America. Marsupials largely disappeared from the other continents as well, except for Australia, where they survived by virtue of continued isolation. South American flora and fauna moved northward through tropical Central America. This process of continental movement explains the similarity of fossil flora and fauna among the Gondwanaland-derived continents and among biogeographic realms, e.g., *Nothofagus* forests in southern continents versus *Quercus* forests in northern continents.

Continental movements result from the stresses placed on the Earth's crust by planetary motion. Fractures appear along lines of greatest stress and are the basis for volcanic and seismic activity, two powerful forces that lead to displacement of crustal masses. The midoceanic ridges and associated volcanism mark the original locations of the continents and preserve evidence of the direction and rate of continental movements. Rift valleys and fault lines typically provide depressions for development of aquatic ecosystems. Mountain ranges develop along lines of collision and subsidence between plates and create elevational gradients and boundaries to dispersal. Volcanic and seismic activity represent a continuing disturbance in many ecosystems.

B. Regional Patterns

Within biogeographic provinces, a variety of biomes can be distinguished on the basis of their characteristic vegetation or aquatic characteristics (see Chapter 2). Much of the variation in environmental conditions that produce biomes at the regional scale is the result of global circulation patterns and topography. Mountain ranges and large rivers may be impassable barriers that limit the distribution of many species. Furthermore, mountains show relatively distinct elevational zonation of biomes (life zones). The area available as habitat becomes more limited at higher elevations. Mountaintops resemble oceanic islands in their degree of isolation within a matrix of lower elevation environments and are most vulnerable to climate changes that shift temperature and moisture combinations upward (Fig. 5.2).

Geographic ranges for many, perhaps most, species are restricted by geographic barriers or environmental conditions beyond their tolerance limits. Species with large geographic ranges often show considerable genetic variation among subpopulations, reflecting adaptations to regional environmental factors. For example, Istock (1981) reported that northern and southern populations of a transcontinental North American pitcher-plant mosquito, *Wyeomyia smithii*, show distinct genetically based life history patterns. The proportion of third instars entering diapause increases with latitude, reflecting adaptation to seasonal changes in habitat or food availability. Controlled crosses between northern and southern populations yielded high proportions of diapausing progeny from northern × northern crosses, intermediate proportions from northern × southern crosses, and low proportions from southern × southern crosses, for larvae subjected to conditions simulating either northern or southern photoperiod and temperature.

C. Island Biogeography

Ecologists have been intrigued at least since the time of Hooker (1847, 1853, 1860) by the presence of related organisms on widely separated oceanic islands. Darwin (1859) and Wallace (1911) later interpreted this phenomenon as evidence of natural selection and speciation of isolated populations following colonization from distant population sources. Simberloff (1969), Simberloff and Wilson (1969), and Wilson and Simberloff (1969) found that many arthropod species were capable of rapid colonization of experimentally defaunated islands.

Although the **theory of island biogeography** originally was developed to explain patterns of equilibrium species richness among oceanic islands (MacArthur and Wilson, 1967), the same factors and processes that govern colonization of oceanic islands explain rates of species colonization and metapopulation dynamics (see following) among isolated landscape patches, e.g., Simberloff (1974), Hanski and Simberloff (1997), and Soulé and Simberloff (1986). Critics of this approach have argued that oceanic islands clearly are surrounded by habitat unsuitable for terrestrial species, whereas terrestrial patches may be surrounded by relatively more suitable patches. Some terrestrial habitat patches may be more similar to oceanic islands than others, e.g., alpine tundra on mountaintops may represent substantially isolated habitats, whereas disturbed patches in grassland may be less distinct. A second issue concerns the extent to which the isolated populations constitute distinct species or metapopulations of a single species (Hanski and Simberloff, 1997). The resolution of this issue depends on the degree of genetic drift among isolated populations over time.

D. Landscape and Stream Continuum Patterns

Within terrestrial biomes, gradients in climate and geographic factors interacting with the patch scale of disturbances across landscapes produces a shifting mosaic of habitat types that affects the distribution of populations. Local extinction of demes must be balanced with colonization of new habitats as they appear in order for species to survive. However, colonists can arrive in terrestrial patches from various directions and distances. By contrast, distribution of aquatic species is more constrained by the linear (single dimension) pattern of water flow. Colonists are more likely to come from upstream (if movement is governed by water flow) or downstream (flying adults), with terrestrial patches between stream systems being relatively inhospitable. Population distributions often are relatively distinct among drainage basins (watersheds), depending on the ability of dispersants to colonize new headwaters or tributaries. Hence, terrestrial and aquatic ecologists have developed different approaches to studying spatial dynamics of populations, especially during the 1980s when **landscape ecology** became a paradigm for terrestrial ecologists and **stream continuum** a paradigm for stream ecologists.

Distribution of populations in terrestrial landscapes, stream continua, and oceanic islands is governed to a large extent by probabilities of extinction versus colonization in particular sites (Fig. 7.2; see Chapter 5). The dispersal ability of the species, the size of the patch, island, or stream habitat, and its distance from the population source determine the probability of colonization by a dispersing individual (Fig. 5.5). Island or patch size and distance from population sources influence the likelihood that an insect able to travel a given distance in a given direction will contact that island or patch.

Individual capacity for sustained travel and for detection of cues that facilitate orientation determine dispersal ability. Species that fly can travel long distances and traverse obstacles in an aquatic or terrestrial matrix better than can flightless species. Many small insects, including flightless species, catch air

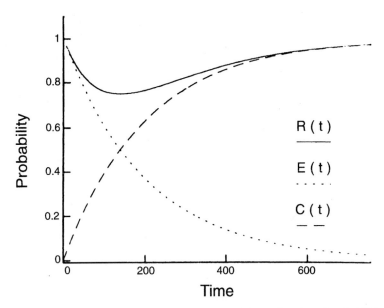

FIG. 7.2 Probability of species presence in an ecosystem (R), as a function of probabilities of local extinction (E) and colonization (C) over time, for specified values of v = probability of colonization over time and λ = probability of extinction over time. From Naeem (1998), reprinted by permission of Blackwell Science, Inc.

currents and are carried long distances at essentially no energetic cost to the insect. Edwards and Sugg (1990) reported that a variety of insects could be collected on montane glaciers far from the nearest potential population sources. Torres (1988) reported deposition, by hurricanes, of insect species from as far away as Africa on Caribbean islands.

On the other hand, many small, flightless species have limited capacity to disperse. Any factor that increases the time to reach a suitable habitat increases the risk of mortality due to predation, extreme temperatures, desiccation, or other factors. Distances of a few meters, especially across exposed soil surfaces, can effectively preclude dispersal by many litter species sensitive to heat and desiccation. Some aquatic species, e.g., Ephemeroptera, have limited life spans as adults to disperse among stream systems. Courtney (1985, 1986) reported that short adult life span was a major factor influencing the common selection of less suitable larval food plants for oviposition (see Chapter 3). Clearly, the distance between an island or habitat patch and the source population is inversely related to the proportion of dispersing individuals able to reach it (Fig. 5.5).

Island or patch size and complexity also influence the probability of successful colonization. The larger the patch (or the shorter its distance from the source population), the greater the proportion of the horizon it represents and the more likely a dispersing insect will be to contact it. Similarly, the distribution of microsites within landscape or watershed patches affects the ability of dispersing insects to perceive and reach suitable habitats. Basset (1996) reported that the presence of arboreal insects is influenced more strongly by local factors in complex habitats, such as tropical forests, and more strongly by regional factors in less complex habitats, such as temperate forests.

The increasing rate of dispersal during rapid population growth increases the number of insects moving across the landscape and the probability that some will travel sufficient distance in a given direction to discover suitable patches. Therefore, population contribution to patch colonization and genetic exchange with distant populations is maximized during population growth.

II. SPATIAL DYNAMICS OF POPULATIONS

As populations change in size, they also change in spatial distribution of individuals. Population movement (epidemiology) across landscapes and watersheds (stream continuum) reflects integration of physiological and behavioral attributes with landscape or watershed structure. Growing populations tend to spread across the landscape as dispersal leads to colonization of new habitats, whereas declining populations tend to constrict into more or less isolated refuges. Isolated populations of irruptive or cyclic species can coalesce during outbreaks, facilitating genetic exchange.

Insect populations show considerable spatial variation in densities in response to geographic variation in habitat conditions and resource quality (Fig. 7.3). Variation can occur over relatively small scales because of the small size of insects and their sensitivity to environmental gradients (e.g., Heliövaara and Väisänen, 1993; Lincoln *et al.*, 1993). The spatial representation of populations can be described across a range of scales from microscopic to global. The pattern of population distribution can change over time as population size and environmental conditions change. Two general types of spatial variation are represented by the expansion of growing populations and by the discontinuous pattern of fragmented populations, or metapopulations.

A. Expanding Populations

Growing populations tend to spread geographically as density-dependent dispersal leads to colonization of nearby resources. This spread occurs in two ways. First, diffusion from the origin, as density increases, produces a gradient of decreasing density toward the fringe of the expanding population. Grilli and Gorla (1997) reported that leafhopper, *Delphacodes kuscheli*, density was highest within the epidemic area and declined toward the fringes of the population. The difference in density between pairs of sampling points increased as the distance between the sampling points increased. Second, long-distance dispersal leads to colonization of vacant patches and "proliferation" of the population (Hanski and Simberloff, 1997). Subsequent growth and expansion of these new

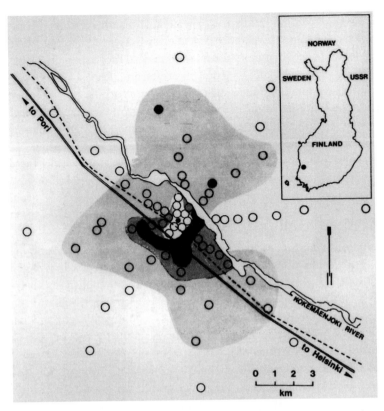

FIG. 7.3 Gradient in pine bark bug, *Aradus cinnamomeus*, densities with distance from the industrial complex (*) at Harjavalta, Finland. White circles, 0–0.50 bugs 100 cm^{-2}; light gray circles, 0.51–1.75 bugs 100 cm^{-2}; dark gray circles, 1.76–3.50 bugs 100 cm^{-2}; and black circles, 3.51–12.2 bugs 100 cm^{-2}. From Heliövaara and Väisänen (1986) by permission from Blackwell Wissenschafts-Verlag GmbH.

demes can lead to population coalescence, with local "hot spots" of superabundance that eventually may disappear as resources in these sites are depleted.

The speed at which a population expands likely affects the efficiency of density-dependent regulatory factors. Populations that expand slowly may experience immediate density-dependent negative feedback in zones of high density, whereas induction of negative feedback may be delayed in rapidly expanding populations, as dispersal slows increase in density. Therefore, density-dependent factors should operate with a longer time lag in populations capable of rapid dispersal during irruptive population growth.

The speed, extent, and duration of population spread are limited by the duration of favorable conditions and the homogeneity of the patch or landscape. Populations can spread more rapidly and extensively in homogeneous patches or landscapes, such as agricultural and silvicultural systems, than in heterogeneous systems in which unsuitable patches limit spread (Schowalter, 1996). Insect species with annual life cycles often show incremental colonization and population expansion. Disturbances can terminate the spread of sensitive populations. Frequently disturbed systems, such as crop systems or streams subject to annual scouring, limit population spread to the intervals between recolonization and subsequent disturbance. Populations of species with relatively slow dispersal may expand only to the limits of a suitable patch during the favorable period. Spread beyond the patch depends on the suitability of neighboring patches (Liebhold and Elkinton, 1989).

The direction of population expansion depends on several factors. The direction of population spread often is constrained by environmental gradients, by wind or water flow, and by unsuitable patches. Gradients in temperature, moisture, or chemical concentrations often restrict the directions in which insect populations can spread, based on tolerance ranges to these factors. Even relatively homogeneous environments, such as enclosed stored grain, are subject to gradients in internal temperatures that affect spatial change in granivore populations (Flinn et al., 1992). Furthermore, direction and flow rate of wind or water have considerable influence on insect movement. Insects with limited capability to move against air or water currents move primarily downwind or downstream, whereas insects capable of movement toward attractive cues move primarily upwind or upstream. Insects that are sensitive to stream temperature, flow rate, or chemistry may be restricted to spread along linear stretches of the stream. Jepson and Thacker (1990) reported that recolonization of agricultural fields by carabid beetles dispersing from population centers was delayed by extensive use of pesticides in neighboring fields.

Schowalter et al. (1981b) examined the spread of southern pine beetle, *Dendroctonus frontalis*, populations in east Texas (Fig. 7.4). They described the progressive colonization of individual trees or groups of trees through time by computing centroids of colonization activity on a daily basis (Fig. 7.5). A centroid is the center of beetle mass (numbers) calculated from the weighted abundance of beetles among the x,y coordinates of colonized trees at a given time.

The distances between centroids on successive days was a measure of the rate of population movement (Fig. 7.4). Populations moved at a rate of 0.9 m/day, primarily in the direction of the nearest group of available trees. However, since southern pine beetle populations generally were sparse during the period of this study, indicating relatively unfavorable conditions, this rate may be near the minimum necessary to sustain population growth.

The probability that a tree would be colonized depended on its distance from currently occupied trees. Trees within 6 m of sources of dispersing beetles had a 14–17% probability of being colonized, compared to <4% probability for trees further than 6 m from sources of dispersing beetles. Population spread, in most cases, ended at canopy gaps where no trees were available within 6 m. However, one population successfully crossed a larger gap encountered at peak abundance (Fig. 7.4), indicating that a sufficiently large number of beetles dispersed across the gap to ensure aggregation on suitable trees and sustained population spread.

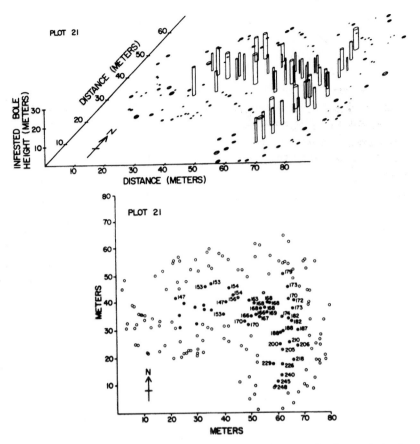

FIG. 7.4 Spatial and temporal pattern of spread of a southern pine beetle population in east Texas during 1977. In the upper figure, cylinders are proportional in size to size of colonized trees; ellipses represent uncolonized trees within 10 m of colonized trees. In the lower figure, Julian dates of initial colonization are given for trees colonized (black circles) after sampling began. Open circles represent uncolonized trees within 10 m of colonized trees. Reprinted from *Forest Science* (Vol. 27, no. 4, p. 840) published by the Society of American Foresters, 5400 Grosvenor Lane, Bethesda, MD 20814-2198. Not for further reproduction.

Population spread in this species may be facilitated by colonization experience and cooperation between cohorts of newly emerging beetles and beetles "reemerging" from densely colonized hosts. Many beetles reemerge after laying some eggs, especially at high colonization densities under outbreak conditions, and seek less densely colonized trees in which to lay remaining eggs. The success of host colonization by southern pine beetles depends on rapid attraction of sufficiently large numbers to overwhelm host defenses (see Chapter 3). For a given day, the centroid of colonization was, on average, twice as far from the centroid of new adults dispersing from brood trees as from the centroid of reemerging beetles (Fig. 7.5). This pattern suggested that reemerging beetles se-

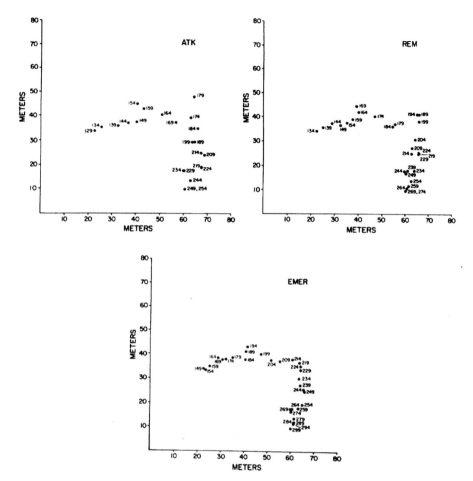

FIG. 7.5 Centroids of colonization (ATK), re-emergence (REM) and emergence (EMER), by Julian date, for the southern pine beetle population in Figure 7.4. Reprinted from *Forest Science* (Vol. 27, no. 4, p. 843) published by the Society of American Foresters, 5400 Grosvenor Lane, Bethesda, MD 20814-2198. Not for further reproduction.

lect the next available trees and provide a focus of attraction for new adults dispersing from further away.

Related research has reinforced the importance of host tree density for population spread of southern pine beetle and other bark beetles (Amman *et al.*, 1988; M. Brown *et al.*, 1987; Mitchell and Preisler, 1992; Sartwell and Stevens, 1975). Schowalter and Turchin (1993) demonstrated that patches of relatively dense pure pine forest are essential to growth and spread of southern pine beetle populations from experimental refuge trees (Fig. 6.5). Populations did not spread from refuge trees surrounded by sparse pines or pine/hardwood mixtures.

A critical aspect of population spread is the degree of continuity of hospitable resources or patches on the landscape. As described above for the southern pine beetle, unsuitable patches can interrupt population spread unless population growth is sufficient to maintain high dispersal rates across inhospitable patches. Heterogeneous landscapes composed of a variety of patch types force insects to expend their acquired resources detoxifying less suitable resources or searching for more suitable resources. Therefore, heterogeneous landscapes tend to limit population spread, whereas more homogeneous landscapes, such as large areas devoted to plantation forestry, pasture grasses, or major crops, provide conditions more conducive to acquisition of surplus energy and nutrients and to sustained population spread.

Corridors of suitable habitat or resources can facilitate population spread across otherwise unsuitable patches. For example, populations of the western harvester ant, *Pogonomyrmex occidentalis*, do not expand across patches subject to frequent anthropogenic disturbance (specifically, soil disruption through agricultural activities) but are able to expand along well-drained, sheltered roadside ditches (DeMers, 1993). Roads provide a disturbed habitat with conditions suitable for dispersal of weedy vegetation and associated insects. Roadside conditions also may increase plant suitability for herbivorous insects and facilitate movement across landscapes fragmented by roads (Spencer and Port, 1988; Spencer *et al.*, 1988).

Population expansion for many species depends on the extent or duration of suitable climatic conditions. Kozár (1991) reported that several insect species showed sudden range expansion northward in Europe during the 1970s, likely reflecting warming temperatures during this period. Population expansion of spruce budworm, western harvester ants, and grasshoppers during outbreaks are associated with warmer, drier periods (Capinera, 1987; DeMers, 1993; Greenbank, 1963).

An important consequence of rapid population growth and dispersal is the colonization of marginally suitable resources or patches where populations could not persist in the absence of continuous influx. Whereas small populations of herbivores, such as locusts or bark beetles, may show considerable selectivity in acceptance of potential hosts, rapidly growing populations often eat all potential hosts in their path. Dense populations of the range caterpillar, *Hemileuca oliviae*, disperse away from population centers, as grasses are depleted, and form an expanding ring, leaving denuded grassland in their wake. Landscapes that are conducive to population growth and spread, because of widespread homogeneity of resources, facilitate colonization of surrounding patches and more isolated resources because of the large numbers of dispersing insects. Epidemic populations of southern pine beetles, generated in the homogenous pine forests of the southern Coastal Plain during the drought years of the mid-1980s, produced sufficient numbers of dispersing insects to discover and kill most otherwise resistant pitch pines in the southern Appalachian Mountains.

B. Metapopulation Dynamics

A metapopulation is a population composed of relatively isolated demes maintained by some degree of dispersal among suitable patches (Hanski and Simberloff, 1997; Harrison and Taylor, 1997; Levins, 1970). Metapopulation structure can be identified at various scales, depending on the scale of distribution and the dispersal ability of the population (Fig. 7.6). For example, metapopulations of some sessile, host-specific insects, such as scale insects (Edmunds and Alstad, 1978), can be distinguished among host plants on a local level, although the insect occurs commonly over a wide geographic range. Local populations of black flies (Simuliidae) can be distinguished at the scale of isolated stream sections characterized by particular substrate, water velocity, temperature, proximity to lake outlets, etc., whereas many species occur over a broad geographic area (e.g., Adler and McCreadie, 1997; Hirai *et al.*, 1994). Conversely, many litter-feeding species occur throughout patches of a particular vegetation type, but populations associated with that particular vegetation type are fragmented at the landscape scale.

Metapopulation structure is most distinct where patches of suitable habitat or food resources are distinct and isolated due to natural variability created by natural environmental heterogeneity (e.g., desert or montane landscapes)

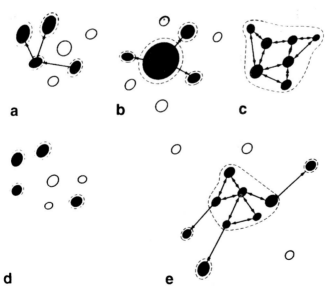

FIG. 7.6 Diagrammatic representation of different metapopulation models. Filled circles are occupied patches; open circles are unoccupied patches; dotted lines are boundaries of local populations; arrows represent dispersal. (a) classic (Levins) model of dispersal among demes, (b) island biogeography model with the mainland providing a source of colonists, (c) a network of interacting demes, (d) a nonequilibrium metapopulation with little capacity for recolonization of vacant patches, and (e) an intermediate case combining features of a–d. From Harrison and Taylor (1997).

or anthropogenic fragmentation. The spatial pattern of metapopulation dynamics reflects a number of interacting factors and largely determines gene flow, species viability and, perhaps, evolution of life-history strategies (e.g., Colegrave, 1997). Hence, interest in spatially structured populations has increased rapidly in recent years.

Metapopulation structure can develop in a number of ways (Fig. 7.6). One is through the colonization of distant resources, and subsequent population development, that occurs during expansion of the source population. A second is through the isolation of population remnants during population decline. A third way represents a stable population structure in which new patches are colonized at the same rate that local extinction occurs in old patches.

The colonization of new patches as dispersal increases during population growth is an important mechanism for initiating new demes and facilitating population persistence on the landscape. The large number of dispersants generated during rapid population growth maximizes the probabilities that suitable resources will be colonized over a considerable area (e.g., Schowalter et al., 1981b) and that more founders will infuse the new demes with greater genetic heterogeneity (Hedrick and Gilpin, 1997). Species with ruderal life histories generally exhibit considerable dispersal capacity and often arrive at sites quite remote from their population sources (Edwards and Sugg, 1990). Such species quickly find and colonize disturbed sites and represent a widely occurring "weedy" fauna. By contrast, species with competitive strategies show much slower rates of dispersal and may travel shorter distances consistent with their more stable population sizes and adaptation to more stable habitats. Such species can be threatened by rapid changes in environmental conditions that exterminate demes more rapidly than new demes are established (Hanski, 1997; Hedrick and Gilpin, 1997).

If conditions for population growth continue, the outlying demes may grow and coalesce with the expanding source population. This process contributes to more rapid expansion of growing populations than would occur only as diffusive spread at the fringes of the population. A well-known example of this is seen in the pattern of gypsy moth, *Lymantria dispar*, population expansion during outbreaks in eastern North America. New demes appear first on ridgetops in the direction of the prevailing wind, because of the wind-driven dispersal of ballooning larvae. These demes grow and spread downslope, merging in the valleys. Similarly, swarms of locusts may move great distances to initiate new demes beyond the current range of the population (Lockwood and DeBrey, 1990).

As a population retreats during decline, subpopulations often persist in isolated refuges, establishing the post-outbreak metapopulation structure. Refuges are characterized by relatively lower population densities that escape the density-dependent decline of the surrounding population. These surviving demes may remain relatively isolated until the next episode of population growth. The existence and distribution of refuges is extremely important to population persistence. For example, bark beetle populations typically persist as scattered demes in isolated lightning-struck, diseased, or injured trees which can be colonized by small numbers of beetles (Flamm *et al.*, 1993). Such trees appear on the landscape with sufficient frequency and proximity to beetle refuges that endemic populations are maintained (Coulson *et al.*, 1983). Croft and Slone (1997) and Strong *et al.* (1997) reported that predaceous mites quickly find colonies of spider mites. New leaves on expanding shoots provide important refuges for spider mite colonists by increasing their distance from predators associated with source colonies.

If suitable refuges are unavailable, too isolated, or of limited persistence, a population may decline to extinction. Under these conditions, the numbers and low heterozygosity of dispersants generated by remnant demes are insufficient to ensure viable colonization of available habitats (Fig. 5.6). For most species, life-history strategies represent successful adaptations that balance population processes with natural rates of patch dynamics, i.e., the rates of appearance and disappearance of suitable patches across the landscape. However, anthropogenic activities have dramatically altered these natural rates and landscape pattern of patch turnover and put many species at risk of extinction (Fielding and Brusven, 1993; Lockwood and DeBray, 1990; Vitousek *et al.*, 1997).

Lockwood and DeBray (1990) suggested that loss of critical refuges as a result of anthropogenically altered landscape structure led to the extinction of a previously widespread and periodically irruptive grasshopper species. The Rocky Mountain grasshopper, *Melanoplus spretus*, occurred primarily in permanent breeding grounds in valleys of the northern Rocky Mountain region but was considered to be one of the most serious agricultural pests in western North America prior to 1900. Large swarms periodically migrated throughout the western United States and Canada during the mid-1800s, destroying crops over areas as large as 330,000 km^2 before declining precipitously. The frequency and severity of outbreaks declined during the 1880s, and the last living specimen was collected in 1902. Macroscale changes during this period (e.g., climate changes, reduced activity of Native Americans and bison, and introduction of livestock) do not seem adequate by themselves to explain this extinction. However, the population refuges for this species during the late 1800s were riparian habitats where agricultural activity, e.g., tillage, irrigation, trampling by cattle, introduction of nonnative plants and birds, was concentrated. Hence, competition between humans and grasshoppers for refugia with suitable oviposition and nymphal development sites may have been the factor leading to extinction of *M. spretus* (Lockwood and DeBrey, 1990).

III. ANTHROPOGENIC EFFECTS ON SPATIAL DYNAMICS

The disappearance of *M. spretus* indicates the vulnerability to extinction of even cyclically abundant species when populations decline to near or below their extinction thresholds (see Chapter 6). Populations always have been vulnerable

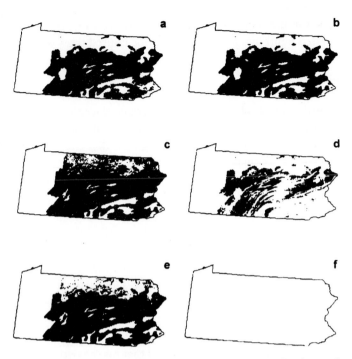

FIG. 7.7 Potential outbreak areas of gypsy moth in Pennsylvania under climate change scenarios. (a) current temperature and precipitation, (b) a 2°C increase, (c) a 2°C increase and 0.5 mm d^{-1} precipitation increase, (d) a 2°C increase and 0.5 mm d^{-1} precipitation decrease, (e) GISS model, and (f) GFDL model. From Williams and Liebhold (1995) by permission from the Entomological Society of America.

to local extinctions as a result of disturbances or habitat loss during environmental changes. Species persist to the extent that dispersal capabilities are adapted to the frequency and scale of these changes. Species adapted to relatively unstable habitats typically have greater dispersal capabilities than do species adapted to more stable habitats.

Human activities affect spatial distribution of populations in several ways. Climate changes eventually will force many species to shift their geographic ranges or face extinction as changing temperatures and humidities exceed tolerance ranges or alter energy balance in their current ranges (Franklin *et al.*, 1992; Kozár, 1991; Rubenstein, 1992) (Fig. 5.2). Changing conditions may favor range expansion for other species. Williams and Liebhold (1995) found that some climate change scenarios predicted larger areas of defoliation, whereas other scenarios predicted smaller areas of defoliation (Fig. 7.7).

Fragmentation of terrestrial ecosystems, alteration and pollution of aquatic ecosystems, and redistribution of species arguably are the most serious and immediate threats to ecosystems worldwide (Samways, 1995). Patch scale, distribution, and abruptness of edges have been altered as a result of habitat fragmentation. This has been particularly evident for wetlands and grasslands. Wetlands historically occupied large portions of floodplains but have been virtual-ly eliminated as a result of draining and filling and stream channelization for urban and agricultural developments. Grasslands have been fragmented se-verely worldwide because of their suitability for agricultural uses. Reservoirs have altered drainage characteristics and reduced the distances between lake ecosystems. Industrial and agricultural pollution threatens many aquatic species. A large number of vagrant species (including various crops and "weeds," rodents, and livestock, as well as insects) have been transported, in-tentionally and unintentionally, far beyond their natural ranges by human ac-tivities. These exotic species have significantly altered the structure and func-tion of their new ecosystems.

A. Fragmentation

Habitat fragmentation is especially deleterious to species adapted to relatively stable ecosystems (e.g., Samways, 1995). Such species typically are less adapt-ed to rapid or long-distance dispersal and may be less able to recolonize vacant or new habitats (resulting from disturbance or climate change) across large in-hospitable patches, compared to ruderal species adapted to long-distance col-onization of disturbed habitats (Powell and Powell, 1987; see Chapter 5). New habitats resulting from changing climatic patterns will be colonized at varying rates by different species, leading to unpredictable patterns of community de-velopment. Insects will not be able to colonize these new habitat patches suc-cessfully until their hosts are established. Schowalter (1995a) found that 70% of arboreal arthropod species occurring in old-growth conifer forests in west-ern Oregon did not occur in adjacent young conifer plantations. Predators and detritivores were particularly affected. Given that 75% of the forested land-scape in this region was clear-cut harvested over a 50-year period (1940–1990), a significant proportion of species associated with old-growth forest now exist as relatively small and isolated populations in a matrix of apparently inhos-pitable young forest. Similarly, Powell and Powell (1987) found that flower vis-itation by male euglossine bees declined following forest fragmentation, even in the 100-ha fragment size, and was proportional to fragment size, indicating that very large areas of forest are necessary for some species.

Isolated patches of natural habitats in fragmented landscapes have rela-tively high ratios of edge to area. These patches are highly vulnerable to influx of nonindigenous species from neighboring patches that may compete with, or prey upon, indigenous species (Punttila et al., 1994). Environmental conditions within habitat patches are further compromised by the characteristics of the edges between neighboring patches in fragmented landscapes. Natural gra-dients of climate and geology interacting with disturbances produce relative-ly large patches, with broad transition zones (ecotones) between patches that dampen interference by one patch on environmental conditions of anoth-er. By contrast, human land use practices tend to produce smaller patches with abrupt edges, e.g., distinct agricultural monocultures within fenced boundaries, plowed edges against grasslands, harvested and regenerating plantations against mature forests. These distinct edges substantially influence environmental conditions of the adjacent patches. For example, an edge of tall trees along an abrupt boundary with an adjacent plantation of short trees is exposed to much greater insolation and airflow, depending on edge orientation, leading to higher temperatures and lower humidities under the canopy than occurred when the edge trees were buffered by neighbors. Chen *et al.* (1995) discovered that microclimatic gradients extended 180–480 m into old-growth Douglas-fir, *Pseudotsuga menziesii*, forests from clear-cut edges, affecting habitat conditions for associated organisms. They concluded that forest patches <64 ha would be completely compromised by ambient environmental conditions, i.e., would be characterized entirely as edge habitat rather than as interior forest habitat. Similarly, grasslands overgrazed by livestock within fenced boundaries expose soil to desiccation, leading to death of surrounding vegetation and an increasing area of desertification (e.g., Schlesinger *et al.*, 1990).

Insects are sensitive to these edge effects. Roland and Kaupp (1995) found that transmission of nuclear polyhedrosis virus was reduced along forest edges, prolonging outbreaks of the forest tent caterpillar, *Malacosoma disstria*. Ozanne *et al.* (1997) documented lower abundances of Psocoptera, Lepidoptera, Coleoptera, Hymenoptera, Collembola, and Araneae and higher abundances of Homoptera and Thysanoptera at forest edges compared to interior forest habitats. Schowalter (1994a, 1995a) reported that these two groups of taxa generally characterize undisturbed and disturbed forests, respectively.

Fragmentation of natural ecosystems typically is associated with homogenization of landscape patterns. Widespread planting of commercial crops and suppression of natural disturbances have eliminated much of the patch pattern of natural landscapes that provided barriers to spread of rapidly growing populations (Schowalter, 1996). In a diverse landscape, outbreaks of particular demes most often would be confined at the patch scale. Agricultural and forested landscapes have become more conducive to expansion and regionwide outbreaks of adapted species.

B. Disturbances to Aquatic Ecosystems

Stream channelization and impoundment have reduced heterogeneity in channel morphology and flow characteristics. Channelization constrains channel morphology, removes obstacles to flow, and shortens stream length. These modifications eliminate habitats in overflow areas (such as wetlands and side channels) and in logs and other impediments and speed drainage in the channeled sections. Impoundments replace a sequence of turbulent sections and pools behind logs and other obstacles (characterized by rocky substrates and high oxygen contents) with deep reservoirs (characterized by silty substrates and stratification of oxygen content and temperature). These changes in stream conditions eliminate habitat for some species (such as species associated with high flow and oxygen concentrations) and increase habitat availability for others (such as species found at low oxygen concentrations).

The linear configuration of stream systems (i.e., the stream continuum concept; see Vannote *et al.*, 1980) makes them particularly vulnerable to disturbances that occur upstream. For example, heavy precipitation in the watershed is concentrated in the stream channel, scouring the channel and redistributing materials and organisms downstream. Fire can expose streams to increased sunlight, raising temperatures and increasing primary production, affecting temperature and biotic resources downstream. Human activities significantly alter water chemistry in ways that influence aquatic insects. Industrial effluents, runoff of agricultural materials (e.g., fertilizers), or accidental inputs of toxic materials (e.g., pesticides) can affect habitat suitability for aquatic species for long distances downstream until sufficient dilution has occurred. Eutrophication, resulting from addition of limiting nutrients, substantially alters the biological and chemical conditions of aquatic systems.

Lake Balaton (Europe's largest lake) in Hungary has experienced incremental eutrophication since the early 1960s, when lake chemistry was relatively uniform (Somlyódy and van Straten, 1986). Since that time, phosphorus inputs from agricultural runoff and urban development have increased, starting at the west end where the Zala River enters the lake. The division of Lake Balaton into four relatively distinct basins draining distinct subwatersheds facilitated documentation of the progression of eutrophication from west to east (Somlyódy and van Straten, 1986). Dévai and Moldován (1983) and Ponyi *et al.* (1983) found that the abundance and species composition of chironomid larvae were correlated with this longitudinal gradient in water quality. The original species characterizing oligo-mesotrophic conditions have been replaced by species characterizing eutrophic conditions in a west to east direction. Similarly, sedimentation resulting from erosion of croplands or clear-cut forests or from trampling of streambanks by livestock alter substrate conditions and habitat suitability for organisms downstream.

Pringle (1997) reported that disturbances and anthropogenic modification of downstream areas (e.g., urbanization, channelization, impoundment) also affect conditions for organisms upstream. Degraded downstream areas may be more vulnerable to establishment of exotic species that are more tolerant of stream degradation. These species subsequently invade upstream habitats. Degradation of downstream areas may restrict movement of upstream species within the watershed, thereby isolating headwater populations and limiting gene flow between watersheds. Finally, degradation downstream may prevent movement of anadromous or catadromous species.

C. Species Introductions

Human transportation of exotic species across natural barriers to their dispersal has altered dramatically the structure and function of natural ecosystems across the globe (Samways, 1995; Wallner, 1996). Examples include the devastation of island vegetation by pigs and goats introduced intentionally by explorers, destruction of grasslands globally by introduced livestock, disruption of aquatic communities by introduced mollusks (e.g., zebra mussel in North America), disruption of grassland and forest communities by introduced plants (e.g., spotted knapweed in North America), mammals (e.g., rabbits in Australia), reptiles and amphibians (brown tree snake in Oceania, African clawed frog in North America), insects (e.g., gypsy moth in North America, the European wood wasp, *Sirex noctulio*, in Australia), and pathogens (e.g., chestnut blight and white pine blister rust in North America, Dutch elm disease in North America and Europe, pinewood nematode in Japan). Exotic species, especially of insects, can be found in virtually all "natural" ecosystems on all continents. Many herbivorous insects and mites have arrived on agricultural or forestry products and become plant pests in agroecosystems or forests. Some herbivorous and predaceous arthropods have been introduced intentionally for biological control of exotic weeds or plant pests (e.g., Croft, 1990; Kogan, 1998; McEvoy *et al.*, 1991). Despite evaluation efforts, these biological control agents, especially arthropod predators, compete with native species and have the potential to colonize native hosts related to the exotic host and develop new biotypes. Indigenous herbivore species also can colonize exotic hosts and develop new biotypes (Strong *et al.*, 1984), with unknown consequences for long-term population dynamics and community structure. Samways *et al.* (1996) found that different invertebrate assemblages were found on exotic vegetation, compared to indigenous vegetation, in South Africa.

Urban areas represent increasingly large and interconnected patches on regional landscapes and are particularly important ports for the spread of exotic species into surrounding ecosystems. Urban centers are the origin or destination for commercial transport of a wide variety of materials, including forest and agricultural products. Urban areas are characterized by a wide variety of exotic species, especially ornamental plants and their associated exotic insects and pathogens. Exotic or native ornamental species typically are stressed by soil compaction, air and water pollutants, elevated urban temperatures, etc. Arriving exotics often have little difficulty finding suitable hosts and becoming established in urban centers and subsequently spreading into surrounding ecosystems.

Road systems connecting urban centers and penetrating natural ecosystems represent major corridors that facilitate spread of exotic species. Roadsides typically are highly disturbed by road maintenance, other human activities, and air pollution from vehicles and provide suitable habitat for a variety of invasive species. Gypsy moth is particularly capable of spreading via human transportation (of pupae or egg masses attached to vehicles, outdoor equipment, or commercial products) between urban centers. Stiles and Jones (1998) demonstrated that population distribution of the imported fire ant, *Solenopsis invicta*, was significantly affected by width and disturbance frequency of road and power-line corridors through forests in the southeastern U.S. (Fig. 7.8). Mound den-

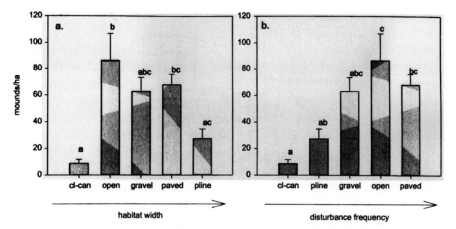

FIG. 7.8 Mean (+ standard error) density of fire ant, *Solenopsis invicta*, mounds along roads under various canopy and substrate conditions in order of increasing corridor width (a) and disturbance frequency (b) at the Savanna River Site in South Carolina. cl-can, closed canopy; pline, powerline cut; and open, gravel, and paved, open canopy roads with dirt, gravel, or paved surfaces, respectively. $N = 10$ for each treatment. Bars with different letters are significantly different at $p < .05$. Reproduced from Stiles, J. H. and R. H. Jones (1998), "Distribution of the red imported fire ant, *Solenopsis invicta*, in road and powerline habitats," *Landscape Ecology* **13**: 335–346, Fig. 3, with kind permission from Kluwer Academic Publishers and J. H. Stiles.

sities were significantly highest along dirt roads not covered by forest canopy and lowest along roads covered by forest canopy. Powerline corridors and graveled or paved roads not covered by forest canopy supported intermediate densities of mounds. These trends suggest that canopy openings of intermediate width and high disturbance frequency are most conducive to fire ant colonization.

IV. CONSERVATION BIOLOGY

A growing number of species is becoming vulnerable to extinction as populations shrink and become more isolated in disappearing habitats (Boecklen, 1991; Wilson *et al.*, 1997) or are displaced by exotic competitors. Examples include a number of butterfly species, the American burying beetle, *Necrophorus americanus*, and a number of aquatic and cave-dwelling species (e.g., Boecklen, 1991; Hanski and Simberloff, 1997; Thomas and Hanski, 1997; Wilson *et al.*, 1997). All of these species are vulnerable to extinction because of their rarity and the increasing fragmentation and isolation of their habitats. Maintenance or recovery of endangered species requires attention to the size and distribution of nature reserves for remnant populations.

The theory of island biogeography was a dominating paradigm in conservation biology during the 1970s and 1980s and continues to shape perspectives of nature reserves as habitat islands (e.g., Diamond and May, 1981; Harris,

1984). One of the important early applications of this theory was to the development of rules for refuge design. The most widely debated of these rules was the SLOSS (single large or several small) rule, based on the likelihood of colonization and persistence of large versus small islands or patches. Diamond and May (1981) noted that the value of various options for species viability depended on the habitat area required by a species and its dispersal capability. Small organisms such as insects could persist in smaller reserves than could larger organisms such as vertebrates. In fact, insects often can persist undetected on rare hosts in relatively small, isolated patches, as was the case for Fender's blue butterfly, *Icaricia icarioides fenderi*. This species was last seen in 1936 before being rediscovered in 1989 in small remnant patches of its host lupine, *Lupinus sulphureus kincaidii*, in western Oregon (Wilson *et al.*, 1997). Nevertheless, species in disappearing habitats remain vulnerable to extinction, as in the case of the Rocky Mountain grasshopper (Lockwood and DeBrey, 1990).

Island biogeography theory has largely been supplanted by models of metapopulation dynamics. Metapopulation models are based on the landscape pattern of demes and gene flow among demes in a nonequilibrium landscape (Hanski and Simberloff, 1997; Harrison and Taylor, 1997). Small demes are most vulnerable to local extinction due to disturbances, but their presence may be critical to recolonization of vacant patches or gene exchange with nearby demes. Dispersal among patches is critical to maintaining declining populations and preventing or delaying local extinction. Clearly, population recovery for such species depends on restoration or replacement of habitats.

Principles of metapopulation dynamics may be particularly important for conservation and restoration of populations of entomophagous predators and parasites in landscapes managed for ecosystem commodities (e.g., forestry and agricultural products). Predators and parasitoids are recognized as important natural agents of crop pest regulation but, as a group, appear to be particularly vulnerable to habitat fragmentation (Kruess and Tscharntke, 1994; Schowalter, 1995a) and pesticide application (Sherratt and Jepson, 1993). Hassell *et al.* (1991) and Sherratt and Jepson (1993) suggested that predator and parasite persistence in agroecosystems depends on the metapopulation dynamics of their prey, as well as on the frequency and distribution of pesticide use, and that connectivity between locally unstable predator–prey interactions could allow their mutual persistence. Thomas *et al.* (1992) found that creation of islands of grassland habitats in agricultural landscapes increased the abundance of several groups of entomophagous arthropods.

Corridors connecting otherwise isolated habitat patches have been identified as critical needs for conservation biology. Just as roads and other disturbed corridors facilitate movement of invasive species among disturbed habitats (DeMers, 1993; Spencer and Port, 1988; Spencer *et al.*, 1988), corridors of undisturbed habitat connecting undisturbed patches can facilitate movement of species characterizing these habitats. Riparian corridors have been a focus of many conservation efforts, but these may be insufficient to conserve upland species in areas with steep elevational gradients.

V. MODELS

The most significant advance in population dynamics research in recent years has been the development of spatially explicit models of population dynamics. A number of approaches have been used to model spatial dynamics. As with temporal dynamics, spatial dynamics can be modeled using deterministic, stochastic, or chaotic functions (Hassell *et al.*, 1991; Matis *et al.*, 1994; Sherratt and Jepson, 1993). Different spatial dynamics result from using these different types of functions.

The earliest attempts to model spatial dynamics either applied diffusion models to describe insect dispersal and population spread from population centers (Rudd and Gandour, 1985; Skellam, 1951; Turchin, 1998) or modeled population dynamics independently among individual landscape patches, based on local conditions within each patch and linked patches by dispersal processes (e.g., Clark, 1979). Diffusion models assume that the environment is homogeneous and that individuals disperse independently and with equal probability in any direction. The diffusion approach is useful for modeling spatial dynamics of insects in stored grain or relatively homogenous crop systems but less useful in most natural landscapes where patchiness interrupts diffusion.

Advances in spatial modeling have been facilitated by development of more powerful computers that can store and manipulate large data sets. Concurrent development of geographical information systems (GIS) and geostatistical software has been a key to describing insect movement (Turchin, 1998) and population epidemiology (Liebhold *et al.*, 1993) across landscapes.

A GIS is an integrated set of programs that facilitate collection, storage, manipulation, and analysis of geographically referenced data, such as topography, vegetation type and density, and insect population densities. Data for a particular set of coordinates can be represented as a value for a cell, and each cell in the matrix is given a value. This method is called the **raster method**. A second method, which requires less storage space, is the **vector method** in which only data representing the vertices of polygons containing data must be stored. Various matrices representing different map layers can be superimposed to analyze interactions. For example, a map layer representing insect population distribution can be superimposed on map layers representing the distribution of host plants, predator abundances, climatic conditions, disturbances, or topography to evaluate the effects of patchiness or gradients in these factors on the spatial dynamics of the insect population.

Geostatistics are a means of interpolating the most probable population densities between sample points, in order to improve representation of spatial distribution over landscapes. Early attempts to characterize spatial patterns were based on modifications of s^2/x, Taylor's Power Law, Lloyd's Patchiness Index, and Iwao's patchiness regression coefficients (Liebhold *et al.*, 1993). These indices focus on frequency distributions of samples and are useful for identifying dispersion patterns (see Chapter 5) but ignore the spatial locations of samples. Modeling spatial dynamics across landscapes requires information on the location of sampling points as well as population density data. The locations of population aggregations affect densities in adjacent cells (Coulson *et al.*, 1996; Liebhold and Elkinton, 1989). Development of geographic positioning systems (GPS) has facilitated incorporation of precise sample locations in GIS databases.

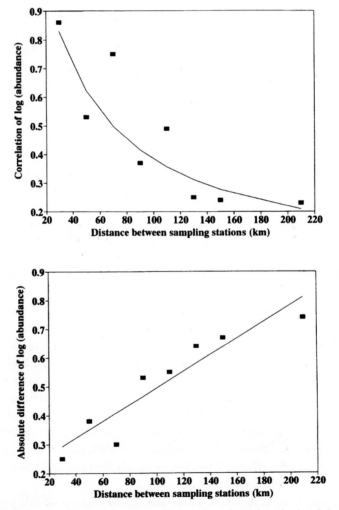

FIG. 7.9 Relationships between the temporal correlation of *Delphacodes kuscheli* density and the distance between sampling stations (top) and between the mean absolute difference between densities of pairs of sampling stations and the distance between sampling stations (bottom) in Argentina. From Grilli and Gorla (1997) by permission from CAB International.

An underlying assumption of geostatistics is that the degree of similarity between sample points is correlated with their proximity (Fig. 7.9) (Coulson *et al.*, 1996; Grilli and Gorla, 1997; Liebhold *et al.*, 1993). Population structure in a given cell is influenced by the population structures in neighboring cells more than by distant cells. An autocorrelation matrix can be developed from data for different distance classes, i.e., x and y coordinates differing by a given distance (Liebhold and Elkinton, 1989). This spatial autocorrelation can be used to interpolate values for unsampled locations by taking a weighted linear average of available samples, a technique known as kriging (Gribko *et al.*, 1995; Grilli and Gorla, 1997; Hohn *et al.*, 1993; Liebhold *et al.*, 1993). Kriging represents an advance over traditional methods of interpolation in several ways, but its most important provision is incorporation of several forms of information simultaneously. The joint spatial dependence of population density and factors such as climate, soil conditions, and vegetation can be integrated to provide more accurate estimates than would be possible with any single variable. These techniques permit prediction of the spatial distribution of insect population densities under potential future climates (Fig. 7.7)

VI. SUMMARY

Factors affecting the geographic distributions of populations have intrigued ecologists for at least the past two centuries. Distributions can be described at different geographic scales. Six distinctive floral and faunal associations (biogeographic realms) can be identified, conforming roughly to continental boundaries, but also reflecting the history of continental movement (plate tectonics). Topography also creates gradients in environmental conditions on mountains and temperature stratification with depth in aquatic ecosystems.

The distribution of species among islands also intrigued early ecologists. The ability of populations to colonize oceanic islands was found to reflect the dispersal capacity of the species, the size of the island, and its distance from the population source. Although controversial, principles of island biogeography have been applied to colonization of terrestrial habitat islands, e.g., mountaintops and patches of unique habitat in otherwise inhospitable landscapes.

At more local scales, the spatial distribution of populations changes with population size. Growing populations expand over a larger area as individuals move from high density patches to the fringe of the population. Rapidly expanding populations generate large numbers of dispersing individuals that maximize the colonization of new patches. Under favorable conditions, these satellite demes expand and coalesce with the main population, affecting ecosystem processes over large areas. Declining populations shrink into isolated refuges that maintain distinct demes of a metapopulation. The extent of movement of individuals among these demes determines genetic heterogeneity and ability to recolonize patches following local extinctions.

All populations are vulnerable to local extinctions due to changing environmental conditions and disturbances. Populations survive to the extent that their dispersal strategies facilitate recolonization and population movement over landscapes. Anthropogenic activities alter spatial distribution in several ways. Climate changes affect the geographic distribution of suitable habitats. However, the most serious anthropogenic effects on spatial patterns are habitat fragmentation, alteration and pollution of aquatic ecosystems, and redistribution (intentionally or unintentionally) of various species. Fragmentation increases isolation of demes and places many species at risk of extinction. At the same time, predators and parasites appear to be most vulnerable to fragmentation and habitat disturbances, often increasing opportunities for population growth by prey species. Humans also are responsible for the introduction of a large and growing number of plant and animal species to new regions as a result of transportation of commercial species and forest and agricultural products. Urban areas represent centers of commercial introductions and provide opportunities for exotic ornamental and associated species to become established and move into surrounding ecosystems. These species affect various ecosystem properties, often dramatically altering vegetation structure and competing with, or preying on, native species.

Modeling of spatial distribution patterns has been facilitated by recent development of geographic information systems (GIS) and geostatistical techniques. Early models represented population expansion as a simple diffusion process. Application of GIS techniques to the patch dynamics of metapopulations permits integration of data on population dynamics with data on other spatially varying factors across landscapes. Geostatistical techniques, such as kriging, permit interpolation of density data between sampling stations to improve mapping and projecting of population distributions. These techniques are improving our ability to evaluate population contributions to ecosystem properties across landscapes.

III

COMMUNITY ECOLOGY

SPECIES CO-OCCURRING AT A SITE INTERACT TO VAR-
ious degrees, both directly and indirectly, in ways that have
intrigued ecologists since earliest times. These interactions
represent mechanisms that control population dynamics, and, hence,
community structure, and also control rates of energy and matter
fluxes, and, hence, ecosystem function. Some organisms engage in close,
direct interactions, as consumers and their hosts, whereas others interact
more loosely and indirectly. For example, predation on mimics depends
on the presence of their models, and herbivores are affected by their
host's chemical or behavioral responses to other herbivores. Direct
interactions, i.e., competition, predation, and symbioses, have been the
focus of research on factors controlling community structure and dynamics, but indirect interactions also control community organization.
Species interactions are the focus of Chapter 8.

A community is composed of the plant, animal, and microbial species
occupying a site. Some of these organisms are integral and characteristic
components of the community and help define the community type,
whereas others occur by chance as a result of movement across a landscape or through a watershed. For example, certain combinations of

species (ruderal, competitive, or stress-tolerant) distinguish desert, grassland, or forest communities. Different species assemblages are found in turbulent water (stream) versus standing water (lake) or eutrophic versus oligotrophic systems. The number of species and their relative abundances define species diversity, a community attribute that is the focus of a number of ecological issues. Chapter 9 addresses the various approaches to describing community structure and factors determining geographic patterns of community structure.

Communities change through time as populations respond differentially to changing environmental conditions, especially to disturbances. Just as population dynamics reflect the net effects of individual natality, mortality, and dispersal interacting with the environment, community dynamics reflect the net effects of species population dynamics interacting with the environment. Severe disturbance or environmental changes can lead to drastic changes in community structure. Changes in community structure through time are the subject of a vast literature summarized in Chapter 10.

Community structure largely determines the biotic environment affecting individuals (Section I) and populations (Section II). The community modifies the environmental conditions of a site. Vegetation cover increases albedo (reflectance of solar energy), reduces soil erosion, modifies temperature and humidity within the boundary layer, and alters energy and biogeochemical fluxes, compared to nonvegetated sites. Species interactions, including those involving insects, modify vegetation cover and affect these processes, as discussed in Section IV. Different community structures affect these processes in different ways.

Species Interactions

JUST AS INDIVIDUALS INTERACT IN WAYS THAT AFFECT POPULATION structure, species populations in a community interact in ways that affect community structure. Species interactions vary considerably in their form and effect and often are quite complex. One species can influence the behavior or abundance of another species directly, e.g., a predator feeding on its prey, or indirectly through effects on other associated species, e.g., a herbivore inducing production of host plant chemicals that attract predators or deter feeding by herbivores arriving later. The web of interactions, both direct and indirect and with positive or negative feedbacks, determines the structure of the community (see Chapter 9) and controls rates of energy and matter fluxes through ecosystems (see Chapter 11).

Insects have provided rich fodder for studies of species interactions. Insects are involved in all types of interactions, as competitors, prey, predators, parasites, commensals, mutualists, and hosts. For example, the complex and elaborate interactions between insect pollinators and their hosts have been among the most widely studied. Our understanding of plant–herbivore, predator–prey, animal–fungus, and various symbiotic interactions is derived largely from models involving insects. This chapter describes the major classes of interactions, factors that affect these interactions, and consequences of interactions for community organization.

I. CLASSES OF INTERACTIONS

Species can interact in various ways and with varying degrees of intimacy. For example, individuals may compete with, prey on, or be prey for, random associations of other species, as well as have more specific interactions with particular species (i.e., symbiosis). Categories of interactions generally have been distinguished on the basis of the sign of their direct effects, such as positive, neutral, or negative effects on growth or mortality of each species. However, the complexity of indirect effects on interacting pairs of species by other associated species has become widely recognized. Furthermore, interactions often have multiple effects on the species involved, requiring consideration of the net effects of the interaction in order to understand its origin and consequences.

A. Competition

Competition is the struggle for use of shared, limiting resources. Resources can be limiting at various amounts and for various reasons. For example, water or nutrient resources may be largely unavailable and support only small populations or a few species in certain habitats (e.g., desert and oligotrophic lakes) but be abundant and support larger populations or more species in other habitats (e.g., rain forest and eutrophic lakes). Newly available resources may be relatively unlimited until sufficient colonization has occurred to reduce per capita availability. Any resource can be an object of interspecific competition, e.g., basking or oviposition sites, food resources, etc.

Although competition for limited resources has been a major foundation for evolutionary theory (Malthus, 1789; Darwin, 1859), its role in natural communities has been controversial (e.g., Connell, 1983; Lawton, 1982; Lawton and Strong, 1981; Schoener, 1982; Strong *et al.*, 1984). Denno *et al.* (1995) and Price (1997) attribute the controversy over the importance of interspecific competition to three major criticisms that arose during the 1980s. First, early studies were primarily laboratory experiments or field observations. Few experimental field studies were conducted prior to the late 1970s. Second, Hairston *et al.* (1960) argued that food must rarely be limiting to herbivores because so little plant material is consumed under normal circumstances (see also Chapter 3). As a result, most field experiments during the late 1970s and early 1980s focused on effects of predators, parasites and pathogens on herbivore populations. Third, many species assumed to compete for the same resource(s) cooccur and appear not to be resource limited. In addition, many communities apparently were unsaturated, i.e., many niches were vacant (e.g., Kozár, 1992b; Strong *et al.*, 1984). The controversy during this period led to more experimental approaches to studying competition. Some (but not all) experiments in which one competitor is removed have demonstrated increased abundance or resource use by the remaining competitor(s), indicative of competition (Denno *et al.*, 1995; Istock, 1973, 1977; Pianka, 1981). However, many factors affect interspecific competition (Colegrave, 1997), and Denno *et al.* (1995) and Pianka (1981) suggested that competition may operate over a gradient of intensities, depending on the degree of niche partitioning (see following).

Denno *et al.* (1995) reviewed studies involving 193 pairs of phytophagous insect species. They found that 76% of these interactions demonstrated competition, whereas only 18% indicated no competition, although they acknowledged that published studies might be biased in favor of species expected to compete. The strength and frequency of competitive interactions varied considerably. Generally, interspecific competition was more prevalent, frequent, and symmetric among haustellate (sap-sucking) species than among mandibulate (chewing) species or between sap-sucking and chewing species. Competition was more prevalent among species feeding internally, e.g., miners and seed-, stem-, and wood-borers (Fig. 8.1), than among species feeding externally. Competition was observed least often among free-living, chewing species, i.e., those generally emphasized in earlier studies that challenged the importance of competition.

Most competitive interactions (84%) were asymmetric, i.e., one species was a superior competitor and suppressed the other (Denno *et al.*, 1995). For example, Istock (1973) demonstrated experimentally that competition between two waterboatmen species is asymmetrical (Fig. 8.2). Population size of *Hesperocorixa lobata* was significantly reduced when *Sigara macropala* was present, but population size of *S. macropala* was not significantly affected by the presence of *H. lobata*. Root feeders were consistently out-competed by folivores, although this, and other, competitive interactions may be mediated by host plant factors (see following under Indirect Effects of Other Species).

Competition generally is assumed to have only negative effects on both (all) competing species (but see following). As discussed in Chapter 6, competition among individuals of a given population represents a major negative feedback mechanism for regulation of population size. Similarly, competition among species represents a major mechanism for regulation of the total abundance of multiple species-populations. As the total density of all individuals of competing species increases, each individual has access to a decreasing share of the resource(s). If the competition is asymmetric, the superior species may competitively suppress other species, leading over sufficient time to **competitive exclusion** (Denno *et al.*, 1995; Park, 1948; Strong *et al.*, 1984). However, Denno *et al.* (1995) found evidence of competitive exclusion in <10% of the competitive interactions they reviewed. Competitive exclusion normally may be prevented by various factors that limit complete preemption of resources by any species. For example, predators or competitors that curb population growth of the most abundant competing species can reduce its ability to competitively exclude other species (Paine, 1966, 1969a,b).

Interspecific competition can take different forms and have different possible outcomes. **Exploitation competition** occurs when all individuals of the competing species have equal access to the resource. A species that can find or exploit a resource more quickly, develop or reproduce more rapidly, or increase

FIG. 8.1 Competition: evidence of interference between Douglas-fir bark beetle, *Dendroctonus pseudotsugae*, larvae (horizontal mines radiating from the vertical parent mine) and cooccurring cerambycid, *Monochamus scutellatus*, larvae (larger mines to the lower left and upper right) preserved in bark from a dead Douglas-fir. The larger cerambycid larvae often remove phloem resources in advance of bark beetle larvae and/or consume bark beetle larvae in their path.

its efficiency of resource utilization will be favored under such circumstances. **Interference competition** involves preemptive use of, and often defense of, a resource that allows a more aggressive species to increase its access to and share of the resource, to the detriment of other species.

Many species avoid resources that have been marked or exploited previously, thereby losing access. Interestingly, males of territorial species typically compete with conspecific males for mates and often do not attack males of other species competing for food resources. Foraging ants may attack other predators and preempt prey resources. For example, Halaj *et al.* (1997) reported that

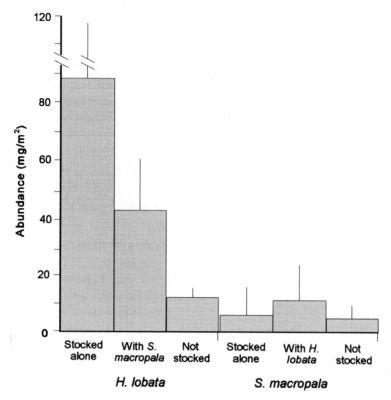

FIG. 8.2   Results of competition between two waterboatmen species, *Hesperocorixa lobata* and *Sigara macropala*, in 1.46 m² enclosures in a 1.2 ha pond. Enclosures were stocked in June with adult *H. lobata* and/or *S. macropala* and final abundance measured after 2 months. Waterboatmen in unstocked enclosures provided a measure of colonization. Vertical bars represent 1 *S.D. N* = 4–8. Data from Istock (1973).

exclusion of foraging ants in young conifer plantations increased abundances of arboreal spiders > 1.5-fold. However, Gordon and Kulig (1996) reported that foragers of the harvester ant, *Pogonomyrmex barbatus*, often encounter foragers from neighboring colonies but relatively few encounters (about 10%) involved fighting and fewer (21% of fights) resulted in death of any of the participants.

Because competition is costly, in terms of lost resources, time, and energy expended in defending resources (see Chapter 4), evolution should favor strategies that reduce competition. Hence, species competing for a resource might be expected to minimize their use of the contested portion and maximize use of the noncontested portions. This results in partitioning of resource use, a strategy referred to as **niche partitioning**. Over evolutionary time, sufficiently consistent partitioning might become fixed as part of the species' adaptive strategies, and the species would no longer respond to changes in the abundance of the former competitor(s). In such cases, competition is not evident, although niche partitioning may be evidence of competition in the past (Connell, 1980). Congeners also typically partition a niche as a result of specialization and divergence into unexploited niches or portions of niches, not necessarily as a result of competition (Fox and Morrow, 1981).

Niche partitioning is observed commonly in natural communities. Species competing for habitat, food resources, or oviposition sites, tend to partition thermal gradients, time of day, host species, host size classes, etc. Several examples are noteworthy.

Granivorous ants and rodents frequently partition available seed resources. Ants specialize on smaller seeds and rodents on larger seeds when the two compete. Brown *et al.* (1979) reported that both ants and rodents increased in abundance in the short term when the other taxon was removed experimentally. However, Davidson *et al.* (1984) found that ant populations in rodent-removal plots declined gradually but significantly after about 2 years. Rodent populations did not decline over time in ant-removal plots. These results reflected a gradual displacement of small-seeded plants (on which ants specialize) by large-seeded plants (on which rodents specialize) in the absence of rodents. Ant removal led to higher densities of small-seeded species but these species could not displace large-seeded plants.

Predators frequently partition resources on the basis of prey size. Predators must balance the higher resource gain against the greater energy expenditure (for capture and processing) of larger prey (e.g., Ernsting and van der Werf, 1988). Generally, predators should select the largest prey that can be handled efficiently (Holling, 1965; Mark and Olesen, 1996), but prey size preference also depends on hunger level and prey abundance (Ernsting and van der Werf, 1988) (see following).

Most bark beetle (Scolytidae) species can colonize extensive areas of dead or dying trees when other species are absent. However, given the relative scarcity of dead or dying trees and the narrow window of opportunity for colonization (the first year after tree death), these insects are adapted to finding such trees rapidly (see Chapter 3) and typically several species co-occur in suitable trees. Under these circumstances, the beetle species tend to partition the subcortical resource on the basis of beetle size, because each species shows the highest survival in phloem that is thick enough to accommodate growing larvae and because larger species are capable of repulsing smaller species (e.g., Flamm *et al.*, 1993). Therefore, the largest species, turpentine beetles, typically occur around the base of the tree, and progressively smaller species occupy successively higher portions of the bole, with the smallest species colonizing the upper bole and branches. However, other competitors, such as wood-boring cerambycids and buprestids often excavate through bark beetle mines and reduce bark beetle survival (Fig. 8.1) (Coulson *et al.*, 1980).

Many competing species partition resource use in time. Partitioning may be by time of day, e.g., nocturnal versus diurnal Lepidoptera (Schultz, 1983) and nocturnal bat and amphibian versus diurnal bird and lizard predators (Reagan *et al.*, 1996), or by season, e.g., asynchronous occurrence of 12 species of waterboatmen (Heteroptera: Corixidae) which breed at different times (Istock, 1973). However, temporal partitioning does not preclude competition through preemptive use of resources or induced host defenses (see Indirect Effects of Other Species).

In addition to niche partitioning, other factors also may obscure or prevent competition. Resource turnover in frequently disturbed ecosystems may prevent species saturation on available resources and prevent competition. Similarly, spatial patchiness in resource availability may hinder resource discovery and prevent species from reaching abundances at which they would compete. Finally, other interactions, such as predation, can maintain populations below sizes at which competition would occur (Paine, 1966, 1969a, b; see Indirect Effects of Other Species).

Competition has proven to be rather easily modeled (see Chapter 6). The Lotka–Volterra equation generalized for n competitors is

$$N_{i(t+1)} = N_{it} + r_i N_{it}(K - N_{it} - \sum_{j>1}^{n} \alpha_{ij} N_{jt})/K \qquad (8.1)$$

where N_i and N_j are species abundances, and α_{ij} represents the per capita effect of N_j on the growth of the ith population and varies for different species. For instance, species j might have a greater negative effect on species i than species i has on species j, i.e., asymmetric competition.

Istock (1977) evaluated the validity of the Lotka–Volterra equations for cooccurring species of waterboatmen, *Hesperocorixa lobata* (species 1) and *Sigara macropala* (species 2), in experimental exclosures (Fig. 8.2). He calculated the competition coefficients, α_{12} and α_{21}, as

$$\alpha_{12} = (K_1 - N_1)N_2 = 3.67 \text{ and } \alpha_{21} = (K_2 - N_2)N_1 = -0.16 \quad (8.2)$$

The intercepts of the zero isocline $(dN/dt = 0)$ for *H. lobata* were $K_1 = 88$ and $K_1/\alpha_{12} = 24$; the intercepts for *S. macropala* were $K_2 = 6$ and $K_2/\alpha_{21} = -38$. The negative K_2/α_{21} and position of the zero isocline for *S. macropala* indicate that the competition is asymmetric, consistent with the observation that *S. macropala* population growth was not affected significantly by the interaction (Fig. 8.2). Although niche partitioning by these two species was not clearly identified, the equations correctly predicted the observed coexistence.

B. Predation

Predation has been defined in various ways, as a general process of feeding on other (prey) organisms (e.g., May, 1981) or as a more specific process of killing and consuming prey (e.g., Price, 1997). Parasitism (and the related parasitoidism), the consumption of tissues in a living host, may or may not be included (e.g., Price, 1997). Both predation and parasitism have positive effects for the predator or parasite but negative effects for the prey. In this section, pre-

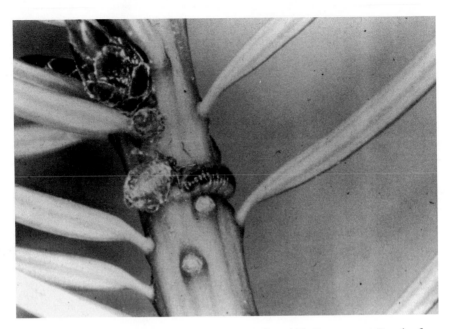

FIG. 8.3   Predation: syrphid larva preying on a conifer aphid, *Cinara* sp., on Douglas-fir.

dation is treated as the relatively opportunistic capture of multiple prey during a predator's lifetime. The following section will address the more specific parasite–host interactions.

Although typically considered in the sense of an animal killing and eating other animals (Fig. 8.3), predation applies equally well to carnivorous plants that kill and consume insect prey and to herbivores that kill and consume plant prey, especially those that feed on seeds and seedlings. Predator–prey and herbivore–plant interactions represent similar foraging strategies and are affected by similar factors (prey density and defensive strategy, predator ability to detect and orient toward various cues, etc.; see Chapter 3).

Insects and related arthropods represent major predators in terrestrial and aquatic ecosystems. Predaceous species occur in many insect orders, especially the Odonata, Mantodea, Heteroptera, Thysanoptera, Mecoptera, Neuroptera, Coleoptera, Diptera, and Hymenoptera. The importance of many arthropods as predators of insects has been demonstrated widely through biological control programs and experimental studies (e.g., Price, 1997; Strong *et al.*, 1984; van den Bosch *et al.*, 1982; Van Driesche and Bellows, 1996). However, many arthropods prey on vertebrates, as well. Predaceous aquatic dragonfly larvae, water bugs, and beetles include fish and amphibians as prey. Terrestrial spiders and centipedes often feed on amphibians, reptiles, and nestling birds (e.g., Reagan *et al.*, 1996).

Insects also represent important predators of plants or seeds. Some bark beetles might be considered to be predators to the extent that they kill multiple trees. Seed bugs (Heteroptera), weevils (Coleoptera), and ants (Hymenoptera) often compete significantly with rodent and avian seed predators and may be capable of preventing plant reproduction under some conditions (e.g., Davidson *et al.*, 1984; Turgeon *et al.*, 1994; see Chapter 13).

Insects are an important food source for a variety of other organisms. Carnivorous plants generally are associated with nitrogen-poor habitats and depend on insects for adequate nitrogen (Juniper *et al.*, 1989; Krafft and Handel, 1991). A variety of mechanisms for entrapment of insects has evolved among carnivorous plants, including water-filled pitchers (pitcher plants), triggered changes in turgor pressure that alter the shape of capture organs (flytraps and bladderworts), and sticky hairs (e.g., sundews). Some carnivorous plants show conspicuous ultraviolet patterns that attract insect prey (Joel *et al.*, 1985), similar to floral attraction of some pollinators (see Chapter 3). Insects also are prey for other arthropods (e.g., predaceous insects, spiders, mites) and vertebrates. Many fish, amphibian, reptile, bird, and mammal taxa feed largely or exclusively on insects (e.g., Dial and Roughgarden, 1995; Gardner and Thompson, 1998; Tinbergen, 1960). Aquatic insects provide the food resource for major freshwater fisheries, including salmonids.

Predation has been widely viewed as a primary means of controlling prey population density. Appreciation for this lies at the heart of predator control policies designed to increase abundances of commercial or game species by alleviating population control by predators. However, mass starvation and declining genetic quality of populations protected from nonhuman predators have demonstrated the importance of predation to maintenance of prey population vigor, or genetic structure, through selective predation on old, injured, or diseased individuals. As a result of these changing perceptions, predator reintroduction programs are being implemented in some regions. At the same time, recognition of the important role of entomophagous species in controlling populations of insect pests has justified augmentation of predator abundances, often through introduction of exotic species, for biological control purposes (van den Bosch *et al.*, 1982; Van Driesche and Bellows, 1996). As discussed in Chapter 6, the relative importance of predation to population regulation, compared to other regulatory factors, has been a topic of considerable discussion.

Just as coevolution between competing species has favored niche partitioning for more efficient resource use, coevolution between predator and prey has produced a variety of defensive strategies balanced against predator foraging strategies. Selection favors prey that can avoid or defend against predators and predators that can efficiently acquire suitable prey. Prey defenses include speed, predator detection and alarm mechanisms, spines or horns, chemical defenses, cryptic, aposematic, disruptive or deceptive coloration, and behaviors (such as aggregation or warning displays) that enhance these defenses (see Chapter 4). Prey attributes that increase the energy cost of capture will restrict the number of predators able to exploit that prey.

Selection also favors predator attributes that increase the efficiency of prey capture. Predator attributes that increase their efficiency in immobilizing and acquiring prey include larger size, detection of cues that indicate vulnerable prey, speed, claws or sharp mouthparts, venoms, and behaviors (such as ambush or attacking the most vulnerable body parts) that compensate for or circumvent prey defenses and reduce the effort necessary to capture the prey. For example, a carabid beetle, *Promecognathus laevissimus*, straddles its prey, polydesmid millipedes, and quickly moves toward the head. It then pierces the neck and severs the ventral nerve cord with its mandibles, thereby paralyzing its prey and circumventing its cyanide spray defense (Parsons *et al.*, 1991).

Predators are relatively opportunistic with respect to prey taxa, compared to parasites, although prey frequently are selected on the basis of factors determining foraging efficiency. For example, chemical defenses of prey affect attractiveness to nonadapted predators (e.g., Bowers and Puttick, 1988; Stamp *et al*, 1997; Traugott and Stamp, 1996). Prey size affects the resource gained per foraging effort expended. Predators generally should select prey sizes within a range that provides sufficient energy and nutrient rewards to balance the cost of capture (Ernsting and van der Werf, 1988; Iwasaki, 1990, 1991; Richter, 1990; Streams, 1994; Tinbergen, 1960). Within these constraints, foraging predators should attack suitable prey species in proportion to their probability of encounter, i.e., more abundant prey types are encountered more frequently than are less abundant prey types (e.g., Tinbergen, 1960).

Predators exhibit both functional (behavioral) and numeric responses to prey density. The functional response reflects predator hunger, handling time required for individual prey, increased discovery of prey, handling efficiency resulting from learning, etc. (Holling, 1959, 1965; Tinbergen, 1960). For many invertebrate predators, the percentage of prey captured is a negative binomial function of prey density, Holling's (1959) type 2 functional response. The ability of type 2 predators to respond individually to increased prey density is limited by their ability to capture and consume individual prey. Vertebrates, and some invertebrates, are capable of increasing their efficiency of prey discovery (e.g., through development of a search image that enhances recognition of appropriate prey; Tinbergen, 1960) and prey processing time through learning, up to a point. The percentage of prey captured initially increases as the predator learns to find and handle prey more quickly but eventually approaches a peak and subsequently declines as discovery and handling time reach maximum efficiency, Holling's (1959) type 3 functional response. The type 3 functional response is better able than the type 2 response to regulate prey population size(s) because of its capacity to increase the percentage of prey captured as prey density increases, at least initially.

Various factors affect the relationship between prey density and proportion of prey captured. The rate of prey capture tends to decline as a result of learned avoidance of distasteful prey, and the maximum rate of prey capture depends on how quickly predators become satiated and on the relative abundances of palatable and unpalatable prey (Holling, 1965). Some insect species, such as the periodical cicadas, apparently exploit the functional responses of their major predators by appearing en masse for brief periods following long periods of inaccessibility. Predator satiation maximizes the success of such mating aggregations (Williams and Simon, 1995). Palatable species experience greater predation when associated with less palatable species than when associated with equally or more palatable species (Holling, 1965).

In addition to these functional responses, predator growth rate and density tend to increase with prey density. Fox and Murdoch (1978) reported that growth rate and size at molt of the predaceous heteropteran, *Notonecta hoffmanni*, increased with prey density in laboratory aquaria. Numeric response reflects predator orientation toward, and longer residence in, areas of high prey density, and subsequent reproduction in response to food availability. However, increased predator density also may increase competition, and conflict, among predators. The combination of type 3 functional response and numeric response (total response) make predators more effective in cropping abundant prey and maintaining relatively stable populations of various prey species. However, the tendency to become satiated and to reproduce more slowly than their prey limits the ability of predators to regulate irruptive prey populations released from other controlling factors.

The importance of predator–prey interactions to population and community dynamics has generated considerable interest in modeling this interaction. The effect of a predator on a prey population was first incorporated into the logistic model by Lotka (1925) and Volterra (1926). As described in Eq. (6.11), their model for prey population growth was

$$N_{1(t+1)} = N_{1t} + r_1 N_{1t} - p_1 N_{1t} N_{2t},$$

where N_2 is the population density of the predator and p_1 is a predation constant. Lotka and Volterra modeled the corresponding predator population as

$$N_{2(t+1)} = N_{2t} + p_2 N_{1t} N_{2t} - d_2 N_{2t}, \tag{8.3}$$

where p_2 is a predation constant and d_2 is per capita mortality of the predator population. The Lotka–Volterra equations describe prey and predator populations oscillating cyclically and out of phase over time. Small changes in parameter values lead to extinction of one or both populations after several oscillations of increasing amplitude.

Pianka (1974) proposed modifications of the Lotka–Volterra competition and predator–prey models to incorporate competition among prey and among predators for prey. Equation (6.12) represents the prey population

$$N_{1(t+1)} = N_{1t} + r_1 N_{1t} - r_1 N_{1t}^2/K_1 - r_1 N_{1t}\alpha_{12}N_{2t}/K_1,$$

where α_{12} is the per capita effect of the predator on the prey population. The corresponding model for the predator population is

$$N_{2(t+1)} = N_{2t} + \alpha_{21}N_{1t}N_{2t} - \beta_2 N_{2t}^2/N_{1t}, \tag{8.4}$$

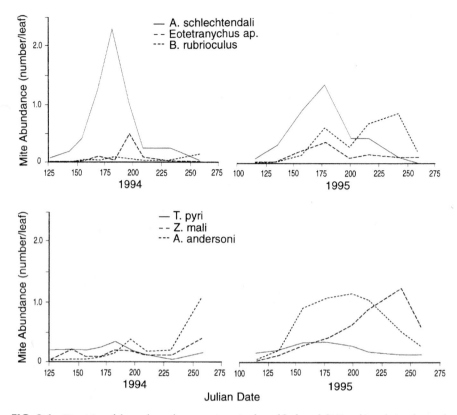

FIG. 8.4 Densities of three phytophagous mites, *Aculus schlechtendali*, *Bryobia rubrioculus*, and *Eotetranychus* sp. (prey), and three predaceous mites, *Amblyseius andersoni*, *Typhlodromus pyri*, and *Zetzellia mali*, in untreated apple plots (*N* = 2) during 1994 and 1995. Data from Croft and Slone (1997).

where α_{21} is the negative effect of predation on the prey population and β_2 incorporates the predator's carrying capacity as a function of prey density (Pianka, 1974). This refinement provides for competitive inhibition of the predator population as a function of the relative densities of predator and prey. The predator–prey equations have been modified further to account for predator and prey distributions (see Begon and Mortimer, 1981) and for functional responses and competition among predators for individual prey (Holling, 1959, 1966). Other models have been developed primarily for parasitoid–prey intractions (see following). However, these modeling approaches have focused on paired predator and prey. Real communities are composed of multiple predator species exploiting multiple prey species, resulting in complex interactions (Fig. 8.4).

C. Symbiosis

Symbiosis involves an intimate association between two species. Three types of interactions are considered symbiotic, although symbiosis often has been used as a synonym for only one of these, mutualism. Parasitism describes interactions in which the symbiont derives a benefit at the expense of the host, as in predation. Commensalism occurs when the symbiont derives a benefit without significantly affecting its partner. Mutualism involves both partners benefitting from the interaction. Insects have provided some of the most interesting examples of symbiosis.

1. Parasitism

Parasitism affects the host (prey) population in ways that are similar to predation and can be described using predation models. However, whereas predation involves multiple prey killed and consumed during a predator's lifetime, parasites feed on living prey. Parasitoidism is unique to insects, especially flies and wasps, and combines attributes of both predation and parasitism. The adult parasitoid typically deposits eggs or larvae on, in, or near multiple hosts, and the larvae subsequently feed on their living host and eventually kill it. Parasites must be adapted to long periods of exposure to the defenses of a living host. Therefore, parasitic interactions tend to be relatively specific associations between coevolved parasites and their particular host species and may involve modification of host morphology, physiology, or behavior to benefit parasite development or transmission. Because of this specificity, parasites and parasitoids tend to be more effective than predators in responding to and controlling population irruptions of their host populations and, therefore, have been primary agents in biological control programs (Hochberg, 1989).

Parasitic interactions can be quite complex. Parasites can be assigned to several categories (van den Bosch *et al.*, 1982). **Ectoparasites** feed externally, by inserting mouthparts into the host (e.g., lice, fleas, mosquitoes, ticks), and **endoparasites** feed internally, within the host's body (e.g., bacteria, nematodes, bot flies, wasps). A **primary parasite** develops on or in a nonparasitic host, whereas a **hyperparasite** develops on or in another parasite. Some parasitic species parasitize other members of the same species (**autoparasitism** or **adelphoparasitism**), as is the case for the hymenopteran, *Coccophagus scutellaris*. The female of this species parasitizes scale insects and the male is an obligate hyperparasite of the female (van den Bosch *et al.*, 1982). **Superparasitism** refers to more individuals of a parasitoid species occuring in the host than can develop to maturity. **Multiple parasitism** occurs when more than one parasitoid species is present in the host simultaneously. In most cases of superparasitism and multiple parasitism, one dominant individual competitively suppresses the others and develops to maturity. In a special case of multiple parasitism, some parasites preferentially attack hosts parasitized by other species (**cleptoparasitism**). The cleptoparasite is not a hyperparasite but typically kills and consumes the original parasite as well as the host.

FIG. 8.5   Parasitism: a nymphalid caterpillar feeding on cecropia foliage in Puerto Rico.

Insects are parasitized by a number of organisms, including viruses, bacteria, fungi, protozoa, nematodes, flatworms, mites, and other insects (Hajek and St. Leger, 1994; Tanada and Kaya, 1993). Some parasites are sufficiently effective that they have been exploited as agents of biological control (van den Bosch *et al.*, 1982). Epidemics of parasites often are responsible for termination of host outbreaks (Hajek and St. Leger, 1994; Hochberg, 1989).

Some parasites alter the physiology or behavior of their hosts in ways that enhance parasite development or transmission. For example, parasitic nematodes often destroy the host's genital organs, sterilizing the host (Tanada and Kaya, 1993). Parasitized insects frequently show prolonged larval development (Tanada and Kaya, 1993). Flies, grasshoppers, and ants infected with fungal parasites often climb to high places where they cling following death, facilitating transmission of wind-blown spores (Tanada and Kaya, 1993).

Insects have evolved various defenses against parasites (see Chapter 3). Hard integument, hairs and spines, and antibiotics secreted by metapleural glands prevent attachment or penetration by some parasites (e.g., Hajek and St. Leger, 1994; Peakall *et al.*, 1987). Ingested antibiotics or gut modifications prevent penetration by some ingested parasites (Tallamy *et al.*, 1998; Tanada and Kaya, 1993). Endocytosis is the infolding of the plasma membrane by a phagocyte engulfing and removing viruses, bacteria, or fungi from the hemocoel. When the foreign particle is too large to be engulfed by phagocytes, aggregation and adhesion of hemocytes can form a dense covering around the particle, encapsulating and destroying the parasite (Tanada and Kaya, 1993). However, some parasitic wasps inoculate the host with a virus that inhibits the encapsulation of their eggs or larvae (Edson *et al.*, 1981; Godfray, 1994). Ants, *Pheidole dentata*, cease foraging or other exposed activities when the presence of the parasitic phorid fly, *Apocephalus feeneri*, is detected (Feener, 1981).

Many insects and other arthropods function in the capacity of parasites. Although parasitism generally is associated with animal hosts, most insect herbivores can be viewed as parasites of living plants (Fig. 8.5). Some herbivores, such as sap-suckers and gall-formers, appear more analogous to blood-feeding or internal parasites of animals. Virtually all arthropods and most vertebrates are parasitized by insect or mite species. The majority of insect parasites of animals are wasps, flies, fleas, and lice (e.g., Price, 1997). Parasitic wasps are a highly diverse group that differentially parasitize the eggs, juveniles, pupae, or adults of various arthropods. Spider wasps, e.g., tarantula hawks, provision burrows with paralyzed spiders for their parasitic larvae. Flies parasitize a wider variety of hosts. Mosquitoes and other biting flies are important blood-sucking ectoparasites of vertebrates. Oestrids, tachinids, and others are important endoparasites of vertebrates and insects. Fleas and lice are ectoparasites of vertebrates. Mites, chiggers, and ticks parasitize a wide variety of hosts.

Generally, parasitoids attack only other arthropods, but a sarcophagid fly, *Anolisomyia rufianalis*, is a parasitoid of *Anolis* lizards in Puerto Rico. Dial and Roughgarden (1996) found a slightly higher rate of parasitism of *Anolis evermanni* compared to *Anolis stratulus*. They suggested that this difference in parasitism may be due to black spots on the lateral abdomen of *A. stratulus* that resemble the small holes made by emerging parasites. Host-seeking flies may tend to avoid lizards showing signs of prior parasitism.

Nicholson and Bailey (1935) proposed a model of parasitoid–prey interactions that assumed that prey are dispersed regularly in a homogeneous environment, that parasitoids search randomly within a constant area of discovery, and the ease of prey discovery and parasitoid oviposition do not vary with prey density. The number of prey in the next generation (u_s) was calculated as

$$pa = log_e(u_i/u_s), \quad (8.5)$$

where p = parasitoid population density, a = area of discovery, and u_i = host density in the current generation.

Hassell and Varley (1969) showed that the area of discovery (a) is not constant for real parasitoids. Rather log a is linearly related to parasitoid density (p) as

$$log\ a = log\ Q - (m\ log\ p), \quad (8.6)$$

where Q is a quest constant and m is a mutual interference constant. Hassell and Varley (1969) modified the Nicholson–Bailey model to incorporate density limitation (Q/p^m). By substitution,

$$pa = log_e(u_i/u_s) = Qp^{1-m}. \tag{8.7}$$

As m approaches Q, model predictions approach those of the Nicholson–Bailey model.

2. Commensalism

Commensalism benefits the symbiont without significantly affecting the host. This is a relatively rare type of interaction, because few hosts can be considered to be completely unaffected by their symbionts. Epiphytes, plants that benefit by using their hosts for aerial support but gain their resources from the atmosphere, and cattle egrets, that eat insects flushed by grazing cattle, are well-known examples of commensalism. However, epiphytes may capture and provide nutrients to the host (a benefit) and increase the likelihood that overweight branches will break during high winds (a detriment). Some interactions involving insects may be largely commensal.

Phoretic or vector interactions (Fig. 2.13) benefit the hitchhiker or pathogen, especially when both partners have the same destination, and normally may have little or no effect on the host. However, hosts can become overburdened when the symbionts are numerous, inhibiting dispersal, resource acquisition, or escape. Schowalter and Crossley (1982) found that a mite thought to be phoretic caused an increased egestion rate in the host cockroaches (see Fig. 4.11). In some cases, the phoretic partners may be mutualists, with predaceous hitchhikers reducing competition or parasitism for their host at their destination (Kinn, 1980). Examples of commensalism often may be seen to exemplify other interaction types as additional information becomes available.

A number of insect and other arthropod species function as nest commensals in ant or termite colonies. Such species are called myrmecophiles or termitophiles, respectively. These symbionts gain shelter, and often detrital food, from their host colonies with little, if any, effect on their hosts. This relationship is distinguished from interactions involving species that intercept host food (e.g., through trophallaxis) and, therefore, function as colony parasites. Some vertebrate species also are commensals of termite castles in the tropics. These termite nests may reach several meters in height and diameter and provide critical shelter for reptile, bird, and mammal species in tropical savannas (see Chapter 14).

Bark beetle galleries provide habitat and resources for a variety of invertebrate and microbial commensals, most of which have little or no effect on the bark beetles (e.g., Stephen et al., 1993). Many of the invertebrate species are fungivores or detritivores that depend on penetration of the bark by bark beetles in order to exploit resources provided by the microbial decay of wood (Fig. 8.6).

3. Mutualism

Mutualistic interactions benefit both partners (positive effects on each) and, therefore, represent cooperative or mutually exploitative relationships. One

FIG. 8.6 Commensalism: an unidentified mite in an ambrosia beetle, *Trypodendron lineatum*, mine in Douglas-fir, *Pseudotsuga menziesii*. A variety of predaceous and detritivorous mites exploit resources in bark and ambrosia beetle mines.

member of a mutualism provides a resource that is exploited by the other (the symbiont). The symbiont, in turn, unintentionally provides a service to its host. For example, plants expend resources to attract pollinators, ants (for defense), or mycorrhizal fungi, which perform a service to the plant in the process of exploiting plant resources. Similarly, bark beetles provide nourishment to their symbiotic microorganisms that improve resource suitability for their host as a consequence of being transported to new resources. Gut symbionts of many insects, and other animals, provide nourishment as a consequence of exploiting resources in the host gut. Some mutualisms require less sacrifice of resources by either member of the pair. For example, aphids attract ants to their waste product, honeydew, and benefit from the protection the ants provide.

Mutualisms have received considerable attention, and much research has focused on examples such as pollination (see Chapter 13), ant–plant, mycorrhizae–plant interactions, and other conspicuous mutualisms. Nevertheless, Price (1997) argued that ecologists have failed to appreciate mutualism as equal in importance to predation and competition, at least in temperate communities (e.g., Goh, 1979; May, 1981; Williamson, 1972). As cooperative relationships, mutualism can contribute greatly to the presence and ecological function of the partners. However, the extent to which mutualism stabilizes or destabilizes interacting species populations is not clear (e.g., May, 1981).

As with parasitism, mutualistic interactions tend to be relatively specific associations between coevolved partners and often involve modification of host morphology, physiology, or behavior to provide habitat or food resources for the symbiont. In return, the symbiont provides necessary resources or protection from competitors or predators. Although the classic examples of mutualism often involve mutually dependent (obligate) partners, i.e., disappearance of one leads to demise of the other, some mutualists are less tightly coupled. To some degree, herbivores on plants often may function as mutualists, pruning and permitting reallocation of resources to more productive plant parts in return for their resources. Many insect species engage in mutualistic interactions with other organisms, including plants, microorganisms, and other insects.

Among the best-known mutualisms are those involving pollinator and ant associations with plants (Feinsinger, 1983; Huxley and Cutler, 1991; Jolivet, 1996). The variety of obligate relationships between pollinators and their floral hosts in the tropics perhaps has contributed to the perception that mutualism is more widespread and important in the tropics. As discussed in Chapter 13, the prevalence of obligate mutualisms between plants and pollinators in the tropics, compared to temperate regions, largely reflects the high diversity of plant species that precludes wind pollination between nearest neighbors. Rare or understory plants in temperate regions also tend to have mutualistic association with pollinators. Other mutualistic associations (e.g., insect–microbial association; see following) may be more prominent in temperate than in tropical regions. Protection of plants by ants or predaceous mites that either are housed in or attracted to resources provided by plants appears to be a widely distributed example of mutualism. Many plants provide nest sites or shelters (domatia), e.g., hollow stems or pilose vein axils, for ants or predaceous mites that protect a plant from herbivores (O'Dowd and Willson, 1991). Other plant species provide extrafloral nectaries rich in amino acids and lipids that attract ants (e.g., Dreisig, 1988; Jolivet, 1996; Oliveira and Brandâo, 1991; Rickson, 1971; Schupp and Feener, 1991; Tilman, 1978).

Clarke and Kitching (1995) discovered an unusual example of a mutualistic interaction between an ant and a carnivorous pitcher plant in Borneo. The ant, *Camponotus* sp., nests in hollow tendrils of the plant and is capable of swimming in pitcher-plant fluid, where it feeds on large prey items caught in the pitcher. Through ant-removal experiments, Clarke and Kitching found that accumulation of large prey (but not small prey) in ant-free pitchers led to putrefaction of the pitcher contents and disruption of prey digestion by the plant. By removing large prey, the ants prevent putrefaction and accumulation of ammonia. Hence, in return for food and nesting sites from the plant, the ants prevent the damaging accumulation of large prey in the pitchers.

Seed-feeding ants often benefit plants by assisting dispersal of nonconsumed seeds. This mutualism is exemplified by myrmecochorous plants that provide a nutritive body (elaiosome) attached to the seed to attract ants. The elaiosome typically is rich in lipids (Jolivet, 1996). The likelihood that a seed will be discarded in or near an ant nest following removal of the elaiosome increases with elaiosome size, perhaps reflecting increasing use by seed disperser, rather than seed-predator, species with increasing elaiosome size (Mark and Olesen, 1996; Westoby et al., 1991). The plants benefit primarily through seed dispersal by ants (Horvitz and Schemske, 1986; Ohkawara et al., 1996), not necessarily from seed relocation to more nutrient-rich microsites (Horvitz and Schemske, 1986; Westoby et al., 1991; see Chapter 13). This interaction has been implicated in the rapid invasion of new habitats by myrmecochorous species (Smith, 1989).

Gressitt et al. (1965, 1968) reported that large phytophagous weevils (Coleoptera: Curculionidae), primarily in the genera *Gymnopholus* and *Pantorhytes*, host diverse communities of cryptogamic plants, including fungi, algae, lichens, liverworts, and mosses, on their backs. These weevils have specialized scales or hairs and produce a thick waxy secretion from glands around depressions in the elytra that appear to foster the growth of these symbionts. In turn, the weevils benefit from the camouflage provided by this growth and, possibly, from chemical protection. Predation on these weevils appears to be rare.

Insects exhibit a wide range of mutualistic interactions with microorganisms. A notable example of a mutualistic interaction between insects and viruses involves parasitoid wasps that inoculate their host with a virus that prevents cellular encapsulation of the parasitoid larva (Edson et al., 1981; Godfray, 1994; see Chapter 3). Intestinal bacteria may synthesize some of the pheromones used by bark beetles to attract mates (Byers and Wood, 1981).

Virtually all wood-feeding species interact mutualistically with some cellulose-digesting microorganisms. Among the best-known examples are the obligate mutualisms between ambrosia beetles (Scolytidae and Platypodidiae) and ambrosia (mold) fungi, between siricid wasps and *Amylostereum* (decay) fungi, and between termites and gut bacteria or protozoa. In the first case, the beetles are the only means of transport for the fungus, carrying hyphae in specialized invaginations of the cuticle (mycangia) that secrete lipids for fungal nourishment, and require the nutrition provided by the fungus. The adult beetles carefully cultivate fungal gardens in their galleries, removing competing fungi. Their offspring feed exclusively on the fungus, which derives its resources from the wood surrounding the gallery, and collect and transport fungal hyphae when they disperse (Batra, 1966; French and Roeper, 1972).

The siricid wasp also is the only means of dispersal for its associated fungus, and its larvae die in the absence of the fungus. The adult female wasp collects fungal hyphae from its gallery prior to exiting. The wasp stores and nourishes the fungus in a mycangium at the base of the ovipositor, then introduces the fungus during oviposition in the wood. The fungus decays the wood around the larva, that feeds on the fungal mycelium, destroying it in the gut, and pass-

FIG. 8.7   Mutualism: weaver ants tending honeydew-producing scale insects in Darwin, Australia.

es decayed wood fragments around the body to combine posteriorly with its frass (Morgan, 1968). Termites similarly depend on mutualistic bacteria or protozoa in their guts for digestion of cellulose (Breznak and Brune, 1994).

Many mutualistic interactions involve insects and other arthropods. A well-known example is the mutualism between honeydew-producing Homoptera and ants (Fig. 8.7). Homoptera excrete much of the carbohydrate solution (honeydew) that composes plant sap in order to concentrate sufficient nutrients (see Chapter 3). Aphid species are particularly important honeydew producers. A variety of species are tended by ants that harvest this carbohydrate resource and protect the aphids from predators and parasites (Bristow, 1991; Dixon, 1985; Dreisig, 1988). This mutualism involves only about 25% of aphid species and varies in its strength and benefits, perhaps reflecting plant chemical influences or the relative costs of defending aphid colonies (Bristow, 1991). Ant species show different preferences among aphid species, and the efficiency of protection often varies inversely with aphid and ant densities (Bristow, 1991; Cushman and Addicott, 1991; Dreisig, 1988).

Dung beetles (Scarabaeidae) and bark beetles often have mutualistic association with phoretic predaceous mites. The beetles are the only means of long-distance transport for the mites, and the mites feed on the competitors or parasites of their hosts (Kinn, 1980; Krantz and Mellott, 1972).

Although mutualism typically is viewed from the perspective of mutual benefits, this interaction also can be viewed as mutual exploitation or manipulation. The structures and resources necessary to maintain the mutualism represent costs to the organisms involved. For example, the provision of domatia or extrafloral nectaries by ant-protected plants represents a cost in terms of energy and nutrient resources that otherwise could be allocated to growth and reproduction. Therefore, plants may lose ant-related traits when the benefit from the ants is removed (Rickson, 1977).

Models of mutualistic interactions have lagged behind models for competitive or predator–prey interactions, largely because of the difficulty of simultaneously incorporating negative (density limitation) and positive (cooperation) feedback. The Lotka–Volterra equations are inadequate for extension to mutualism because they lead to unbounded exponential growth of both populations (May, 1981). May asserted that minimally realistic models for mutualists must allow for saturation in the magnitude of at least one of the reciprocal benefits, leading to a stable equilibrium point, with one (most often both) of the two equilibrium populations being larger than that sustained in the absence of the mutualistic interaction. However, recovery from perturbations to this equilibrium may take longer than in the absence of the mutualistic interaction, leading to instability (May, 1981). May presented a simple model for two mutualistic populations

$$N_{1(t+1)} = N_{1t} + r_1 N_{1t}[1 - (N_{1t} + \alpha N_2)/K_1] \tag{8.8}$$

$$N_{2(t+1)} = N_{2t} + r_2 N_{2t}[1 - (N_{2t} + \beta N_1)/K_2], \tag{8.9}$$

in which the carrying capacity of each population is increased by the presence of the other, with α and β representing the beneficial effect of the partner, $K_1 \rightarrow K_1 + \alpha N_2$, $K_2 \rightarrow K_2 + \beta N_1$, and $\alpha\beta < 1$ to limit uncontrolled growth of the two populations. The larger the product, $\alpha\beta$, the more tightly coupled the mutualists. For obligate mutualists, a threshold effect must be incorporated to represent the demise of either partner if the other becomes rare or absent. May (1981) concluded that mutualisms are stable when both populations are relatively large and increasingly unstable at lower population sizes, with a minimum point for persistence.

Dean (1983) proposed a biologically realistic model that incorporates density dependence as the means by which two mutualists can reach a stable equilibrium. As a basis for this model, Dean developed a model to describe the relationship between population carrying capacity (k_y) and an environmental variable (M) that limits k_y

$$dk_y /dM = a(K_y - k_y)/K_y, \tag{8.10}$$

where K_y is the maximum value of k_y and the constant a is reduced by a linear function of k_y. This equation can be integrated as

$$k_y = K_y(1 - e^{(-aM + C_y)/K_y}), \tag{8.11}$$

where C_y is the integration constant. Equation 8.11 describes the isocline where $dY/dt = 0$.

For species Y exploiting a replenishable resource provided by species X, Eq. (8.11) can be rewritten as

$$k_y = K_y(1 - e^{(-aN_x + C_y)/K_y}),\tag{8.12}$$

where N_x is the number of species X. The carrying capacity of species X depends on the value of Y and can be described as

$$k_x = K_x(1 - e^{(-bN_y + C_x)/K_x}),\tag{8.13}$$

where N_y is the number of species Y. Mutualism will be stable when the number of one mutualist (N_y) maintained by a certain number of the other mutu-

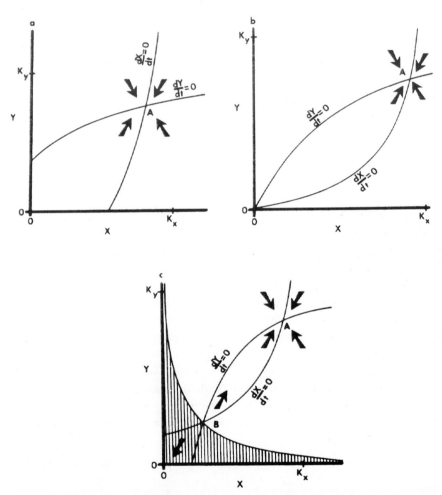

FIG. 8.8 The effect of integration constants in Dean's (1983) model on the form of mutualism (see text for equations) over a range of densities (X and Y) for two interacting species. (a) when C_x and $C_y > 0$, the interacting species are facultative mutualists, (b) when C_x and $C_y = 0$, both species are obligate mutualists, and (c) when C_x and $C_y < 0$, both species are obligate mutualists and have extinction thresholds at densities of B. Reproduced from Dean (1983), "A simple model of mutualism," *American Naturalist* **121**: 409–417, by permission from the University of Chicago.

alist (N_x) is greater than the N_y necessary to maintain N_x. When this condition is met, both populations grow until density effects limit the population growth of X and Y, so that isoclines defined by Eqs. (8.12) and (8.13) inevitably intersect at a point of stable equilibrium. Mutualism cannot occur when the isoclines do not intersect and is unstable when the isoclines are tangential. This condition is satisfied when any value of N_x or N_y can be found to satisfy either of the following equations

$$K_y(1 - e^{(-aN_x + C_y)/K_y}) > -(C_x + K_x[ln(K_x - N_x) - lnK_x])/b \quad (8.14)$$

$$K_x(1 - e^{(-bN_y + C_x)/K_x}) > -(C_y + K_y[ln(K_y - N_y) - lnK_y])/a \quad (8.15)$$

The values of the constants, C_x and C_y, in Eqs. (8.14) and (8.15) indicate the strength of mutualistic interaction. When C_x and $C_y > 0$, the interacting species are facultative mutualists; when C_x and $C_y = 0$, both species are obligate mutualists; when C_x and $C_y < 0$, both species are obligate mutualists and their persistence is determined by threshold densities (Fig. 8.8).

The growth rates of the two mutualists can be described by modified logistic equations as

$$N_{y(t+1)} = N_{y(t)} + (r_y N_{y(t)}[k_y - N_{y(t)}])/k_y \quad (8.16)$$

$$N_{x(t+1)} = N_{x(t)} + (r_x N_{x(t)}[k_x - N_{x(t)}])/k_x \quad (8.17)$$

where r_y and r_x are the intrinsic rates of increase for species Y and X, respectively. However, k_y and k_x are not constants but are determined by Eqs. (8.12) and (8.13).

II. FACTORS AFFECTING INTERACTIONS

The strength, and even type, of interaction, can vary over time and space depending on biotic and abiotic conditions (e.g., Tilman, 1978). Interactions can change during life history development or differ between sexes. For example, immature butterflies (caterpillars) are herbivores, but adult butterflies are pollinators. Several orders of insects (Odonata, Ephemeroptera, Plecoptera, and Trichoptera) have aquatic associations as immatures but terrestrial associations as adults. Immature males of the strepsipteran family Myrmecolacidae parasitize ants, whereas immature females parasitize grasshoppers (de Carvalho and Kogan, 1991). Herbivores and host plants often interact mutualistically at low population densities, with the herbivore benefitting from plant resources and the plant benefitting from limited pruning, but the interaction becomes increasingly predatory as plant condition declines. The strength of nonobligate interactions depends on the proximity of the two species, their ability to perceive each other, their relative densities, and their motivation to interact. These factors, in turn, are affected by abiotic conditions, resource availability, and indirect effects of other species.

A. Abiotic Conditions

Relatively few studies have addressed the effects of abiotic conditions on species interactions. Chase (1996) experimentally manipulated temperature and solar radiation in experimental plots containing grasshoppers and wolf spiders in a grassland. When temperature and solar radiation were reduced by shading during the morning, grasshopper activity was reduced, but spider activity was unaffected and spiders reduced grasshopper density. In contrast, grasshopper activity remained high in unshaded plots, and spiders did not reduce grasshopper density. Stamp and Bowers (1990) also noted that temperature affects the interactions between plants, herbivores, and predators.

Hart (1992) studied the relationship between crayfish, their caddisfly (Trichoptera) prey, and the algal food base in a stream ecosystem. He found that crayfish foraging activity was impaired at high flow rates, limiting predation on the caddisfly grazers and altering the algae–herbivore interaction.

Abiotic conditions that affect host growth or defensive capability influence predation or parasitism. Increased exposure to sunlight can increase plant production of defensive compounds and reduce herbivory (Dudt and Shure, 1994; Niesenbaum, 1992). Stamp *et al.* (1997) reported that the defensive chemicals sequestered by a caterpillar had greater negative effects on a predator at higher temperatures. Light availability to plants may affect their relative investment in toxic compounds versus extrafloral nectaries and domatia to facilitate defense by ants (Davidson and Fisher, 1991). Fox *et al.* (1999) reported that drought stress did not affect growth of St. John's wort, *Hypericum perforatum*, in the U.K. directly but increased plant vulnerability to herbivores.

Altered atmospheric conditions, e.g., CO_2 enrichment or pollutants, affect interactions between many herbivores and their hosts (Alstad *et al.*, 1982; Arnone *et al.*, 1995; Brown, 1995; Heliövaara and Väisänen, 1986, 1993; Kinney *et al.*, 1997; Roth and Lindroth, 1994; Salt *et al.*, 1996). For example, Hughes and Bazzaz (1997) reported that elevated CO_2 significantly increased C to N ratio and decreased percentage nitrogen in milkweed tissues, resulting in lower densities but greater per capita leaf damage by the western flower thrips, *Frankliniella occidentalis*. However, increased plant growth at elevated CO_2 levels more than compensated for leaf damage. Salt *et al.* (1996) reported that elevated CO_2 did not affect the competitive interaction between shoot- and root-feeding aphids. Coûteaux *et al.* (1991) found that elevated CO_2 affected litter quality and decomposer food web interactions.

Disturbances affect species interactions in several ways. First, disturbances act like predators for intolerant species and reduce their population sizes, thereby affecting their interactions with other species. Second, disturbances contribute to landscape heterogeneity, thereby providing potential refuges from negative interactions (e.g., Denslow, 1985). For example, disturbances often reduce abundances of predators, perhaps facilitating population growth of prey populations in disturbed patches (Kruess and Tscharntke, 1994; Schowalter and Ganio, 1999).

B. Resource Availability and Distribution

Resource availability affects competition and predation. If suitable resources (plants or animal prey) become more abundant, resource discovery becomes easier and populations of associated consumers grow. The probability of close contact and competition among consumers increases, up to a point at which the superior competitor(s) suppress or exclude inferior competitors. As a result, the intensity of interspecific competition may peak at intermediate levels of resource availability, although the rate of resource use may continue to rise with increasing resource availability (depending on functional and numerical responses). Population outbreaks sufficient to deplete resources also reduce populations of competing species.

Interactions are affected by the heterogeneity of the landscape. Potential competitors, or predators and their prey, often may not occur simultaneously in the same patches, depending on their respective dispersal and foraging strategies. Sparse resources in heterogeneous habitats tend to maintain small, low-density populations of associated species. The energetic and nutrient costs of detoxifying current resources or searching for more suitable resources limits growth, survival, and reproduction (see Chapter 4). Under these conditions, potentially interacting species are decoupled in time and space, co-occurring infrequently on a particular resource. Hence, competition is minimized, and predator-free space is maximized, in patchy environments. In contrast, more homogeneous environments facilitate population spread of associated species and maximize the probability of co-occurrence.

C. Indirect Effects of Other Species

Ecologists traditionally have focused on pairs of species that interact directly, i.e., through energy or material transfers, as described above. However, indirect interactions, such as reduced predation on mimics when the models are present, have received less attention. Only recently has tri-trophic level interaction been recognized as a key to understanding both herbivore–plant and predator–prey interactions (e.g., Boethel and Eikenbary, 1986; Price *et al.*, 1980). Even tri-trophic level interaction represents a highly simplified model of communities (Gutierrez, 1986) in which species interactions with many other species are affected by changing environmental conditions (see Chapters 9 and 10). The tendency for multiple interactions to stabilize or destabilize species populations and community structure has been debated (May, 1973, 1983; Price, 1997). May (1973) proposed that community stability depends on predator–prey interactions being more common than mutualistic interactions. Because multispecies interactions control rates of energy and nutrient fluxes through ecosystems, resolution of the extent to which interactions contribute to stability of community structure will contribute significantly to our understanding of ecosystem stability.

Associated species affect particular interactions in a variety of ways. For

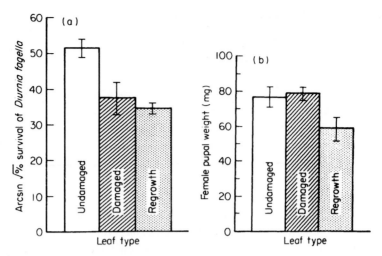

FIG. 8.9 Differential survival to pupation (a) and mean female pupal weight (b) of *Diurnea flagella* on foliage that was undamaged, naturally damaged by folivores, and produced following damage. Vertical lines represent standard errors of the mean. *D. flagella* larvae feeding on regrowth foliage show both reduced survival to pupation and reduced pupal weight. From Hunter (1987) by permission from Blackwell Scientific.

example, much research has addressed the effects of herbivore induction of plant defenses on later colonists (Fig. 8.9) (e.g., Harrison and Karban, 1986; Hunter, 1987; Kogan and Paxton, 1983; Moran and Whitham, 1990; Sticher *et al.*, 1997; Wold and Marquis, 1997) and on decomposers (Grime *et al.*, 1996). Baldwin and Schultz (1983) and Rhoades (1983) found evidence that damage by herbivores can be communicated chemically among plants, leading to induced defenses in plants in advance of herbivory (see also Zeringue, 1987). Although their hypothesis that plants may communicate chemically with each other has been challenged (e.g., Fowler and Lawton, 1985), Sticher *et al.* (1997) reviewed a number of studies demonstrating the effect of volatile chemical elicitors on acquired resistance in plants. In particular, some volatile compounds (jasmonic acid, methyl jasmonic acid, and ethylene) have been shown to induce production of proteinase inhibitors and other defenses against insects and pathogens when applied at low concentrations to some plant species.

Plant defenses can be augmented by endophytic fungi (see Chapter 3) and other organisms. Differential colonization of foliage by microbial species can affect patterns of foliage quality for folivores (Fig. 8.10) (Carroll, 1988; Clay, 1990). Carroll (1988) reported that mycotoxins produced by mutualistic endophytic fungi complement host defenses in deterring herbivores. Clay *et al.* (1993) documented complex effects of insect herbivores and endophytic fungi on the competitive interactions among grass species. For example, tall fescue, *Festuca arundinacea*, competed poorly with orchard grass, *Dactylis glomerata*, when herbivores were absent, but fescue infected with its fungal endophyte,

FIG. 8.10 Indirect effects of associated species. The light-colored foliage at the ends of shoots is new grand fir, *Abies grandis*, foliage produced during 1995, a dry year, in western Washington; the blackened 1993 foliage was colonized by sooty mold during a wet year; normal foliage prior to 1993 was produced during extended drought. Sooty mold exploits moist conditions, especially honeydew accumulations and, in turn, may affect foliage quality for folivores.

Acremonium spp., competed better than either orchard grass or uninfected fescue when herbivores were present.

Volatile defenses of plants induced by defoliators often attract the herbivore's predators and parasites (e.g., Price, 1986; Turlings *et al.*, 1990). At the same time, however, plant defenses sequestered by herbivores can affect herbivore–predator and herbivore–pathogen interactions (Brower *et al.*, 1968; Stamp *et al.*, 1997; Tallamy *et al.*, 1998; Traugott and Stamp, 1996). Inflorescence spiders preying on pollinators affect the pollinator–plant interaction (Louda, 1982). Herbivores feeding above ground frequently deplete root resources, through compensatory translocation, and negatively affect root-feeding herbivores (e.g., Masters *et al.*, 1993; Rodgers *et al.*, 1995; Salt *et al.*, 1996).

Ants interact in a variety of ways with virtually all other organisms in the community. Among their many interactions, ants frequently visit floral resources but have little importance as pollinators. Peakall *et al.* (1987) suggested that antibiotic secretions produced by most ants, to inhibit infection by entomophagous fungi in a subterranean habitat, also inhibit germination of pollen. Ants lacking these secretions are known to function as pollinators. Hence, pathogens apparently influence this interaction between ants and plants.

Ants affect interactions among other organisms, as well. Because ant activity is nest-focused, interactions with other organisms vary in strength with proximity to the nest. For example, the degree of competition for foraging territories among species depends on proximity of nests and foraging distances. Herbivory by leaf-cutter ants is inversely related to distance from the ant nest. Ants attracted to domatia, to floral or extrafloral nectories, or to aphid honeydew commonly affect herbivore–plant interactions (Cushman and Addicott, 1991; Fritz, 1983; Jolivet, 1996; Oliveira and Brandâo, 1991; Tilman, 1978). Attraction of ants to honeydew or extrafloral nectaries also varies inversely with distance from the ant nest. Tilman (1978) reported that ant visits to extrafloral nectaries declined with the distance between cherry trees and ant nests. The associated predation on tent caterpillars by nectar-foraging ants also declined with distance from the ant nest.

Complex interactions among a community of invertebrates and fungi affect bark beetle interactions with host trees. Southern pine beetles, *Dendroctonus frontalis*, once were thought to have a mutualistic association with blue stain fungi, with beetles providing transport and the fungus contributing to tree death and beetle reproduction. However, several studies have shown that this beetle can colonize trees in the absence of the fungus (Bridges *et al.*, 1985), that the blue stain fungus is, in fact, detrimental to beetle development and is avoided by the mining beetles (Barras, 1970; Bridges, 1983; Bridges and Perry, 1985), and that other mycangial fungi are necessary for optimal beetle development (Bridges and Perry, 1985). Subsequent research demonstrated that phoretic tarsonemid mites collect spores of the blue-stain fungus in specialized structures, sporothecae (Fig. 8.11) (Bridges and Moser, 1983; Moser, 1985). Beetles carrying these mites transport the blue stain fungus significantly more often than do mite-free beetles (Bridges and Moser, 1986). This beetle–host interaction is affected further by phoretic predaceous mites that prey on nematode parasites of the beetle (Kinn, 1980).

Termite interaction with its mutualistic gut symbionts is affected by host wood and associated fungi. Using forced feeding and preference tests involving combinations of several conifer species and fungi, Mankowski *et al.* (1998) found that termite preferences for wood–fungal combinations generally reflected the suitability of the resource for the gut fauna, as indicated by changes in faunal densities when termites were forced to feed on wood–fungus combinations.

Competitive interactions between a pair of species may be modified by the presence of additional competitors. Pianka (1981) proposed a model in which two species with modest competitive overlap over a range of resource values could become "competitive mutualists" with respect to a third species that could compete more strongly for intermediate resource values. The two species benefit each other by excluding the third species from both sides of its resource spectrum (niche).

Competitive interactions among several species also can be modified by

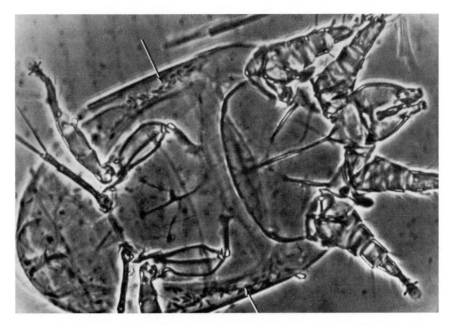

FIG. 8.11   Ascospores of *Ceratocystis minor* in sporothecae (arrows) formed by tergite 1 on the ventral–lateral sides of a *Tarsonemus ips* female, phoretic on the southern pine beetle, *D. frontalis*. From Moser (1985) by permission from the British Mycological Society.

predators. A predator that preys indiscriminately on several competing prey species, as these are encountered, will tend to prey most often on the most abundant prey species, thereby preventing that species from competitively suppressing others. Paine (1966, 1969a,b) introduced the term **keystone species** to refer to top predators that maintain balanced populations of competing prey species. However, this term has become used more broadly to include any species whose effect on community and ecosystem structure or function is disproportionately large compared to its abundance (Bond, 1993; Power *et al.*, 1996). Some insect species play keystone roles. For example, many herbivorous insects affect plant competitive interactions by selectively reducing the density of abundant host species and providing additional space and resources for nonhost plants (Louda *et al.*, 1990a; Schowalter and Lowman, 1999).

III. CONSEQUENCES OF INTERACTIONS

A given species interacts with many other species in a variety of ways (competing for various food, habitat and other resources, preying, or being preyed, on, and cooperating with mutualists) with varying degrees of positive and negative effects on abundance. Therefore, the population status of species in the community represents the net effects of these positive and negative effects.

A. Population Regulation

As discussed in Chapter 6, competition and predation have been recognized as two primary mechanisms, along with resource quality, for limiting population growth of a given species (e.g., May, 1983). Any particular species typically interacts with at least 2–5 other species as prey (see Chapter 9) and with additional species as a competitor. Life table analysis often is used to identify key factors, especially predators or other interactions, that contribute most to population change, but the combination of interactions provides some "redundant" control of population growth. If the major regulating species should disappear, other predators, parasites, or competitors might compensate.

As noted above, mutualistic interactions may reduce the probability of either species declining to extinction. Mutualistic species often are closely associated, especially in obligate relationships, and enhance each other's resource acquisition or energy and nutrient balance. Although mutualism is likely to become unstable at low population densities of either partner, depending on the degree of obligation (May, 1983), mutualism could help to maintain the two populations above extinction thresholds.

The combination of various interactions involving a particular species should maintain its population levels within a narrower range than would occur in the absence of these various interactions. Croft and Slone (1997) found that three predaceous mite species maintained populations of the European red mite, *Panonychus ulmi*, at lower equilibrium levels than did fewer predator species. However, few studies have documented the importance of species diversity or food web structure to stability of population levels.

B. Community Regulation

The extent to which the network of regulatory interactions maintains stable community structure (see Chapters 9 and 10) has been a topic of considerable debate. Although some irruptive species show wide amplitude in population size over time and space, the range in population size is narrower, and the duration of deviations shorter, in the presence of regulatory interactions than when those interactions are disrupted through habitat alteration or introduction into new habitats (see Chapters 6 and 7).

The capacity for the network of interactions to stabilize species populations may be enhanced by compensatory interactions and changes in the nature or strength of interaction with changing environmental conditions. For example, the diversity of plant species at a site can, at the same time, compete for resources, share nutrients via mycorrhizae, be growth-limited by herbivores, and limit herbivore populations through the mingling of attractive host odors and repellent (or unattractive) nonhost odors (Allen and Allen, 1990; Hunter and Arssen, 1988; Visser, 1986). The net result of these negative and positive effects of interaction may be balanced coexistence. Ants gain more by preying on aphids when the value of honeydew rewards is low (e.g., scattered individuals or individuals dispersing from dense colonies) and by tending aphids when the value of honeydew rewards is high (Bristow, 1991; Cushman and Addicott, 1991). Commensal interactions can become parasitic, e.g., when dense populations of phoretic mites overburden their host. Finally, competitive interactions could become mutualistic if two competitors mutually exclude a third, more competitive, species from the intermediate region of the shared niche (Pianka, 1981). Such flexibility in species interactions may facilitate regulation in a variable environment. If the various species in the community respond to changes in each other's population densities in ways that are neutral or beneficial at low densities and increasingly negative at higher densities (see Chapters 12 and 15), then some stability of community structure should result. Stabilization of community structure has substantial implications for the stability of ecosystem processes (see Chapter 15).

Interactions strongly affect energy or nutrient balances, survival and reproduction of the associated species and, therefore, represent major selective forces. Strongly negative interactions should select for adaptive responses that minimize the negative selection, e.g., niche partitioning among competitors or prey defenses. Therefore, negative interactions should evolve toward more neutral or mutualistic interactions (Carroll, 1988; Price, 1997).

IV. SUMMARY

Species interact in a variety of ways with the other species that cooccur at a site. These interactions produce combinations of positive, neutral, or negative effects for species pairs. However, other species may alter the nature or strength of particular pairwise interactions, e.g., predators can reduce the intensity of competition among prey species by maintaining their populations below levels that induce competition.

Some species compete for a shared resource, with the result that the per capita share of the resource is reduced. This interaction has negative effects on both species. Competition can be exploitative, when all individuals have equal access to the resource, or interfering, when individuals of one species preempt use of, or defend, the resource. In cases of asymmetric competition, the superior competitor can exclude inferior competitors over a period of time (competitive exclusion), unless the inferior competitor can escape through dispersal or survival in refuges where superior competitors are absent.

Predator–prey interactions involve a predator's killing and eating prey and, therefore, have a positive effect on the predator but a negative effect on the prey. Predators and parasites affect prey populations similarly, but predators generally are opportunistic with respect to prey taxa, whereas parasites generally are more specialized for association with particular host species. Predators show preferences for prey size or defensive capability that maximize capture and utilization efficiency.

Symbiosis involves an intimate association between a symbiont and its host species, often coevolved to maximize the probability of association and to mitigate any host defense against the symbiont. Symbiosis includes parasitism, commensalism, and mutualism. Parasitism is beneficial to the parasite but detrimental to its host. Although parasitism typically is considered to involve animal hosts, insect herbivores have a largely parasitic association with their host plants. Parasitoidism is unique to insects and involves an adult female ovipositing on or in a living host, with her offspring feeding on and eventually killing the host. Most hosts of parasitoids are other arthropods, but at least one sarcophagid fly is a parasitoid of tropical lizards. Commensalism benefits the symbiont but has neutral effects for the host. Typically the symbiont uses its host or its products as habitat or as a means of transport with negligible effects on the host. Mutualism benefits both partners and is exemplified by pollinator–plant, ant–plant, ant–aphid, and detritivore–fungus interactions.

A variety of factors influence the nature and intensity of interaction. Abiotic factors that affect the activity or condition of individuals of a species may alter their competitive, predatory, or defensive ability. Resource availability, particularly the quality and patchiness of resources, may mitigate or exacerbate competition or predation by limiting the likelihood that competitors or predators and their prey co-occur in time and space. Other species can influence pairwise interactions indirectly. For example, predators often reduce populations of various prey species below sizes that would induce competition. Induced plant responses can influence predator–herbivore interactions and competition among herbivores in time and space. Species whose presence significantly affects diversity or community structure have been considered keystone species. A number of insect species function as keystone species.

Competition and predation/parasitism have been recognized as important mechanisms of population regulation and have been amenable to mathematical modeling. Mutualism has been viewed largely as a curiosity, rather than an important regulatory interaction, and modeling efforts have been more limited. However, mutualism may promote both populations and reduce their risk of decline to unstable levels. The network of interactions affecting a particular species may maintain population size within a narrower range with less frequent irruptions than occur when populations are released from their regulatory network. The extent of mutual regulation (stabilization) of populations through this network of interactions has been widely debated, but has significant implications for the stability of community structure and ecosystem processes governed by these interactions.

Community Structure

A COMMUNITY IS COMPOSED OF ALL THE ORGANISMS OCCUPYING A site. The extent to which these organisms are coevolved to form a consistent and recurring integrated community or represent ad hoc assemblages of loosely interacting species remains a topic of much discussion. Considerable research has been directed toward identifying spatial and temporal patterns in community structure and evaluating factors that determine community composition. Such efforts have become increasingly important to conservation efforts, with recognition of the dependence of many species on the presence of associated species. However, comparison of community structures within or among broadly distributed community types that share few, if any, species requires approaches that are independent of the taxonomic composition of the community per se.

Ecologists have developed a variety of nontaxonomic approaches to describing community structure, providing different types of information to meet different objectives. The diversity of approaches has hindered comparison of community structures described in different terms. Nevertheless, distinct geographic patterns can be seen in community structure, and some community types characterize particular habitat conditions. A number of factors determine community composition, distribution, and dynamics. This chapter focuses on approaches to describing community structure, and on biogeographic patterns and underlying factors contributing to community structure. Temporal patterns in community structure are the focus of the next chapter.

I. APPROACHES TO DESCRIBING COMMUNITIES

Although the community is understood to include all organisms at a site, most studies have addressed subsets of this community, for practical reasons. Hence, the literature on communities includes references to the plant community, the arthropod community, the bird community, consumer communities associated with different plant species, tritrophic interactions, etc. Insects are addressed to varying degrees in studies of communities, although insects represent the majority of species in most terrestrial and freshwater aquatic communities (Table 9.1) and clearly are integral to community structure and dynamics, as pollinators or herbivores of vegetation, as resources for vertebrates, etc. Description of particular subsets of the community involves further differentiation in approaches.

Three general approaches to describing community structure can be identified: species diversity, species interactions, and functional organization. Although the "ideal" approach is a topic of intense ecological debate (e.g., Polis 1991a), each approach provides useful information, and the choice largely reflects objectives and practical considerations.

A. Species Diversity

Insects represent the vast majority of species in most terrestrial and aquatic ecosystems. For example, in systems where diversity of insect or arthropod species has been inventoried, along with plants and vertebrates, arthropods account for 70–80% of the total number of recorded species (Table 9.1), roughly the same proportion as the total number of described species of organisms. Given that plant and vertebrate inventories are relatively complete, whereas currently described insect species represent only a fraction of the estimated total number of species (May, 1988; Wilson, 1992), the proportional representation of invertebrates likely will increase.

Species diversity is a central theme in ecology. An enormous amount of research has addressed how diversity develops under different environmental conditions, how anthropogenic changes are affecting diversity, and how diversity affects the stability of natural communities (see Chapters 10 and 15). Clearly, the measurement of diversity is fundamental to meeting these objectives.

Diversity can be represented in various ways (Magurran, 1988). The simplest representation is a catalog of species, or the total number of species (richness), a measure that indicates the variety of species in a community (α diversity). Species richness can be standardized for various ecosystems by measuring the number of species per unit area or per 1000 individuals. This measure of

TABLE 9.1 Numbers of Species of Vascular Plants, Vertebrates, and Arthropods in Desert, Grassland, Forest and Marshland Ecosystems

	Desert			Grassland/Savanna			Forest		Marshland
	USSR[a]	USA1[a]	USA2[a]	Hungary[b]	Hungary[c]	USA[d]	USA[e]	Puerto Rico[f]	Hungary[g]
Vascular plants	125	174	>600	1311	1762	521	600	470	804
Vertebrates	198	145	201	347	289	355	88	78	118
Arthropods	>1360	>1100	>2640	8496	7095	>1750[b]	>3500	>1500	5332
% Arthropods	75	77	77	93	78	67	84	73	85

[a]Data from Polis (1991b).
[b]Data from Mahunka (1986, 1987) and Szujko-Lacza and Kovacs (1993).
[c]Data from Mahunka (1981, 1983) and Szujko-Lacza (1982).
[d]Data from Hazlett (1998) and Lavigne et al. (1991).
[e]Data from Parsons et al. (1991).
[f]Data from Reagan et al. (1996).
[g]Data from Mahunka (1991).
[b]Insects only.

diversity accounts for the typical increase in number of species with increasing sample area or number of individuals. Species richness for many plant and animal groups increases from high latitudes to lower latitudes and from smaller, more isolated islands to larger islands near continents (MacArthur and Wilson, 1967; Magurran, 1988; Stiling, 1996). Diversity also increases from harsh or frequently disturbed ecosystems that restrict richness to more productive ecosystems that provide a greater number of niches, but typically declines again in very productive ecosystems (Tilman and Pacala, 1993). Species diversity appears generally to peak at intermediate levels of disturbance (the **Intermediate Disturbance Hypothesis**) due to a combination of sufficient resources and insufficient time for competitive exclusion (Connell, 1978; Huston, 1979; Lubchenco, 1978; Pickett and White, 1985; Sousa, 1979). Insect diversity may reflect primarily the diversity of plants, which affects diversity of host resources and habitat structure (Curry, 1994; Magurran, 1988; Schowalter, 1995b; Stiling, 1996).

The various species in a community are not equally abundant. Typically, a few species are abundant and many species are represented by only one or a few individuals. The distribution of numbers of individuals among species (evenness) is a measure of each species' importance. Rank–abundance curves are a commonly used method of presenting species abundance data (Magurran, 1988). Four rank–abundance patterns are most commonly used for comparison among different communities (Fig. 9.1). The **geometric model** (or **niche-preemption hypothesis**) describes a community in which successively less abundant species use the same proportion of resources available after preemption by the more abundant species and is predicted to occur when species arrive in an unsaturated community at regular time intervals and exploit a fraction of the remaining resources. The **log series model** is closely related to the geometric model but is predicted to occur when the time intervals between species arrival are random rather than regular. The **log normal model** has been shown to be widely applicable, because this distribution results mathematically from random variation among a large number of factors producing a normal distribution. In natural communities, the large number of environmental factors affecting species abundances fulfils this condition. The **broken stick model** reflects relatively uniform use of resources among species in the community. Generally, as richness and evenness increase, the rank–abundance pattern shifts from a geometric pattern to a log pattern and finally to a broken stick pattern. Disturbances and other environmental changes can alter rank–abundance patterns (Figs. 9.2 and 9.3) (Bazzaz, 1975; Kempton, 1979).

Richness and evenness have been combined mathematically in various ways to calculate diversity indices based on proportional abundances of species (e.g., Magurran, 1988; Stiling, 1996). Two indices have been used widely, the Shannon–Wiener, or Shannon (often incorrectly referred to as the Shannon–Weaver) index, and Simpson's index. The two indices differ in their emphasis on species richness (Shannon–Wiener) or abundance (Simpson's).

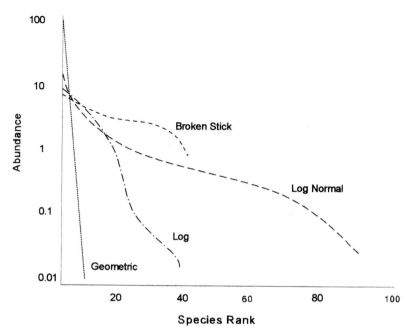

FIG. 9.1 Typical shapes of four rank abundance models. Species are ranked from most to least abundant. Adapted from Magurran (1988).

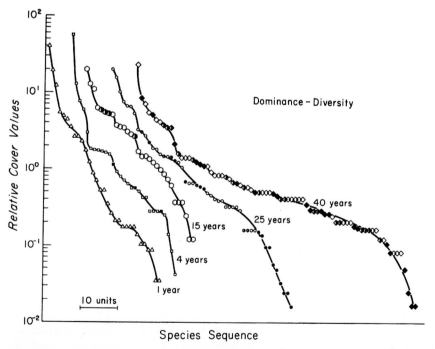

FIG. 9.2 Rank-abundance curves for old fields representing five post-abandonment ages in southern Illinois. Open symbols are herbs, half-open symbols are shrubs and closed symbols are trees. From Bazzaz (1975) by permission from the Ecological Society of America.

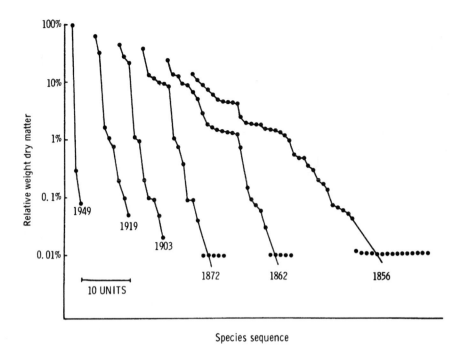

FIG. 9.3 Change over time in rank abundance of plant species in an experimental plot of permanent pasture at Rothamsted, UK, following continuous application of nitrogen fertilizer since 1856. Species with abundances < 0.01% were recorded as 0.01%. From Kempton (1979) by permission from the International Biometric Society.

The Shannon–Wiener index assumes that individuals are randomly sampled from an effectively infinite population and that all species are represented in the sample. Diversity (H') is calculated as

$$H' = -\sum_{i=1}^{n} p_i \ln p_i, \tag{9.1}$$

where p_i is the proportion of individuals found in the ith species. Values generally fall in the range 1.5–3.5, rarely surpassing 4.5. If the rank–abundance pattern follows a log normal model, 10^5 species are necessary to produce a value of $H' > 5$. If the index is calculated for a number of samples, the indices will be normally distributed and amenable to use of parametric statistics, including ANOVA, to compare diversities among sets of samples (Margurran, 1988), e.g., for different ecosystems (Schowalter, 1995a). If all species were equally abundant, a maximum diversity (H_{max}) can be calculated as *ln S*, where *S* is the total number of species. The ratio of observed to maximum diversity is a measure of evenness.

When randomness cannot be assured (e.g., data from light trapping, that samples insects based on differential attraction to light), the Brillouin index is a more appropriate measure of diversity (Margurran, 1988). This index (*HB*) is calculated as

$$HB = (ln\ N! - \Sigma\ ln\ n_i!)/N \qquad (9.2)$$

where N is the total number of individuals, and n_i is the number of individuals in the ith species. Values of this index rarely exceed 4.5 and generally are correlated with, but lower than, Shannon indices for the same data.

Simpson's index differs from the Shannon–Wiener and Brillouin indices in being weighted toward the abundances of the commonest species, rather than species richness (Margurran, 1988). This index (*D*) is calculated as

$$D = \sum_{i=1}^{n}(n_i(n_i - 1))/(N(N - 1)), \qquad (9.3)$$

where n_i is the number of individuals in the ith species and N is the total number of individuals. Diversity decreases as D increases, so Simpson's index generally is expressed as $1 - D$ or $1/D$. Once the number of species exceeds 10, the underlying rank–abundance pattern is important in determining the value of D.

Diversity indices have been a tool for comparing taxonomically distinct communities, based on their rank–abundance patterns. However, important information is lost when species diversities are reduced to an index (Magurran, 1988). For example, increased abundances of invasive or exotic species may indicate increased diversity without conveying important information about the change in community integrity or function. Very different communities can produce the same diversity index. The large number of species represented by single individuals poses a dilemma: should these be included in the diversity calculation or not? Their presence may be accidental or reflect inadequate or biased sampling. Furthermore, ecologically unique communities are not necessarily diverse and would be lost if conservation decisions were made on the basis of diversity alone (Magurran, 1988).

Diversity also can be measured as variation in species composition among communities or areas (β diversity). Several techniques have been developed to compare communities, based on their species compositions and rank–abundance patterns, across environmental gradients or between areas (Magurran, 1988).

The simplest of these similarity measures are indices based on species presence or absence in the communities being compared. The Jaccard index (C_J) is calculated as

$$C_J = j/(a + b - j) \qquad (9.4)$$

and the Sorenson index (C_S) as

$$C_S = 2j/(a + b), \qquad (9.5)$$

where j is the number of species found in both sites, a is the number of species in the first site, and b is the number of species in the second site. Neither of these indices accounts for species abundances.

Two quantitative similarity indices have been used widely. A modified version of the Sorenson index (C_N) is calculated as

$$C_N = 2_{jN}/(aN + bN), \tag{9.6}$$

where jN is the sum of the lower of the two abundances for each species found in both sites, aN is the total number of individuals in the first site, and bN is the total number of individuals in the second site. Most quantitative similarity indices are influenced strongly by species richness and sample size. The Morisita–Horn index (C_{mH}) is less influenced by species richness and sample size but is sensitive to the abundance of the dominant species. Nevertheless, it may be generally the most satisfactory similarity index (Magurran, 1988). This index is calculated as

$$C_{mH} = 2\Sigma(an_i bn_i)/(da + \text{db})aN \cdot bN, \tag{9.7}$$

where aN is the total number of individuals in the first site, an_i is the number of individuals of the ith species in the first site, and $da = \Sigma an_i^2/aN^2$.

More recently, multivariate statistical techniques have been applied to comparison of communities. Cluster analysis can be performed using either presence–absence or quantitative data. Each pair of sites is evaluated on the degree of similarity, then combined sequentially into clusters to form a dendrogram with the branching point representing the measure of similarity (Figs. 9.4 and 9.5). Ordi-

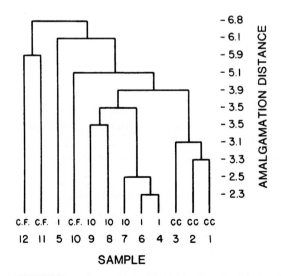

FIG. 9.4 Dendrogram of similarity for dung beetles (Scarabaerdae) in clear-cuts, 1 ha and 10 ha forest fragments, and contiguous forest. From Klein (1989) by permission from the Ecological Society of America.

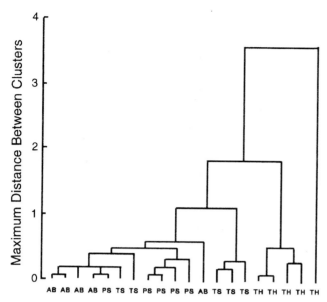

FIG. 9.5   Dendrogram of arthropod community similarity in canopies of four old-growth conifer species at the Wind River Canopy Crane Research Facility in southwestern Washington. AB, *Abies grandis* (grand fir); PS, *Pseudotsuga menziesii* (Douglas-fir); TS, *Tsuga heterophylla* (western hemlock); and TH, *Thuja plicata* (western redcedar). Data from Schowalter and Ganio (1998).

nation techniques include principal components analysis (PCA), detrended correspondence analyses, and nonmetric multidimensional scaling. Ordination compares sites on their degree of similarity, then plots them in Euclidian space, with the distance between points representing their degree of dissimilarity (Figs. 9.6

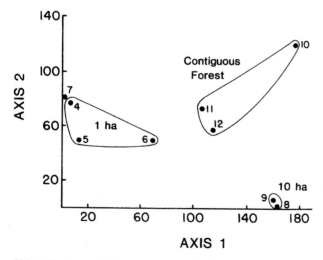

FIG. 9.6   Detrended Correspondence Analysis ordination of dung beetle assemblages in 1 ha and 10 ha forest fragments, and contiguous forest. From Klein (1989) by permission from the Ecological Society of America.

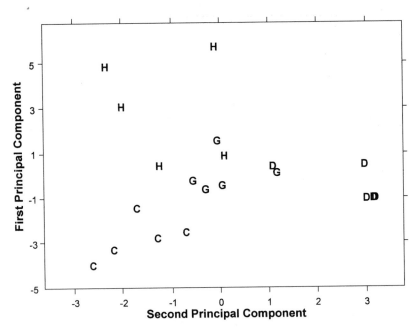

FIG. 9.7 Principal Components Analysis ordination of arthropod communities in canopies of four old-growth conifer species at the Wind River Canopy Crane Research Facility in southwestern Washington. G, grand fir (*Abies grandis*); D, Douglas-fir (*Pseudotsuga menziesii*); H, western hemlock (*Tsuga heterophylla*); and C, western redcedar (*Thuja plicata*). From Schowalter and Ganio (1998) by permission from CAB International.

and 9.7). Both cluster and ordination techniques also can indicate which species (or other components) contribute most to the discrimination.

B. Species Interactions

Communities can be characterized in terms of the relationships among species, most commonly trophic (feeding) interactions, i.e., food webs. Clearly, the most complete description of the community would include all possible interactions (including indirect interactions) among the total number of species (e.g., Polis, 1991a). In practice, this is difficult to accomplish, even in relatively species-poor communities (Camilo and Willig, 1995; Polis, 1991a,b; Reagan *et al.*, 1996), because of the largely unmanageable number of arthropod species (Table 9.1) and lack of complete information on their interactions. More commonly, research focuses on subsets or simplified representations of the community.

The simplest approach to community description emphasizes interactions between only a few species, e.g., plant–herbivore or predator–prey interactions. In particular, many studies have addressed the relatively distinct assemblages of arthropods based on individual plant species (e.g., Richerson and Boldt, 1995; Schowalter, 1994a; Schowalter and Ganio, 1998) or soil/litter resources (e.g., Moore and Hunt, 1988; Seastedt *et al.*, 1989). This approach maximizes description of interactions among a manageable number of relatively resource-specific herbivores or detritivores and their associated predators and parasites. Detailed descriptions at this level have been useful for identifying and comparing factors affecting these trophic interactions (e.g., chemical defenses, see Chapters 3 and 8), for evaluating the coevolutionary patterns of speciation between insects and their hosts (e.g., Becerra, 1997) and for comparing trophic interactions among community types, e.g., comparing phenological responses of insect herbivores to leaf emergence in tropical and temperate forests (Coley and Aide, 1991). However, this approach emphasizes relatively linear trophic relationships (food chains) and does not address linkages among members of different component communities.

Broader subcommunities can be identified. For example, Hunt *et al.* (1987) described the trophic interactions among arthropod and microbial species composing the litter subcommunity of a grassland ecosystem. Moore and Hunt (1988) subsequently noted that relatively discrete component communities supported by particular resource bases (bacteria, fungi, or plant roots) could be distinguished within this broader subcommunity (Table 9.2). Similarly, individual plant species represent resource bases for relatively discrete component communities of associated arthropods and other organisms in the above-ground subcommunity (Curry, 1994). Resource-based component communities are linked to each other by generalist herbivores and predators. Similarly, the canopy and soil/litter subcommunities are linked by species that feed above-ground but

TABLE 9.2   The Proportion of Energy and Nitrogen Derived from the Bacterial, Fungal, and Root (Including Mycorrhizal Fungi) Resource Channels by Different Faunal Groups in the North American Shortgrass Steppe

Faunal group	Resource channel		
	Bacteria	Fungi	Roots
Protozoa			
Flagellates	100	0	0
Amoebae	100	0	0
Ciliates	100	0	0
Nematodes			
Bacteriovores	100	0	0
Fungivores	0	90	10
Root-feeders	0	0	100
Omnivores	100	0	0
Predators	69	3	28
Microarthropods			
Mycophagous Collembola	0	90	10
Mycophagous oribatid mites	0	90	10
Mycophagous prostigmatid mites	0	90	10
Nematophagous mites	67	4	30
Predaceous mites	40	39	22

From Moore and Hunt (1988) by permission from Nature, (©) 1988 Macmillan Magazines, Ltd.

pupate in the soil or feed on litter resources but disperse and bask on foliage and by predators and detritivores that move among substrates in search of resources.

The most inclusive approach to community description is represented by interaction webs, in which all species are connected by arrows indicating interactions. Relatively few communities are composed of sufficiently few species to depict all interactions conveniently. Hot springs and other communities subject to extreme abiotic conditions typically are composed of a few tolerant algal and invertebrate species (Collins *et al.*, 1976). Communities composed of relatively few invertebrate and vertebrate species characterize some aquatic ecosystems, e.g., vernal pools and riffles. However, even the desert communities described by Polis (1991a) were composed of $>10^3$ arthropod species, most of which had not been studied sufficiently to provide complete information on interactions. A number of studies have addressed trophic interactions, i.e., food webs, although even trophic interactions are poorly known for many species, especially insects.

Several properties have appeared to characterize food webs (see Briand and Cohen, 1984; Cohen and Palka, 1990; Cohen *et al.*, 1990; Martinez, 1992; May, 1983; Pimm, 1980, 1982; Pimm and Kitching, 1987; Pimm and Lawton, 1977, 1980; Pimm and Rice, 1987; Pimm *et al.*, 1991; Polis, 1991b; Reagan *et al.*, 1996). However, food web analysis typically has been based on combination of all insects (often all arthropods) into a single category, in contrast to resolution of individual species of plants and vertebrates. Polis (1991b) and Reagan *et al.* (1996) increased the resolution of arthropod diversity to individual "kinds," based on taxonomy and similar phylogeny or trophic relationships, for evaluation of food web structure in desert and tropical rainforest communities, respectively. They found that the structure of their food webs differed from that of food webs in which arthropods were combined. Goldwasser and Roughgarden (1997) analyzed the effect of taxonomic resolution on food web structure and found that food web properties reflected the degree of taxonomic resolution. The following discussion evaluates the proposed properties of food webs, based on analyses with insects or arthropods as a single category, in view of challenges based on greater resolution of arthropod diversity.

1. Food Chain Length

The length of food chains within food webs should be relatively short, at most 3–5 links (May, 1983; Pimm and Kitching, 1987; Pimm and Lawton, 1977), because the laws of thermodynamics predict energy limitation at higher trophic levels. Therefore, energy gain can be maximized by feeding lower on the food chain. At the same time, competition for prey is most severe at lower levels, perhaps restricting energy gains. Consequently, the trophic level selected by predators represents a trade-off between maximizing energy availability and minimizing competition. However, Polis (1991b) and Reagan *et al.* (1996)

found chain lengths of 6–19 links, using food webs with greater resolution in arthropod taxonomy. Reagan *et al.* (1996) reported a mean chain length of 8.6, double the length of chains found when arthropods are treated as a single category.

2. Trophic Loops

Loops, or reciprocal predation, in which two species feed on each other or a third species feeds on one and is eaten by the other, should be rare or absent because the size range of prey is constrained by physical limits and because loops potentially reduce population recovery following disturbance (Pimm, 1982; Pimm and Rice, 1987). Cannibalism is considered a "self-loop" (see Fox, 1975a).

Polis (1991b) and Reagan *et al.* (1996) reported the occurrence of a substantial number of loops, especially involving arthropods. In most cases, each species in the loop preys on juveniles of the other species. For example, in a tropical forest in Puerto Rico, adult centipedes prey on young toads, whereas adult toads prey on young centipedes. Polis (1991b) reported that several species of desert ants regularly prey on each other. Other predators constituted 9% of the overall diet of the aquatic heteropteran, *Notonecta hoffmanni*, studied by Fox (1975b). Longer loops involving up to four species have been observed (Reagan *et al.*, 1996). Reagan *et al.* (1996) found that 35% of 19,800 observed chains (corrected to exclude loops) include at least one species involved in at least one loop. Reciprocal predation may be more common than previously recognized and may complicate measurement of food chain length.

3. Food Web Connectance

Community connectance, the proportion of potential feeding relationships that actually occurs in the community (Pimm, 1982), has been found to increase with increasing species richness as

$$L = 0.14S^2, \tag{9.8}$$

where L is the number of links and S is the number of species (Martinez, 1992). This **constant connectivity hypothesis** predicts that, on average, each species will be involved in predator–prey interactions with 14% of the other species in the community. Havens (1992) analyzed 50 pelagic food webs with species richness ranging from 10–74 and found that the number of links per species increased four-fold over this range. Reagan *et al.* (1996) reported that the food web in a tropical forest in Puerto Rico supported constant connectance at low taxonomic resolution but that connectance dropped quickly as taxonomic resolution was increased. Polis (1991b) and Reagan *et al.* (1996) also found that the prediction that each species interacts with only 2–5 other species greatly underestimates the actual number of linkages per species and concluded that these properties are sensitive to taxonomic resolution.

4. Food Web Compartmentalization

Pimm and Lawton (1980) proposed that food webs should be compartmentalized between, but not within, habitats. Whereas the relatively distinct communities representing disturbed versus undisturbed patches within an ecosystem represent compartmentalization, the communities within habitat patches should not be compartmentalized. This property largely follows from the constant connectivity hypothesis, i.e., compartmentalization is inconsistent with equal linkage among species.

The vague definition of habitat complicates assessment of compartmentalization. For example, does soil/litter constitute a habitat or a subunit of the site habitat? Soil/litter subcommunities tend to be relatively distinct from plant-based above-ground subcommunities.

Nevertheless, compartmentalization can be identified within recognized habitats. Reagan *et al.* (1996) found distinct diurnal and nocturnal compartments in a tropical forest food web. Moore and Hunt (1988), Polis (1991b) and Reagan *et al.* (1996) found distinct compartmentalization within the community of a single patch when arthropod species or "kinds" were distinguished (Table 9.2). Compartmentalization reflects the development of component communities composed of specialists feeding on particular resources and the resulting channels of energy and material transfer. Host specificity appears to occur more frequently and at a finer spatial scale among herbivorous and detritivorous arthropods, based on their small size, short life spans, and intricate biochemical interactions (see Chapter 3) that facilitate rapid adaptation for utilization of particular resources, even within individual leaves (e.g., Mopper and Strauss, 1998; Parsons and de la Cruz, 1980). Many parasitoids also are host specific, so that compartmentalization is maintained at higher trophic levels among arthropods. Of course, generalists at all trophic levels connect compartments and maintain the web of interactions. Moore and Hunt (1988) found that compartmentalized models of food webs were more stable than noncompartmentalized webs.

5. Omnivory

Omnivores (defined as species feeding on more than one trophic level) should be rare (Pimm, 1982; Pimm and Rice, 1987). Pimm and Rice (1987) found that omnivory reduced the stability of food web interactions. However, Polis (1991b) and Reagan *et al.* (1996) reported that omnivory is common in food webs when arthropods are resolved to species or kinds. In fact, they found that most species fed at more than one trophic level, often from nonadjacent trophic levels, in desert and tropical rain forest communities.

6. Ratio of Basal to Top Species

Finally, ratios of species and links from basal to intermediate to top trophic levels (where basal species are prey only, intermediate species are prey and predators, and top predators have no predators) are expected to be constant

(Briand and Cohen, 1984). This implies a large proportion of top predators. Top predators are expected to comprise 29% of all species in a given community, and prey to predator ratios should be <1.0 (Briand and Cohen, 1984).

As shown for the properties discussed above, this property reflects poor resolution of arthropod diversity. Top predators appear to be common because they are easily distinguished vertebrate species, whereas poor taxonomic resolution at basal and intermediate levels underrepresents their diversity. Reagan *et al.* (1996) reported that in a rain forest food web that distinguished kinds of arthropods representation of basal and intermediate species was 30 and 70% of all species, respectively, and the proportion of top predators was < 1%. Polis (1991b) also reported that top predators are rare or absent in desert communities. Both Polis (1991b) and Reagan *et al.* (1996) reported that ratios of prey species to predator species are much greater than 1.0 when the true diversity of lower trophic levels is represented.

Although the properties of food webs identified by early theorists may be flawed to the extent that arthropod diversity has not been resolved adequately, they represent hypotheses that have stimulated considerable research into community organization. Future advances in food web theory will reflect efforts to treat arthropods at the same level of taxonomic resolution as other taxa.

C. Functional Organization

A third approach to community description is based on the guild, or functional group, concept (Cummins, 1973; Hawkins and MacMahon, 1989; Körner, 1993; Root, 1967; Simberloff and Dayan, 1991). The guild concept was originally proposed by Root (1967), who defined a guild as a group of species, regardless of taxonomic affiliation, that exploit the same class of environmental resources in a similar way. This term has been useful for studying potentially coevolved species that compete for, and partition use of, a common resource. The largely equivalent term "functional group" was proposed by Cummins (1973) to refer to a group of species having a similar ecological function. Insects, as well as other organisms, have been combined into guilds or functional groups based on similarity of response to environmental conditions (e.g., Coulson *et al.*, 1986; Fielding and Brusven, 1993; Grime, 1977; Root, 1973) or of effects on resources or ecosystem processes (e.g., Romoser and Stoffolano, 1998; Schowalter and Lowman, 1999; Siepel and de Ruiter-Dijkman, 1993). This method of grouping is one basis for pooling kinds of organisms, as discussed above.

Pooling species in this way has been attractive for a number of reasons (Root, 1967; Simberloff and Dayan, 1991). First, it reflects the compartmentalization of natural communities (see previous) and focuses attention on sympatric species that share an ecological relationship, e.g., competing for a resource or affecting a particular ecological process, regardless of taxonomic relationship. Second, it helped resolve multiple usage of the term "niche" to refer to both the functional role of a species and the set of conditions that determines its presence in the community. Use of guild or functional group to refer to species' ecological role(s) permits limitation of the term niche to refer to the conditions that determine species presence. Third, this concept facilitates comparative studies of communities which may share no taxa but do share functional groupings, e.g., herbivores, pollinators, and detritivores. Guild or functional groupings permit focus on a particular group, with specific functional relationships, among community types. Hence, researchers avoid the necessity of cataloging and studying all species represented in the community, a nearly impossible task, before comparison is possible. Functional groupings are particularly useful for simplifying ecosystem models to emphasize effects of functional groups with particular patterns of carbon and nutrient use on fluxes of energy and matter. Nevertheless, this method for describing communities has been used more widely among aquatic ecologists than among terrestrial ecologists.

The designation of functional groupings is largely a matter of convenience and depends on research objectives (e.g., Hawkins and MacMahon, 1989; Körner, 1993; Simberloff and Dayan, 1991). For example, defining "same class of resources" or "in a similar manner" is ambiguous. Each species represents a unique combination of abilities to respond to environmental conditions and to affect ecosystem processes, i.e., species within functional groups are similar only on the basis of the particular criteria used to distinguish the groups. Characterization of functional groups based on response to climate change, response to a disturbance gradient, effect on carbon flux, or effect on biogeochemical cycling would involve different combinations of species. Insects are particularly difficult to categorize because functional roles can change during maturation (sedentary larvae becoming mobile adults, aquatic larvae becoming terrestrial adults, herbivorous larvae becoming pollinating adults, etc.), and many species are too poorly known to assign functional roles. All Homoptera can be assigned to a plant sap-sucking functional group, but various species would be assigned to different functional groups on the basis of the plant part(s) affected (e.g., foliage, shoots, or roots, xylem or phloem). Clearly, functional groups can be subdivided to represent a diversity of responses to different gradients or subtle differences in ecological effects. For example, a ruderal or stress-adapted "functional group" could be divided into subgroups that tolerate desiccation, physiologically prevent desiccation, or avoid desiccation by feeding on plant fluids. Similarly, a foliage-feeder guild can be divided into subgroups that fragment foliage, mine foliage, or suck cellular fluids, feed on different plant species, etc., each subgroup affecting energy and matter fluxes in a different manner.

Species included in a particular functional group should not be considered redundant (Beare *et al.*, 1995; Lawton and Brown, 1993) but rather complementary, in terms of ensuring ecological functions. Schowalter *et al.* (1999) reported that each functional group defined on the basis of feeding type included species that responded positively, negatively, or nonlinearly to moisture availability. Species replacement within functional groups maintained functional organization over a moisture gradient.

Changes in the relative abundances or biomass of functional groups can signal changes in the rate and direction of ecological processes. For example, changes in the relative proportions of filter-feeder versus shredder functional groups in aquatic ecosystems affect the ways in which detrital resources are processed within the stream community and their contribution to downstream communities. Similarly, changes in the relative proportions of folivores versus sap-suckers affect the flux of nutrients as solid materials versus liquid (e.g., honeydew) and their effect on the detrital community (e.g., Schowalter and Lowman, 1999).

The functional group concept permits a convenient compromise in dealing with diversity, i.e., sufficient grouping to simplify taxonomic diversity while retaining an ecologically relevant level of functional diversity. Therefore, the functional group approach has become widely used in ecosystem ecology.

II. PATTERNS OF COMMUNITY STRUCTURE

A central theme of community ecology has been identification of patterns in community structure across environmental gradients in space and time (see also Chapter 10). The diversity of community types at landscape and regional scales has been a largely neglected aspect of biodiversity, but is important to the maintenance of regional species pools and metapopulation dynamics for many species. In addition, the mosaic of community types on a landscape may confer conditional stability to the broader ecosystem, in terms of relatively consistent proportions of community types over time (see Chapters 10 and 15).

Identification of patterns in community organization has become increasingly important to population and ecosystem management goals. Introduction of exotic insects to combat noxious pests (weeds or other insects) requires attention to the ability of the biocontrol agent to establish itself within the community and to its potential effects on nontarget components of the community. Efforts to conserve or restore threatened species require consideration and maintenance of the underlying community organization.

Depending on the descriptive approach taken (see previous), patterns have been sought in terms of species diversity, food web structure, or guild or functional group composition. Unfortunately, comparison of data among communities has been hampered by the different approaches used to describe communities, compounded by the variety of sampling techniques, with their distinct biases, used to collect community data. For example, sweep netting, light trapping, interception trapping, pitfall trapping, soil coring, canopy fumigation, and branch bagging are among the techniques commonly used to sample terrestrial arthropods. These techniques differ in their representation of nocturnal versus diurnal flying insects, arboreal versus soil/litter species, and sessile versus mobile species, etc. (e.g., Blanton, 1990; Majer and Recher, 1988; Southwood, 1978). Relatively few studies have used the same, or similar techniques, to provide comparative data among community types or locations. Some proposed patterns have been challenged, as subsequent studies provided more directly comparable data or increased resolution of arthropod taxonomy (e.g., Hawkins and MacMahon, 1998; Polis, 1991b; Reagan *et al.*, 1996). Disturbance history, or stage of postdisturbance recovery, also affects community structure (e.g., Wilson, 1969, see Chapter 10). However, the history of disturbance at sampled sites often is unknown, potentially confounding interpretation of differences in community structure. Nevertheless, apparent patterns identified at a variety of spatial scales may serve as useful hypotheses to guide future studies.

A. Global Patterns

Communities can be distinguished more easily on a taxonomic basis at a global scale than at smaller spatial scales, largely because of the distinct faunas among biogeographic realms (Wallace, 1876). However, global patterns of species richness, food web structure, and functional group organization have been related to effects of environmental suitability. In particular, a number of studies have indicated patterns related to latitudinal gradients in temperature and moisture and to the ecological history of adaptive radiation of particular taxa in particular places.

Latitudinal gradients in temperature and precipitation establish a global template of habitat suitability, as discussed in Chapters 2 and 7. Equatorial areas, characterized by high sun angle and generally high precipitation, provide favorable conditions of light, temperature, and moisture, although seasonal patterns of precipitation in some tropical areas create periods of adverse conditions for many organisms. The strongly seasonal climate of temperate zones requires specific adaptations for survival during seasonally unfavorable conditions, thereby limiting species diversity. The harsh conditions of temperate deserts and high latitude zones generally restrict the number of species that can be supported or that can adapt to these conditions.

Species richness generally decreases with latitude for a wide variety of taxa (Price, 1997; Stout and Vandermeer, 1975; Willig and Lyons, 1998). This gradient may be particularly steep for insects, which would be expected to show increasing species richness toward warmer latitudes. Although arctic and temperate arthropods are relatively well known, tropical arthropod faunas are poorly known. Several studies suggest that the tropics may support several million new species (Erwin, 1995; May, 1988; Wilson, 1992). This trend may not be reflected by all taxa (e.g., aphids; Dixon, 1985) or component communities (Vinson and Hawkins, 1998). Although diversity may be high in the tropics, densities may be low and make detection of many species difficult.

Vinson and Hawkins (1998) reviewed literature for stream communities and concluded that species richness is highly variable, and no strong latitudinal trends are apparent. Schowalter (1995b) compared arthropod diversities in wet temperate coniferous and wet tropical broadleaved forests and concluded that the greater species richness and diversity in the tropical forest was based on the greater number of tree species represented. The diversity of arthropod species computed for each tree species was comparable between the two sites.

Terborgh (1973) showed that the apparent trend in species richness with latitude may reflect increasing land area toward the equator. He noted that climate is relatively constant across a wide belt between 20° N and S latitudes but shows a distinct latitudinal gradient above 20° N and S latitudes. Combining climate and surface area gradients yielded a latitudinal gradient in habitat area available within each climate class, with a preponderance of global surface area in tropical habitat. These data support an alternative hypothesis that gradients in species richness reflect habitat area available for within-habitat speciation (see discussion following).

Latitudinal gradients in species richness may reflect greater primary productivity in the tropics (Rosenzweig and Abramsky, 1993; Tilman and Pacala, 1993; Waide *et al.*, 1999; see following). Willig and Lyons (1998) showed that latitudinal gradients also can result from chance.

Superimposed on the latitudinal gradients are the relatively distinct biogeographic realms identified by Wallace (1876). These biogeographic realms reflect the history of continental breakup, with southern floras and faunas largely distinct from northern floras and faunas. However, the southern continents show a varied history of reconnection with the northern continents that has resulted in invasion primarily by northern species. The proximity of North America and Eurasia has facilitated movement of species between these land masses, leading to development of a Holarctic species component, especially within the arctic and boreal biomes. Whereas many genera, and even some species, occur throughout the Holarctic realm, the flora and fauna of Australia have remained relatively distinct as a result of continued isolation.

Species richness also may be related to geological time. Wilson (1969) suggested that coevolution should improve the efficiency of total resource exploitation and lead to further increase in coexisting species over time. In other words, a habitat or resource that has persisted for a longer period of time would acquire more species than would a more recently derived habitat or resource. Birks (1980) found that the residence time of each tree species in Britain was strongly correlated with the diversity of its associated insect species. Tree species that had a longer history of occurrence in Britain hosted a larger number of species than did tree species with shorter residence times. Again, because residence time is correlated with area of occurrence (habitat area), the effects of these two factors cannot be distinguished easily (Price, 1997, see following).

B. Biome and Landscape Patterns

Patterns in species richness, food web structure and functional organization have been observed among biomes and across landscape patches. To some extent, patterns may reflect variation in occurrence or dominance of certain taxa in different biomes. Regional species pools may obscure effects of local habitat conditions on species richness (Kozár 1992a), especially in temperate ecosystems (Basset, 1996), but few ecologists have addressed the extent to which the regional species pool may influence local species richness. Various hypotheses have been proposed to account for apparent patterns at the biome and landscape level (e.g., Price, 1997; Tilman and Pacala, 1993). However, as more data have become available, some patterns have become equivocal.

General functional groups are common to all biomes, e.g., grazing herbivores (depending on degree of autochthonous primary production in streams), predators, parasites, and detritivores, whereas other functional groups depend on particular resources being present, e.g., sap-suckers require vascular plants and wood borers require wood resources. Proportions of the fauna representing the different functional groups vary among biomes. Low order streams have primarily detrital-based resources and their communities are dominated by detritivores and their associated predators and parasites. Other communities represent various proportions of autotroph functional groups (e.g., chemoautotrophs, ruderal, competitive, and stress-tolerant vascular versus nonvascular plants) and heterotroph functional groups (herbivores, predators, detritivores).

Different species compose these functional groups in different biomes. For example, the folivore functional group is composed primarily of moths, beetles, and tree crickets in broadleaved forests, moths and sawflies in coniferous forests (Schowalter, 1994a, 1995a; Schowalter et al., 1981c), grasshoppers in grasslands and shrublands (Curry, 1994), and caddisflies and flies in aquatic communities (e.g., Hart, 1992). The predator functional group in terrestrial arthropod communities is dominated by a variety of arachnids, beetles, flies, and wasps, whereas in aquatic arthropod communities this functional group is dominated by dragonflies, true bugs, and beetles.

Among terrestrial biomes, species richness generally is assumed to increase from harsh biomes (e.g., tundra and desert) to grassland to forest, again reflecting differences in physical complexity, suitability, and stability of the habitat (Bazzaz, 1975; Tilman and Pacala, 1993). However, this trend is not apparent for arthropods among communities where extensive species inventories are available (Table 9.1). Species richness is not always linearly related to primary productivity and patterns likely depend on scale (Rosenzweig and Abramsky, 1993; Tilman and Pacala, 1993; Waide et al., 1999). Species richness often declines above intermediate levels of productivity, perhaps because more productive communities are dominated by larger individuals that reduce habitat heterogeneity or because more productive and stable communities favor competitive exclusion of some species by the best adapted species (Tilman and Pacala, 1993). For example, continuous fertilization of permanent pasture at Rothamsted, UK, since 1856 resulted in changes in species rank-abundance pattern from a log normal curve in 1856 to progressively more geometric curves by 1949 (Fig. 9.3) (Kempton, 1979).

Functional group composition has not shown consistent differences among biomes (Hawkins and MacMahon, 1989; Stork, 1987). Detritivores represent a relatively greater proportion of the community in boreal forests and other biomes characterized by accumulated organic material and a lower proportion in tropical forests, deserts, and other biomes with little organic matter (Seastedt, 1984). Wood borers occur only in forest or large shrub ecosystems with abundant wood resources. Pollinators are more diverse in tropical forests where plant diversity has led to greater reliance on insect and vertebrate pollinators, compared to temperate and arctic biomes. Proportional representation of species and individuals among functional groups varies widely among canopy arthropod communities in temperate and tropical forests, depending on tree species composition (Fig. 9.8) (Moran and Southwood, 1982; Schowalter and Ganio, 1998, 1999; Stork, 1987).

At the landscape or drainage basin scale, patterns in species richness and functional group organization can be related to local variation in physical conditions. The history and geographic pattern of disturbance may be particularly important factors affecting variation in community structure. Polis *et al.* (1997) concluded that the movement of organisms and resources among the interconnected community types composing a landscape can contribute to the organization of the broader landscape community by subsidizing more resource-limited local communities. However, Basset (1996) found that diversity in tropical rain forest trees was related to five factors: numbers of young leaves available throughout the year, ant abundance, leaf palatability, leaf water content, and altitudinal range. These data suggested that local factors may be more important determinants of local species diversity and community structure in complex ecosystems, such as tropical forests, than in less complex ecosystems, such as temperate forests.

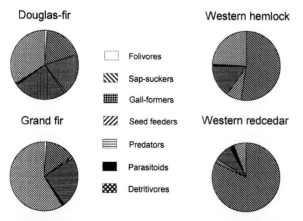

FIG. 9.8 Functional group organization of arthropod communities in canopies of four old-growth conifer species at the Wind River Canopy Crane Research Facility in southwestern Washington. Data from Schowalter and Ganio (1998).

Vinson and Hawkins (1998) found six studies that compared species richness of stream insects over drainage basins. Species diversity varied with elevation, which covaried with a number of important factors, such as stream morphology, flow rate and volume, riparian cover, and agricultural or urban land use. In one study, Carter *et al.* (1996) used multivariate analyses (TWINSPAN) to compare species composition among 60 sites representing first-order (characterized by narrow V-shaped channel, steep gradient, nearly complete canopy cover) to sixth-order (characterized by wide channel, low gradient, little canopy cover) streams over a 15,540 km^2 drainage basin. They identified five communities distinguished largely by elevation. The highest species richness occurred in mid-order, mid-elevation streams that included species groups characterizing valley or montane sites.

Similarly, the transition zones (ecotones) between terrestrial community types typically have higher species richness because they include species from each of the neighboring communities. Ecotones can move across the landscape as environmental conditions change. For example, the northern edge of Scots pine, *Pinus sylvestris*, forest in Scotland moved rapidly 70–80 km northward about 4000 years BP, then retreated southward again about 400 years later (Gear and Huntley, 1991). Sharp edges between community types, such as result from land use practices, reduce the value of this ecotone as a transition zone.

Patches representing different stages of postdisturbance recovery show distinct patterns of species richness, food web structure, and functional group organization (see Chapter 10). Species richness typically increases during community development up to an equilibrium, perhaps declining somewhat prior to reaching equilibrium (e.g., MacArthur and Wilson, 1967; Wilson, 1969). As the number of species increases, the number of species interactions increases. Food chains that characterize simpler communities develop into more complex food webs (Wilson, 1969). Schowalter (1994a, 1995a) and Schowalter *et al.* (1981c) found that patches of recently disturbed temperate and tropical forests were characterized by higher sap-sucker/folivore ratios than were patches of undisturbed forests, even when data were reported as biomass.

Shure and Phillips (1991) reported that species richness and functional group composition are modified by patch size (Fig. 6.4). Species richness was lowest in midsized canopy openings (0.08–0.4 ha). Herbivore guilds generally had lowest biomass in midsized canopy openings; omnivore biomass peaked in the smallest openings (0.016 ha) and then declined as opening size increased; predator biomass was highest in the control forest and smallest openings and lowest in the midsized openings; and detritivore biomass was similar among most openings but much lower in the largest openings (10 ha). This pattern may indicate the scale that distinguishes communities characterizing closed-canopy and open-canopy forest. Smaller openings were influenced by surrounding forest, whereas larger openings favored species more tolerant of exposure and altered plant conditions, e.g., early successional species and higher phenolic concentrations (Dudt and Shure, 1994; Shure and Wilson, 1993). Intermediate-sized openings may be too exposed for forest species but insufficiently exposed for earlier successional species. However, species richness generally increases with habitat area (Fig. 9.9) (Johnson and Simberloff, 1974; MacArthur and Wilson, 1967), for reasons that will be discussed.

III. DETERMINANTS OF COMMUNITY STRUCTURE

A number of factors affect community structure (e.g., Price, 1997). Factors associated with habitat area, resource availability, and species interactions appear to have the greatest influence.

A. Habitat Area and Complexity

The relationship between number of species and sampling effort, in time or space, has been widely recognized. The increase in number of species with increasing number of samples reflects the greater representation of the community. Similarly, a larger habitat area will "sample" a larger proportion of a potential species pool. Increasing habitat area also tends to represent increasing heterogeneity of habitat conditions (e.g., Johnson and Simberloff, 1974; Strong *et al.*, 1984), providing an increasing number of niches.

In developing the **Theory of Island Biogeography**, MacArthur and Wilson (1967) emphasized the relationship between species richness (*S*) and island area (*A*), expressed as

$$S = Ca^z \tag{9.9}$$

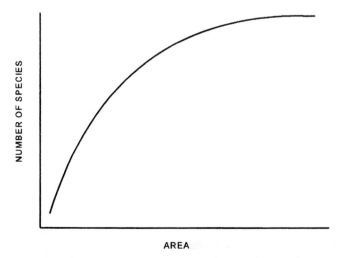

FIG. 9.9   Relationship between species richness and geographic area.

where C depends on the taxon and biogeographic region and z is a parameter that varies little among taxa or biogeographic regions, generally falling in the range 0.20–0.35 (Fig. 9.9). The value of z increases with habitat heterogeneity and proximity to the mainland. For nonisolated sample areas within islands or within continental areas, the relationship between species number and sample area is similar, but z is smaller, generally 0.12–0.17 (MacArthur and Wilson, 1967).

Habitat area has continued to be viewed as a primary factor affecting species richness, likely influencing apparent gradients in species richness with latitude and host residence time (e.g., Birks, 1980; Price, 1997; Terborgh, 1973), as discussed previously. However, habitat area also is a surrogate for habitat heterogeneity. Larger islands are more likely than smaller islands to represent a wider range in elevation, soil types, aspects, etc. Similarly, larger continental areas are more likely than smaller areas to represent a range of habitat conditions. Because relatively distinct component communities develop on particular resources, such as plant or microbial species (e.g., Moore and Hunt, 1988; Schowalter, 1995b), species richness increases exponentially as representation of resource diversity increases. Furthermore, habitat heterogeneity provides for refuges from competition and/or predation, i.e., local patches of competition- or predator-free space. The architectural complexity of individual plants also can affect the diversity of associated fauna (Lawton, 1983).

B. Habitat Stability

Habitat stability determines the length of time available for community development (see Chapter 10). Wilson (1969) proposed four stages in community development. The **noninteractive stage** occurs on newly available habitat or immediately following a disturbance, when numbers of species and population sizes are low. As species number rises during the **interactive stage**, competition and predation influence community structure, with some species disappearing and new species arriving. The **assortative stage** is characterized by persistence of species that can coexist and utilize resources most efficiently, facilitating species packing. Finally, the **evolutionary stage** is characterized by coevolution that increases the efficiency of overall utilization and species packing. Community development in frequently disturbed habitats cannot progress beyond earlier stages, whereas more stable habitats permit advanced community development and increased species richness. However, the most stable habitats also allow the most adapted species to preempt resources from other species, leading to a decline in species richness (see following). This trend has led to the development of the **Intermediate Disturbance Hypothesis**, which predicts that species richness peaks at intermediate levels of disturbance (e.g., Connell, 1978; Sousa, 1985; but see Reice, 1985). Community recovery from disturbance is described more fully in Chapter 10.

C. Resource Availability

As discussed above, the availability of particular resources determines the presence of associated species. If a limiting resource (host) becomes more abundant, then associated species also become more abundant, until some other factor(s) become limiting, but some species compete more effectively for a resource when it occurs at higher abundance.

Limiting resources may preclude any single adaptive strategy from becoming dominant and thereby maintain high species richness. Rosenzweig and Abramsky (1993), Tilman and Pacala (1993), and Waide *et al.* (1999) concluded that species richness is not always linearly related to productivity. Intermediate levels of productivity often support the highest diversity because higher productivity favors dominance by the most competitive species.

A number of studies have compared species richness between relatively homogeneous and heterogeneous environments (e.g., Cromartie, 1975; Risch, 1980, 1981; Root, 1973; Strong *et al.*, 1984; Tahvanainen and Root, 1972). Because organisms have greater difficulty maintaining energy and nutrient balance when resources are scattered (see Chapter 4), the abundance of individual species generally decreases with increasing resource heterogeneity, precluding exclusive use of the niche and permitting species richness to increase. By contrast, homogeneous resources facilitate rise to competitive dominance by the best-adapted species, leading to reduced species richness. Extensive planting of agricultural or silvicultural monocultures establishes the conditions necessary for some species to reach epidemic population levels across landscapes (see Chapter 7), reducing availability of resources shared with other species but providing prey resources for predators (Polis *et al.*, 1997; see also Chapter 8).

D. Species Interactions

Species interactions can enhance or preclude persistence of some species, as discussed in Chapter 8. As has been noted, species populations cannot persist where their host species are absent. However, the presence of competitors, predators, and mutualists also affects persistence of associated species.

Some species can have particularly profound effects on community structure. Their presence in a community leads to a different community structure than occurs in their absence. A top predator that preferentially preys on the most abundant of several competing prey species can prevent any single species from competitively suppressing others. Paine (1966, 1969a,b) considered such species to be **keystone species**. Bond (1993) and Power *et al.* (1996) applied this term to any species that have effects on ecosystem structure or function disproportionate to their abundance or biomass.

Some insect species could be considered to be keystone species to the extent that their abundance greatly alters diversity, productivity, rates of energy or nutrient flux, etc. Many herbivorous insects increase the diversity of plant species by selectively reducing the density of abundant host species and providing space and resources for nonhost plants (Lawton and Brown, 1993; Schowalter and Lowman, 1999). The southern pine beetle, *Dendroctonus frontalis*, is capable, at high population densities, of killing pine trees and increasing the availability of woody resources that maintain populations of other xylophagous species (Flamm *et al.*, 1993). Termites and ants affect soil structure and fertility in ways that determine vegetation development (see Chapters 13 and 14).

As discussed in Chapter 6, the combination of bottom-up (resource supply) and top-down (trophic cascades) factors tends to stabilize population levels. Changes in abundance of any trophic level, however, affect abundances at other trophic levels. Generally, increased abundance at one trophic level increases resources available to the next higher trophic level, increasing abundance at that level, but reducing abundance at the next lower level. Reduced abundance at the lower trophic level reduces its control over the second lower trophic level, which increases in abundance and reduces abundance at the third lower trophic level. Trophic cascades are commonly observed in lake ecosystems (Carpenter and Kitchell, 1984, 1987, 1988; Letourneau and Dyer, 1998; Vanni and Layne, 1997). Although examples of trophic cascades controlled by top predators are rare in terrestrial ecosystems, Letourneau and Dyer (1998) found a potential example in neotropical rain forest communities where a beetle, *Phyllobaenus* sp., was an effective top predator. Beetle predation on ants reduced ant abundance and increased herbivore abundance on *Piper* ant-plants. Where this beetle was absent and spiders were a less effective top predator, ant abundance was higher and reduced herbivore abundance.

IV. SUMMARY

Communities are composed of the species occupying a site. Identification of patterns in community structure has been a major goal of ecological research. However, no standard approach for delimiting a site and describing or comparing community structure has been adopted. Indices of species diversity, food web structure and functional group organization are three methods used to facilitate comparison among communities.

Species diversity has two components: richness and evenness. Richness is the number of species in the community, whereas evenness is a measure of relative abundances. These two components can be represented by rank–abundance curves and by diversity indices. Geometric rank–abundance curves characterize harsh or disturbed habitats with a limited number of adapted species and strong dominance hierarchy, whereas log and broken stick models characterize more stable habitats with higher species accumulation and greater evenness in abundance among species. A number of diversity indices and similarity indices have been developed to integrate richness and evenness in a variable that can be compared among community types.

Food web structure represents the network of pairwise interactions among the species in the community. A number of food web attributes have been proposed, based on limited taxonomic resolution of insects and other arthropods, that are being challenged as greater resolution of arthropod taxonomy reveals networks of interactions within this diverse group.

Functional group organization reflects combination of species on the basis of functional responses to environmental variables or effects on ecological processes, regardless of taxonomic affiliation. This approach has become popular because it simplifies species diversity in an ecologically meaningful way. However, the allocation of species to functional groups is based on particular objectives and is, therefore, arbitrary to the extent that each species represents a unique combination of functional responses or effects.

The noncomparable descriptions of communities based on these three approaches, compounded by the variety of arthropod sampling techniques, each with its unique biases, has hindered comparison of community structure among habitat types. Biogeographic patterns based on limited data often are inconclusive, and the apparent underlying factors often are correlated. Nevertheless, many taxa show latitudinal gradients in abundance, with species richness increasing toward the equator. Some groups are more diverse within biogeographic realms of origin or where resources have been available over longer time periods. Some functional groups are more abundant in certain biomes, e.g., pollinators in diverse tropical habitats, detritivores and wood borers in habitats with greater organic matter or wood accumulation.

Habitat area and stability, resource availability, and species interactions are major factors that affect community structure. Habitat area affects the pool of species available and the heterogeneity of habitat conditions and resources. Habitat stability determines the length of time available for species accumulation, assortment, and species packing. Species richness generally increases with resource availability, up to a point at which the most adapted species competitively suppress other species. Species interactions often affect persistence in a particular habitat. Colonists cannot survive unless their host resources are available. Competition, predation, and mutualism also affect species' ability to persist. Keystone species have effects on community structure or ecosystem processes that are disproportionate to their numbers or biomass. Keystone species include predators that focus on the most abundant prey species, thereby reducing competition among prey species and maintaining more species than would coexist in the absence of the predator. Many herbivorous insects function as keystone species by selectively reducing abundance of dominant host species and facilitating persistence of nonhosts.

Community Dynamics

COMMUNITY STRUCTURE CHANGES THROUGH TIME AS SPECIES ABUN-
dances change, altering the network of interactions. Short-term (e.g., seasonal
or annual) changes in community structure represent responses to environmental changes that favor some species or affect interaction strength (see Chapter 8).
Longer-term (e.g., successional) changes in community structure often reflect
relatively predictable trends during community development on newly available
or disturbed sites. Finally, changes in community structure over evolutionary
time reflect responses to long-term trends in environmental conditions.

Among the major environmental issues facing governments worldwide is
the effect of anthropogenic activities (e.g., altered atmospheric or aquatic chemistry, land use, species redistribution) on the composition of natural communities and the ecosystem services they provide to humans. How stable is community structure, and how sensitive are communities and ecosystems to changes in
species composition? Our perception of communities as self-organizing entities
or random assemblages has significant implications for our sensitivity to species
loss and our approach to management of ecosystem resources.

As with population dynamics, study of changes in community structure require long periods of observation. Few studies have continued over sufficiently
long time periods to evaluate many of the factors presumed to affect community structure. However, paleoecological evidence and studies of community recovery following disturbance have provided useful data. Research on factors af-

fecting community structure over a range of temporal scales can enhance understanding of the degree of stability in community structure and anticipation of responses to environmental changes.

I. SHORT-TERM CHANGE IN COMMUNITY STRUCTURE

Community structure changes over relatively short time periods. Short-term variation in community structure reflects interactions among species responding differentially to fluctuating abiotic conditions and species interactions.

Fluctuating climatic conditions can cause appreciable changes in arthropod species abundances, interactions, and community structure. For example, Schowalter *et al.* (1999) found that particular arthropod species, as well as the entire arthropod community, associated with a desert shrub, creosote bush, *Larrea tridentata*, in southern New Mexico showed distinct trends over an experimental gradient in precipitation volume. Abundances of many species increased with moisture availability, whereas abundances of a few species declined with moisture availability, and some species showed nonlinear or nonsignificant responses. Multivariate analysis indicated distinct community structures on plants subjected to different amounts of precipitation.

Similarly, as shown in Fig. 10.1, the relative abundances of the dominant folivore and sap-sucker species changed over time in conifer canopies in western Oregon. In particular, western spruce budworm, *Choristoneura occidentalis*, sawflies, *Neodiprion abietis*, and aphids, *Cinara* spp., were abundant only during a drought period, 1987–1993. The budmoth, *Zeiraphera hesperiana*, was virtually absent during this period but was a dominant species during wet years.

Factors that increase competition or predation can reduce population sizes of particular species. Some species may become locally extinct, whereas others show population irruptions. Changes in species abundances affect species interactions. Interaction strength can change greatly. Herbivores that interact with their hosts in a commensal, or even mutualistic, manner at low abundances interact in a more predatory manner at high abundances. Reduced abundance of one mutualist can jeopardize the persistence of the other.

Changes in species composition and abundance alter species diversity, food web structure and species function. Change in abundance of species at one trophic level can affect the diversity and abundance of species at lower trophic levels through trophic cascades. For example, reduced predator abundance typically increases herbivore abundance, thereby decreasing plant abundance (Carpenter and Kitchell, 1987, 1988; Letourneau and Dyer, 1998).

II. SUCCESSIONAL CHANGE IN COMMUNITY STRUCTURE

Relatively predictable changes in community structure occur over periods of decades to centuries as a result of succession on newly exposed or disturbed

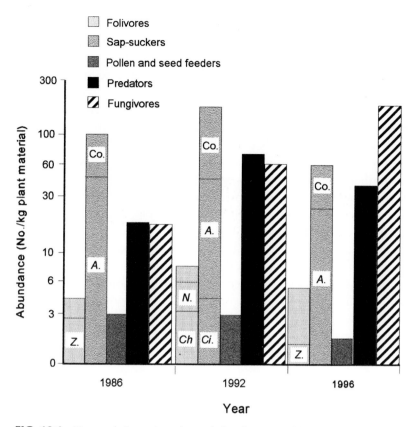

FIG. 10.1 Temporal change in arthropod abundances in old-growth Douglas-fir, *Pseudotsuga menziesii*, canopies at the H.J. Andrews Experimental Forest in western Oregon; 1989 and 1996 were relatively wet years; 1992 was in the middle of an extended drought period (1987–1993). Z., *Zeiraphera hesperiana*; Ch., *Choristoneura occidentalis*; N., *Neodiprion abietis*; Ci., *Cinara* spp.; A., *Adelges cooleyi*; Co., Coccoidea (4 spp.). Note the log scale of abundance. Data from Schowalter (1989, 1995a, and unpublished data).

sites. New habitats become available for colonization as a result of tectonic activity, glacial movement, sea level change, and sediment deposition or erosion. Species colonizing newly exposed surfaces typically are small in stature, tolerant of exposure or able to exploit small shelters, and able to exploit nonorganic or exogenous resources. Disturbances to existing communities affect each species differently, depending on its particular tolerances to disturbance or postdisturbance conditions (see Chapter 2). Often, legacies from the predisturbance community (such as buried rhizomes, seed banks, woody litter, and animal survivors in protected microsites) remain following disturbance and influence the trajectory of community recovery.

The process of community development on disturbed or newly exposed sites is called **ecological succession**. The succession of populations and communities on disturbed or newly exposed sites has been a unifying concept in ecology since the time of Cowles (1911) and Clements (1916). These early ecologists viewed succession as analogous to the orderly development of an organism (ontogeny). Succession progressed through a predictable sequence of stages (seres), driven by biogenic processes, that culminated in a self-perpetuating community (the climax) determined by climatic conditions. Succession is exemplified by the sequential colonization and replacement of species: weedy annual to perennial grass to forb, to shrub, to shade-intolerant tree and, finally, to shade-tolerant tree stages on abandoned cropland. Succession following fire or other disturbances shows a similar sequence of stages (Fig. 10.2).

Although the succession of species and communities on newly exposed or disturbed sites is one of the best-documented phenomena in ecology, the nature of the community and mechanisms driving species replacement have been debated intensely from the beginning. Gleason (1917, 1926, 1927) argued that succession is not directed by autogenic processes, but reflects population dynamics of individual species based on their adaptations to changing environmental conditions. Egler (1954) further argued that succession could proceed along many potential pathways, depending on initial conditions and initial species pools. Odum (1969) integrated the Clementsian model of succession with ecosystem processes by proposing that a number of ecosystem properties,

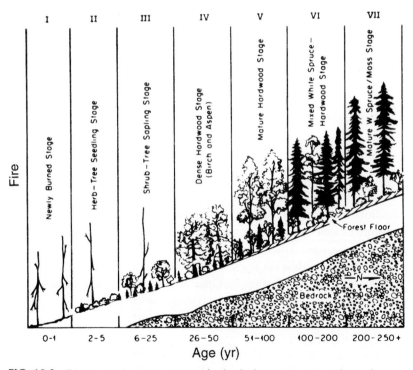

FIG. 10.2 Diagrammatic representation of upland white spruce, *Picea glauca*, forest succession in Alaska following fire. From Van Cleve and Viereck (1983) by permission from Springer-Verlag.

including species diversity, primary productivity, biomass, and efficiency of energy and nutrient use, increase during succession. Drury and Nisbet (1973) viewed succession as a temporal gradient in community structure, similar to the spatial gradients discussed in Chapter 9, and argued that species physiological tolerances to environmental conditions were sufficient to explain species replacement. More recently, the importance of disturbances and heterotroph activity in determining successional processes and preventing ascension to the climatic climax has been recognized (e.g., MacMahon, 1981; Ostfeld *et al.*, 1997; Pickett and White, 1985; Schowalter, 1981, 1985; Willig and Walker, 1999).

The concept of succession as goal-oriented toward a climax has succumbed to various challenges, especially recognition that succession can progress along various pathways to nonclimatic climaxes under different environmental conditions (Whittaker, 1953). Furthermore, the mechanism of species replacement is not necessarily facilitated by the replaced community (e.g., Botkin, 1981; Connell and Slatyer, 1977; Horn, 1981; McIntosh, 1981; Peet and Christensen, 1980; Whittaker, 1953, 1970). Nevertheless, debate continues over the integrity of the community, the importance of autogenic factors that influence the process, and the degree of convergence on particular community composition (Bazzaz, 1990; Peet and Christensen, 1980; Glenn-Lewin *et al.*, 1992; West *et al.*, 1981).

A. Patterns of Succession

Two types of succession can be recognized. **Primary succession** occurs on newly exposed substrates, e.g., lava flows, uplifted marine deposits, dunes, newly deposited beaches. Primary succession typically involves a long period of soil formation and colonization by species requiring little substrate modification. **Secondary succession** occurs on sites where the previous community was disturbed and is influenced by remnant substrate and surviving individuals. Although most studies of succession have dealt with trends in vegetation, heterotrophic successions, including successions dominated by insects or other arthropods, have contributed greatly to perspectives on the process. Insects and other arthropods dominate the development of freshwater communities and litter (especially woody litter and carrion) communities, and succession in these habitats occurs over shorter time scales than does succession involving longer-lived plant species.

Succession varies in duration from weeks for communities with little biomass (e.g., carrion feeders) to centuries for communities with abundant biomass (e.g., forests). Shorter successions are amenable to study by individual researchers. However, forest or desert succession spans decades to centuries and has not been studied adequately throughout its duration (Fig. 10.2). Rather, forest succession typically has been studied by selecting plots of different age since disturbance or abandonment of management to represent various seres (i.e., the chronosequence approach). Although this approach has proven convenient for comparing and contrasting various seres, it fails to account for effects of differences in initial conditions on subsequent species colonization and turnover processes (e.g., Egler, 1954; Schowalter *et al.*, 1992). Even Clements (1916) noted that comparison of the successional stages is less informative than is evaluation of the factors eliciting the transitions between stages. However, this approach requires establishment of long-term plots protected from confounding activities and a commitment by research institutions to continue studies beyond the usual confines of individual careers. Characterization of succession is a major goal of the network of U.S. and International Long-Term Ecological Research (LTER) sites (e.g., Van Cleve and Martin, 1991). Long-term and comparative studies will improve understanding of successional trajectories and their underlying mechanisms.

A number of trends have been associated with vegetation succession. Generalists or r-strategists generally dominate early successional stages, whereas specialists or K-strategists dominate later successional stages (Table 10.1; Fig. 10.2) (Boyce, 1984; Brown, 1984, 1986; Brown and Hyman, 1986; Brown and Southwood, 1983; Grime, 1977; Janzen, 1977; Strong *et al.*, 1984; see Chapter 5). Species richness typically increases during early to mid succession but reaches a plateau or declines during late succession (Peet and Christensen, 1980; Whittaker, 1970), a pattern similar to the spatial gradient in species richness across ecotones (Chapter 9).

Wilson (1969), based in part on data from Simberloff and Wilson (1969), suggested that community organization progresses through four stages: noninteractive, interactive, assortative, and evolutionary. The noninteractive stage occurs early during succession (first decade), when species richness and population densities are too low to induce density-dependent competition, predation, or parasitism. As species number and densities increase, interaction strength increases and produces a temporary decline or equilibrium in species number, as some species are excluded by competition or predation. The assortative stage occurs overs long disturbance-free time periods, as a result of species persistence in the community on the basis of efficient resource use and coexistence. Niche partitioning allows more species to colonize and persist. Finally, coevolution over very long time periods increases the efficiency of interaction and permits further increase in species number. However, most communities are disturbed before reaching the assortative stage. Species richness typically peaks at intermediate stages of succession, when both early and late successional species are represented. Indeed, the Intermediate Disturbance Hypothesis predicts that species richness is maximized through intermediate levels of disturbance that maintain a combination of early and late successional species (Connell, 1978; Sousa, 1985).

Arthropod communities also change during vegetative succession (Table 10.1) (Brown, 1984; Shelford, 1907; Weygoldt, 1969). Schowalter (1994a, 1995a) and Schowalter and Crossley (1988) reported that sap-sucking insects (primarily Homoptera) dominate early successional temperate and tropical

TABLE 10.1 Life History Strategies of Insects from Different Successional Stages

Characteristic	Successional stage				Source
	Ruderal 0–1 years	Early 1–5 years	Mid 7–11 years	Late 60+ years	
Mobility (% fully winged species)	94	84	80	79	Heteroptera (Brown, 1982)
Generation time (% species >1 generation/yr)	43	50	33	3	Exopterygote herbivores (Brown and Southwood, 1983)
	41	37	10	12	Heteroptera (Brown, 1982)
Size (mean body length, mm, ± SEM)	3.68 ± 0.57	3.59 ± 0.63	3.86 ± 0.63	4.14 ± 0.67	All insect species (Brown, 1986)
Reproductive potential (mean number of embryos ± SEM)		70.0 ± 4.4[a]		50.2 ± 2.0[b]	Aphids (Brown and Llewellyn, 1985)
Niche breadth (scale 1–5; 1, highly specialized)	3.35	3.10	2.87	1.79	Sap feeders (Brown and Southwood, 1983)
	1.60	1.29	1.33	3.05	Weevils (Brown and Hyman, 1986)

[a]On herbaceous plants.
[b]On woody plants.
Updated from Brown (1984) by permission from V.K. Brown and the American Institute of Biological Sciences, © 1984 American Institute of Biological Sciences.

forests, whereas folivores, predators, scavengers, and fungivores dominate later successional forests. This trend likely reflects the abundance of young, succulent tissues with high translocation rates that favor sap-suckers and tending ants during early regrowth.

Brown and Southwood (1983) reported a similar trend toward increased representation of predators, scavengers, and fungivores in later successional stages. They noted, in addition, that species richness of herbivorous insects and plants were highly correlated during the earliest successional stages but not later successional stages, whereas numbers of insects and host plants were highly correlated at later stages but not the earliest successional stages. Brown and Southwood (1983) suggested that early colonization by herbivorous insects depends on plant species composition, but that population increases during later stages depend on the abundance of host plants (see also Chapters 6 and 7). Torres (1992) reported that several species of Lepidoptera reached outbreak levels on early successional host plants in the months following Hurricane Hugo (1989) in Puerto Rico, but disappeared after depleting their resources. Schowalter (unpublished data) observed this process repeated following Hurricane Georges (1998). Schowalter (1981) and Schowalter and Lowman (1999) suggested that insect outbreaks reflect elevated abundances and suitability of hosts that are increasingly stressed by competition. Insect-induced mortality accelerates plant replacement by nonhost plants, i.e., the next sere.

Punttila *et al.* (1994) reported that the diversity of ant species declined during forest succession in Finland. Most ant species were found in early successional stages, but only the three species of shade-tolerant ants were common in old (>140 years old) forests. They noted that forest fragmentation favored species that require open habitat by reducing the number of forest patches with sufficient interior habitat for more shade-tolerant species.

Starzyk and Witkowski (1981) examined the relationship between bark- and wood-feeding insect communities and stages of oak-hornbeam forest succession. They found the highest species richness in older forest (>70 years old) with abundant dead wood and in recent clear-cuts with freshly cut stumps. Densities of mining larvae also were highest in the older forest and intermediate in the recent clear-cut. Intermediate stages of forest succession supported fewer species and lower densities of bark- and wood-feeding insects. These trends reflected the decomposition of woody residues remaining during early stages and the accumulation of woody debris again during later stages.

Heterotrophic successions have been studied in decomposing wood and animal carcasses. Both processes can be divided into distinct stages characterized by relatively discrete heterotrophic communities.

In general, succession in wood is initiated by the penetration of the bark barrier by bark and ambrosia beetles (Scolytidae and Platypodidae) at, or shortly after, tree death (Ausmus, 1977; Dowding, 1984; Savely, 1939; Swift, 1977; Zhong and Schowalter, 1989). These beetles inoculate galleries in fresh wood (decay class I) with a variety of symbiotic organisms (e.g., Schowalter *et al.*,

1992; Stephen *et al.*, 1993; see Chapter 8) and provide access to interior substrates for a diverse assemblage of saprotrophs and their predators. The bark
and ambrosia beetles remain only for the first year but are instrumental in penetrating bark, separating bark from wood, and facilitating drying of wood (initiating decay class II). These insects are followed by wood boring beetles, wasps,
and their associated saprophytic microorganisms (Chapter 8). Powderpost and
other beetles, carpenter ants, *Canponotus* spp., or termites dominate the later
stages of wood decomposition (decay classes III–IV), depending on wood conditions (especially moisture content) and proximity to population sources.
Wood becomes increasingly soft and porous, and holds more water, as decay
progresses. These insects and associated bacteria and fungi complete the decomposition of wood and incorporation of recalcitrant humic materials into the
forest floor (decay class V).

Insect species composition follows characteristic successional patterns in
decaying carrion (Fig. 10.3 and 10.4), with distinct assemblages of species occurring during each stage: fresh, bloated, decay, dry, and remains (Payne, 1965;
Tantawi *et al.*, 1996; Tullis and Goff, 1987). For small animals, various carrion
beetles initiate the successional process by burying the carcass prior to oviposi-
sition. For all animal carcasses, the fresh, bloated, and decay stages are dominated by various Diptera, especially calliphorids, whereas later stages are dominated by Coleoptera, especially dermestids. The duration of each stage depends
on environmental conditions that affect the rate of decay (compare Figs. 10.3
and 10.4) (Tantawi *et al.*, 1996) and on predators, especially ants (Tullis and
Goff, 1987; Wells and Greenberg, 1994). This distinct sequence of insect community types, as modified by local environmental factors, has been applied by
forensic entomologists to determine time since death.

B. Factors Affecting Succession

Succession should progress toward the community type characteristic of the
biome within which it occurs, e.g., toward deciduous forest within the deciduous forest biome or toward chaparral within the chaparral biome (e.g., Whittaker 1953, 1970). However, succession can progress along various alternative
pathways and reach alternative endpoints [such as beech (*Fagus*) forest, maple
(*Acer*) forest, or hemlock (*Tsuga*) forest within the eastern deciduous forest of
the northeastern U.S.], depending on a variety of local abiotic and biotic factors. Substrate conditions represent an abiotic factor that selects a distinct sub-
set of the regional species pool determined by climate. Distinct initial communities reflecting disturbance conditions or unique conditions of local or regional
populations can affect the success of subsequent colonists. These initial conditions can guide succession into alternative pathways leading to distinct self-perpetuating endpoints (Whittaker, 1953; Egler, 1954). Recurrence of disturbance
can truncate community development, and the sequential pattern of disturbance types can direct succession along alternative pathways.

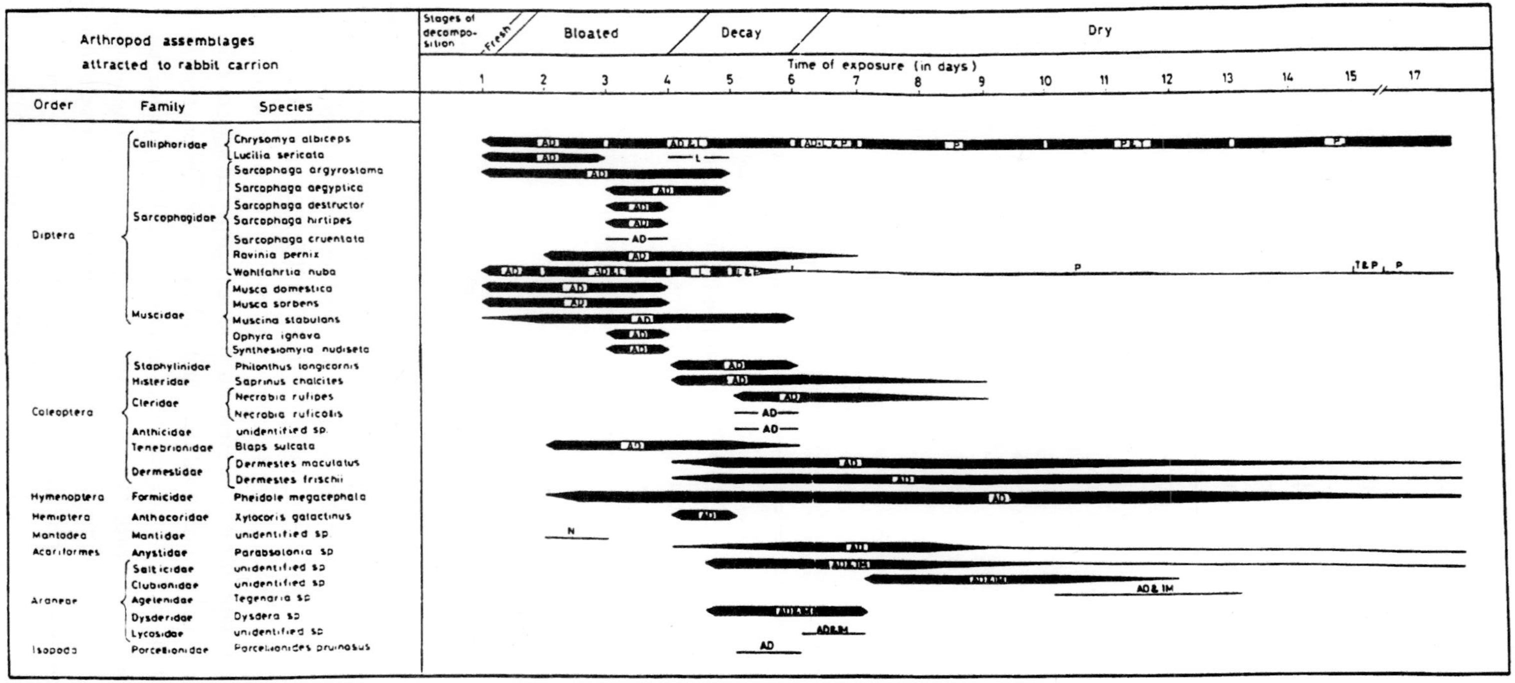

FIG. 10.3 Succession of arthropods on rabbit carrion during summer in Egypt. From Tantawi *et al.* (1996) by permission from the Entomological Society of America.

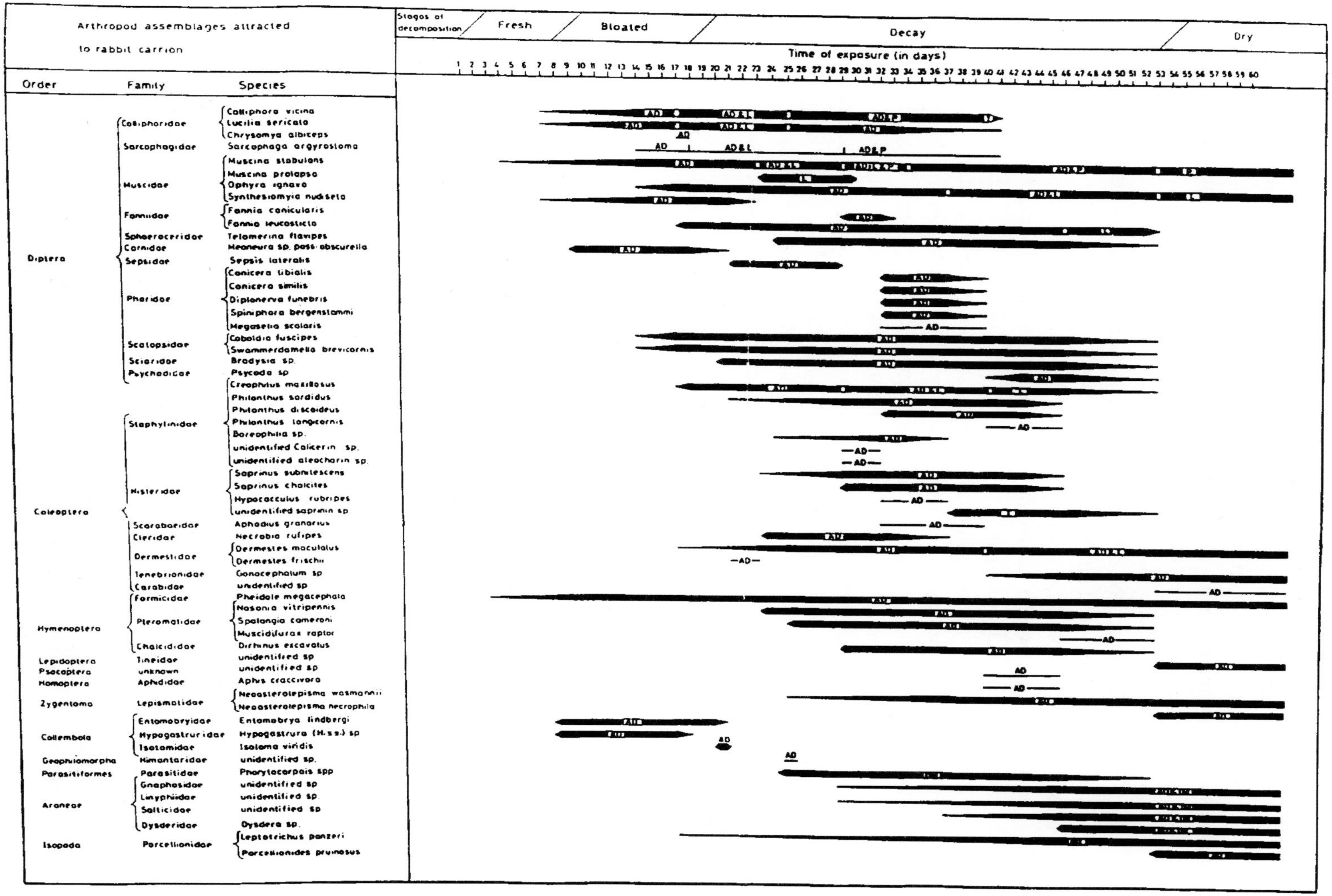

FIG. 10.4 Succession of arthropods on rabbit carrion during winter in Egypt. From Tantawi *et al.* (1996) by permission from the Entomological Society of America.

Substrate conditions affect the ability of organisms to settle, become established, and derive necessary resources. Some substrates restrict species representation, e.g., serpentine soils, gypsum dunes, and lava flows. Relatively few species can tolerate such unique substrate conditions or the exposure resulting from limited vegetative cover. In fact, distinct subspecies often characterize the communities on these and the surrounding substrates. Different communities characterize cobbled or sandy sections of streams, because of different exposure to water flow and filtration of plant or detrital resources. Finally, sites with a high water table support communities distinct from the surrounding communities, e.g., swamp or marsh communities imbedded within desert, grassland, or forested landscapes.

Successional pathways are affected by the composition of initial colonists and survivors from the previous community. The initial colonists of a site represent regional species pools and their composition can vary, depending on proximity to population sources. A site is more likely to be colonized by abundant species than by rare species. Rapidly growing and expanding populations are more likely to colonize even marginally suitable sites than are declining populations. For example, trees dying during a period of minimal bark beetle abundance would undergo a delay in initiation of heterotrophic succession, dominated by a different assemblage of insect species associated with different microorganisms (e.g., Schowalter *et al.*, 1992). Wood initially colonized by decay fungi, such as inoculated by wood-boring beetles, wasps, and termites, decays more rapidly, thereby affecting subsequent colonization, than does wood initially colonized by mold fungi, such as inoculated by bark and ambrosia beetles (Käärik, 1974; Schowalter *et al.*, 1992).

Many individuals survive disturbance, depending on their tolerance to (or protection from) disturbance, and affect subsequent succession (Egler, 1954). Disturbance scale also affects the rate of colonization. Succession initiated primarily by ruderal colonists will differ from succession initiated by a combination of ruderal colonists and surviving individuals and propagules, such as seed banks. Such legacies from the previous community contribute to the early appearance and advanced development of later successional species. These may preclude establishment of some ruderal species that would lead along a different successional pathway. Large-scale disturbances promote ruderal species that can colonize a large area rapidly, whereas small-scale disturbances may expose too little area for shade-intolerant ruderal species and be colonized instead by later successional species expanding from the edge (Brokaw, 1985; Denslow, 1985; Shure and Phillips, 1991).

The sequence of disturbances during succession determines the composition of successive species assemblages. For example, fire followed by drought would filter the community through a fire-tolerance sieve, then a drought-tolerance sieve, whereas flooding followed by fire would produce a different sequence of communities. Disturbance also can truncate community development. Grasslands and pine forests often dominate sites with climatic conditions that could support mesic forest, but succession is arrested by topographic or seasonal factors that increase the incidence of lightning-ignited fires and preclude persistence of mesic trees.

Similarly, environmental changes (including anthropogenic suppression of disturbances) affect community structure. Ironically, fire suppression to "protect" natural communities often results in successional replacement of fire-dominated communities, such as pine forests and grasslands. The replacing communities also may be more vulnerable to some disturbances. For example, fire suppression in the intermountain region of western North America has caused major change in community structure from relatively open, pine/larch woodland maintained by frequent ground fires to closed-canopy pine/fir forest that has become increasingly vulnerable to stand replacing crown fires (Agee, 1993; Schowalter and Lowman, 1999; Wickman, 1992).

The importance of animal activity to successional transitions has not been recognized widely, despite obvious effects of many herbivores on plant species composition (e.g., Louda *et al.*, 1990a; see Chapter 12). Vegetation changes caused by animal activity often have been attributed to plant senescence. Animals affect succession in a variety of ways (Davidson, 1993; MacMahon, 1981; Schowalter and Lowman, 1999; Willig and McGinley, 1999). Animals that construct burrows or mounds or that wallow or compact soils can kill all vegetation in small (several m^2) patches and provide suitable germination habitat and other resources for ruderal plant species (Andersen and MacMahon, 1985; MacMahon, 1981; see also Chapter 14), thereby reversing succession. Jonkman (1978) reported that the collapse of leaf-cutter ant, *Atta vollenweideri*, nests following colony abandonment provides small pools of water that facilitate plant colonization and accelerate development of woodlands in South American grasslands.

Herbivorous species can delay colonization by host species (Tyler, 1995; Wood and Andersen, 1990) and can suppress or kill host species and facilitate their replacement by nonhosts over areas as large as 10^6 ha during outbreaks (Schowalter and Lowman, 1999). Bullock (1991) reported that the scale of disturbance can affect animal activity, thereby influencing colonization and succession. Generally, herbivory during early seres halts or advances succession (Brown, 1984; Schowalter, 1981; Torres, 1992), whereas herbivory during later seres halts or reverses succession (Davidson, 1993; Schowalter and Lowman, 1999). Similarly, Tullis and Goff (1987) and Wells and Greenberg (1994) reported that predaceous ants affected colonization and activity of carrion feeders and affected succession of the carrion community.

Granivores tend to feed on the largest seeds available, which most often represent later successional plant species, and thereby inhibit succession (Davidson, 1993). Herbivores and granivores can interact competitively to affect local patterns of plant species survival and succession. For example, Ostfeld *et al.* (1997) reported that voles dominated interior portions of old fields, fed preferentially on hardwood seedlings over white pine seedlings, and competitively displaced mice, which fed preferentially on white pine seeds over hardwood seeds near the forest edge. This interaction favored growth of hard-wood seedlings in the ecotonal zone and favored growth of white pine seedlings in the old field interior.

A number of studies have demonstrated that species richness peaks at intermediate stages of succession and tends to decline during later successional stages (e.g., Peet and Christensen, 1980; Wilson, 1969). Intermediate levels of disturbance (severity, frequency, or scale) may permit community development up to stages that maximize representation of both early and late successional species, the Intermediate Disturbance Hypothesis (Connell, 1978; Sousa, 1985).

Relatively few studies have evaluated community development experimentally. Patterns of arthropod colonization of new habitats represent a relatively short-term succession amenable to analysis. Strong *et al.* (1984) considered the unwitting movement of plants around the world by humans to represent a natural experiment for testing hypotheses about development of phytophage assemblages on a new resource. They noted that relatively few arthropod colonists on exotic plants were associated with the plant in its native habitat. Most arthropods associated with exotic plants are new recruits derived from the native fauna of the new habitat. Most of the insects that colonize introduced plants are generalists that feed on a wide range of hosts, often unrelated to the introduced plant species, and most are external folivores and sap-suckers (Kogan, 1981; Strong *et al.*, 1984). Miners and gall-formers represent higher proportions of the associated fauna in the region of plant origin, likely because of the higher degree of specialization required for feeding internally. For example, endophages represent 10–30% of the phytophages associated with two species of thistles in native European communities but represent only 1–5% of phytophages associated with these thistles in southern California, where they were introduced (Strong *et al.*, 1984). These results indicate that generalists are better colonists than are specialists, but adaptation over ecological time increases exploitation efficiency (Kogan, 1981; Strong *et al.*, 1984).

In one of the most ambitious studies of community development, Simberloff and Wilson (Simberloff, 1969; Simberloff and Wilson, 1969; Wilson and Simberloff, 1969) defaunated (using methyl bromide fumigation) six small mangrove islands formed by *Rhizophora mangle* in Florida Bay and monitored the reestablishment of the arthropod community during the following year. Simberloff and Wilson (1969) reported that by 250 days after defaunation, all but the most distant island had species richness and composition similar to those of untreated islands, but densities were lower on treated islands. Initial colonists included both strong and weak fliers, but weak fliers, especially psocopterans, showed most rapid population growth. Ants, which dominate the mangrove fauna, were among the later colonists but showed the highest consistency in colonization among islands. Simberloff and Wilson (1969) found that colonization rates for ant species were related to island size and distance from population sources. The ability of an ant species to colonize increasingly smaller islands was similar to its ability to colonize increasingly distant islands.

Species richness initially increased, declined gradually as densities and interactions increased, then reached a dynamic equilibrium with species colonization balancing extinction (see also Wilson, 1969). Calculated species turnover rates were > 0.67 species per day (Simberloff and Wilson, 1969), consistent with the model of MacArthur and Wilson (1967).

These studies explain why early successional stages are dominated by r-selected species with wide tolerances (generalists) and rapid reproductive rates, whereas later stages are dominated by K-selected species with narrower tolerances for coexistence with more specialized species (see Chapter 5). The first arthropods to appear on newly exposed or denuded sites (also glaciated sites) typically are generalized detritivores and predators, that exploit residual or exogenous dead organic material, and dying colonists unable to survive. These arthropods feed on less toxic material than do herbivores or on material in which the defensive compounds have decayed. Herbivores appear only as their host plants appear, and their associated predators similarly appear as their prey appear.

C. Models of Succession

Clements (1916) noted that comparison of successional stages is less useful than is understanding processes affecting the transitions from one sere to another. Nevertheless, few studies have continued over sufficient periods to evaluate the mechanism(s) producing successional transitions. Rather, a number of non-mutually exclusive models have been proposed, all of which may affect particular transitions to varying degrees, and have been widely debated (e.g., Connell and Slatyer, 1977; Horn, 1981; McIntosh, 1981; Peet and Christensen, 1980). The debate involves competing views of succession as (a) resulting from population dynamics or emergent ecosystem processes and (b) as stochastic or converging on an equilibrial community structure (Horn, 1981; McIntosh, 1981).

The **facilitation model** was proposed by Clements (1916), who viewed communities as a single entity that showed progressive (facilitated) development similar to the ontogeny of individual organisms. According to this model, also called **relay floristics** (Egler, 1954), successive stages cause progressive changes in environmental conditions that facilitate their replacement by the subsequent stage, and later successional species cannot appear until sufficient environmental modification by earlier stages has occurred. For example, soil development or increased plant density during early stages makes the environment less suitable for recruitment of additional early species but more suitable for recruitment of later successional species. Fire-dominated ecosystems (in which nitrogen is volatilized during fire) typically are colonized following fire by symbiotic nitrogen fixers such as alders (*Alnus* spp.), ceanothus (*Ceanothus* spp.), or cherries (*Prunus* spp.). These species are relatively shade intolerant, and increasing density eventually suppresses their photosynthesis and nitrogen-fixation, facilitating replacement by shade-tolerant species growing in the understory and exploiting the replenished organic nitrogen in the soil (e.g., Boring *et al.*, 1988). The increasing porosity and altered nutrient content of decomposing wood, resulting from heterotroph activity, precludes further recruitment of early successional species, e.g., bark beetles and anaerobic or microaerophilic microorganisms, and facilitates replacement by later successional wood borers and more aerobic microorganisms (e.g., Schowalter *et al.*, 1992).

This model was challenged early. Gleason (1917, 1926, 1927), Whittaker (1953, 1970), and, more recently, Drury and Nisbet (1973), argued for a reductionist view of species colonization and turnover on the basis of individual life history attributes. Connell and Slatyer (1977), Horn (1981), and MacMahon (1981) have proposed a broader view of succession as reflecting multiple pathways and mechanisms.

Egler (1954) argued that secondary succession often may reflect differential longevity of colonizing species. Most of the eventual dominants colonize early when competition is low. Failure of species to become established at this early stage reduces the probability of future dominance. Juveniles of later species grow to maturity over a longer period, tolerating the early dominance of ruderal species, and eventually exclude the early successional species (through shading, preemptive use of water, etc.). Connell and Slatyer (1977) referred to this model as the **tolerance model**. This model is represented best in ecosystems dominated by species that sprout from roots or stumps, germinate from seed banks, or colonize rapidly from adjacent sources. These attributes ensure early appearance along with more ruderal species. However, many large-seeded trees, flightless insects, and other animals characterizing later successional stages of forest ecosystems require a long period of establishment and achieve dominance only during late succession, especially in large areas of disturbed habitat.

A third model proposed by Connell and Slatyer (1977) to explain at least some successional transitions is the antithesis of facilitation, i.e., inhibition of replacement. According to this **inhibition model**, the initial colonists preempt use of resources and exclude, suppress, or inhibit subsequent colonists for as long as these initial colonists persist. Succession can proceed only as individuals are damaged or killed and thereby release resources (including growing space) for other species. Examples of inhibition are successional stages dominated by allelopathic species, such as shrubs that increase soil salinity or acidity, by species that preempt space, such as many perennial sodforming grasses whose network of rhizomes restricts establishment by other plants, by species whose life spans coincide with the average interval between disturbances, and by species that create a positive feedback between disturbance and regeneration (e.g., Shugart *et al.*, 1981). In decomposing wood, the sequence of colonization by various insects determines initial fungal association; initial colonization by mold fungi can catabolize available labile carbohydrates and inhibit subsequent establishment of decay fungi (Käärik, 1974), restricting further succession (Schowalter *et al.*, 1992). Environmental fluctuation, disturbances, or animal

activity (such as gopher mounds, bison wallows, trampling, and insect outbreaks) often are necessary to facilitate replacement of these stages (MacMahon, 1981; Schowalter et al., 1981a; Schowalter and Lowman, 1999). However, Agee (1993), Schowalter (1985), and Schowalter et al. (1981a) note that
bark beetle outbreaks increase fuel accumulation and the probability of fire,
thereby ensuring the continuity of pine forest (Fig. 10.5).

Horn (1981) developed a model of forest succession as a tree-by-tree replacement process using the number of saplings of various species growing under each canopy species (ignoring species for which this is not a reasonable pre-
dictor of replacement) and correcting for expected longevity. This model
assumes that knowing what species occupies a given position narrows the statistical range of expected future occupants and that the probability of replacement depends only on the species occupying that position and does not change
with time unless the occupant of that position changes. The model is not directly applicable to communities in which recurrent large-scale disturbances are
the primary factor affecting vegetation dynamics. Interestingly, Horn (1981)
found that successive iterations by a given replacement matrix invariably converged on a particular community composition, regardless of the starting composition. This result indicates that convergence is not necessarily a reflection of
biotic processes (Horn, 1981) and should increase attention to the rate of convergence and transition states producing convergence.

Many ecologists consider vegetation changes over time to be no more than
expressions of species life history characteristics. Species distributions in time
reflect their physiological tolerances to changing environmental conditions,
parallel to distributions in space (Botkin, 1981; Drury and Nisbet, 1973). Several major simulation models of forest gap succession are based on species-spe-
cific growth rates and longevities as affected by stochastic mortality (e.g., Doyle,
1981; Shugart et al., 1981; Solomon et al., 1981). However, no current models
address the importance of animals or various disturbances on the successional
process. The variety of successional pathways determined by unique combinations of interacting initial and subsequent conditions may favor models that apply chaos theory.

III. PALEOECOLOGY

Paleoecology provides a context for understanding extant interactions and
community structure. Although most paleoecological study has focused on bio-
geographical patterns (e.g., Price, 1997), fossils also reveal much about prehistoric species interactions and community structure (Labandeira and Sepkoski,
1993). We can infer that similar morphological features of fossil and extant organisms imply similar functions and associated behaviors (Boucot, 1990;
Poinar, 1993; Scott and Taylor, 1983), helping to explain fossil records as well
as to understand long-term patterns of community change.

The fossil record contains abundant evidence of similar functions and be-

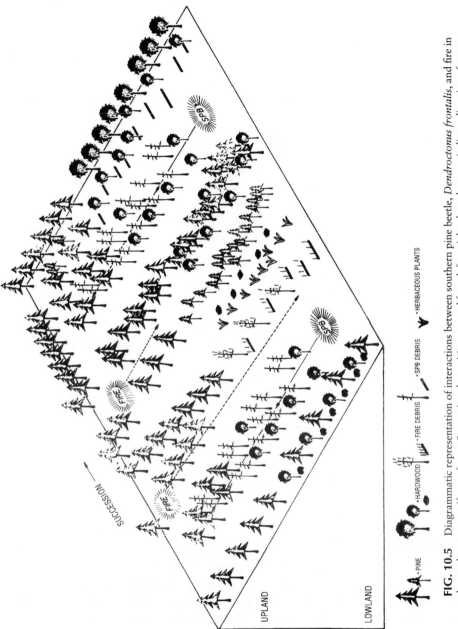

FIG. 10.5 Diagrammatic representation of interactions between southern pine beetle, *Dendroctonus frontalis*, and fire in the southeastern coniferous forest. Successional transitions extend from left to right; dotted arrows indicate direction of movement. Fire is a regular feature of the generally dry uplands but moves into generally moist lowlands where drought or southern pine beetle creates conditions favorable for combustion. Southern pine beetle is a regular feature of both forests but is most abundant where pines occur at high density and stress levels. Fire is necessary for regeneration of pines, especially following succession to hardwoods if fire return is delayed. From Schowalter *et al.* (1981a) by permission from the Entomological Society of America.

haviors. For example, haustellate mouth parts of proto-Hemiptera suggest early appearance of feeding on plant sap (Labandeira and Sepkoski, 1993; Scott and Taylor, 1983). A fossil termite bug, *Termitaradus protera*, in Mexican amber has the same morphological modifications as its extant congeners for surviving in termite colonies and, therefore, can be assumed to have had similar interactions with termites (Poinar, 1993). Dental structure of Upper Carboniferous amphibians suggests that most were predaceous, many insectivorous (Scott and Taylor, 1983).

Evidence of consistent species roles suggests that host selection behaviors and other species associations within communities have been conserved over time, the **Behavioral Fixity Hypothesis** (Boucot, 1990; Poinar, 1993). Association of potentially interacting taxa in the same deposits and anatomical evidence of interaction are common. For example, bark beetle (Scolytid) galleries and termite nests, complete with fecal pellets, in fossil conifers from the early-to mid-Tertiary demonstrate a long evolutionary history of association between these insects and conifers (Boucot, 1990). Evidence of wood boring, perhaps by ancestral beetles, can be found as early as the Upper Carboniferous (Scott and Taylor, 1983). Some vertebrate coprolites from the Upper Carboniferous contain arthropod fragments (Scott and Taylor, 1983). The presence of fig wasps (Agaonidae) in Dominican amber suggests the cooccurrence of fig trees (Poinar, 1993). Many fossil leaves from as early as the Upper Carboniferous show evidence of herbivory similar to that produced by modern insects (Boucot, 1990; Scott and Taylor, 1983).

Boucot (1990) reported a unique example of an extant insect species associated with extant genera in an Upper Miocene deposit in Iceland. The hickory aphid, *Longistigma caryae*, occurred in the same deposit with fossil leaves of *Carya* (or *Juglans*), *Fagus*, *Platanus*, and *Acer*. This aphid species survives on the same tree genera in eastern North America, providing strong evidence for long-term association between this insect and its hosts.

Demonstrated interaction between pairs of species is less common but provides more convincing evidence of behavioral constancy (Fig. 10.6). Gut contents from arthropods in Upper Carboniferous coal deposits indicate herbivorous, fungivorous, or detritivorous diets for most early arthropods (Scott and Taylor, 1983). Mermithid nematodes commonly parasitize chironomid midges, typically castrating males and causing diagnostic changes in antennal morphology. A number of chironomid males from Baltic and Dominican amber show both the altered antennal morphology and the nematode emerging at the time of host death (Boucot, 1990; Poinar, 1993). Parasitic mites frequently are found attached to their hosts in amber. Phoretic mites associated with their beetle or fly hosts are relatively rare, but occur in Dominican amber (Poinar, 1993). Similarly, staphylinid beetles commensal in termite nests have been found with their termite hosts in Dominican amber (Poinar, 1993).

Surprisingly few examples of demonstrated mutualistic interactions are preserved in the fossil record. Scott and Taylor (1983) noted that spores of Up-

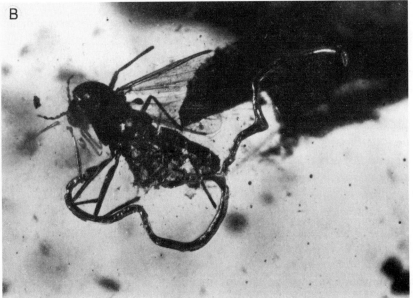

FIG. 10.6 Evidence of parasitism of extinct insects. (A) fungal synnema (spore-bearing structure) protruding from the body of a *Troctopsocopsis* sp. (Psocoptera) in Dominican amber, (B) two allantonematid nematodes emerging from a chironomid midge in Dominican amber, and (C) parasitic mites attached to the abdomen of a chironomid midge in Dominican amber. From Poinar (1993) with permission from the *Annual Review of Entomology*, Vol. 38, © 1993, by Annual Reviews.

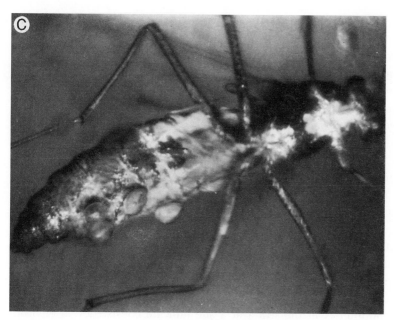

FIG. 10.6 (*Continued*)

per Carboniferous plants had a resistant sporoderm capable of surviving passage through animal guts, suggesting that herbivores may have served as agents of spore dispersal. An Upper Carboniferous arthropod, *Arthropleura armata*, was found with pollen grains of a medullosan seed fern attached along its posterior edge at the base of its legs. This species could have been an early pollinator of these seed ferns whose pollen was too large for wind transport. Furthermore, some Upper Carboniferous plants produced glandular hairs that might have been an early type of nectary to attract pollinators (Scott and Taylor, 1983).

Fossil data permit limited comparison of diversity and species interactions between taxonomically distinct fossil and extant communities (see also Chapter 9). Insect diversity has increased at a rate of about 1.5 families per million years since the Devonian; the rise of angiosperms during the Cretaceous contributed to diversification within families but did not increase the rate of diversification at the family level (Labandeira and Sepkoski, 1993). Arthropod diversity was high in the communities recorded in Upper Carboniferous coal deposits and Dominican and Mexican ambers (Poinar, 1993; Scott and Taylor, 1983). Similar interactions, as discussed previously, indicate that virtually all types of interactions represented by extant communities (e.g., herbivore–plant, arthropod–fungus, predator–prey, pollinator, wood-borer, and detritivore) were established as early as the Upper Carboniferous.

The Behavioral Fixity Hypothesis permits reconstruction of prehistoric communities, to the extent that organisms associated in coal, amber, or other deposits represent prehistoric communities (e.g., Fig. 10.7) (Poinar, 1993). The Upper Carboniferous coal deposits indicate a diverse, treefern-dominated, swamp ecosystem. The fossils in Dominican amber indicate a tropical, evergreen angiosperm rain forest. Some insect specimens indicate the presence of large buttress-based host trees, whereas other specimens indicate the presence of palms in forest openings (Poinar, 1993). The presence of fig wasps indicates that fig trees also occurred. Baltic amber contains a combination of warm temperate and subtropical groups, suggesting a number of possible community structures. The temperate elements could have originated at a higher elevation, or Baltic amber may have formed during a climate change from subtropical to temperate conditions (Poinar, 1993). Diversity, food web structure, and functional group organization were similar between these extinct communities and extant communities (Poinar, 1993; Scott and Taylor, 1983), suggesting that broad patterns of community structure are conserved through time, even as species composition changes.

The fossil record also records changes in community structure at a site through time. The degree to which particular community types are continuous across discontinuities in the strata at a site indicate consistency of environmental conditions and community structure (Boucot, 1990). Boucot (1990) noted that, although a particular fossilized community (taxonomic association) rarely persists long in a local stratigraphic section, communities typically recur over larger areas for 10^6–10^7 years, indicating a high degree of stability with-

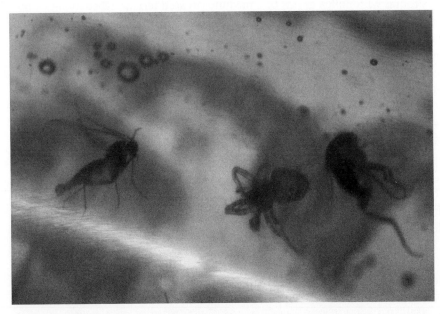

FIG. 10.7   Sciarid and phorid flies (Diptera) and spider from Colombian amber. From a sample containing > 12 species of insects (4 orders) and spiders.

in environmental constraints. Pollen or other fossil records often indicate relatively rapid changes in distribution of particular plant species and, presumably, of associated heterotrophs. For example, Gear and Huntley (1991) reported that dating of fossilized Scots pine, *Pinus sylvestris*, stumps in northern Scotland indicate that pine forest expanded rapidly northward 70–80 km about 4000 years BP and persisted for about 400 years before retreating southward again, suggesting a 400-year period of warmer climate and community change. However, they noted that even this remarkably rapid rate of species movement would be insufficient (by an order of magnitude) to accomplish range change necessary under future climate change scenarios, especially if impeded by landscape fragmentation.

IV. DIVERSITY VERSUS STABILITY

The relationship between biodiversity and community stability remains a controversial issue (e.g., de Ruiter *et al.*, 1995; Grime, 1997; Hooper and Vitousek, 1997; Schulze and Mooney, 1993; Tilman *et al.*, 1997; see Chapter 15). An early assumption that diversity conferred stability on communities and ecosystems was challenged, beginning in the 1970s, by modeling efforts that indicated increasing vulnerability to perturbation of more complex systems (e.g., May, 1973, 1981; Yodzis, 1980). However, new studies have addressed the importance of biodiversity for variability of ecosystem processes (e.g., de Ruiter *et al.*, 1995; Tilman and Downing, 1994; Tilman *et al.*, 1997). Among these are studies of "pest" dynamics and their effects on community structure in diverse ecosystems versus simple ecosystems (e.g., Schowalter and Turchin, 1993).

Fundamental to our understanding of this relationship are definitions and measurements of diversity and stability. As noted in Chapter 9, the variety of methods for measuring diversity has complicated comparison of communities, including assessment of community change. Should diversity be measured as species richness, functional group richness, or some diversity index using species or functional groups (de Ruiter *et al.*, 1995; Grime, 1997; Hooper and Vitousek, 1997; Tilman and Downing, 1994; Tilman *et al.*, 1997)? Stability can be defined as reduced variability in system behavior. However, ecologists have disagreed over how best to measure stability. Stability has been shown to have multiple components, one representing capacity to resist change (perturbation) and the other representing ability to recover following a change (i.e., succession), which indicate different degrees of stability for a given ecosystem. The variable(s) chosen to measure stability also can indicate different degrees of stability.

Traditionally, stability was measured by population and community ecologists as the constancy of species composition and community structure (e.g., Grime, 1997; May, 1973, 1983). Ecosystem ecologists have emphasized the variability of ecosystem processes such as primary productivity, energy flux, and biogeochemical cycling, especially as variability changes during succession (e.g., de Ruiter *et al.*, 1995; Kratz *et al.*, 1995; Odum, 1969; Tilman and Downing, 1994). Species diversity may stabilize some variables but not others, leading to different conclusions. The extent to which diversity contributes to ecosystem integrity will be addressed in Chapter 15.

A. Components of Stability

Holling (1973) originally defined stability as the ability of a community to withstand disturbance with little change in structure, whereas resilience was the capacity of the community to recover following perturbation. Webster *et al.* (1975) subsequently refined the definition of stability to incorporate both **resistance** to change and **resilience** following perturbation. Succession is the expression of resilience. However, the criteria for measuring stability remain elusive. What degree of change can be accommodated before resistance is breached? Does resilience require the recovery of a predisturbance community structure or of ecosystem functions that produce certain species combinations and over what period of time?

Webster *et al.* (1975) developed a functional model to evaluate the relative stability of ecosystems based on the lowest turnover rates, i.e., the longest time constraint, and damping factors, i.e., factors that reduce amplitude of fluctuation, in the system. The system has not fully recovered from displacement until the slowest component of the response has disappeared. They concluded that ecosystems with greater structure and amounts of resource storage were more resistant to disturbance, whereas ecosystems with greater turnover (e.g., via consumers) were more resilient. From a community standpoint, resistance depends on the level of tolerance of the dominant species to characteristic disturbances or other environmental changes, e.g., through protected meristems or propagules, or resource storage; resilience is conferred by species with rapid recolonization and growth rates. Overall, temperate forests, with high biotic and abiotic storage and slow turnover, appear to be most resistant to disturbance, and stream systems, with low biotic and abiotic storage and high turnover, appear to be least resistant. Resistance and resilience also were found to be related inversely, with their relative contributions to stability in a given ecosystem determined by the proportions of K and r specialists (see Chapter 5). Succession appears to represent a trend from more resilient to more resistant communities.

Resistance and resilience are affected by regional species abundance and distribution. Resistance can be compromised by fragmentation, which increases community exposure to external factors. For example, trees in interior forest communities typically are buffered from high temperatures and high wind speeds by surrounding trees and typically have less buttressing than do open-grown trees. Fragmentation increases the proportion of trees exposed to high temperatures and wind speeds and thereby vulnerable to moisture stress or toppling (Chen *et al.*, 1995; Franklin *et al.*, 1992). Fragmentation also interferes with the adapted abilities of species in the regional pool to recolonize disturbed sites. Species are adapted to levels of dispersal and colonization sufficient to maintain populations within the characteristic habitat matrix of the landscape. If the rate of patch turnover is increased through fragmentation, the colonization rates for many species may be insufficient to provide the necessary level of resilience for community recovery.

B. Stability of Community Variables

A number of community variables can be examined from the standpoint of their variability with respect to diversity. Among these are species composition and food web structure. Simpler communities, in terms of species composition and food web structure, often appear to be more stable than complex communities (e.g., May, 1973, 1983). Boucot (1990) noted that simple marine communities in the fossil record continue across sedimentary discontinuities more often than do complex marine communities. Boucot (1990) also noted that particular taxonomic associations typically recur over larger areas for 10^6–10^7 years, indicating a high degree of stability within environmental constraints. The variety of successional pathways leading to multiple endpoints (Horn, 1981; Whittaker, 1953) has indicated that many communities do not necessarily recover their predisturbance composition or food web structure, although some mechanisms lead to positive feedback between disturbance and community organization (Schowalter, 1985; Schowalter *et al.*, 1981a; Shugart *et al.*, 1981). May (1973, 1983) and Yodzis (1980) concluded, using modeling approaches, that more complex communities are more vulnerable to disruption by perturbations in any particular species population because of their propagation through the network of interactions involving that species. However, de Ruiter *et al.* (1995) incorporated the patterning of interaction strengths in real communities and found that simultaneous occurrence of strong top-down regulation of lower trophic levels and strong bottom-up regulation of higher trophic levels imposed stabilizing patterns on interaction strengths.

To some extent, lack of a clear correlation between diversity and stability of community variables may be an artifact of the duration of succession or the number of intermediate stages that can generate alternative pathways. More frequently disturbed communities may appear to be more stable than infrequently disturbed communities because the ecological attributes of ruderal species favor rapid recovery, whereas longer time periods and more intervening factors affect recovery of tree species composition. Furthermore, if maximum species diversity occurs at intermediate levels of disturbance (the Intermediate Disturbance Hypothesis), then the lower species diversity of earlier and later successional communities is associated with both high and low stability, in terms of frequency and amplitude of departure from particular community structure.

A major source of diversity is the variety of community types and the regional species pool maintained in a shifting landscape mosaic of patch types. Although the community of any particular site may appear unstable because of multiple factors interacting to affect its response to perturbation, the landscape pattern of local communities minimizes the distance between population sources and sinks and ensures proximity of colonists for species packing and assortment during site recovery. Even if the community does not recover to the same endpoint, that predisturbance endpoint likely appears in other patches.

Our perspective on the effect of diversity on stability may differ at a larger spatial scale. Diverse, or complex, communities can be compartmentalized spatially as a result of heterogeneity of resources and population growth patterns (see Chapter 7). As Boucot (1990) noted, fossil communities rarely persist long at a site but recur over larger areas for long time periods, indicating a high degree of stability at a landscape scale. Hence, species composition and community structure may be conditionally stable at the landscape scale but not at the site scale, with the landscape-scale community represented as a shifting mosaic of local component communities.

V. SUMMARY

Community structure changes over a range of time scales, from annual to decadal to millennial time periods. Temporal patterns of community organization and their sensitivity to environmental changes can indicate their stability to anthropogenic changes.

Community structure changes on annual time scales as population sizes respond to environmental conditions. Changes in resource quality, competition, and predation lead to population irruptions of some species and local extinction of others, thereby affecting their interactions with other species and leading to changes in community structure.

Ecological succession, the sequential stages of community development on newly exposed or disturbed sites, is one of the best documented ecological phenomena and has provided a unifying concept that integrates species life history strategies, population behavior, community dynamics, and ecosystem processes. Early successional communities typically are dominated by relatively generalized ruderal species with high mobility and rapid reproductive rates. Later successional stages are increasingly dominated by species that are more specialized, are less mobile, and have lower reproductive capacities. Although most studies of succession have focused on plants, insects show successional patterns associated with changes in vegetation, and relatively rapid heterotrophic succession in decomposing wood and animal carcasses has contributed much to successional theory.

A number of factors influence successional pathways. Local substrate conditions can restrict initial colonists to those from the surrounding species pool that can become established on distinct substrates, such as serpentine, volcanic, or water-saturated soils. The composition of the initial community, including survivors of the previous disturbance and colonists, can affect the success of subsequent colonists. Subsequent disturbances and animal activity can affect successional pathways. Animals, including insects, create germination sites for colonists and suppress some host species, thereby facilitating, inhibiting, or reversing succession. In fact, animal activity often may account for vegetation changes that have been attributed to plant senescence.

Several models of succession have augmented the early model of succession as a process of facilitated community development, in which earlier stages create conditions more conducive to successive stages. In some cases, all the eventual dominants are present in the initial community, and succession reflects differential development time and longevity among species, i.e., the tolerance model. Some successional stages are able to competitively exclude later colonists, the inhibition model. Succession may advance beyond such stages as a result of plant injury or death from subsequent disturbances or animal activity.

Paleoecological research indicates that species interactions and community structures have been relatively consistent over evolutionary time. However, the communities occupying particular sites have changed over these time periods as the environmental conditions of the site have changed.

The relationship between species or functional diversity and community or ecosystem stability has been highly controversial. Much of the discussion reflects different definitions of diversity and stability. Stability can be seen to have two major components, resistance to change, and resilience following perturbation. Succession is the expression of resilience. Although much evidence indicates that a particular community composition or structure may not be replaced at a site, indicating instability at the local level, the structure of communities at a landscape scale ensures that disturbed sites are near population sources and that component communities are maintained within a shifting landscape mosaic, indicating stability at the landscape or regional level.

ECOSYSTEM LEVEL

THE ECOSYSTEM LEVEL OF ORGANIZATION INTEGRATES
species interactions and community structure with their responses to, and effects on, the abiotic environment. Interactions
among organisms is the mechanism governing energy and nutrient fluxes through ecosystems. The rates and spatial patterns in
which individual organisms and populations acquire and allocate
energy and nutrients determine the rate and direction of these fluxes (see
Chapters 4 and 8).

Communities vary in their ability to modify their abiotic environment.
The relative abundance of various nutrient resources affects the efficiency
with which they are cycled and retained within the ecosystem. Increasing
biomass confers increased storage capacity and buffering against changes
in resource availability. Community structure also can modify climatic
conditions, by controlling albedo and hydric fluxes, buffering individuals
against changing environmental conditions.

A major issue at the ecosystem level is the extent to which communities are organized to maintain optimal conditions for the persistence of
the community. Species interactions and community structures may
represent adaptive attributes at the supraorganismal level, which stabilize

ecosystem properties near optimal levels for the various species. If so, anthropogenic interference with community organization (e.g., species redistribution, pest control, overgrazing, deforestation) may disrupt stabilizing mechanisms and contribute to ecosystem degradation.

Insects affect virtually all ecosystem properties, especially through their effects on vegetation, detritus, and soils. Insects clearly affect primary productivity, and, hence, the capture and flux of energy and nutrients. In fact, insects are the dominant pathway for energy and nutrient flow in many aquatic and terrestrial ecosystems. They affect vegetation density and porosity, and hence, albedo and the penetration of light, wind, and precipitation. They affect accumulation and decomposition of litter, and mixing and porosity of soil and litter, thereby affecting soil fertility and water flux. They often determine disturbance frequency, succession, and associated changes in efficiency of ecosystem processes over time. Their small size and rapid and dramatic responses to environmental changes are ideal attributes for regulators of ecosystem processes, through positive and negative feedback mechanisms. Ironically, effects of detritivores (largely ignored by insect ecologists) on decomposition have been addressed by ecosystem ecologists, whereas effects of herbivorous insects (the focus of insect ecology) on ecosystem processes have been all but ignored until recently.

Chapter 11 summarizes key aspects of ecosystem structure and function, including energy flow, biogeochemical cycling, and climate modification. Chapters 12–14 cover the variety of ways in which insects affect ecosystem structure and function. The varied effects of herbivores are addressed in Chapter 12. Although not often viewed from an ecosystem perspective, pollination and seed predation affect patterns of primary production, as described in Chapter 13. The important effects of detritivores on organic matter turnover and soil development are the focus of Chapter 14. Finally, the potential roles of these organisms as regulators of ecosystem processes are explored in Chapter 15.

Ecosystem Structure and Function

TANSLEY (1935) COINED THE TERM "ECOSYSTEM" TO RECOGNIZE THE INtegration of the biotic community and its physical environment as the fundamental unit of ecology, within a hierarchy of physical systems that spanned the range from atom to universe. Shortly thereafter, Lindeman's (1942) study of energy flow through an aquatic ecosystem introduced the modern concept of an ecosystem as a feedback system capable of redirecting and reallocating energy and matter fluxes. More recently, during the 1950s–1970s, concern over the fate of radioactive isotopes from nuclear fallout generated considerable research on biological control of elemental movement through ecosystems (Golley, 1993). Recognition of anthropogenic effects on atmospheric conditions, especially greenhouse gas and pollutant concentrations, has renewed interest in how natural and altered communities control fluxes of energy and matter and modify abiotic conditions.

Delineation of ecosystem boundaries can be problematic. Ecosystems can be described at various scales. At one extreme, the diverse flora and fauna living on the backs of rain forest beetles (Gressitt *et al.*, 1965, 1968) or the aquatic communities in water-holding plant structures (Fig. 11.1) constitute an ecosystem. At the other extreme, the interconnected terrestrial and marine

FIG. 11.1   The community of aquatic organisms, including microflora and invertebrates, that develops in water-holding structures of plants, such as *Heliconia* flowers, represents a small-scale ecosystem with measurable inputs of energy and matter, species interactions that determine fluxes and cycling of energy and matter, and outputs of energy and matter.

ecosystems constitute a global ecosystem (Golley, 1993; Lovelock, 1988; Tansley, 1935). Generally, ecosystems have been described at the level of the landscape patch composed of a relatively distinct community type. However, increasing attention has been given to the interconnections among patches that compose a broader landscape-level or watershed-level ecosystem (e.g., Polis *et al.*, 1997; Vannote *et al.*, 1980).

Ecosystems can be characterized by their structure and function. Structure reflects the way in which the ecosystem is organized, e.g., species composition, distribution of energy and matter (biomass), and trophic or functional organization in space. Function reflects the biological modification of abiotic conditions, including energy flow, biogeochemical cycling, and climate. This chapter describes the major structural and functional parameters of ecosystems in order to provide background for description of insect effects on these parameters in Chapters 12–14. Insects affect ecosystem structure and function in a number of ways and are primary pathways for energy and nutrient fluxes.

I. ECOSYSTEM STRUCTURE

Ecosystem structure represents the various pools (both sources and sinks) of energy and matter and their relationships to each other, i.e., directions of matter or information flow (e.g., Fig. 1.3). The size of these pools (i.e., storage capacity) determines the buffering capacity of the system. Ecosystems can be compared on the basis of the sizes and relationships of various biotic and abiotic compartments for storage of energy and matter. Major characteristics for comparing ecosystems are their trophic or functional group structure, biomass distribution, or spatial and temporal variability in structure.

A. Trophic Structure

Trophic structure represents the various feeding levels in the community. Organisms generally can be classified as **autotrophs** (or primary producers), which synthesize organic compounds from abiotic materials, and several kinds of **heterotrophs** (or secondary producers), including insects, which ultimately derive their energy and resources from autotrophs (Fig. 11.2).

Autotrophs are those organisms capable of fixing (acquiring and storing) inorganic resources in organic molecules. Photosynthetic plants, responsible for fixation of abiotic carbon into carbohydrates, are the sources of organic molecules. This chemical synthesis is powered by solar energy. Free-living and symbiotic bacteria and cyanobacteria are an important means of fixing inorganic N_2 into ammonia, the source of most nitrogen available to plants. Other chemoautotrophic bacteria oxidize ammonia into nitrite or nitrate (the form of nitrogen available to most green plants) or oxidize inorganic sulfur into organic compounds. Production of autotrophic tissues must be sufficient to compensate for amounts consumed by heterotrophs.

Heterotrophs can be divided into several trophic levels depending on their

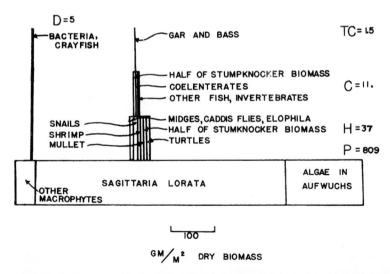

FIG. 11.2 Biomass pyramid for the Silver Springs ecosystem. P, primary producers; H, herbivores; C, predators; TC, top predators; D, decomposers. From Odum (1957) by permission from the Ecological Society of America.

source of food. Primary consumers (herbivores) eat plant tissues. Secondary consumers eat primary consumers, tertiary consumers eat secondary consumers, etc. Omnivores feed on more than one trophic level. Finally, reducers (including detritivores and decomposers) feed on dead plant and animal matter (Whittaker, 1970). Detritivores fragment organic material and facilitate colonization by decomposers, which reduce the chemical compounds.

Each trophic level can be subdivided into functional groups, based on the way in which organisms gain or use resources. For example, autotrophs can be subdivided into photosynthetic, nitrogen-fixing, nitrifying, and other functional groups. The photosynthetic functional group can be subdivided further into ruderal, competitive, and stress-tolerant functional groups (Grime, 1977), or into C-3 and C-4, nitrogen-accumulating, calcium accumulating, high-lignin or low-lignin functional groups, etc., to represent their different strategies for resource use and growth. Similarly, primary consumers can be subdivided into migratory grazers (e.g., many ungulates and grasshoppers), sedentary grazers (various leaf-chewing insects), leaf miners, gall-formers, sap-suckers, root feeders, parasitic plants, plant pathogens, etc., to reflect different modes for acquiring and affecting their plant resources.

The distribution of biomass in an ecosystem is an important indicator of storage capacity, a characteristic that influences ecosystem stability (Webster *et al.*, 1975). Harsh ecosystems, such as tundra and desert, restrict autotrophs to a relatively few, small plants with relatively little biomass to represent storage of energy and matter. Dominant species are adapted to retain water, but water storage capacity is limited. By contrast, wetter ecosystems permit development of large producers with greater storage capacity in branch and root systems. Accumulated detritus represents an additional pool of stored organic matter that buffers the ecosystem from changes in resource availability. Tropical, and other warm, humid ecosystems generally have relatively low detrital biomass because of rapid decomposition and turnover. Stream and tidal ecosystems lose detrital material as a result of export in flowing water. Detritus is most likely to accumulate in cool, moist ecosystems, especially boreal forest, in which detritus decomposes slowly. Biomass of heterotrophs is small in most terrestrial ecosystems but may be larger than primary producer biomass in some aquatic ecosystems, as a result of high production and turnover by low biomass of algae (Whittaker, 1970).

Trophic structure can be represented by numbers, mass (biomass), or energy content of organisms in each trophic level (Fig. 11.2). Such representations are called numbers pyramids, biomass pyramids, or energy pyramids (see Elton, 1939) because the numbers, mass, or energy content of organisms generally decline at successively higher trophic levels. However, the form of these pyramids differs among ecosystems. Terrestrial ecosystems typically have large numbers or biomasses of primary producers that supports progressively smaller numbers or biomasses of consumers. Many stream ecosystems are supported primarily by allochthonous material (detritus entering from the adjacent terrestrial ecosystem) and have few primary producers (e.g., Oertli, 1993; Wallace *et al.*, 1997). Numbers pyramids for terrestrial ecosystems may be inverted because individual plants can support numerous invertebrate consumers. Biomass pyramids for some aquatic ecosystems are inverted because a small biomass of plankton with a high rate of reproduction and turnover can support a larger biomass of organisms with low rates of turnover at higher trophic levels (Whittaker, 1970).

B. Spatial Variability

At one time, the ecosystem was considered to be the interacting community and abiotic conditions of a site. This view gradually has expanded to incorporate the spatial pattern of interacting component communities at a landscape or watershed level (see Chapter 9). Patches within a landscape or watershed are integrated by disturbance dynamics and interact through the movement of organisms, energy, and matter (Schowalter, 1996; see Chapter 7). For example, the Stream Continuum Concept (Vannote *et al.*, 1980) integrates the various stream sections that mutually influence each other. Downstream ecosystems are influenced by inputs from upstream, but the upstream ecosystems are influenced by organisms returning materials from downstream (e.g., Pringle, 1997). Riparian zones (floodplains) are affected by both the terrestrial and aquatic ecosystems, receiving inputs from periodic flooding by the aquatic ecosystem and runoff from the terrestrial ecosystem.

The structure of ecosystems at a stream continuum or landscape scale may have important consequences for recovery from disturbances, by increasing proximity of population sources and sinks. Patches representing various stages of recovery from disturbance provide the sources of energy and matter (including colonists) for succession of any particular patch. Terrestrial ecosystems are major sources of energy and material for aquatic ecosystems. Important members of some trophic levels, especially migratory herbivores and birds, often are concentrated seasonally at particular locations along migratory routes. Social insects may forage long distances from their colonies, integrating patches through pollination, seed dispersal, or other interactions. Such aggregations add spatial complexity to trophic structure.

II. ENERGY FLOW

Life represents a balance between the tendency to increase entropy (Second Law of Thermodynamics) and the decreased entropy, through continuous energy inputs, necessary to maintain basic metabolism and to expand life through growth and reproduction. All energy for life on Earth ultimately comes from solar radiation, which powers the chemical storage of energy through photosynthesis. Given the First and Second Laws of Thermodynamics, the energy flowing through ecosystems, including resources harvested for human use, can be no greater, and typically is much less, than the amount of energy stored in carbohydrates.

Organisms have been compared to thermodynamic machines powered by the energy of carbohydrates to generate maximum power output, in terms of progeny (Lotka, 1925; Odum and Pinkerton, 1955; Wiegert, 1968). Just as organisms can be studied in terms of their energy acquisition, allocation, and energetic efficiency (Chapter 4), so ecosystems can be studied in terms of their energy acquisition, allocation, and energetic efficiency (Odum, 1969; Odum and Pinkerton, 1955). Energy acquired from the sun powers the chemical synthesis of carbohydrates, which represents storage of potential energy that is then channeled through various trophic pathways, each with its own power output, and eventually is dissipated completely as heat through the combined respiration of the community (Lindeman, 1942; Odum, 1969; Odum and Pinkerton, 1955).

The study of ecosystem energetics was pioneered by Lindeman (1942), whose model of energy flow in a lacustrine ecosystem ushered in the modern concept of the ecosystem as a thermodynamic machine. Lindeman noted that the distinction between the community of living organisms and the nonliving environment is obscured by the gradual death of living organisms and conversion of their tissues into abiotic nutrients that are reincorporated into living tissues.

The rate at which available energy is transformed into organic matter is called productivity. This energy transformation at each trophic level (as well as by each organism) represents the storage of potential energy that fuels metabolic processes and power output at each trophic level. Energy flow reflects the integration of productivity by all trophic levels.

A. Primary Productivity

Primary productivity is the rate of conversion of solar energy into plant matter. The total rate of solar energy conversion into carbohydrates (total photosynthesis) is **gross primary productivity**. However, a portion of gross primary productivity must be expended by the plant through metabolic processes necessary for maintenance, growth, and reproduction, and is lost as heat through respiration. The net rate at which energy is stored as plant matter is **net primary productivity**. The energy stored in net primary production (NPP) becomes available to heterotrophs.

Primary productivity, turnover, and standing crop biomass are governed by a number of factors that differ among successional stages and between terrestrial and aquatic ecosystems. NPP is correlated with foliar standing crop biomass (Fig. 11.3). Hence, reduction of foliar standing crop biomass by herbivores can affect NPP. Often, only above-ground NPP is measured, although below-ground production typically exceeds above-ground production in grassland and desert ecosystems (Webb et al., 1983). Among major terrestrial biomes, total (aboveground + belowground) NPP ranges from 2000 g m^{-2} year^{-1}

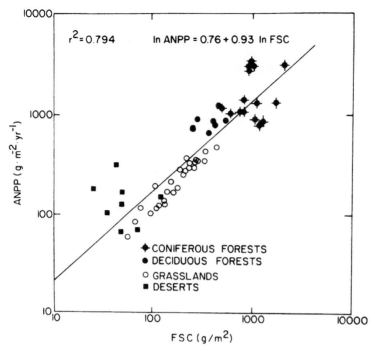

FIG. 11.3 Relationship between above-ground net primary production (ANPP) and peak foliar standing crop (FSC) for forest, grassland, and desert ecosystems. From Webb *et al.* (1983) by permission from the Ecological Society of America.

in tropical forests, swamps and marshes, and estuaries to < 200 g m^{-2} year^{-1} in tundra and deserts (Fig. 11.4) (Brown and Lugo, 1982; Waide *et al.*, 1999; Webb *et al.*, 1983; Whittaker, 1970).

Photosynthetic rates and NPP are sensitive to environmental conditions. Net primary production increases with annual precipitation and temperature,

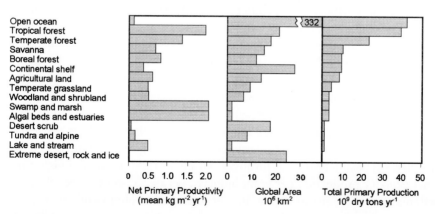

FIG. 11.4 Net primary production, total area, and contribution to global net primary production of the major biomes. Data from Whittaker (1970).

but the relationships are not simply linear (Whittaker, 1970). Low temperatures inhibit photosynthesis, and foliage loss in seasonally cold biomes reduces the period during which photosynthesis can occur. Photosynthetic rates and NPP also reflect access to photosynthetically active radiation, carbon dioxide, and water (Whittaker, 1970).

Photosynthetically active radiation occurs within the range 400–700 nm. The energy content of NPP divided by the supply of short-wave radiation, on an annual basis, provides a measure of photosynthetic efficiency (Webb *et al.*, 1983). Photosynthetic efficiency generally is low, ranging from 0.065 to 1.4% for ecosystems with low to high productivities (Sims and Singh, 1978; Whittaker, 1970).

Photosynthetically active radiation can be limited as a result of latitude, topography, cloud cover, or dense vegetation, which restrict penetration of short-wave radiation. Terborgh (1985) discussed the significance of differences in tree geometries among forest biomes. Boreal tree crowns are tall and narrow to maximize interception of lateral exposure to sunlight filtered through a greater thickness of atmosphere, whereas tropical tree crowns are umbrella-shaped to maximize interception of sunlight filtered through the thinner layer of atmosphere overhead. Solar penetration through tropical tree canopies, but not boreal tree canopies, is sufficient for development of multiple layers of understory plants.

The relationship between precipitation and potential evapotranspiration is an important factor affecting photosynthesis. Water limitation can result from insufficient precipitation, relative to evapotranspiration. Plants respond to water deficits by closing stomata, thereby reducing O_2 and CO_2 exchange with the atmosphere. Plants subject to frequent water deficits must solve the problem of acquiring CO_2, when stomatal opening facilitates water loss. Many desert and tropical epiphyte species are able to take up and store CO_2 as malate at night (when water loss is minimal) through crassulacean acid metabolism (CAM), then carboxylate the malate (to pyruvate) and refix the CO_2 through normal photosynthesis during the day (Winter and Smith, 1996; Woolhouse, 1981). Although CAM plants require high light levels to provide the energy for fixing CO_2 twice (Woolhouse, 1981), desert plants often have high photosynthetic efficiencies relative to foliage biomass (Webb *et al.*, 1983).

Air circulation is necessary to replenish CO_2 within the uptake zone neighboring the leaf surface. Although atmospheric concentrations of CO_2 may appear adequate, high rates of photosynthesis, especially in still air, can deplete CO_2 in the boundary area around the leaf, reducing photosynthetic efficiency.

Ruderal plants in terrestrial ecosystems and phytoplankton in aquatic ecosystems typically have high turnover rates (short life spans) and high rates of net primary production per gram biomass, because resources are relatively nonlimiting and the plants are composed primarily of photosynthetic tissues. Net primary production by all vegetation is low, however, because of the small biomass available for photosynthesis. By contrast, later successional plant species have low turnover rates (long life spans) and lower rates of net primary production per gram, because shading reduces photosynthetic efficiency and large portions of biomass necessary for support and access to sunlight are non-photosynthetic but still respire, e.g., wood and roots.

Typically, the NPP that is consumed by herbivores on an annual basis is low, an observation that prompted Hairston *et al.* (1960) to conclude that herbivores are not resource limited and must be controlled by predators. Early studies of energy content of plant material involved measurement of change in enthalpy (heat of combustion) rather than free energy (Wiegert, 1968). However, we now know that the energy initially stored as carbohydrates is incorporated, through a number of metabolic pathways, into a variety of compounds varying widely in their digestibility by herbivores. The energy stored in plant compounds often costs more to digest than the free energy it provides (see Chapters 3 and 4). Many of these herbivore-deterring compounds require energy expenditure by the plant, reducing the free energy available for growth and reproduction (e.g., Coley, 1986). The methods used to measure herbivory often greatly underestimate the turnover of NPP (Risley and Crossley, 1993; Schowalter and Lowman, 1999; see Chapter 12).

B. Secondary Productivity

Net primary production provides the energy for all heterotrophic activity. Consumers capture the energy stored within the organic molecules of their food sources. Therefore, each trophic level captures the energy represented by the biomass consumed from the lower trophic level. The rate of conversion of NPP into heterotroph tissues is **secondary productivity**. As with primary productivity, we can distinguish the total rate of energy consumption by secondary producers (gross secondary productivity) from the energy incorporated into consumer tissues (net secondary productivity) after expenditure of energy through respiration. Secondary productivity is limited by the amount of net primary production, because only the net energy stored in plant matter is available for consumers, secondary producers cannot consume more matter than is available, and energy is lost during each transfer between trophic levels.

Not all food energy removed by consumers is ingested. Consumer feeding often is wasteful. Scraps of food are dropped or damaged plant parts are abscissed (Risley and Crossley, 1993), making this material available to decomposers. Of the energy contained in ingested material, some is not assimilable and is egested, becoming available to reducers. A portion of assimilated energy must be used to support metabolic work, for maintenance, food acquisition, and various other activities, and is lost through respiration (see Chapter 4). The remainder is available for growth and reproduction (secondary production).

Secondary production can vary widely among heterotrophs and ecosystems. Herbivores generally have lower efficiencies of food conversion (ingestion/GPP < 10%) than do predators (<15%) because the chemical composition of animal food is more digestible than is plant food (Whittaker, 1970). Heterotherms have higher efficiencies than do homeotherms because of the greater respiratory losses associated with maintaining constant body temperature (Golley, 1968; see also Chapter 4). Therefore, ecosystems dominated by invertebrates or heterothermic vertebrates (e.g., most freshwater aquatic ecosystems dominated by insects and fish) will have higher rates of secondary production, relative to net primary production, than will ecosystems with greater representation of homeothermic vertebrates.

Eventually, all plant and animal matter enters the detrital pool as organisms die. The energy in detritus then becomes available to reducers (detritivores and decomposers). Detritivores fragment detritus and inoculate homogenized detritus with microbial decomposers during gut passage. Detrital material consists primarily of lignin and cellulose, but detritivores often improve their efficiency of energy assimilation by association with gut microorganisms or by reingestion of feces (coprophagy) following microbial decay of cellulose and lignin (Schowalter and Crossley, 1982).

C. Energy Budgets

Energy budgets can be developed from measurements of available solar energy, primary productivity, secondary productivity, decomposition, and respiration. Comparison of budgets and conversion efficiencies among ecosystems can indicate factors affecting energy flow and contributions to global energy budget. Development of energy budgets for agricultural ecosystems can be used to evaluate the efficiency of human resource production.

Lindeman (1942) was the first to demonstrate that ecosystem function can be represented by energy flow through a trophic pyramid or food web. He accounted for the energy stored in each trophic level, transferred between each pair of trophic levels, and lost through respiration. Odum (1957) and Teal (1957, 1962) calculated energy storage and rates of energy flow among trophic levels in several aquatic and wetland ecosystems (Fig. 11.5). Odum and Smalley (1959) and Smalley (1960) calculated energy flow through consumer populations. The International Biological Programme (IBP) focused attention on the energy budgets of various ecosystems (e.g., Bormann and Likens, 1979; Misra, 1968; Odum, 1969; Petrusewicz, 1967; Sims and Singh, 1978), including energy flow through insect populations (Kaczmarek and Wasilewski, 1977; McNeill and Lawton, 1970; Reichle and Crossley, 1967).

More recently, the energy budgets of agricultural ecosystems have been evaluated from the standpoint of energetic efficiency and sustainability. Whereas the energy available to natural communities comes from the sun, additional energy inputs are necessary to maintain agricultural productivity. These include energy from fossil fuels (used to produce fertilizers and pesticides and to power machinery) and from human and animal labor (Bayliss-Smith, 1990; Schroll, 1994). These additional inputs of energy have been difficult to quantify (Bayliss-

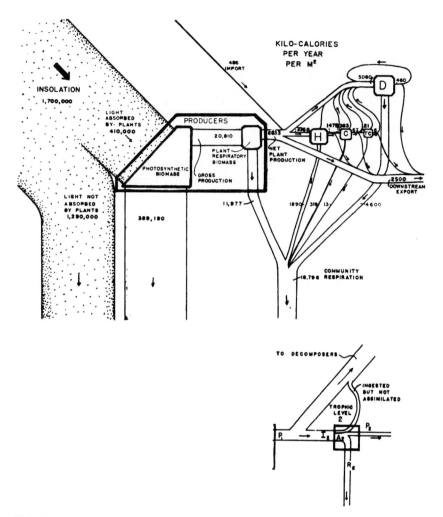

FIG. 11.5 Energy flow (kcal m^{-2} yr^{-1}) in the Silver Springs ecosystem. H, herbivores; C, predators; TC, top predators; D, decomposers. From Odum (1957) by permission from the Ecological Society of America.

Smith, 1990). Although the amount and value of food production is well known, the efficiency of food production (energy content of food produced per unit of energy input) is poorly known but critical to sustainability and economic development (Patnaik and Ramakrishnan, 1989). Promotion of predaceous insects to control pests, as an alternative to energy-expensive pesticides, and of soil organisms (including insects) to reduce loss of soil organic matter, as an alternative to fertilizers, has been proposed as a means to increase efficiency of agricultural production (Elliott *et al.*, 1984; Ostrom *et al.*, 1997).

Costanza *et al.* (1997), Daily (1997), Myers (1996), and Odum (1996) attempted to account for all energy used to produce and maintain the goods and services that support human culture. In addition to the market and energy value of current ecosystem resources, energy was expended in the past to produce those resources (fossil fuel). Additional energy is expended for transportation of resources to population centers and development of societal infrastructures. For example, given that NPP is proportional to biomass (Webb *et al.*, 1983), the energy used in the past to produce biomass must be included in the energy value of the system. When forests are harvested, the energy or resources derived from the timber can be replaced only by cumulative inputs of solar energy to replace the harvested biomass. Solar energy also generates tides and evaporates water necessary for maintenance of intertidal and terrestrial ecosystems and their resources.

Odum (1996) proposed the term **emergy** to denote the total amount of energy used to produce resources and cultural infrastructures. Costanza *et al.* (1997), Daily (1997), and Odum (1996) note that ecosystems provide a variety of "free" services, such as filtration of air and water and fertilization of floodplains, with energy from the sun and from topographic gradients, that must be replaced at the cost of fossil fuel expenditure when these services are lost as a result of environmental degradation (e.g., channelization and impoundment of streams). Sustainability of systems based on ecosystem resources thus depends on the energy derived from the ecosystem relative to the total emergy required to produce the resources. Consequently, many small-scale subsistence agricultural systems are far more efficient and sustainable than larger scale, industrial agricultural systems that could not be sustained without massive inputs from nonrenewable energy sources. Unfortunately, these more sustainable agroecosystems may not provide sufficient production to feed the growing world population.

III. BIOGEOCHEMICAL CYCLING

Organisms use the energy available to them as currency to acquire, concentrate, and organize chemical resources for growth and reproduction (see Chapter 4). Even sedentary organisms living in or on their material resources must expend energy to acquire resources against chemical gradients or to make these resources usable (e.g., through oxidation and reduction reactions necessary for digestion and assimilation). If energy gains do not at least equal energy expenditures, then energy is insufficient for further resource acquisition, growth, and reproduction.

Energy and matter are transferred from one trophic level to the next through consumption but, whereas energy is dissipated ultimately as heat, matter is conserved and can be reused. Conservation and reuse of nutrients within the ecosystem buffer organisms against resource limitation and contribute to ecosystem stability. The efficiency with which limiting elements are recycled varies among ecosystems. Biogeochemical cycling results from fluxes among biotic and abiotic storage pools.

Biogeochemical cycling occurs over a range of spatial and temporal scales. Cycling occurs within ecosystems as a result of trophic transfers and recycling of biotic materials made available through decomposition. Rapid cycling by microbial components is coupled with slower cycling by larger, longer-lived organisms within ecosystems. However, nutrients exported from one ecosystem become inputs for another. Detritus washed into streams during storms often is the major source of nutrients for the stream ecosystem. Nutrients moving downstream are major sources for estuarine and marine ecosystems. Nutrients lost to marine sediments become the source of nutrients from uplifted sediments for terrestrial ecosystems. Materials from storage in these long-term abiotic pools become available for extant ecosystems through weathering and erosion.

A. Abiotic and Biotic Pools

The sources of all elemental nutrients necessary for life are abiotic pools, the atmosphere, oceans, and sediments. The atmosphere is the primary source of nitrogen and of carbon (as carbon dioxide) and water for terrestrial ecosystems. Sediments are a major pool of carbon (as calcium carbonate), as well as the primary source of mineral elements, such as phosphorus and cations such as Na, K, Ca, and Mg released through chemical weathering. The ocean is the primary source of water but also is a major source of carbon (from carbonates) for marine organisms and of terrestrial cations that enter the atmosphere when winds > 20 kph lift water and dissolved minerals from the ocean surface.

Resources from abiotic pools are not available to all organisms but must be transformed into biologically useful compounds by autotrophic organisms. Photosynthetic plants acquire atmospheric or dissolved carbon dioxide and synthesize carbohydrates, which then are stored in biomass. Nitrogen-fixing bacteria and cyanobacteria acquire atmospheric or dissolved N_2 and convert it into ammonia, which they and some plants can incorporate directly into amino acids and nucleic acids. Nitrifying bacteria oxidize ammonia into nitrite and nitrate, the form of nitrogen available to most plants. These autotrophs also acquire other essential nutrients in dissolved form. The living and dead biomass of these organisms represents the pool of energy and nutrients available to heterotrophs.

The size of biotic pools represents storage capacity that buffers the organisms representing these pools against reduced availability of nutrients from abiotic sources. Larger organisms have a greater capacity to store energy and nutrients for use during periods of limited resource availability. Many plants can mobilize stored nutrients from tubers, rhizomes, or woody tissues to maintain metabolic activity during unfavorable periods. Similarly, larger animals can store energy, such as in the fat body of insects, and can retrieve nutrients from muscle or other tissues during periods of inadequate resource acquisition. Detritus represents a major pool of organic compounds. The nutrients from detritus become available to organisms through decomposition. Ecosystems with greater nutrient storage in living or dead biomass tend to be more resistant to certain environmental changes than are ecosystems with more limited storage capacity (Webster *et al.*, 1975).

B. Major Cycles

The biota modifies chemical fluxes. In the absence of biota, the rate and direction of chemical fluxes would be controlled solely by the physical and chemical factors determining exchanges between abiotic pools. Chemicals would be retained at a site only to the extent that chelation or concentration gradients restricted leaching or diffusion. Exposed nutrients would continue to move with wind or water (erosion). Biotic uptake and storage of chemical resources creates a biotic pool that reduces chemical storage in abiotic pools, altering rates of exchange among abiotic pools, and restricting movement of nutrients across chemical and topographic gradients. For example, the uptake and storage of atmospheric CO_2 by plants (including long-term storage in fossil biomass, i.e., coal, oil, and gas) and the uptake and storage of calcium carbonate by marine animals (and deposition in marine sediments) control concentration gradients of CO_2 available for exchange between the atmosphere and ocean (Keeling *et al.*, 1995; Sarmiento and Le Quéré, 1996). Conversely, fossil fuel combustion, deforestation and desertification, and destruction of coral reefs is reducing CO_2 uptake by biota and releasing CO_2 from biotic storage, thereby increasing global CO_2 available for exchange between the atmosphere and ocean. Biotic uptake of various sedimentary nutrients retards their movement from higher elevations back to marine sediments.

   Consumers, including insects, affect the rate at which nutrients are acquired and stored (see Chapters 12–14). Consumption reduces the biomass of the lower trophic level, thereby reducing nutrient uptake at that trophic level, and moves nutrients from consumed biomass into biomass at the higher trophic level (through secondary production) or into the environment (through secretion and excretion) where nutrients become available to detritivores, soil microorganisms, or are exported via water flow to aquatic food webs. Nutrients are recycled through decomposition of dead plant and animal biomass, which releases simple organic compounds or elements into soil solution for reacquisition by autotrophs.

   Some nutrients are lost during trophic transfers. Carbon is lost (exported) from ecosystems as CO_2 during respiration. Gaseous or dissolved CO_2 remains available to organisms in the atmospheric and oceanic pools. Organic biomass can be blown or washed away. Soluble nutrients are exported as water percolates through the ecosystem and enters streams. The efficiency with which nutrients are retained within an ecosystem reflects their relative availability. Nutrients such as nitrogen and phosphorus often are limiting and tend to be cycled and retained in biomass more efficiently than are nutrients that are more consistently available, such as K and Ca. The following four examples exemplify the processes involved in biogeochemical cycling.

1. Hydric Cycle

Water availability, as discussed in Chapters 2 and 9, is one of the most important factors affecting the distribution of terrestrial organisms. Many organisms are modified to optimize their water balances, e.g., through their adaptations for acquiring and retaining water (Chapter 2). Water available to plants is a primary factor affecting photosynthesis and ecosystem energetics. Water absorbs solar energy, with little change in temperature, thereby buffering humid ecosystems against large changes in temperature. At the same time, water use by organisms significantly affects its passage through terrestrial ecosystems.

The primary source of water for terrestrial ecosystems is water vapor from evaporation over the oceans (Fig. 11.6). The availability of water to terrestrial ecosystems is controlled by a variety of factors, including the rate of evaporation from the ocean, the direction of prevailing winds, atmospheric and topographic factors that affect convection and precipitation, temperature, relative humidity, and soil texture. Water enters terrestrial ecosystems as precipitation and condensation and as subsurface flow and groundwater derived from precipitation or condensation at higher elevations. Condensation may be a major avenue for water input to arid ecosystems. Many plants in arid regions are

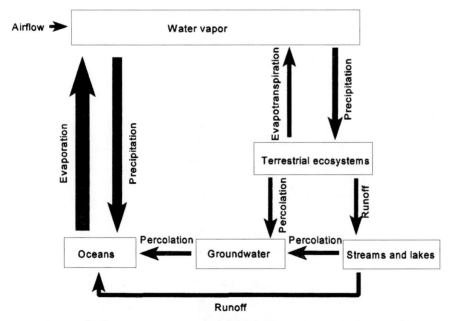

FIG. 11.6 The hydric cycle. Net evaporation over the oceans is the source of water vapor carried inland by air currents. Water precipitated into terrestrial ecosystems eventually is returned to the ocean.

adapted to acquire water through condensation. Some desert insects also acquire water through condensation on specialized hairs or body parts (Chapman, 1982). Vegetation intercepts up to 50% of precipitation, depending on crown structure and plant surface area (Parker, 1983). Most intercepted water evaporates. The remainder penetrates the vegetation as throughfall (water dripping from foliage) and stemflow (water funneled to stems).

Vegetation takes up water primarily from the soil, using some in the synthesis of carbohydrates. Vascular plants conduct water upward and transpire much of it through the stomata. Evapotranspiration is the major mechanism for maintaining the upward capillary flow of water from the soil to the canopy. This active evaporative process greatly increases the amount of water moving back into the atmosphere, rather than flowing downslope, and may increase the availability of water for a particular site, as will be discussed.

Vegetation stores large amounts of water intra- and extracellularly and controls the flux of water to the atmosphere. Accumulation of organic material increases soil water storage capacity and further reduces downslope flow. Soil water storage mediates plant acquisition of other nutrients in dissolved form. Food passage through arthropods and earthworms, together with materials secreted by soil microflora, binds soil particles together, forming soil aggregates (Hendrix *et al.*, 1990; Setälä *et al.*, 1996). These aggregates increase water and nutrient storage capacity and reduce erosibility. Burrowing organisms, such as earthworms and wood borers, increase the porosity and water storage capacity of soil and decomposing wood (e.g., Eldridge, 1994). Macropore flow increases the rate and depth of water infiltration.

Some organisms also control water movement in streams. Swamp and marsh vegetation restricts water flow in low gradient ecosystems. Trees falling into stream channels impede water flow. Similarly, beaver dams impede water flow and store water in ponds. However, water eventually evaporates or reaches the ocean, completing the cycle.

2. Carbon Cycle

The carbon cycle (Fig. 11.7) is particularly important because of its intimate association with energy flow, via the transfer of chemical energy in carbohydrates, through ecosystems. Carbon is stored globally both as atmospheric carbon dioxide and as sedimentary and dissolved carbonates (principally, calcium carbonate). The atmosphere and ocean mediate the global cycling of carbon among terrestrial and aquatic ecosystems. The exchange of carbon between atmosphere and dissolved or precipitated carbonates is controlled by temperature, carbonate concentration, salinity, and biological uptake that affects concentration gradients (Keeling *et al.*, 1995; Sarmiento and Le Quéré, 1996).

Carbon enters ecosystems primarily as a result of photosynthetic fixation of CO_2 in carbohydrates. The chemical energy stored in carbohydrates is used to synthesize all the organic molecules used by plants and animals. Carbon en-

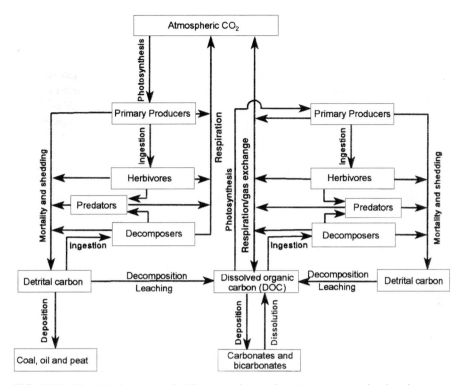

FIG. 11.7   The global carbon cycle. The atmosphere is the primary source of carbon for terrestrial ecosystems (left), whereas dissolved carbonates and bicarbonates are the primary source of carbon for marine ecosystems (right). Exchange of carbon between atmosphere, hydrosphere, and geosphere is regulated largely by biotic uptake and deposition.

ters many aquatic ecosystems, especially those with limited photosynthesis, primarily as allochthonus inputs of exported terrestrial materials (e.g., terrestrial organisms captured by aquatic animals, detritus, and dissolved organic material entering with runoff or leachate). Carbon is transferred among trophic levels through consumption, converted into an astounding diversity of compounds for a variety of uses, and eventually is returned to the atmosphere as CO_2 from respiration, especially during decomposition of dead organic material, completing the cycle. However, loss of carbon from an ecosystem is minimized by rapid acquisition and immobilization of soluble and fine particulate carbon by soil organisms and aquatic filter feeders, from which carbon becomes available for transfer within soil and aquatic food webs (de Ruiter *et al.*, 1995; Wallace and Hutchens, 2000).

However, some carbon compounds (especially complex polyphenols, such as lignin) decompose very slowly, if at all, and are stored for long periods as soil organic matter, peat, coal, or oil. Humic compounds are phenolic polymers that are resistant to chemical decomposition and constitute long-term carbon storage in terrestrial soils. These compounds contribute to soil water and nutrient holding capacities because of their large surface area and numerous binding sites. Plants produce organic acids that are secreted into the soil through roots. These acids facilitate extraction of mineral nutrients from soil exchange sites, maintain ionic balance (with mineral cations), reduce soil pH, and inhibit decomposition of organic matter. Similarly, peat accumulates in bogs where low pH inhibits decomposition and eventually may be buried, contributing to formation of coal or oil. Coal and oil represent long-term storage of accumulated organic matter that decomposed incompletely as a result of burial, anaerobic conditions, and high pressure. The carbon removed from the atmosphere by these fossil plants is now reentering the atmosphere rapidly, as a result of fossil fuel combustion, leading to increased atmospheric concentrations of CO_2.

3. Nitrogen Cycle

Nitrogen is a critical element for synthesis of proteins and nucleic acids and is available in limited amounts in most ecosystems. The atmosphere is the reservoir of elemental nitrogen, making nitrogen an example of a nutrient with an atmospheric cycle (Fig. 11.8). Most organisms cannot use gaseous nitrogen and many other nitrogen compounds. In fact, some nitrogen compounds are toxic in small amounts to most organisms (e.g., ammonia). Nitrogen cycling is mediated by several groups of microorganisms that transform toxic or unavailable forms of nitrogen into biologically useful compounds.

Gaseous N_2 from the atmosphere becomes available to organisms through fixation in ammonia, primarily by nitrogen-fixing bacteria and cyanobacteria. These organisms are key components of most ecosystems, but are particularly important in ecosystems subject to periodic massive losses of nitrogen, such as through fire. Many early successional plants, especially in fire-dominated ecosystems, have symbiotic association with nitrogen-fixing bacteria in root nodules. These plants can use the ammonia produced by the associated bacteria, but most plants require nitrate (NO_3) as their source of nitrogen.

Ammonium compounds also are produced by lightning and volcanic eruptions. Nitrifying bacteria oxidize ammonia to nitrite (NO_2) and nitrate, which then is available to plants, for synthesis of amino acids and nucleic acids, and is transferred to higher trophic levels through consumption. The nitrogen compounds in dead organic matter are decomposed to ammonium by ammonifying bacteria. Organic nitrogen enters aquatic ecosystems as exported terrestrial organisms, detritus, or runoff and leachate solutions. Nitrogen in freshwater ecosystems similarly is transferred among trophic levels through consumption, eventually reaching marine ecosystems. Under anaerobic conditions, which occur naturally as well as a result of eutrophication or soil compaction, the cycle can be disrupted by anaerobic denitrifying bacteria that convert nitrate to gaseous nitrogen, which is lost to the atmosphere, thereby completing the cycle. However, nitrogen loss is minimized by soil organisms that aerate the soil through excavation and by the rapid acquisition and immobilization of soluble

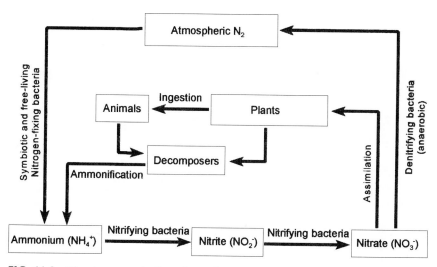

FIG. 11.8 The nitrogen cycle. Bacteria are the primary organisms responsible for transforming elemental nitrogen into forms available for assimilation by plants. Note that the return of nitrogen to the atmospheric pool occurs almost exclusively under anaerobic conditions.

nitrogen by soil microorganisms and aquatic filter feeders, from which nitrogen becomes available to plants and soil and aquatic food webs.

4. Sedimentary Cycles

Many nutrients, including phosphorus and mineral cations, are available only from sedimentary sources. These nutrients are cycled in similar ways, as exemplified by phosphorus (Fig. 11.9). Phosphorus is biologically important in molecules that mediate energy exchange during metabolic processes (ATP and ADP) and in phospholipids. Like nitrogen, it is available to organisms only in certain forms and is in limiting supply in most ecosystems. Phosphorus and mineral cations become available to terrestrial ecosystems as a result of chemical weathering or erosion of geologically uplifted, phosphate-bearing sediments.

Phosphate enters an ecosystem from weathered bedrock and moves among terrestrial ecosystems through materials washed downslope or filtered from the air. Phosphorus is highly reactive but available to plants only as phosphate that often is bound to soil particles. Plants extract phosphorus (and mineral cations) from cation exchange and sorption sites on soil particles and from soil solution. Phosphorus then is synthesized into biological molecules and transferred to higher trophic levels through consumption; it eventually is returned to the soil as dead organic matter and is decomposed. Phosphorus enters aquatic ecosystems largely in particulate forms exported from terrestrial ecosystems. It is transferred between aquatic trophic levels through consumption, eventually being deposited in deep ocean sediments, completing the cycle. Phosphorus loss is minimized by soil organisms and aquatic filter feeders which rapidly acquire

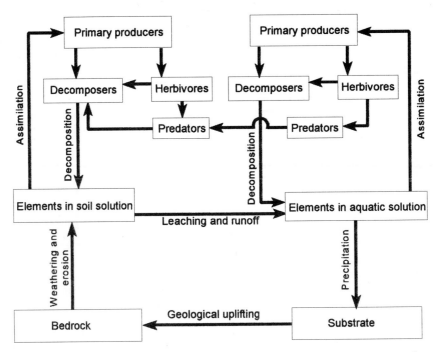

FIG. 11.9 Sedimentary cycle. Phosphorus and other nongaseous nutrients precipitate from solution and are stored largely in sediments of marine origin. These nutrients become available to terrestrial ecosystems primarily through chemical weathering of uplifted sediments.

and immobilize soluble phosphorus and make it available for plant uptake and exchange among soil and aquatic organisms.

C. Factors Influencing Cycling Processes

A number of factors alter the rates and pathways of biogeochemical fluxes. Variation in fluxes reflects the chemical properties and source of the nutrient, interactions with other cycles, and the composition of the community, especially the presence of specialized organisms that control particular fluxes. Hence, changes in community composition resulting from disturbance and recovery alter the rates and pathways of chemical fluxes.

The chemical properties of various elements and compounds, especially their solubility and susceptibility to pH changes, as well as biological uses, affect cycling behavior. Some elements, such as Na and K, form compounds that are readily soluble over normal ranges of pH. These elements generally have high rates of input to ecosystems via precipitation but also high rates of export via runoff and leaching. Other elements, such as Ca and Mg, form compounds that are not as soluble over usual ranges of pH and have lower rates of input and export. Elements such as nitrogen and phosphorus are necessary for all organisms, relatively limiting, and generally conserved within organisms. For example, deciduous trees typically resorb nitrogen from senescing foliage prior to leaf fall. Sodium has no known function in plants and is not retained in plant tissues, but is required by animals for osmotic balance and for muscle and nerve function. Consequently, it is conserved tightly by these organisms. In fact, animals often seek mineral sources of sodium (e.g., Seastedt and Crossley, 1981b). Many decay fungi accumulate sodium (Cromack *et al.*, 1975; Schowalter *et al.*, 1998), despite absence of apparent use in fungal metabolism, perhaps to attract animal vectors of fungal spores.

Biogeochemical cycles interact with each other in complex ways. For example, some plants respond to increased atmospheric CO_2 by reducing stomatal opening, thereby acquiring sufficient CO_2 while reducing water loss. Hence, increased size of the atmospheric pool of CO_2 may alter transpiration, permitting some plant species to colonize more arid habitats. Similarly, the calcium cycle interacts with cycles of several other elements. Calcium carbonate generally accumulates in arid soils as soil water evaporates. Acidic precipitation, such as resulting from industrial emission of nitrous oxides and sulfur dioxide into the atmosphere, dissolves and leaches calcium carbonate from soils and sediments. Soils with high content of calcium carbonate are relatively buffered against pH change, whereas soils depleted of calcium carbonate become acidic, increasing export (through leaching) of other cations as well.

Some biogeochemical fluxes are controlled by particular organisms. The nitrogen cycle depends on several groups of microorganisms that control the transformation of nitrogen among various forms that are available or unavailable to other organisms (see previous). Soil biota secrete substances that bind soil particles into aggregates that facilitate retention of soil water and nutrients. Some plants [e.g., western redcedar (*Thuja plicata*) and dogwoods (*Cornus* spp.)] accumulate calcium in their tissues (Kiilsgaard *et al.*, 1987) and generally increase pH and buffering capacity of surrounding soils. Their presence or absence thereby affects retention of other nutrients, as well.

Changes in community composition following disturbance or during succession affect rates and pathways of biogeochemical fluxes. Early successional communities frequently are inefficient because of limited competition for resources by the small biomass, and early successional species have little selective pressure to retain nutrients. For example, the early successional tropical tree, *Cecropia* spp., has large, thin leaves that transpire water more rapidly than do the smaller, more sclerotized leaves of later successional species. Although later successional communities are not always efficient, declining resource supply relative to growing biomass promotes efficiency of nutrient retention within the ecosystem (Odum, 1969; Schowalter, 1981). Agricultural and silvicultural systems are inefficient largely because communities composed of a single or few plant species cannot acquire or retain all available forms of matter effectively. For example, nitrogen fixation often is controlled by noncommercial species, such as symbiotic nitrogen-fixing lichens, herbs and shrubs, or structures, such as large decomposing woody litter, that are suppressed or eliminated by management activities. Necessary nitrogen must be supplied anthropogenically, often in excess amounts that leach into groundwater and streams.

IV. CLIMATE MODIFICATION

Although most previous studies have emphasized the effect of climate on survival and distribution of organisms (see Chapters 2 and 7), communities of organisms also modify local and regional climatic conditions, perhaps influencing global climatic gradients (Chase *et al.*, 1996; Parker, 1995; Pielke and Vidale, 1995). Climate modification largely reflects the capacity of vegetation to shade and protect the soil surface, abate airflow, and control water fluxes (Fig. 11.10). Hence, biomes and successional stages vary widely in ability to modify climate.

When vegetation development or moisture is limited, as in deserts, or reduced, as in desertified regions, the soil surface is exposed fully to sunlight and contains insufficient water to restrict temperature change (Lewis, 1998). Sur-

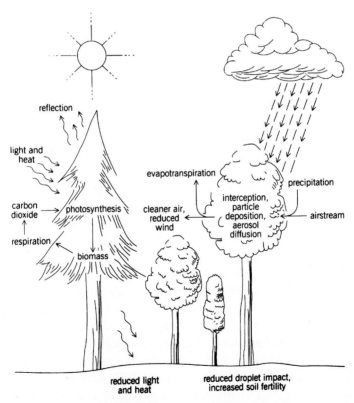

FIG. 11.10 Diagrammatic representation of the effects of vegetation on climate and atmospheric variables. The capacity of vegetation to modify climate depends on vegetation density and vertical height and complexity. Reproduced from Schowalter (1991), "Forest ecology," in *1992 Yearbook of Science and Technology*, McGraw-Hill, New York, by permission from The McGraw Companies.

face temperatures can reach 60–70°C during the day (e.g., Seastedt and Crossley, 1981a), but fall rapidly at night as a result of long-wavelength radiation. Deforested areas also show lower precipitation, compared to forested areas (Meher-Homji, 1991). Surface warming increases the convective rise of water vapor, increasing the likelihood of condensation and precipitation, but also increasing export of moisture-laden air in ecosystems exposed to high winds. Where biotic capacity to take up and impede movement of water over and through soils is limited, water flows downslope at a rate determined primarily by topography, soil texture, and sedimentary properties, eventually returning to the ocean via streamflow.

Vegetation reduces diurnal surface temperatures (Lewis, 1998) by reflecting much solar energy, absorbing energy stored in photosynthates, and using the remainder to evaporate transpired water (Parker, 1995). The canopy absorbs reradiated energy from the forest floor at night, maintaining warmer nocturnal temperatures. Canopy cover reduces the impact of raindrops on the soil surface and impedes the downslope movement of water, thereby reducing erosion and loss of soil (Fig. 11.11). Soil organic matter retains water, increasing soil moisture capacity and reducing temperature change. Salati (1987) reported that 30% of precipitation in the Amazon basin was generated locally by evapotranspiration. This recycling of water may be most pronounced in montane areas, where steep vertical temperature gradients condense rising evapotranspired water. Reduced air flow and increased turbulence caused by an irregular vegetation height profile reduce wind erosion and facilitate filtering of materials from the airstream. Exposure of individual organisms to damaging or lethal wind speeds is reduced as a result of buffering by surrounding individuals.

Modification of climatic conditions depends on vegetation density and vertical structure. Sparse vegetation has a lower capacity to modify temperature, water flow, and wind speed than does dense vegetation. Albedo is related inversely to vegetation height and "roughness," declining from 0.25 for vegetation < 1.0 m in height to 0.10 for vegetation > 30 m height, and generally reaches highest values in vegetation with a smooth canopy surface (Monteith, 1973). Compared to taller vegetation, short vegetation traps less radiation between multiple layers of leaves and stems and modifies climatic conditions within a shallower column of air.

Parker (1995) demonstrated that the canopy above 30 m was affected most by rising temperatures during midday in a temperate forest (Fig. 11.12). Temperature between 40 and 50 m height ranged from 16°C at night to 38°C during midafternoon (a diurnal range of 22°C); relative humidity in this canopy zone declined from > 95% at night to 50% during midafternoon. Below 10 m, temperature range was only 10°C and relative humidity was constant at > 95%. Windsor (1990) reported similar gradients in canopy environment in a lowland tropical forest.

Tall, multicanopied forests have the greatest capacity to modify local and regional climate because the stratified layers of foliage and denser understory

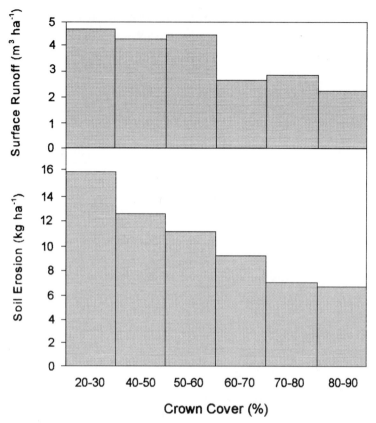

FIG. 11.11 Effect of canopy cover on average runoff and soil erosion, based on 41 runoff-producing storms totaling 1128 mm in northern Thailand. Data from Ruangpanit (1985).

successively trap filtered sunlight, intercept precipitation and throughfall, and impede airflow in the deepest column of air. Soils with high organic content have lower albedo (0.10) than does desert sand (0.30) (Monteith, 1973). However, the thin (3 mm) biological crusts, composed of cyanobacteria, green algae, lichens and mosses, on the surface of soils in arid and semiarid regions are capable of substantially reducing wind erosion (Belnap and Gillette, 1998).

Insects and other organisms (including humans) alter vegetation and soil structure (Fig. 11.13) and thereby affect biotic control of local and regional climate (see Chapters 12–14). Reduction in vegetative cover increases surface temperatures and reduces relative humidity (Lewis, 1998; Salati, 1987). Schlesinger *et al.* (1990) reported that desertification results in a destabilizing positive feedback, whereby initial vegetation removal causes surface warming and drying that stresses and kills adjacent vegetation, leading to an advancing arc of desertified land. Forest fragmentation increases wind fetch and penetration of air from surrounding crop or pasture zones into forest fragments (Chen *et al.*, 1995). Belnap and Gillette (1998) found that disturbance of the brittle bi-

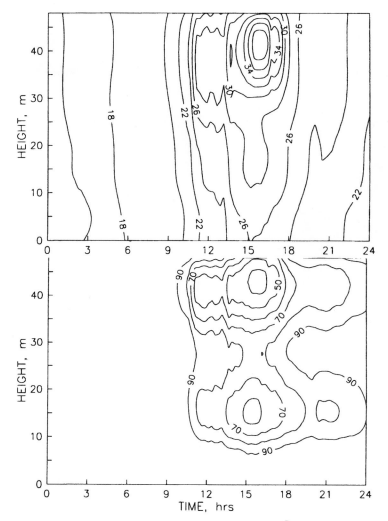

FIG. 11.12 Height–time profiles of air temperature and relative humidity in mixed-hardwood forest in Maryland. Temperature contours are 2°C; relative humidity contours are 10% units. Nocturnal temperature gradients are weak, but a hot spot develops in the upper canopy in midafternoon. Humidity declined in the upper canopy in midafternoon, coincident with peak temperatures, and was near saturation (> 95%) outside the marked contours. From Parker (1995).

ological crusts on desert soils greatly increases the effect of wind on soil loss. The contributions of insect outbreaks to changes in climatic conditions have not been evaluated.

V. MODELING

Modeling has become a useful tool for testing hypotheses concerning behavior and self regulation of complex systems (e.g., Camilo and Willig, 1995; Patten,

FIG. 11.13   Deforestation in Panama. Removal of tropical rain forest cover has exposed soil to solar heating and severe erosion, leading to continued ecosystem deterioration and, potentially, to altered regional temperature and precipitation patterns.

1995; Ulanowicz, 1995) and for predicting ecosystem responses to environmental changes, as well as ecosystem contributions to environmental change, especially carbon flux (e.g., Rastetter *et al.*, 1991; Sarmiento and Le Quéré, 1996). The logistical difficulty of measuring and manipulating all ecosystem components and processes for experimental purposes has placed greater emphasis on modeling to simulate experimental conditions and to identify critical components and processes for further study.

Modeling at the ecosystem level necessarily starts with conceptual models of linkages among components and reflects the perception of individual modelers of the importance of particular components and interactions (e.g., Figs. 1.3, 11.1, 11.6–11.9). Models differ in the degree to which species are distinguished in individual submodels or combined into functional group submodels (e.g., de Ruiter *et al.*, 1995; Naeem, 1998; Polis, 1991b; Reagan *et al.*, 1996) and to which light, water, and nutrient availability are integrated simultaneously with changes in ecosystem structure and composition (e.g., Waring and Running, 1998). Obviously, conceptualizing the integration of the many thousands of species in a given ecosystem is virtually impossible. On the other hand, some global-scale models distinguish the biota only at the community level, if at all. The degree to which individual species are distinguished influences the representation of the variety of interactions and feedbacks that influence ecosystem parameters (Naeem, 1998; Polis, 1991b; Reagan *et al.*, 1996). Similarly, models based on a limited set of variables to predict a single type of output (e.g., carbon flux) may fail to account for effects of other variables (e.g., effects of limiting nutrients, such as nitrogen, on carbon flux) (Waring and Running, 1998). More general models require simplifying assumptions to expand their application and may lose accuracy as a consequence.

After the conceptual organization of the model has been determined, inter-action strengths are quantified (Figs. 11.14 and 11.15), based on available data, or subjected to sensitivity analysis to identify the range of values that represent observed interaction (e.g., Benke and Wallace, 1997; de Ruiter *et al.*, 1995; Parton *et al.*, 1993; Rastetter *et al.*, 1991; Running and Gower, 1991). Direct and indirect interactions can be represented in transition matrix form, e.g.,

	N_1	N_2	N_3	N_4	.	.	N_i
N_1	α_{11}	α_{21}	α_{31}	α_{41}	.	.	α_{i1}
N_2	α_{12}	α_{22}	α_{32}	α_{42}	.	.	α_{i2}
N_3	α_{13}	α_{23}	α_{33}	α_{43}	.	.	α_{i3}
N_4	α_{14}	α_{24}	α_{34}	α_{44}	.	.	α_{i4}
.	.	.	.	.	.	.	.
.	.	.	.	.	.	.	.
N_j	α_{1j}	α_{2j}	α_{3j}	α_{4j}	.	.	α_{ij}

where N_j is the *j*th ecosystem component, and α_{ij} is the relative effect (direct + indirect) of N_j on N_i. When $N_i = N_j$, α_{ij} represents intrinsic effects on numbers or mass. Differential equations of the general form

$$N_{i(t+1)} = N_{it} + \Sigma(\alpha_{ij}N_{jt}) \tag{11.1}$$

are used to calculate the transitional states of each component as input conditions change. Note the application of this inclusive equation to equations for growth of individual populations and interacting species in Chapters 6 and 8. Components must be linked so that changes in the number, mass or energy, or nutrient content of one component have appropriate effects on the numbers, masses or energy, or nutrient contents of other components. Models focused on species will emphasize species interactions and fluxes of energy or matter through food webs. Models focused on energy or matter pools emphasize fluxes of energy and matter among pools, but may include important species that affect flux rates.

Ecosystem models are sensitive to effects of indirect interactions. For example, a direct predator–prey interaction reduces prey abundance, thereby indirectly affecting interactions between the prey and its competitors, hosts, and other predators. Ultimately, indirect effects of this interaction can affect primary production, canopy cover, and resource availability in ways that determine climate, substrate, and resource conditions for the entire ecosystem. Nontrophic interactions are difficult to recognize and measure (Dambacher *et al.*, 1999); quantitative data are available for relatively few potential indirect interactions. Accordingly, the complexity of indirect, as well as direct, interactions is difficult to model, but has important implications for how ecosystems respond to environmental changes (see Chapter 15).

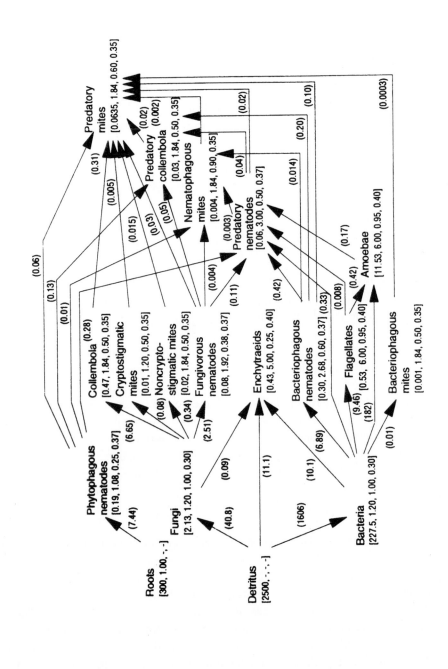

Predatory
mites
[0.0635, 1.84, 0.60, 0.35]

(0.31)

(0.005)

Predatory   (0.02)
collembola   (0.002)
[0.03, 1.84, 0.50, 0.35]

(0.015)

(0.03)

(0.05)

Nematophagous
mites
[0.004, 1.84, 0.90, 0.35]

(0.02)

(0.04)

Predatory
nematodes
[0.06, 3.00, 0.50, 0.37]

(0.003)

(0.014)

(0.10)

(0.20)

(0.06)

(0.13)

(0.01)

Collembola (0.28)
[0.47, 1.84, 0.50, 0.35]

Cryptostigmatic
mites
[0.01, 1.20, 0.50, 0.35]

Noncrypto-
stigmatic mites
[0.02, 1.84, 0.50, 0.35]

Fungivorous
nematodes
[0.08, 1.92, 0.38, 0.37]

(0.08)

(0.34)

(0.004)

(0.11)

Enchytraeids
[0.43, 5.00, 0.25, 0.40]

Bacteriophagous
nematodes
[0.30, 2.68, 0.60, 0.37]

(0.42)

(0.33)

(0.008)

Flagellates
[0.53, 6.00, 0.95, 0.40]
(9.46)
(182)

(0.17)

Amoebae
[11.53, 6.00, 0.95, 0.40]

(0.42)

Bacteriophagous
mites
[0.001, 1.84, 0.50, 0.35]

(0.0003)

Phytophagous
nematodes
[0.19, 1.08, 0.25, 0.37]
(7.44)

Roots
[300, 1.00, -, -]

Fungi
[2.13, 1.20, 1.00, 0.30]
(6.65)
(2.51)

(0.09)

(11.1)

Detritus
[2500, -, -, -]

(40.8)

(1606)

(10.1)

(6.89)

(0.01)

Bacteria
[227.5, 1.20, 1.00, 0.30]

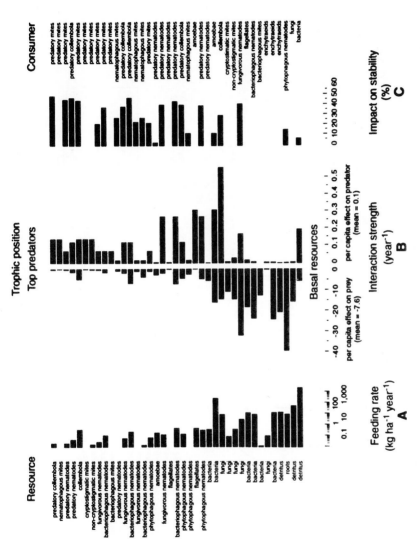

FIG. 11.14 Quantification of feeding rates (top), interaction strengths as per capita effects, and impact of these interactions on soil food web stability (bottom) in conventional agriculture at Lovinkhoeve Experimental Farm, The Netherlands. Reprinted with permission from de Ruiter, P. C., A. M. Neutel, and J. C. Moore. 1995. Energetics, patterns of interaction strengths, and stability in real ecosystems. *Science* **269**: 1257–1260. Copyright 1995 American Association for the Advancement of Science.

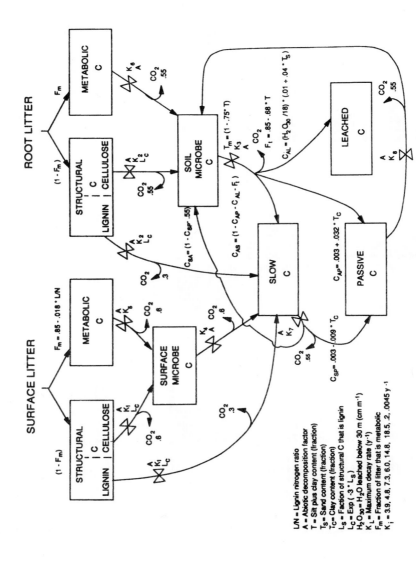

FIG. 11.15 Detail of carbon fluxes in the soil organic carbon submodel of the CENTURY ecosystem model. From Parton *et al.* (1993), courtesy of the American Geophysical Union.

A number of models have been developed to predict fluxes of energy or particular elements, especially carbon or nitrogen, through ecosystems. However, as noted above, interactions among various cycles (e.g., carbon and calcium) may confound predictions based on individual resources.

Comprehensive ecosystem models that integrate energy, carbon, water, and nutrient fluxes include FOREST-BGC/BIOME-BGC (Running and Gower, 1991) and CENTURY (e.g., Fig. 11.15) (Parton *et al.*, 1993), which have been modified to represent a variety of ecosystem types. These models are useful for predicting global biogeochemical processes because they integrate common ecosystem processes in a logical framework and have minimum requirements for detail of inputs on ecosystem characteristics. However, these models emphasize pool sizes and interactions among plant, litter, and soil subsystems. The effects of insects and other invertebrates have been incorporated poorly or not at all in these, or other, existing ecosystem models. At best, insects typically are combined as insects or arthropods, thereby losing valuable information about this diverse group, that can respond dramatically to environmental change and have major effects on ecosystem properties (Chapters 12–14).

VI. SUMMARY

An ecosystem represents the integration of the biotic community and the abiotic environment. The capacity of the community to modify its environment depends on its structure and the degree to which it controls ecosystem processes, especially energy flow, biogeochemical cycling, and climatic conditions.

Ecosystem structure reflects the organization of various abiotic and biotic pools that exchange energy and matter. Abiotic pools are the atmosphere, oceans, and sediments that represent the sources of energy and matter for biotic use. Biotic pools are the various organisms (individuals, species populations, functional groups, or trophic levels) in the community. Autotrophs (or primary producers) are those organisms that can acquire resources from abiotic pools. Heterotrophs (or secondary producers) are those organisms that must acquire their resources from other organisms. Energy and matter storage in these pools can be represented as pyramids of productivity, numbers, or biomass.

Energy available to ecosystems comes primarily from solar radiation, captured and stored in carbohydrates by primary producers through the process of photosynthesis. The total rate at which energy is captured (gross primary productivity) depends on exposure to sunlight, availability of water, and biomass. Some of the energy from gross primary production is expended in plant respiration. The remaining net primary production is stored as plant biomass and is the source of energy and matter for heterotrophs. Primary heterotrophs (herbivores) feed on autotrophs, whereas secondary heterotrophs (predators) feed on other heterotrophs. Consumption transfers the energy stored in consumed biomass to the higher trophic level, with some lost as egestion and consumer respiration. Generally, $< 10\%$ of the energy available at each trophic level is converted into biomass at the next higher trophic level, although predators generally have a higher efficiency of conversion than do herbivores. Energy remaining in organisms at the time of death becomes available to decomposers that release the remaining energy through respiration.

Energy is the currency with which organisms acquire and concentrate material resources necessary for growth and reproduction. Material resources are often available in limited supply, favoring mechanisms that facilitate retention and reuse within the ecosystem. Biogeochemical cycling represents the processes whereby material resources, including water, carbon, nitrogen, and mineral elements, are acquired from abiotic pools and recycled among trophic levels, with eventual return to abiotic pools. The efficiency with which these materials are recycled and conserved, rather than lost to abiotic pools, buffers an ecosystem against resource depletion and reduced productivity. Hence, ecosystems become organized in ways that maximize the capture and storage of resources among organisms. Resources egested or excreted during trophic transfers, as well as dead organisms, become available to decomposers that rapidly acquire and store the nutrients from organic matter. Nutrients in decomposers become available for exchange among soil and aquatic organisms and for plant uptake. Microorganisms are particularly instrumental in making nitrogen available for plant uptake, with different specialists fixing atmospheric nitrogen as ammonia, converting ammonia to nitrate, and organic nitrogen to ammonia. Volatilization by fire and denitrification by anaerobic bacteria complete the cycle by returning elemental nitrogen to the atmosphere.

Ecosystems also modify local and regional climatic conditions. The degree to which vegetation reduces soil warming, evaporation, erosion, altitude at which water vapor condenses, and wind speed depends on density and vertical architecture. Insects and other organisms affect vegetation structure, hence, canopy–atmosphere interactions. Tall, multicanopied forests are most effective at modifying surface temperatures, relative humidities, wind speed, and the altitude at which evapotranspired water condenses, thereby ameliorating local and regional fluctuations in temperature, wind speed, and precipitation.

Models have become important tools for synthesizing complex, and often incomplete, data for prediction of ecosystem responses to, and effects on, global environmental changes. Ecosystem models differ in structure and degree of simplification. Effects of insects on a variety of ecosystem parameters largely have been ignored in ecosystem models.

12

Herbivory

HERBIVORY IS THE RATE OF CONSUMPTION BY ANIMALS OF ANY PLANT parts, including foliage, stems, roots, flowers, fruits, or seeds. Direct effects of insects on plant reproductive parts are addressed in Chapter 13. Herbivory is a key ecosystem process that reduces density of plants or plant materials, transfers mass and nutrients to the soil, and affects habitat and resource conditions for other organisms. Insects are the primary herbivores in many ecosystems, and their effect on primary production can equal or exceed that of more conspicuous vertebrate grazers in grasslands (e.g., Andersen and Lonsdale, 1990; Gandar, 1982; Sinclair, 1975; Wiegert and Evans, 1967).

Loss of plant material through herbivory generally is negligible, or at least inconspicuous, but periodic outbreaks of herbivores have a well-known capacity to reduce growth and survival of host species by as much as 100% and to alter vegetation structure over large areas. A key aspect of herbivory is its variation in intensity among plant species, reflecting biochemical interactions between the herbivore and the various host and nonhost species that comprise the vegetation (see Chapter 3).

Effects of herbivory on ecosystem processes depend on the type of herbivore and pattern of consumption, as well as its intensity (rate). Measurement and comparison of herbivory and its effects among ecosystems and environmental conditions remain problematic due to lack of standardized techniques for measuring or manipulating intensity. Few studies have assessed the effects of herbivory on ecosytem processes other than primary production. Nevertheless, accumulating evidence indicates that effects of herbivory on ecosystem processes, including primary production, are complex. Ecosystem management practices that exacerbate or suppress herbivory may be counterproductive.

I. TYPES AND PATTERNS OF HERBIVORY

A. Herbivore Functional Groups

Herbivorous insects that have similar means of exploiting plant parts for food can be classified into feeding guilds or functional groups. Groups of plant-feeders include **chewers** that consume foliage, stems, flowers, pollen, seeds, and roots, **miners** and **borers** that feed between plant surfaces, **gall-formers** that reside and feed within the plant and induce the production of abnormal growth reactions by plant tissues, **sap-suckers** that siphon plant fluids, and **seed predators** and **frugivores** that consume the reproductive parts of plants (Romoser and Stoffalano, 1998). Some species, such as seed predators, seedling-eaters, and tree-killing bark beetles, are true plant predators, but most herbivores function as plant parasites because they do not kill their hosts but instead feed on the living plant without causing death (Price, 1980). These different modes of consumption affect plants in different ways. For example, **folivores** (species that chew foliage) directly reduce the area of photosynthetic tissue, whereas sap-sucking insects affect the flow of fluids and nutrients within the plant, and **root-feeders** reduce plant capacity to acquire nutrients or remain upright.

Folivory is the best studied aspect of herbivory. In fact, the term herbivory often is used even when folivory alone is measured, because loss of foliage is the most obvious and most easily quantified aspect of herbivory. Folivory represents the direct consumption of photosynthetically active material. Consequently, the loss of leaf area can be used as a relative term to indicate the effect of herbivory. In contrast, other herbivores such as sap-suckers or root-borers cause less conspicuous losses that are more difficult to measure. Nonetheless, Schowalter *et al.* (1981c) reported that calculated loss of photosynthates to sap-suckers greatly exceeded measured foliage loss to folivores in an early successional deciduous forest. Sap-suckers and root-feeders also may have long term effects, e.g., through disease transmission or altered rates of nutrient acquisition.

B. Measurement of Herbivory

Effects of herbivory on ecosystem processes are determined by temporal and spatial variability in the magnitude of consumption. Clearly, evaluation of the effects of herbivory requires robust methods for measuring herbivory as well as various aspects of primary productivity and other ecosystem processes. Measurement of herbivory can be difficult, especially for underground plant parts and forest canopies, and has not been standardized. Several methods commonly used to measure herbivory have been compared by Filip *et al.* (1995), Landsberg (1989), and Lowman (1984).

The simplest and most widely used technique is the measurement of feeding rate by individual herbivores and extrapolation to feeding rate by a population. This technique provides relatively accurate rates of consumption and can be used to estimate per capita feeding rate for sap-suckers as well as folivores (e.g., Gandar, 1982; Schowalter *et al.*, 1981c; Stadler and Müller, 1996). Insect folivores typically consume 50–150% of their dry body mass per day (Blumer and Diemer, 1996; Reichle and Crossley, 1967; Reichle *et al.*, 1973; Schowalter *et al.*, 1981c).

Rates of sap and root consumption are difficult to measure, but a few studies have provided limited information. For example, honeydew production by individual sap-sucking insects can be used as an estimate of their consumption rates. Stadler and Müller (1996) and Stadler *et al.* (1998) reported that individual spruce aphids, *Cinara* spp., produced from 0.1 mg honeydew d^{-1} for 1st instars to 1 mg d^{-1} for adults, depending on aphid species, season, and nutritional status of the host. Schowalter *et al.* (1981c) compiled consumption data from studies of eight herb- and tree-feeding aphids (Auclair, 1958, 1959, 1965; Banks and Macaulay, 1964; Banks and Nixon, 1959; Day and Irzykiewicz, 1953; Llewellyn, 1972; Mittler, 1958, 1970; Mittler and Sylvester, 1961; Van Hook *et al.*, 1980; Watson and Nixon, 1953), a leafhopper (Day and McKinnon, 1951), and a spittlebug (Wiegert, 1964) that yielded an average consumption rate of 2.5 mg dry sap mg^{-1} dry insect d^{-1}.

Several factors affect the rate of sap consumption. Andersen *et al.* (1992) found that leafhopper feeding rate was related to xylem chemistry and fluid tension. Feeding rates generally increased with amino acid concentrations and decreased with xylem tension, ceasing above tensions of 2.1 Mpa when plants were water stressed. Stadler and Müller (1996) reported that aphids feeding on poor quality hosts with yellowing needles produced twice the amount of honeydew as did aphids feeding on high quality hosts during shoot expansion, but this difference disappeared by the end of shoot expansion. Banks and Nixon (1958) reported that aphids tended by ants approximately doubled their rates of ingestion and egestion.

Measurement of individual consumption rate has limited utility for extrapolation to effects on plant growth because more plant material may be lost, or not produced, than actually consumed, as a consequence of wasteful feeding or mortality to meristems (e.g., Blumer and Diemer, 1996; Gandar, 1982). For example, Schowalter (1989) reported that feeding on Douglas-fir, *Pseudotsuga menziesii*, buds by a budmoth, *Zeiraphera hesperiana*, caused an overall loss of < 1% of foliage standing crop, but the resulting bud mortality caused a 13% reduction in production of shoots and new foliage.

Percentage leaf area missing can be measured at discrete times throughout the growing season. This percentage can be estimated visually but is sensitive to observer bias (Landsberg, 1989). Alternatively, leaf area of foliage samples is measured, then remeasured after holes and missing edges have been reconstructed (e.g., Filip *et al.*, 1995; Odum and Ruiz-Reyes, 1970; Reichle *et al.*, 1973; Schowalter *et al.*, 1981c). Reconstruction originally was accomplished using tape or paper cutouts. More recently, computer software has become available to reconstruct leaf outlines and fill in missing portions (Hargrove, 1988). Neither method accounts for expansion of holes as leaves expand, for compensatory growth (to replace lost tissues), for completely consumed or prematurely abscissed foliage, nor for herbivory by sap-suckers (Faeth *et al.*, 1981; Hargrove, 1988; Lowman, 1984; Reichle *et al.*, 1973; Risley and Crossley, 1993; Stiling *et al.*, 1991).

The most accurate method for measuring folivory is detailed life table analysis of marked plant parts (e.g., leaves, branches, roots) at different stages of plant growth (Aide, 1993; Filip *et al.*, 1995; Hargrove, 1988; Lowman, 1984). Continual monitoring permits accounting for consumption at different stages of plant development, with consequent differences in degree of hole expansion, compensatory growth, and complete consumption or loss of damaged parts (Lowman, 1984; Risley and Crossley, 1993). Estimates of herbivory based on long-term monitoring often are 3–5 times the estimates based on discrete measurement of leaf area loss (Lowman, 1984, 1995). Filip *et al.* (1995) compared continual and discrete measurements of herbivory for 12 tree species in a tropical deciduous forest in Mexico. Continual measurement provided estimates 1–5 times higher than those based on discrete sampling. On average, measurements from the two techniques differed by a factor of 2.

Several methods also have been used to measure effects of herbivory on plants or ecosystem processes. A vast literature is available on the effects of herbivory on growth of individual plants or plant populations (e.g., Crawley, 1983; Huntly, 1991).

At the ecosystem level, a number of studies have compared ecosystem processes between sites naturally infested or not infested, during population irruptions. Such comparison confounds herbivore effects with environmental gradients that may be responsible for the discontinuous pattern of herbivory (Chapter 7).

A few studies have involved experimental manipulation of herbivore numbers, especially on short vegetation (e.g., Kimmins, 1972; McNaughton, 1979; Morón-Ríos *et al.*, 1997; Schowalter *et al.*, 1991; Seastedt, 1985; Seastedt *et al.*, 1983; Williamson *et al.*, 1989), but this technique clearly is difficult in mature forests. The most common method has been comparison of ecosystem processes in plots with nominal herbivory versus chemically suppressed herbivory (e.g., Brown *et al.*, 1987, 1988; Gibson *et al.*, 1990; Louda and Rodman, 1996; Seastedt *et al.*, 1983). However, insecticides can provide a source of limiting nutrients that may affect plant growth. Carbaryl, for example, contains nitrogen, frequently limiting and likely to stimulate plant growth. Manipulation of herbivore abundance is the best means for relating effects of herbivory to its intensity (e.g., Schowalter *et al.*, 1991; Williamson *et al.*, 1989), but manipulation of herbivore abundance often is difficult (Baldwin, 1990; Crawley, 1983; Schowalter *et al.*, 1991). Cages constructed of fencing or mesh screening are used to exclude or contain experimental densities of herbivores (e.g., Mc-Naughton, 1985; Palmisano and Fox, 1997). Mesh screening should be installed in a manner that does not restrict air movement or precipitation and thereby alter growing conditions within the cage.

A third option has been to simulate herbivory by clipping or pruning plants or by punching holes in leaves (e.g., Honkanen *et al.*, 1994). This method avoids the problems of manipulating herbivore abundance but may fail to represent important aspects of herbivory, other than physical damage, that influence its effects (e.g., Baldwin, 1990; Crawley, 1983). For example, herbivore saliva may stimulate growth of some plant species (Dyer *et al.*, 1995), and natural patterns of consumption and excretion affect litter condition and decomposition (Hik and Jefferies, 1990; Lovett and Ruesink, 1995; Stadler *et al.*, 1998; Zlotin and Khodashova, 1980).

The choice of technique for measuring herbivory and its effects depends on several considerations. The method of measurement must be accurate, efficient, and consistent with objectives. Measurement of percentage leaf area missing at a point in time is an appropriate measure of the effect of herbivory on canopy porosity, photosynthetic capacity, and canopy-soil or canopy–atmosphere interactions but does not represent the rate of consumption or removal of plant material. Access to some plant parts is difficult, precluding continual monitoring. Hence, limited data are available for herbivory on roots or in forest canopies. Simulating herbivory by removing plant parts or punching holes in leaves fails to represent some important effects of herbivory, such as salivary toxins or stimulants or flux of canopy material to litter as feces, but does overcome the difficulty of manipulating abundances of herbivore species.

Similarly, the choice of response variables depends on objectives. Most studies have examined only effects of herbivory on above-ground primary production, consistent with emphasis on foliage and fruit production. However, herbivores feeding above ground affect root production and rhizosphere processes, as well (Gehring and Whitham, 1991, 1995; Holland *et al.*, 1996; Rodgers, *et al.*, 1995). Effects on some fluxes, such as dissolved organic carbon in honeydew, are difficult to measure (Stadler *et al.*, 1998). Some effects, such as compensatory growth and altered community structure, do not become apparent for long time periods following herbivore outbreaks (Alfaro and Shepherd, 1991; Wickman, 1980).

C. Spatial and Temporal Patterns of Herbivory

All plant species support characteristic assemblages of insect herbivores, although some plants host a greater diversity of herbivores and exhibit higher levels of herbivory than do others (e.g., Coley and Aide, 1991; de la Cruz and Dirzo, 1987). Some plants tolerate continuous high levels of herbivory, whereas other species show negligible loss of plant material (Carpenter and Kitchell, 1984; Lowman and Heatwole, 1992; McNaughton, 1979; Schowalter and Ganio, 1999), and some plant species suffer mortality at lower levels of herbivory than do others. Herbivory typically is concentrated on the most nutritious or least defended plants and plant parts (Chapter 3; Aide and Zimmerman, 1990).

The consequences of herbivory vary significantly, not just among plant–herbivore interactions, but also as a result of different spatial and temporal factors (Huntly, 1991; Maschinski and Whitham, 1989; Schowalter and Lowman, 1999). For example, water or nutrient limitation and ecosystem fragmentation can affect significantly the ability of the host plant to respond to herbivory (e.g., Chapin *et al.*, 1987; Maschinski and Whitham, 1989; Parks, 1993; Schowalter and Lowman, 1999; Webb, 1978). The timing of herbivory with respect to plant development and the intervals between attacks also have important effects on ecosystem processes (Hik and Jefferies, 1990).

Herbivory usually is expressed as daily or annual rates of consumption and ranges from negligible to several times the standing crop biomass of foliage (Table 12.1), depending on ecosystem type, environmental conditions, and regrowth capacity of the vegetation (Lowman, 1995; Schowalter and Lowman, 1999). Herbivory for particular plant species can be integrated at the ecosystem level by weighting rates for each plant species by its biomass or leaf area. When the preferred hosts are dominant plant species, loss of plant parts can be dramatic and conspicuous, especially if these species are slow to replace lost parts (Brown and Ewel, 1987). For example, defoliation of evergreen forests may be visible for months, whereas deciduous forests and grasslands are adapted for periodic replacement of foliage and typically replace lost foliage quickly. Eucalypt forests are characterized by chronically high rates of herbivory (Fox and Morrow, 1992). Some species lose more than 300% of their foliage standing crop annually, based on life table studies of marked leaves (Lowman and Heatwole, 1992).

Comparison of herbivory among ecosystem types (Table 12.1) indicates considerable variation. The studies in Table 12.1 reflect the range of measurement techniques described above. Most are short-term snapshots of folivory, often for only a few plant species, do not provide information on herbivory by sap-suckers or root feeders, and do not address deviation in environmental conditions, plant chemistry, or herbivore densities from long-term means during the period of study. Long-term studies using standardized techniques are necessary for meaningful comparison of herbivory rates.

Cebrián and Duarte (1994) compiled data from a number of aquatic and terrestrial ecosystems and found a significant relationship between percentage plant material consumed by herbivores and the rate of primary production. Herbivory ranged from negligible to $> 50\%$ of photosynthetic biomass removed daily. Rates were greatest in some phytoplankton communities where herbivores consumed all production daily and least in some forests where

TABLE 12.1   Herbivory Measured in Temperate and Tropical Ecosystems (Including Understory)

Location	Ecosystem type	Level of grazing	Technique[a]	Source
Tropical				
Costa Rica	Tropical forest	7.5% (new leaves)	1	Stanton (1975)
	Tropical evergreen forest	30% (old)	1	Stanton (1975)
Panama	Tropical evergreen forest	13%	1	Wint (1983)
Panama (BCI)	Tropical evergreen forest	8% (6% insect; 1–2% vertebrates)	1, 2	Leigh and Smythe (1978)
	Understory only	15%	1, 2	Leigh and Windsor (1982)
		21% (but up to 190%)	3	Coley (1983)
Puerto Rico	Tropical evergreen forest	7.8%	1	Odum and Ruiz-Reyes (1970)
		5.5–16.1%	1	Benedict (1976)
		2–6%	1	Schowalter (1994a)
		2–13%	1	Schowalter and Ganio (1999)
Mexico	Tropical deciduous forest	7–9%	1	Filip et al. (1995)
	Tropical deciduous forest	17%	3	Filip et al. (1995)
Venezuela	Understory only	0.1–2.2%	1	Golley (1977)
New Guinea	Tropical evergreen forest	9–12%	1	Wint (1983)
Australia	Montane or cloud forest	26%	3	Lowman (1984)
	Warm temperate forest	22%	3	Lowman (1984)
	Subtropical forest	14.6%	3	Lowman (1984)
Cameroon	Tropical evergreen forest	8–12%	3	Lowman et al. (1993)

(continued)

TABLE 12.1 *(continued)*

Location	Ecosystem type	Level of grazing	Technique[a]	Source
Tanzania	Tropical grassland	14–38% (4–8% insect; 8–34% vertebrates)	4	Sinclair (1975)
South Africa	Tropical savanna	38% (14% insect; 24% vertebrates)	4	Gandar (1982)
Temperate U.S.	Deciduous forest	2–10%	1	Reichle *et al.* (1973)
		1–5%	1	Schowalter *et al.* (1981c)
	Herbaceous sere	3%	4	Crossley and Howden (1961)
	Coniferous forest	<1%	1	Schowalter (1989)
		1–6%	1	Schowalter (1995a)
	Grassland	5–15%	1	Detling (1987)
Australia	Evergreen forest	15–300%	3	Lowman and Heatwole (1992)
	Dry forest	5–44%	1	Fox and Morrow (1983)
		3–6%	2	Ohmart *et al.* (1983)
Europe	Deciduous forest	7–10%	1	Nielson (1978)
	Alpine grassland	19–30%	1	Blumer and Diemer (1996)

[a]1, leaf area missing; 2, litter or frass collection; 3, turnover of marked foliage; 4, individual consumption rates. Expanded from Lowman (1995).

herbivores removed < 1% of production. Insects are the primary herbivores in forest ecosystems (Janzen, 1981; Wiegert and Evans, 1967) and account for 11–73% of total herbivory in grasslands, where native vertebrate herbivores remove an additional 15–33% of production (Detling, 1987; Gandar, 1982; Sinclair, 1975). Temperate deciduous forests and tropical evergreen forests show similar annual losses of 3–20%, based on discrete sampling of leaf area loss (Coley and Aide, 1991; Landsberg and Ohmart, 1989; Odum and Ruiz-Reyes, 1970; Schowalter and Ganio, 1999; Schowalter et al., 1986). Aquatic ecosystems, evergreen forests, and grasslands, which replace lost photosynthetic tissue continuously, often lose several times their standing crop biomass to herbivores annually, based on loss of marked foliage or on herbivore exclusion (Carpenter and Kitchell, 1984; Cebrián and Duarte, 1994; Crawley, 1983; Landsberg, 1989; Lowman and Heatwole, 1992; McNaughton, 1979).

In addition to the conspicuous loss of photosynthetic tissues, terrestrial plants lose additional material to sap-suckers and root feeders. Schowalter et al. (1981c) compiled data on rates of sap consumption to estimate turnover of 5–23% of foliage standing crop biomass through sap-sucking herbivores, in addition to 1–2% turnover through folivores in a temperate deciduous forest. Brown and Gange (1991) and Morón-Ríos et al. (1997) reported that root-feeding insects can reduce primary production of grasses by 30–50%.

Factors that promote herbivore population growth, e.g., abundant and susceptible hosts, also increase herbivory (see Chapters 6 and 8). Proportional losses of foliage to folivores generally are higher in less diverse ecosystems, compared to more diverse ecosystems (Kareiva, 1983), but the intensity of herbivory also depends on the particular species composition of the vegetation (Moore and Francis, 1991; Moore et al., 1991). Brown and Ewel (1987) demonstrated that ecosystem-level foliage losses per unit ground area were similar among four tropical ecosystems that varied in vegetation diversity, but the proportional loss of foliage standing crop was highest in the less diverse ecosystems. Nevertheless, rare plant species in diverse ecosystems can suffer intense herbivory, especially under conditions that increase their apparency or acceptability (Brown and Ewel, 1987; Schowalter and Ganio, 1999). Fonseca (1994) reported that an Amazonian myrmecophytic canopy tree showed 10-fold greater foliage losses when ants were experimentally removed than when ants were present.

Seasonal and annual changes in herbivore abundance affect patterns and rates of herbivory, but the relationship may not be linear, depending on variation in per capita rates of consumption or wasteful feeding with increasing population density (Crawley, 1983; Stadler et al., 1998). Herbivory in temperate forests typically is concentrated in the spring, during leaf expansion (Feeny, 1970; Hunter, 1987). Hunter (1992) reported that over 95% of total defoliation on Quercus robur in Europe occurs between budburst in April and the beginning of June. Although some herbivorous insects prefer mature foliage (Cates, 1980; Sandlin and Willig, 1993; Volney et al., 1983), most defoliation events are associated with young foliage (Coley, 1980; Jackson et al., 1999;

Lowman, 1985; Hunter, 1992). Herbivory also is highly seasonal in tropical ecosystems. Although tropical plants produce new foliage over a more protracted period than do temperate plants, many plants produce new foliage in response to seasonal variation in precipitation (Aide, 1992; Coley and Aide, 1991; Lowman, 1992; Ribeiro *et al.*, 1994). Young foliage may be grazed more extensively than older foliage in tropical rain forests (Coley and Aide, 1991; Lowman 1984, 1992). Schowalter and Ganio (1999) reported significantly greater rates of leaf area loss during the "wet" season than during the "dry" season in a tropical rain forest in Puerto Rico (Fig. 12.1). However, seasonal peaks of leaf expansion and herbivory are broader in tropical ecosystems than in temperate ecosystems.

Few studies have addressed long-term changes in herbivore abundances or herbivory as a result of environmental changes (see Chapter 6). However, disturbances often induce elevated rates of herbivory at a site. Periods of elevated herbivory frequently are associated with drought (Mattson and Haack, 1987;

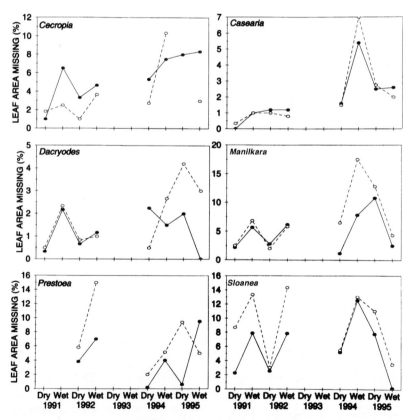

FIG. 12.1   Effects of tree species, hurricane disturbance, and seasonal cycles on leaf area missing in a tropical rain forest in Puerto Rico. *Cecropia*, *Casearia*, and *Prestoea* are early successional trees; *Dacryodes*, *Manilkara*, and *Sloanea* are late successional trees. Solid lines represent intact forest (lightly disturbed); dashed lines represent treefall gaps. From Schowalter and Ganio (1999) by permission from Blackwell Scientific.

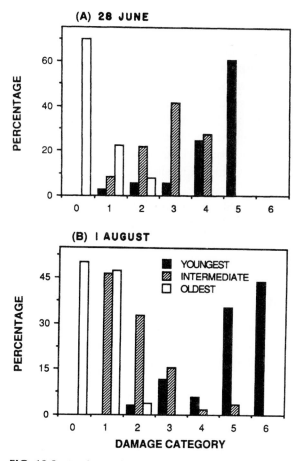

FIG. 12.2　Herbivore damage to plants in young, intermediate, and old successional sites in sand dune vegetation in Michigan in June (A) and August (B) 1988. Percentages are averages for leaves on upper and lower canopy branches by damage category: 0, 0% damage; 1, 1–5%; 2, 6–25%; 3, 26–50%; 4, 51–75%; 5, 76–100%; and 6, no leaves remaining. From Bach (1990) by permission from the Ecological Society of America.

Schowalter and Ganio, 1999; Chapter 6). Other disturbances that injure plants also may increase herbivory, especially by root feeders and stem borers (e.g., Paine and Baker, 1993; Witcosky *et al.*, 1986).

Changes in vegetation associated with disturbance or recovery affect temporal patterns of herbivory. Bach (1990) reported that intensity of herbivory declined during succession in dune vegetation in Michigan (Fig. 12.2). Coley (1980, 1982, 1983), Coley and Aide (1991), and Lowman and Box (1983) found that rapidly growing early successional tree species showed higher rates of herbivory than did slow-growing late successional trees. Schowalter (1995a), Schowalter and Ganio (1999), and Schowalter and Crossley (1988) compared canopy herbivore abundances and folivory in replicated disturbed (harvest or hurricane) and undisturbed patches of temperate deciduous, temperate coniferous, and tropical evergreen forests. In all three forest types, disturbance resulted in greatly increased abundances of sap-suckers and somewhat increased abundances of folivores on abundant rapidly growing early successional plant species. The resulting shift in biomass dominance from folivores to sap-suckers following disturbance resulted in an elevated flux of primary production as soluble photosynthates, relative to fragmented foliage and feces. Schowalter *et al.* (1981c) calculated that loss of photosynthate to sap-suckers increased from 5% of foliage standing crop in undisturbed forest to 20–23% of foliage standing crop during the first two years following clear-cutting, compared to relatively consistent losses of 1–2% to folivores. Torres (1992) reported a sequence of defoliator outbreaks on early successional herbs and shrubs during several months following Hurricane Hugo in Puerto Rico. As each plant species became dominant at a site, severe defoliation facilitated its replacement by other plant species. Continued measurement of herbivory over long time periods will be necessary to relate changes in the intensity of herbivory to environmental changes and to effects on ecosystem processes.

II. EFFECTS OF HERBIVORY

Herbivory affects a variety of ecosystem properties, primarily through differential changes in survival, productivity, and growth form among plant species. It reduces plant density, opens the canopy, stimulates plant growth under some conditions, transfers mass and nutrients to the soil, and affects habitat and resource conditions for other organisms. Herbivory is not evenly distributed among plant species or over time. Rather, some species are subject to greater herbivory than are others, and relative herbivory among plant species varies with environmental conditions (e.g., Coley, 1980; Coley and Aide, 1991; Crawley, 1983; Schowalter and Ganio, 1999). These differential effects on host conditions alter vegetation structure, energy flow, and biogeochemical cycling, and often predispose the ecosystem to various disturbances.

The observed severity of herbivore effects in agroecosystems and some native ecosystems has led to a widespread perception of herbivory as a disturbance (see Chapter 2). This perception raises a number of issues. How can a normal trophic process also be a disturbance? Is predation a disturbance? At what level does herbivory become a disturbance? Do the normally low levels of 5–20% loss of NPP constitute disturbance? Although debate may continue over whether or not herbivory is a disturbance (Veblen *et al.*, 1994; White and Pickett, 1985) rather than simply an ecosystem process (Schowalter, 1985; Schowalter and Lowman, 1999; Willig and McGinley, 1999), herbivory can dramatically alter ecosystem structure and function over large areas.

A. Plant Productivity, Survival, and Growth Form

Traditionally, herbivory has been viewed solely as a process that reduces primary production. As described above, herbivory can remove several times the standing crop of foliage, alter plant growth form, or kill all plants of selected species over large areas during severe outbreaks. However, several recent studies indicate more complex effects of herbivory. The degree to which herbivory affects plant survival, productivity, and growth form depends on the plant parts affected, plant condition, including the stage of plant development, and the intensity of herbivory.

Different herbivore species and functional groups, e.g., folivores, sap-suckers, shoot borers, and root feeders, determine which plant parts are affected. Folivores and leaf miners reduce foliage surface area and photosynthetic capacity, thereby limiting plant ability to produce and accumulate photosynthates for growth and maintenance. In addition to direct consumption of foliage, much nonconsumed foliage is lost due to wasteful feeding by folivores (Risley and Crossley, 1993) and induction of leaf abscission by leaf miners (Faeth *et al.*, 1981; Stiling *et al.*, 1991). Sap-suckers and gall-formers siphon photosynthates from the plant's vascular system and reduce plant ability to accumulate photosynthates for growth and maintenance. Shoot borers and bud feeders damage meristems and growing shoots, altering plant growth rate and form. Root feeders reduce plant ability to acquire water and nutrients. Reduced accumulation of energy often reduces flowering or seed production, often completely precluding reproduction (V. Brown *et al.*, 1987; Crawley, 1989). For example, Parker (1985) and Wisdom *et al.* (1989) reported that flower production by composite shrubs, *Gutierrezia microcephala*, was reduced as much as 80% as a consequence of grazing by the grasshopper, *Hesperotettix viridis*. Many sap-suckers and shoot- and root-feeders also transmit or facilitate growth of plant pathogens, including viruses, bacteria, fungi, and nematodes (e.g., Jones, 1984). Alternatively, folivory may induce resistance to subsequent infection by plant pathogens (Hatcher *et al.*, 1995).

Plant condition is affected by developmental stage and environmental conditions and determines herbivore population dynamics (see Chapters 3 and 6) and plant capacity to compensate for herbivory. Low or moderate levels of herbivory often stimulate plant productivity (e.g., Carpenter and Kitchell, 1984; Carpenter *et al.*, 1985; Carroll and Hoffman, 1980; Detling, 1987, 1988; Dyer *et al.*, 1993; Lowman, 1982; McNaughton, 1979, 1993a; Trumble *et al.*, 1993; Williamson *et al.*, 1989), whereas severe herbivory usually results in mortality or decreased fitness (Detling, 1987, 1988; Marquis, 1984; Williamson *et al.*, 1989). Healthy plants often replace lost foliage, resulting in higher annual primary production, although standing crop biomass of plants often is reduced.

The rapid replacement of primary production lost to herbivores in many aquatic systems is well known (Carpenter and Kitchell, 1984, 1987, 1988; Carpenter *et al.*, 1985; Wallace and O'Hop, 1985). Wallace and O'Hop (1985) reported that new leaves of waterlilies, *Nuphar luteum*, disappeared within 3 weeks as a result of grazing by the leaf beetle, *Pyrrhalta nymphaeae*. A high rate of leaf production was necessary to maintain macrophyte biomass.

Trumble *et al.* (1993) reviewed literature demonstrating that compensatory growth (replacement of consumed tissues) following low to moderate levels of herbivory is a widespread response by terrestrial plants, as well. Increased productivity of grazed grasses, compared to ungrazed grasses, has been demonstrated experimentally in a variety of grassland ecosystems (Detling, 1987, 1988; McNaughton, 1979, 1986, 1993a; Seastedt, 1985; Williamson *et al.*, 1989), but growth enhancement may depend on the presence of herbivore feces (Baldwin, 1990; Hik and Jefferies, 1990) or other herbivore products (Baldwin, 1990). Dyer *et al.* (1995) demonstrated that crop and midgut extracts present in grasshopper regurgitants during feeding stimulate coleoptile growth in grasses, but saliva may not stimulate growth of all plant species (Detling *et al.*, 1980) . Wickman (1980) and Alfaro and Shepherd (1991) reported that short-term growth losses by defoliated conifers were followed by several years, or even decades, of growth rates that exceeded predefoliation rates (Fig. 12.3). Romme *et al.* (1986) reported that annual wood production in pine forests in western North America reached or exceeded preattack levels within 10–15 years following mountain pine beetle, *Dendroctonus ponderosae*, outbreaks.

Detling (1987, 1988), Dyer *et al.* (1993, 1995), McNaughton (1979, 1986, 1993a), and Paige and Whitham (1987) have argued that herbivory may benefit some plants, to the extent that species adapted to replace consumed tissues often disappear in the absence of grazing. Net primary productivity of some grasslands declines when grazing is precluded, due to smothering of shoots as

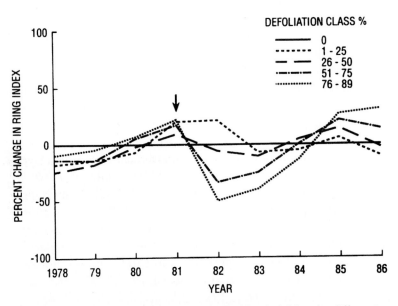

FIG. 12.3 Changes in ring width indices for Douglas-fir defoliated at different intensities by the Douglas-fir tussock moth, *Orgyia pseudotsugata*, in 1981 (arrow). The horizontal line at 0% represents ring width index for nondefoliated trees. From Alfaro and Shepherd (1991). Reprinted from *Forest Science* (Vol. 37, no. 3, p. 963) published by the Society of American Foresters, 5400 Grosvenor Lane, Bethesda, MD 20814-2198. Not for further reproduction.

standing dead material accumulates (Kinyamario and Imbamba, 1992; Knapp and Seastedt, 1986; McNaughton, 1979). Inouye (1982) reported that herbivory by several insect and mammalian herbivores had a variety of positive and negative effects on fitness of a thistle, *Jurinea mollis*.

These observations generated the **herbivore optimization hypothesis** (Fig. 12.4), that primary production is maximized at low to moderate levels of herbivory (Carpenter and Kitchell, 1984; Mattson and Addy, 1975; McNaughton, 1979). Although this hypothesis is widely recognized among aquatic ecologists as the basis for inverted biomass pyramids (Carpenter and Kitchell, 1984, 1987, 1988; Carpenter *et al.*, 1985), its application to terrestrial systems remains somewhat controversial (e.g., Belsky, 1986; Painter and Belsky, 1993; Patten, 1993). Nevertheless, the hypothesis is supported by experimental tests for both insect and vertebrate herbivores in grassland (Detling, 1987; Dyer *et al.*, 1993; McNaughton, 1979, 1993b; Seastedt, 1985), salt marsh (Hik and Jefferies, 1990) and forest ecosystems (Lovett and Tobiessen, 1993; Schowalter *et al.*, 1991).

Compensatory growth likely depends on plant adaptation to herbivory and on suitable growing conditions (Trlica and Rittenhouse, 1993; Williamson *et al.*, 1989). Dyer *et al.* (1991) reported that grazing-adapted and nongrazing-adapted clones of an African C_4 grass, *Panicum coloratum*, differed significantly in their responses to herbivory by grasshoppers. After 12 weeks of grazing, the grazing-adapted plants showed a 39% greater photosynthetic rate and

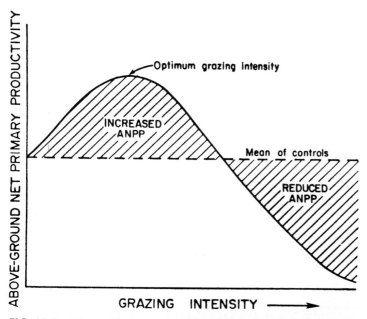

FIG. 12.4 Relationship between intensity of phytophagy and net primary production. Net primary production often peaks at low to moderate intensities of phytophagy, supporting the grazing optimization hypothesis. From Williamson *et al.* (1989) by permission from the Society for Range Management.

26% greater biomass, compared to the nongrazing-adapted plants. Lovett and Tobiessen (1993) reported that defoliation resulted in elevated photosynthetic rates of red oak, *Quercus rubra*, seedlings grown under conditions of low and high nitrogen availability, but that high nitrogen seedlings were able to maintain the high photosynthetic rates for a longer time (Fig. 12.5). Vanni and Layne (1997) found that consumer-mediated nutrient cycling strongly affected phytoplankton production and community dynamics in lakes.

Honkanen *et al.* (1994) artificially damaged needles or buds of Scots pine, *Pinus sylvestris*. Damage to buds increased shoot growth. Damage to needles stimulated or suppressed shoot growth, depending on the degree and timing of damage and the position of the shoot relative to damaged shoots. Growth was significantly reduced by loss of 100%, but not 50%, of needles and was significantly reduced on shoots located above damaged shoots, especially late in the season. Shoots located below damaged shoots showed increased growth. Honkanen *et al.* (1994) suggested that these different effects of damage indicated an important effect of physiological status of the damaged part, i.e., whether it was a sink (bud) or source (needle) for resources.

Morón-Ríos *et al.* (1997) reported that below-ground herbivory by root-feeding scarab beetle larvae, *Phyllophaga* sp., prevented compensatory growth in response to above-ground grazing. Furthermore, salivary toxins or plant pathogens injected into plants by some sap-sucking species can cause necrosis of plant tissues (Jones, 1984; Miles, 1972; Raven, 1983; Skarmoutsos and Millar, 1982), honeydew accumulation on foliage can promote growth of pathogenic fungi and limit photosynthesis (Dik and van Pelt, 1993), and some leaf

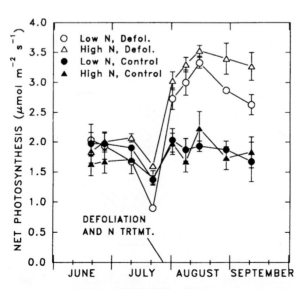

FIG. 12.5 Mean net photosynthetic rate in old leaves from red oak seedlings subjected to four combinations of nitrogen fertilization and defoliation intensity. Defoliation and fertilization treatments began July 26. From Lovett and Tobiessen (1993) by permission from Heron Publishing.

miners induce premature abscission (Chabot and Hicks, 1982; Faeth *et al.*, 1981; Pritchard and James, 1984a, b; Stiling *et al.*, 1991), thereby exacerbating the direct effects of herbivory. However, foliage injury can induce resistance to subsequent herbivory or infection by plant pathogens (Hatcher *et al.*, 1995; Hunter, 1987).

Although primary productivity may be increased by low to moderate intensities of grazing, some plant tissues may be sacrificed by plant allocation of resources to replace lost foliage. Morrow and LaMarche (1978) and Fox and Morrow (1992) reported that incremental growth of *Eucalyptus* stems increased 2–3 times during the year following insecticide application, compared to growth of unsprayed stems.

Root growth and starch reserves are affected significantly by above-ground, as well as below-ground, herbivory. Morón-Ríos *et al.* (1997) noted that root-feeders reduced root to shoot ratios by 40% and live to dead above-ground biomass ratio by 45% through tiller mortality, apparently reducing plant capacity to acquire sufficient nutrients for shoot production. Rodgers *et al.* (1995) observed that starch concentrations in roots were related inversely to the level of mechanical damage to shoots of a tropical tree, *Cedrela odorata* (Fig. 12.6). Gehring and Whitham (1991, 1995) reported that folivory on pinyon pine adversely affected mycorrhizal fungi, perhaps through reduced carbohydrate sup-

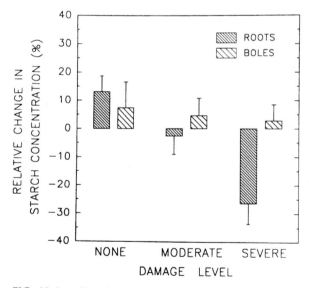

FIG. 12.6 Effect of intensity of artificial herbivory (to simulate terminal shoot damage by a lepidopteran, *Hypsipyla grandella*) on mean relative change (+ standard error) in starch concentrations (percent of initial level) in roots and lower boles of a neotropical hardwood, *Cedrela odorata*, in Costa Rica. In the moderate treatment, 0.2–0.3 cm of terminal shoot was excised; in the severe treatment, 0.5–0.6 cm of terminal was excised. Data represent five sampling dates over a 12-day period beginning 18 days after treatment. From Rodgers *et al.* (1995) by permission from the Association of Tropical Biologists.

ply to roots. However, Holland *et al.* (1996) reported that grasshopper grazing on maize increased carbon allocation to roots. McNaughton (1979, 1993a) and van der Maarel and Titlyanova (1989) concluded that sufficient shoot biomass to maintain root function is critical to plant ability to compensate for losses to herbivores.

Levels of herbivory that exceed plant ability to compensate lead to growth reduction and mortality. Lovett and Tobiessen (1993) reported that resource-limited plants were more likely to succumb to herbivores than were plants with optimal resources. Injured or stressed plants are more vulnerable to herbivores and may have insufficient resources to repair or replace tissues damaged by herbivores. Plant species most stressed by adverse conditions often suffer severe mortality to herbivores (e.g., Crawley, 1983; Painter and Belsky, 1993; Schowalter and Lowman, 1999). Seedlings are particularly vulnerable to herbivores because of their limited resource storage capacity and may be unable to replace tissues lost to herbivores (Wisdom *et al.*, 1989). Clark and Clark (1985) reported that survival of tropical tree seedlings was highly correlated with the percentage of original leaf area present 1 month after germination and with the number of leaves present at 7 months of age.

Herbivory by exotic species may cause more severe or more frequent reduction in productivity and survival, in part because plant defenses may be less effective against newly associated herbivores. The most serious effects of herbivory, however, result from artificially high intensities of grazing by livestock or game (Oesterheld *et al.*, 1992; Patten, 1993). Whereas grazing by native herbivores typically is seasonal, and grasses have sufficient time to replace lost tissues before grazing resumes, grazing by livestock is continuous, allowing insufficient time for recovery (McNaughton, 1993a; Oesterheld and McNaughton, 1988, 1991; Oesterheld *et al.*, 1992).

Continued grazing during stressful periods exacerbates stress. Akiyama *et al.* (1984) reported that primary production was twice the daily energy loss to heavy grazing during the spring, but was only 69% of daily energy loss during the summer and equaled daily energy loss during the fall. Hik and Jefferies (1990) and Thompson and Gardner (1996) reported that regrowth following exposure to herbivory by geese and grasshoppers, respectively, declined during the summer. Grazing early during the growing season had no significant effect on final biomass, but mid- and late-season periods of grazing resulted in significantly reduced biomass.

Herbivory can alter plant architecture significantly. Gall-formers deform expanding foliage and shoots. Repeated piercing during feeding-site selection by sap-sucking species also can cause deformation of foliage and shoots (Miles, 1972; Raven, 1983). Shoot-borers and bud-feeders kill developing shoots and induce growth of lateral shoots (Clark and Clark, 1985; Nielsen, 1978; Reichle *et al.*, 1973; Zlotin and Khodashova, 1980). Severe or repeated herbivory of this type often slows or truncates vertical growth and promotes bushiness. Gange and Brown (1989) reported that herbivory increased variation in plant size. Morón-Ríos *et al*. (1997) found that both above-ground and below-ground herbivory alter shoot-to-root ratios. Suppression of height or root growth restricts plant ability to acquire resources and often leads to plant death. However, pruning also can stimulate growth and seed production (e.g., Inouye, 1982) or improve water and nutrient balance (e.g., Webb, 1978).

B. Community Dynamics

Differential herbivory among plants and plant species in an ecosystem affects both the distribution of individuals of a particular plant species and the opportunities for growth of plant species resistant to or tolerant of herbivory. The intensity of herbivory determines its effects on plant communities. Low to moderate intensities that prevail most of the time generally reduce foliage, root, or plant density and ensure a slow turnover of plant parts or individual plants. High intensities during outbreaks or as a result of management can dramatically reduce the abundance of preferred species and rapidly alter vegetation structure and composition. Overgrazing by domestic livestock has initiated desertification of arid grasslands (by reducing vegetation cover, causing soil desiccation) in many parts of the globe (e.g., Schlesinger *et al*., 1990). Herbivory by exotic insect species (but rarely native species) is capable of eliminating plant species that are unable to compensate (McClure, 1991).

Patterns of herbivory often explain observed geographic or habitat distributions of plant species (Crawley, 1983, 1989; Huntly, 1991; Louda *et al*., 1990a; Schowalter and Lowman, 1999). Herbivory can have a variety of positive and negative effects on plant growth and fitness, even for a particular plant species (Inouye, 1982). Herbivory can prevent successful establishment or continued growth, especially during the vulnerable seedling stage (Hulme, 1994). Louda *et al*. (1990a) reported that patterns of herbivory on two species of goldenbushes, *Happlopappus* spp., explained the significant difference between expected and observed distributions of these species across an environmental gradient from maritime to interior ecosystems in southern California (Fig. 12.7). Louda and Rodman (1996) found that chronic herbivory by insects was concentrated on bittercress, *Cardamine cordifolia*, growing in sunny habitats and largely explained the observed restriction of this plant species to shaded habitats. Schowalter *et al*. (1981a) suggested that differential mortality among pine species (due to southern pine beetle, *Dendroctonus frontalis*) in the southern U.S. largely explained the historic patterns of species distributions over topographic gradients (Fig. 10.5). On the other hand, Inouye (1982) and Paige and Whitham (1987) demonstrated that herbivory can increase seed production.

McEvoy *et al*. (1991) documented changes in plant community structure resulting from herbivore-induced mortality to the exotic ragwort, *Senecio jacobeae*, in western Oregon. Ragwort standing crop declined from > 700 g m^{-2} (representing 90% of total standing crop of vegetation) to 0.25 g m^{-2} over a 2-year period following release of the ragwort flea beetle, *Longitarsus ja-*

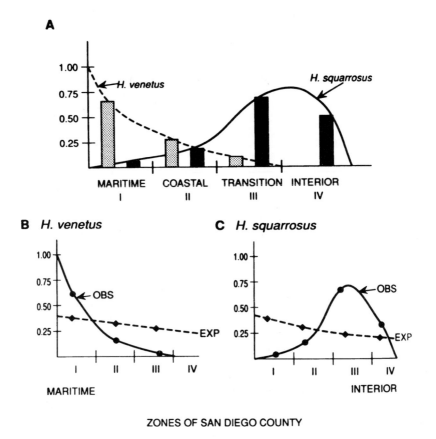

FIG. 12.7 Herbivore effects on plant species distribution. (A) Gradients in observed frequencies of two goldenbushes, *Happlopappus venetus* (stippled) and *H. squarrosus* (black), from maritime to interior montane sites in San Diego County, California. (B and C) Observed frequency accounting for herbivore effects (solid lines) compared to potential distribution in the absence of herbivory (dashed line), based on several measures of performance of control plants when insects were excluded. From Louda *et al.* (1990a).

cobaeae. Grasses responded rapidly to declining ragwort abundance, followed by forbs, resulting in relatively constant vegetation standing crop over the 8 years of measurement.

Herbivory often facilitates successional transitions (see Chapter 10). Selective herbivory among plant species suppresses those on which herbivory is focused and provides space and other resources to others, resulting in altered plant community composition (e.g., Davidson, 1993; McEvoy *et al.*, 1991; Schowalter, 1981a; Schowalter *et al.*, 1986). Brown and Gange (1989), Brown *et al.* (1988) and Gibson *et al.* (1990) reported that chemically-reduced above-ground herbivory resulted in lower plant species richness after two years, whereas Brown and Gange (1989) found that reduced below-ground herbivory resulted in higher plant species richness, largely reflecting differential intensities of herbivory among various grass and forb species. Anderson and Briske

(1995) simulated herbivory in a transplant garden containing mid-seral and late-seral grass species to test alternative hypotheses, that (1) mid-seral species have greater tolerance to herbivory or (2) herbivory is focused on late-seral species, to explain species replacement in intensively grazed grasslands in the southern U.S. They found that late-seral species had greater competitive ability and equivalent or higher tolerance to herbivory, indicating that selective herbivory on the late-successional species is the primary mechanism for reversal of succession, i.e., return to dominance by mid-seral species under intense grazing pressure. Conversely, Bach (1990), Coley (1980, 1982, 1983), Coley and Aide (1991), and Lowman and Box (1983) found that intensities of herbivory were higher in earlier successional stages than in later successional stages. Schowalter *et al.* (1981a) suggested that southern pine beetle is instrumental in advancing succession in the absence of fire by selectively killing early successional pines, thereby favoring their replacement by later successional hardwoods (Fig. 10.5).

Davidson (1993) compiled data indicating that herbivores may retard or reverse succession during early seres but advance succession during later seres. She suggested that herbivory is concentrated on the relatively less defended, but grazing tolerant, midsuccessional grasses, forbs, and pioneer trees (see Bach, 1990). Increased herbivory at early stages of community development tends to retard succession, whereas increased herbivory at later stages advances succession. Environmental conditions may affect this trend. For example, succession from pioneer pine forest to late successional fir forest in western North America can be retarded or advanced, depending primarily on moisture availability and condition of the dominant vegetation. Under conditions of adequate moisture (riparian corridors and high elevations), mountain pine beetles advance succession by facilitating the replacement of host pines by the more shade tolerant, fire-intolerant understory firs. However, inadequate moisture and short fire return intervals at lower elevations concentrate herbivory by several defoliators and bark beetles on water-stressed firs. In the absence of fire during drought periods, understory firs are eliminated and succession truncated at the pine sere. Fire fueled by fir mortality also leads to eventual regeneration of pine forest. Similarly, each plant species that became dominant during succession following Hurricane Hugo in Puerto Rico induced elevated herbivory that facilitated its demise and replacement (Torres, 1992). The direction of succession then depends on which plant species are present and their responses to environmental conditions.

Changes in plant condition, community composition, and structure affect habitat and food for other animals and microorganisms. Changes in nutritional quality or abundance of particular foliage, fruit, or seed resources affects abundances of animals that use those resources. Animals that require or prefer nesting cavities in dead trees may be promoted by tree mortality resulting from herbivore outbreaks.

Grazing on aboveground plant parts can affect litter and rhizosphere processes in a variety of ways (Bardgett *et al.*, 1998). Reduced foliar quality resulting from induced defenses or replacement of palatable by less palatable plant species can reduce the low food quality of detrital material (Fig. 12.8). Seastedt *et al.* (1988) reported that simulation of herbivore effects on throughfall affected litter arthropod communities. Schowalter and Sabin (1991) found that three taxa of litter arthropods were significantly more abundant under experimentally defoliated Douglas-fir saplings, compared to nondefoliated saplings. Altered carbon storage in roots (Filip *et al.*, 1995; Holland *et al.*, 1996) affects resources available for below-ground food webs (Fig. 12.9). Bardgett *et al.* (1997, 1998) reported that microbial biomass, nematode abundance, and soil respiration rates were consistently reduced by removal of sheep grazing (Fig. 12.10). Gehring and Whitham (1991, 1995) documented significantly reduced mycorrhizal activity on roots of piñon pines subject to defoliation by insects, compared to nondefoliated pines.

Insect herbivores or their products constitute highly nutritious resources for insectivores and other organisms. Caterpillars concentrate essential nutrients several orders of magnitude over concentrations in foliage tissues (e.g., Schowalter and Crossley, 1983). Abundances of insectivorous birds and mammals often increase in patches experiencing insect herbivore outbreaks (Barbosa and Wagner, 1989). Arthropod tissues represent concentrations of nutrients for decomposers (Schowalter and Crossley, 1983; Seastedt and Tate, 1981).

A variety of organisms utilize honeydew accumulation. Ants, honey bees, and other animals forage on the carbohydrate-rich honeydew (Fig. 12.11). Stadler and Müller (1996) and Stadler *et al.* (1998) reported that the presence of honeydew significantly increased the growth of a variety of epiphytic bacteria, yeasts, and filamentous fungi on the surface of conifer needles, potentially affecting photosynthetic efficiency of underlying foliage.

C. Water and Nutrient Fluxes

Relatively few studies have addressed effects of insect herbivores on biogeochemical cycling processes, despite the importance of vegetation and litter structure, and turnover of material between these pools, to biogeochemical cycling. Crossley and Howden (1961) pioneered the study of nutrient fluxes from vegetation through arthropod communities. Subsequent research has demonstrated that insect herbivores affect biogeochemical cycling in a number of ways, including altered vegetation composition and structure, direct transfer of material from plants to litter, and associated effects on litter quality.

Altered vegetation composition changes patterns of acquisition and turnover of various nutrients by the vegetation. For example, insects (such as bark beetles) that affect the relative composition of Douglas-fir and western redcedar, *Thuja plicata*, in the northwestern U.S.A. affect calcium dynamics and soil pH, i.e., calcium accumulation and higher pH under western redcedar compared to Douglas-fir (e.g., Kiilsgaard *et al.*, 1987). Similarly, Ritchie *et al.* (1998) re-

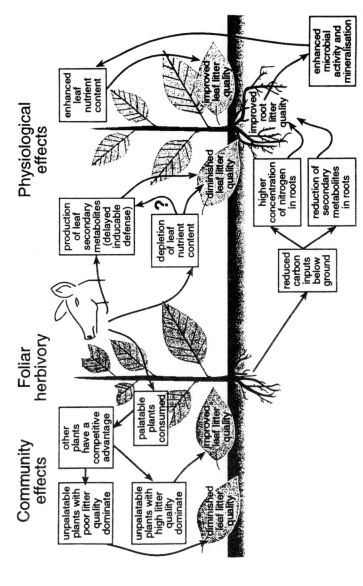

FIG. 12.8 Effects of herbivory on host nutrient allocation and trophic interactions. Reprinted from *Soil Biology and Biochemistry* **30**, Bardgett *et al.*, "Linking above-ground and below-ground interactions: How plant responses to foliar herbivory influence soil organisms," pp. 1867–1878. (1998), with permission from Elsevier Science.

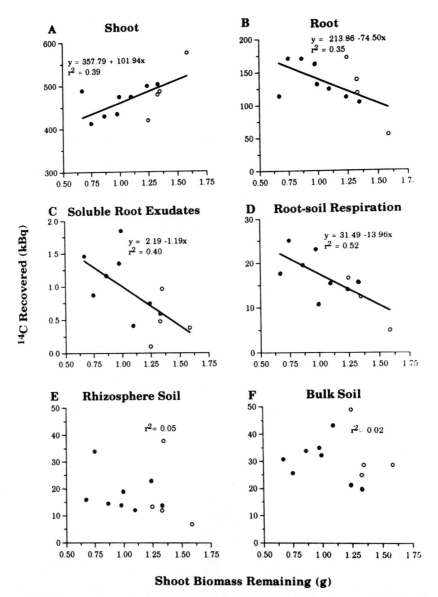

FIG. 12.9   Carbon allocation as a function of intensity of herbivory (measured as shoot biomass remaining) in (A) shoots, (B) roots, (C) soluble root exudates, (D) respiration from roots and soil, (E) rhizosphere soil, and (F) bulk soil. Data were normalized for differences in $^{14}CO_2$ uptake; 1 kBq, 1000 disintegrations sec^{-1}. Shoot biomass was inversely related to leaf area removed by herbivores. Regression lines are shown where significant at $P < 0.05$. Open circles represent ungrazed plants, solid circles grazed plants. From Holland *et al.* (1996) by permission from Springer-Verlag.

ported that herbivory generally reduced the abundance of plant species with N-rich tissues in an oak savanna in the north central U.S.

Reduced metabolic demands by pruned or defoliated plants can reduce water and nutrient uptake (Webb, 1978) and potentially contribute to plant survival dur-

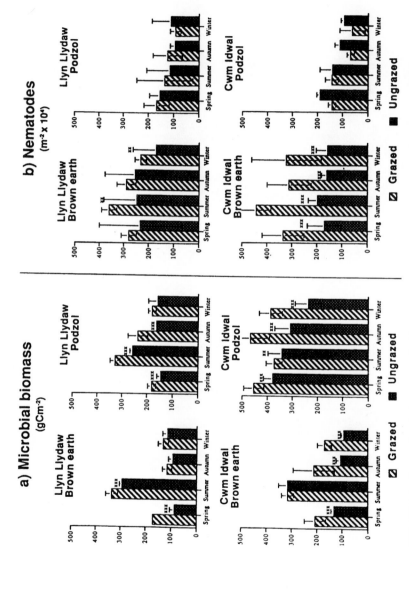

FIG. 12.10 Seasonal variation in microbial biomass and nematode abundance in grazed and ungrazed plots of two grassland types in Ireland. Vertical lines represent standard errors; * = $P < 0.05$, ** = $P < 0.01$, *** = $P < 0.001$. Reprinted from *Soil Biology and Biochemistry* **29**, Bardgett *et al.*, "Seasonality of the soil biota of grazed and ungrazed hill grasslands," pp. 1285–1294. (1997), with permission from Elsevier Science.

FIG. 12.11 Honeydew often collects as droplets on leaves below dense aphid populations (top). This rich source of energy is exploited by ants, other animals, and sooty molds, e.g., honey bee dragging a honeydew tether after feeding on honeydew on an oak leaf (bottom).

ing drought periods. Parks (1993) reported that seedlings of grand fir, *Abies grandis*, defoliated by western spruce budworm, *Choristoneura occidentalis*, showed a higher survival rate during drought stress than did nondefoliated seedlings.

Herbivory affects biogeochemical cycling directly by changing the seasonal timing, amount, and form of nutrients transferred from plants to litter or soil. In the absence of herbivory, litter accumulation may be highly seasonal (i.e., concentrated at the onset of cold or dry conditions) and have low nutrient concentrations (especially of nitrogen which may be resorbed from senescing foliage). Herbivory increases the amount and nutrient content of litter during the growing season, thereby increasing the nutritional value of litter for decomposers. Hollinger (1986) reported that during an outbreak of the California oak moth, *Phryganidia californica*, fluxes of nitrogen and phosphorus to the ground more than doubled, and feces and insect remains accounted for 60–70% of the total nitrogen and phosphorus fluxes.

A major effect of folivory may be the increased flux of nutrients in the form of solutes leached from damaged foliage during precipitation events. Nutrient fluxes from canopy to litter in forests are controlled strongly by foliage area (Lovett *et al.*, 1996; Stachurski and Zimka, 1984). Kimmins (1972), Seastedt *et al.* (1983), Schowalter *et al.* (1991), and Stachurski and Zimka (1984) reported that herbivory greatly increases leaching of nutrients from chewed foliage (Fig. 12.12). However, in ecosystems with high annual precipitation, herbivore-induced nutrient turnover may be masked by nutrient inputs via precipitation (Schowalter *et al.*, 1991).

The contribution of nutrients from honeydew has been a subject of considerable interest. Stadler and Müller (1996) and Stadler *et al.* (1998) documented significant amounts of dissolved organic carbon in aphid honeydew. Most of the honeydew in their studies was immobilized quickly by phylloplane microorganisms before reaching the ground.

Owen (1978) and Owen and Wiegert (1976) suggested that the trisaccharide, melezitose, in aphid honeydew provides a rich, labile carbohydrate resource for free-living, nitrogen-fixing soil bacteria. Petelle (1980) subsequently demonstrated that fructose, also abundant in aphid honeydew, increased nitrogen fixation 9-fold more than did melezitose. However, Grier and Vogt (1990) found that chemical removal of aphids increased available soil nitrogen, nitrogen mineralization rates, net primary production, and nitrogen uptake by red alder, *Alnus rubra*. These data, together with those of Lovett and Ruesink (1995), indicate that folivores and sap-suckers facilitate nitrogen immobilization in soil microorganisms.

The form of nutrients moving from plant parts to litter affects decomposition processes. Senescent foliage, foliage fragments lost via herbivory, and foliage passed through herbivore digestive systems differ in the amount and form of nitrogen and carbon compounds, as well as in the degree of microbial preconditioning. Zlotin and Khodashova (1980) reported that plant material passed through herbivore digestive systems decomposed more rapidly than did raw plant material. Lovett and Ruesink (1995) reported that gypsy moth, *Lymantria dispar*, feces contained much labile carbon and nitrogen. Microbial growth, stimulated by labile carbon (Fig. 12.13), was sufficient to immobilize all the available nitrogen (Fig. 12.14). However, Swank *et al.* (1981) reported that defoliation of hardwood forests in the southern Appalachians of the U.S. increased nitrate export via streamflow. The shift in biomass dominance from folivores to sap-suck-

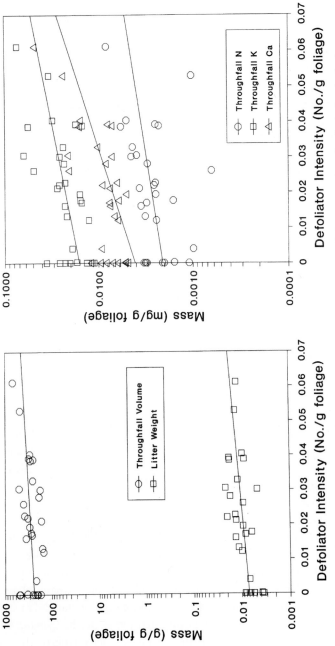

FIG. 12.12 Folivore effects on throughfall, litterfall, and fluxes of N, K, and Ca from young Douglas-fir during the feeding period, April–June, in western Oregon. Reprinted from *Forest Ecology and Management 42*, Schowalter *et al.*, "Phytophage effects on primary production, nutrient turnover, and litter decomposition of young Douglas-fir in western Oregon," pp. 229–243. (1991), with permission from Elsevier Science.

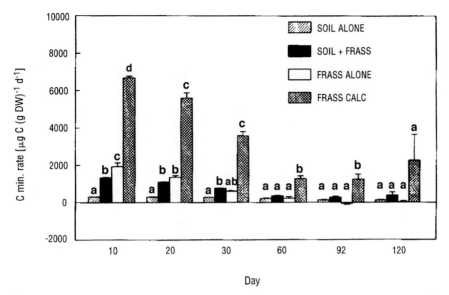

FIG. 12.13 Mean carbon mineralization rate (CO_2 evolution) in soil alone, soil + gypsy moth frass, frass alone, and frass in soil + frass (calculated by subtracting mean net C mineralization in soil alone from that in the soil + frass and expressing the rate per gram dry weight of frass. Vertical lines are standard errors. Within a sample date, bars under the same letter are not significant at $P < 0.05$. From Lovett and Ruesink (1995) by permission from Springer-Verlag.

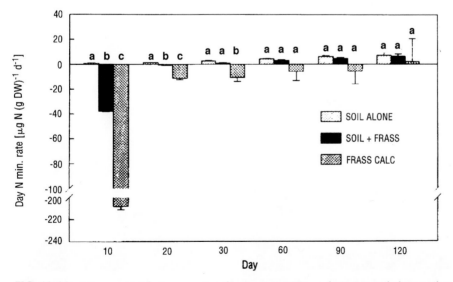

FIG. 12.14 Mean potential nitrogen mineralization rate (CO_2 evolution) in soil alone, soil + gypsy moth frass, frass alone, and frass in soil + frass (see Fig. 12.13). From Lovett and Ruesink (1995) by permission from Springer-Verlag.

ers following disturbance in temperate and tropical forests (Schowalter, 1995a; Schowalter and Ganio, 1999; Schowalter *et al.* 1981c) indicates an accompanying shift in the predominant flux of nutrients. Dominance by folivores in undisturbed forests should transfer nutrients primarily as throughfall (enhanced by leaching from chewed foliage) and fragmented plant parts, whereas dominance by sap-suckers in early successional communities should transfer nutrients predominantly as a labile carbohydrate solution that stimulates microbial growth and immobilization of accompanying nutrients. Holland (1995) reported that soil microbial biomass peaked at intermediate levels of herbivory in no-tillage agricultural systems, perhaps because moderate intensities of herbivory increased root exudates that fuel microbial production (Fig. 12.9) (Holland *et al.*, 1996).

D. Effects on Climate and Disturbance Regime

Changes in vegetation structure affect soil temperature, relative humidity, erosion, soil moisture, and soil fertility (see Chap 11). Changes in litter accumulation can affect ecosystem vulnerability or sensitivity to some disturbances, especially fire.

Herbivory increases vegetation porosity (Fig. 12.15) and penetration of light, precipitation, and wind to the understory and soil surface. Canopy opening greatly affects abiotic conditions in the understory (Chazdon and Fetcher,

FIG. 12.15   Increased canopy porosity resulting from herbivory. Holes chewed by folivorous insects in the large leaves of *Cecropia* increase the penetration of light, water, and airflow to lower strata.

1984; Denslow, 1995, Fernandez and Fetcher, 1991). Increased soil warming due to penetration of sunlight may be offset to some extent by increased penetration of precipitation to the ground. Schowalter *et al.* (1991) reported that 20% loss of foliage mass doubled the amount of water reaching the soil surface. Reduced plant surface area reduces interception of precipitation and evapotranspiration (Parker, 1983). Increased accumulation of litter resulting from herbivory in forest ecosystems may contribute to soil water retention.

Herbivory can increase or decrease the likelihood or severity of future disturbances. Herbivory in grasslands reduces the amount of standing dead material (Knapp and Seastedt, 1986), potentially reducing the severity of fire but increasing soil exposure to desiccation and exacerbating effects of drought. Herbivory in forests increases fuel accumulation in the form of fine and coarse litter material, thereby increasing the likelihood and severity of fire, especially in arid forests where litter decomposes slowly and lightning strikes are frequent (Schowalter, 1985). However, reduced foliage surface area may reduce water demand and mitigate the effects of drought (Parks, 1993).

Herbivory potentially affects climate and disturbance regimes over spatial scales ranging up to thousands of square kilometers. The potential consequences of high intensities of herbivory over large areas for regional climate or disturbance dynamics have not been evaluated. However, McCullough *et al.* (1998) and Schowalter (1985) compiled data indicating that bark beetle outbreaks can modify fire severity, reliability, and scale and modify forest composition and structure in ways that could affect regional climate, especially through changes in water cycling.

III. SUMMARY

Herbivory, the feeding on living plant parts by animals, is a key ecosystem process that has widely recognized effects on primary production and on vegetation structure and composition. The effect of herbivory depends on herbivore feeding type and intensity. Different types of herbivory affect different tissues and the production, translocation, and accumulation of photosynthates to varying degrees.

A number of methods have been used to measure the intensity and effects of herbivory. The most common method for measuring intensity has been estimation of consumption rates by individual herbivores and extrapolation to population size. This method can be used to measure consumption by sap-sucking herbivores as well as folivores. A second method is measurement, by various means, of missing plant biomass. This method does not account for completely consumed (and unobserved) parts or for compensatory growth. Measurement of turnover of marked plant parts is the most accurate, but labor intensive, method for estimating herbivory. Estimates of herbivory can differ by 2–5 times among methods, making standardization a key to comparison among ecosystems.

The intensity of herbivory varies widely but a trend is apparent among ecosystem types. Herbivory generally is lowest (<2% reduction in primary production) in some forests and highest (most primary production consumed daily) in aquatic ecosystems. Insects are the primary herbivores in forest ecosystems and may account for the bulk of herbivory in grasslands, even though vertebrate grazers are more conspicuous.

Herbivory has well-known effects on survival, productivity, and growth form of individual plants. However, the traditional view of herbivory as a negative effect on plants is being replaced by a view that recognizes more complex effects of variable intensity. Moderate intensities of herbivory often stimulate production, through compensatory growth, and flowering, thereby increasing fitness. Herbivory can affect the growth form of plants by terminating shoot growth and initiating branching and by affecting shoot-to-root ratios. Changes in survival, productivity, and growth of individual plant species affect vegetation structure and community dynamics. Herbivores often determine the geographic or habitat patterns of occurrence of plant species and facilitate successional transitions.

Few studies have addressed effects of insect herbivores on biogeochemical cycling or other abiotic conditions. However, herbivores affect, often dramatically, the turnover of plant nutrients to litter as plant fragments, feces and animal tissues, and nutrients leached from chewed surfaces. Folivory alters seasonal patterns of nutrient fluxes by transferring material prior to plant resorption of nutrients from senescing parts. Sap-sucking insects transfer copious amounts of labile carbohydrates (as honeydew) that stimulate growth and nutrient uptake by microbes. Herbivory also may affect climate and the likelihood and intensity of future disturbances. Reducing vegetation cover greatly affects the penetration of light, precipitation, and wind to the understory and soil, affecting soil warming and water content, relative humidity, erosion, transpiration, etc. Reduced vegetation biomass or litter accumulation affects abundance of fuel to support fire and affects water-holding capacity and vegetation demand for water during drought. Therefore, herbivory can influence ecosystem stability substantially (Chapter 15).

13

Pollination, Seed Predation, and Seed Dispersal

INSECTS AFFECT PLANT REPRODUCTION AND ASSOCIATED PROCESSES in a variety of ways. Direct and indirect effects of herbivores on plant production and allocation of resources to reproduction were described in Chapter 12. Pollination, seed predation, and seed dispersal are major processes by which insects (and other animals) affect plant reproduction and distribution. Pollinators control fertilization and reproductive rates for many plant species, especially in the tropics. In fact, some plant species depend on pollinators for successful reproduction and may disappear if their pollinators become rare or extinct (Powell and Powell, 1987; Steffan-Dewenter and Tscharntke, 1997). Seed predators consume seeds and thereby reduce plant reproductive efficiency but often move seeds to new locations and thereby contribute to plant dispersal. Pollinators and seed predators play an important role in seedling recruitment and plant demography.

Insects are the major agents of pollination, seed predation, or seed dispersal in many ecosystems (Bawa, 1990). Pollination and seed dispersal are among the most intricate mutualisms between animals and plants and have been studied widely from the perspective of coevolution. Nonetheless, few studies have evaluated the effects of pollinators, seed predators, and seed dispersers on ecosystem processes, despite their importance to vegetation dynamics. Different functional groups of pollinators and seed-feeders affect seedling recruitment and vegetation dynamics in different ways.

I. MECHANISMS AND PATTERNS OF POLLINATION

Plants exhibit a diversity of reproductive mechanisms. Many reproduce vegetatively, but this mechanism is limited largely to local reproduction. Genetic heterozygosity is increased by outcrossing. Although many plant species are capable of self-fertilization, a large percentage (a vast majority in some ecosystems) are self-incompatible, and many are dioecious (e.g., 20–30% of tropical tree species), with male and female floral structures separated among individual plants to preclude inbreeding (Bawa, 1990; Momose *et al.*, 1998). Mechanisms for transporting pollen between individuals becomes increasingly critical for reproduction with increasing separation of male and female structures and increasing isolation of individual plants (Regal, 1982). Several mechanisms are available for moving pollen among flowering individuals. These mechanisms confer varying degrees of fertilization efficiency, depending on ecosystem conditions.

A. Mechanisms of Pollination

Pollen can be transferred between plants through abiotic and biotic mechanisms (Regal, 1982). Pollen is transported abiotically by means of wind. Biotic transport involves primarily insects, birds, and bats. Insects are the major pollinators for a vast majority of plant species in the tropics (Bawa, 1990), but the proportion of wind-pollinated plants increases toward the poles, reaching 80–100% at northernmost latitudes (Regal, 1982).

Functional groups of pollinators may be more or less restricted to groups of plants based on floral or habitat characteristics (Bawa, 1990). A large number of pollinators are **nonspecialists** with respect to plant species. This functional group includes many beetles, flies, and thrips, etc. that forage on any floral resources available. **Specialist** pollinators often exploit particular floral characteristics that may exclude other pollinators. For example, nocturnally flowering plants with large flowers attract primarily bats, whereas plants with small flowers attract primarily moths. Long, bright red flowers attract birds but are largely unattractive to insects (Johnson and Bond, 1994). Such flowers often are narrow to hinder entry by bees and other insect pollinators (Heinrich, 1979), but may nonetheless be pollinated by some insects (Roubik, 1989). **Pollen feeders** feed primarily on pollen (e.g., beetles and thrips) and are likely to transport pollen acquired during feeding, whereas others are primarily **nec-**

tar-feeders (e.g., beetles, butterflies and moths, and flies) and transport pollen more coincidentally. In fact, many nectar feeders avoid the reproductive organs or, in the case of ants, may reduce pollen viability (Peakall *et al.*, 1987). **Bees, especially** *Apis* spp., feed on pollen and nectar. Functional groupings also reflect attraction to floral odors. For example, **dung- and carrion-feeding flies** are the primary pollinators of plants that emit dung or carrion odors (Appanah, 1990).

Roubik (1989) reviewed studies that distinguished pollinator functional groups on the basis of seasonality. Heithaus (1979) reported that megachilid and anthophorid bees were most active during the dry season in Costa Rica, halictid bees during both wet and dry seasons, and andrenid and colletid bees during the wet season or during both seasons. Social pollinators (e.g., apid bees) require a sequence of floral resources throughout the year to support long-lived colonies and visit a succession of flowering plant species, whereas more ephemeral, solitary species with short life spans can be relatively more specialized on seasonal floral resources (Corbet, 1997; Roubik, 1989).

Pollinator functional groups also can be distinguished on the basis of habitat preferences, such as vegetation stratum (Fig. 13.1). Appanah (1990) distinguished four groups of plant–pollinator associations in a tropical lowland dipterocarp forest. The **forest floor stratum** is characterized by low visibility and limited airflow. Floral rewards are low, reflecting low productivity of light-limited plants and low energy requirements of associated pollinators, and flowering times are extended, increasing the probability of pollination by infrequent visitors. The plant–pollinator association of this stratum is dominated largely by nonselective, low-energetic beetles, midges, and other flies. These pollinators are attracted over short distances by strong olfactory cues, often resembling dung or carrion, which have limited effective range. The **understory stratum** shares many of the environmental features of the forest floor. Plants in this stratum also offer limited visual cues and floral rewards and are pollinated by nonspecific trapliners, i.e., species that revisit particular plants along an established circuit (e.g., trigonid bees, solitary wasps, and butterflies). The **overstory stratum** generally is characterized by brightly colored flowers, held above the canopy to attract pollinators over a wide area, and brief, highly synchronized flowering within plant species. Dominant pollinators are *Apis dorsata* and trapliners such as carpenter bees, birds, and bats. Dipterocarps in the genera *Shorea*, *Hopea*, and *Dipterocarpus* form a separate association based on tiny flowers with limited nectar rewards and nocturnal flowering. Thrips and other tiny, flower-feeding insects are the primary pollinators. Finally, some plant species representing various canopy positions are **cauliflorous**, i.e., they produce flowers along the trunk or main branches. These flowers typically are large, or small and clumped, pale colored, odiferous, and produced during a brief, highly synchronized period. Pollinators include understory and overstory insects, birds, and bats.

Roubik (1993) experimentally manipulated availability of floral resources

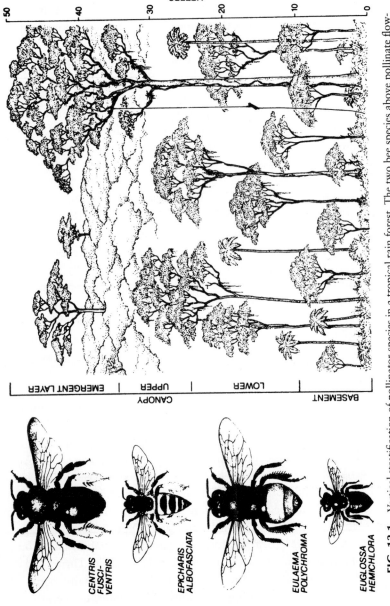

FIG. 13.1 Vertical stratification of pollinator species in a tropical rain forest. The two bee species above pollinate flowers in the upper canopy and the two species below pollinate flowers in the subcanopy. From Perry (1984), © George V. Kelvin/*Scientific American.*

from different canopy strata in tropical forests in Panama. Results indicated that the apparent fidelity of pollinator species to particular canopy strata reflected pollinator preferences for particular floral resources. Most pollinator species were attracted to their preferred floral resources, regardless of their location in the canopy.

B. Pollination Efficiency

A number of factors influence the efficiency of pollen transport between conspecific reproductive structures. The mechanism of pollen transport, proximity of conspecific plants, pollinator attraction to floral structures, adaptations for carrying pollen, fidelity, and thermodynamic constraints determine the probability that a flower will receive conspecific pollen.

Pollination efficiency reflects the probability that pollen reaches a conspecific flower. Wind pollination is highly inefficient. The probability of successful pollen transfer by wind decreases as the cube of distance between plants (Moldenke, 1976). However, plant investment in individual pollen grains is negligible so large numbers can be produced, increasing the probability that some will land on conspecific reproductive structures. Directed transport of pollen by pollinators increases efficiency to the extent that the pollinator visits a conspecific flower before the pollen is lost or contaminated with pollen from other plant species. Hence, animal-pollinated plant species may invest energy and nutrients in adaptations to improve the host-selectivity of the pollinator. These adaptations include nectar rewards to attract pollinators; floral and aromatic advertisements; floral structures that restrict the diversity of pollinators visiting the flowers, synchronized flowering among conspecific individuals, and divergence in time of flowering among plant species to reduce pollen contamination (Heinrich, 1979).

Nectar rewards must be sufficient to compensate the pollinator for the foraging effort. For example, a greater nectar return is necessary to attract bees during cooler periods, when energy allocation to thermoregulation is high, compared to warmer periods (Heinrich, 1979). Heinrich (1979) noted that pollinator fidelity reflects offsetting adaptations. Plants invest the minimum amount of energy necessary to reward pollinators, but pollinators quickly learn to concentrate on flowers offering the greatest rewards. Individual plants in aggregations could attract bees and be pollinated even if they produced no nectar, provided that their neighbors produced nectar. The nonproducers should be able to invest more energy in growth and seed production. However, if these "cheaters" became too common, pollinators would switch to competing plant species that offered greater food rewards (Feinsinger, 1983).

The regularity with which conspecific plants occur in close proximity to each other largely determines their pollination mechanism. Long-lived species that dominate relatively simple ecosystems, i.e., grasslands and temperate forests, are largely wind pollinated. These plant species do not require efficient pollination or frequent reproduction to ensure population survival. Energetically inexpensive wind transport of pollen provides sufficient pollination (and successful reproduction) so that energy need not be diverted to production of expensive nectar rewards and floral displays to advertise availability.

C. Patterns of Insect Pollination among Ecosystems

Pollination by insects is more prevalent in some types of ecosystems than in others. Pollination by animals is more common in angiosperm-dominated ecosystems than in gymnosperm-dominated ecosystems, but, as has been noted, pollination by wind is energetically efficient for dominant species in grasslands and temperate deciduous forests.

Directed transport of pollen by animals becomes critical to reproduction of plant species that are short-lived, sparsely distributed, or occur in habitats with restricted airflow (Appanah, 1990; Moldenke, 1979; Regal, 1982). In contrast to long-lived plants, short-lived plants have limited opportunities for future reproduction and, therefore, depend on more efficient pollination to ensure seed production. Sparsely distributed plants and plants in areas of limited airflow cannot rely on inefficient transport of pollen by wind between distant or inaccessible individuals. Such species include early successional plants dominating ephemeral communities, widely spaced plants in harsh environments (e.g., deserts), scattered forbs in grasslands, subdominant trees, shrubs, and herbs in temperate forests, and most plant species in tropical forests. Regal (1982) reported that fewer than 6% of desert shrub species are wind pollinated. Plant investment in attractants and rewards for pollinators represents an evolutionary trade-off between growth and reproduction (Heinrich, 1979) and may affect the persistence of early successional plant species when herbivory or competition for light reduces acquisition of energy.

II. EFFECTS OF POLLINATION

Pollination contributes to genetic recombination and survival. Many plants can reproduce vegetatively, but this mechanism is not conducive to long-distance colonization or genetic recombination. Species survival and adaptation to changing environmental conditions requires out-crossing and environmental selection among diverse genotypes. Pollinators also contribute to vegetation composition and ecosystem energy and nutrient fluxes.

Pollination is critical to genetic recombination. Wind provides adequate pollination among closely spaced individuals. However, pollination by animals is critical to out-crossing and survival of ephemeral or isolated species. Bawa (1990) reviewed studies that demonstrate long-distance pollen flow and out-crossing for several tropical canopy trees but a high degree of inbreeding for many other tropical taxa, primarily herbs and shrubs.

Pollination also is critical to reproduction of some key plant species. Many animals depend on fruit production. Hence, pollination of fruiting plants has consequences not only for plant reproduction, but also for the survival of frugivores and seed predators (Bawa, 1990).

Differential pollination and reproductive success among plant species affects vegetation dynamics. Plant species that maximize pollination efficiency and increase out-crossing via animal pollinators are able to persist as scattered individuals. However, pollination efficiency by insects is strongly affected by plant spacing. Momose *et al.* (1998) found that pollination by thrips and consequent fruit and seed development of a small (<8 m height) tree species, *Popowia pisocarpa*, in Sarawak declined dramatically when distances between trees exceeded 5 m. Changes in pollinator abundances and pollination efficiency affect plant population dynamics and persistence in communities.

Pollinator-facilitated reproduction is a key factor enhancing the persistence of sparsely distributed plant species. Maintenance of rare plant species or restoration of declining species depends largely on protection or enhancement of associated pollinators (Archer and Pyke, 1991; Corbet, 1997). Pollination may be particularly important to the reproduction and establishment of early successional plant species and to the persistence of early and mid-successional species into later seres.

Steffan-Dewenter and Tscharntke (1997) examined the effects of plant isolation on pollination and seed production in replicate grasslands surrounded by intensively managed farmland. They established small experimental patches of two grassland species, *Sinapsis arvensis* and *Raphanus sativus*, at increasing distances from the grassland boundaries and found that the number and diversity of bees visiting flowers, and seed production, declined with increasing isolation (Fig. 13.2). Number of seeds per plant was reduced by 50% at 260 m from the nearest grassland for *R. sativus* and at 1000 m for *S. arvensis*.

Changes in pollinator abundance, such as those resulting from ecosystem fragmentation, can affect plant reproduction and gene flow (Bawa, 1990; Didham *et al.*, 1996). Powell and Powell (1987) compared attraction of male euglossine bees to floral chemical baits in forest fragments in Brazil. Abundance and species composition did not differ among sites prior to fragmentation. However, after fragmentation, visitation rates for most species declined with fragment size, and the bee species trapped in clearings differed from the species trapped in forests (Fig. 13.3). Powell and Powell (1987) concluded that the reduced abundances of particular pollinators in fragmented forests threaten the viability of their orchid hosts. Aizen and Feinsinger (1994) compared pollinator visitation among replicated blocks containing continuous forest and large (>2.2 ha) and small (< 1 ha) fragments in subtropical dry forest in northwestern Argentina. The diversity and visitation frequency of native pollinators decreased significantly and the visitation frequency of exotic honey bees, *Apis mellifera*, increased significantly with decreasing fragment size (Fig. 13.4).

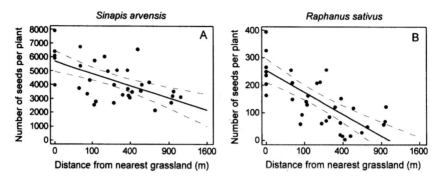

FIG. 13.2   Relationship between the number of seeds per plant and the distance from the nearest chalk grassland supporting populations of pollinators. The regression line is significant at $P <$ 0.001; r = −0.80; dotted lines are 95% confidence limits. From Steffan-Dewenter and Tscharntke (1997) by permission from the International Society for Horticultural Science.

Fragments supported fewer bee species than did continuous forests. Although honey bees from the surrounding agricultural matrix replaced most of the lost visitation by native pollinators, some plant species could be threatened by reduced specificity of pollinators.

Few studies have addressed effects of pollinators on energy and nutrient fluxes. Roubik (1989) calculated the effects of social bees on energy and nitrogen budgets of tropical forests in Central America. He estimated that 600 colonies km^{-2} harvested 1.4×10^7 kJ yr^{-1} and disposed of an equivalent energy value represented by dead bees scattered on the ground within a few dozen meters of each nest. This value exceeds estimates of energy fixed annually by primary producers, indicating that the energetics of flowering are greatly underestimated (Roubik, 1989). The 600 colonies also distribute about 1800 kg of trash (pupal exuviae and feces) ha^{-1} yr^{-1}. At 4% nitrogen content, this represents a flux of 72 kg ha^{-1} yr^{-1} or about 1% of above-ground nitrogen in biomass.

III. MECHANISMS AND PATTERNS OF SEED PREDATION AND DISPERSAL

The fate of seeds is critical to plant reproduction. A variety of animals feed exclusively or facultatively on fruits or seeds, limiting potential germination and seedling recruitment. Many animals, especially frugivores, facilitate seed dispersal. Dispersal of seeds generally is considered to facilitate colonization of new habitats and to permit escape from high mortality near parent plants, but relatively few studies have measured the advantages of seed dispersal to plant fitness (Howe and Smallwood, 1982). As with pollination, a variety of mechanisms contribute to seed dispersal. These mechanisms confer varying degrees of dispersal efficiency and advantages for seedling growth, depending on ecosystem conditions.

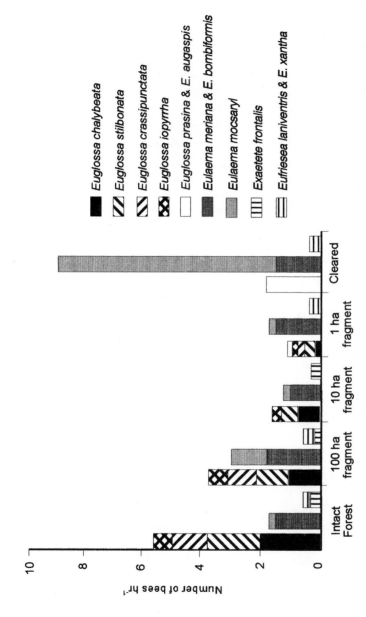

FIG. 13.3 Rates of visitation by male euglossine bees at chemical baits in intact forest, forest fragments of varying size (100 ha, 10 ha, and 1 ha), and recently deforested (500 ha). Modified from Powell and Powell (1987) by permission from the Association for Tropical Biology.

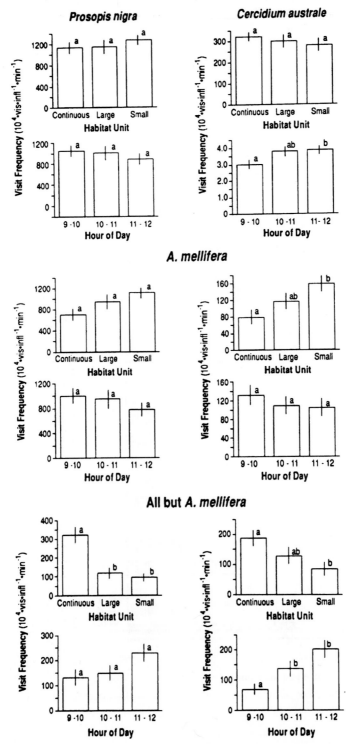

FIG. 13.4 Rates of visitation by all pollinating insects, exotic honey bees (*Apis mellifera*) alone, and native pollinators alone on flowers of two plant species by treatment (continuous forest, and

A. Mechanisms of Seed Predation and Dispersal

Fruits and seeds are highly nutritive food resources as a consequence of plant provision for germination and, often, attraction of dispersal agents. A wide variety of animals feed on fruits or seeds.

Seed dispersal can be accomplished through both abiotic and biotic mechanisms. Abiotic dispersal involves wind and water; biotic dispersal involves autogenic mechanisms, such as explosive fruits, and various animal agents, particularly insects, fish, reptiles, birds, and mammals. Dispersal by animals typically is a consequence of frugivory or seed predation, but some species acquire seeds or spores through external attachment of seeds with various kinds of clinging devices, such as sticky material or barbed spines. Seeds of a majority of plant species in many ecosystems are dispersed by animals (Howe and Smallwood, 1982).

Seed predator and seed disperser functional groups can be distinguished on the basis of consumption of fruits or seeds versus transport of seeds. **Frugivores** feed on fleshy fruits and often terminate fruit or seed development (Sallabanks and Courtney, 1992), but many vertebrate frugivores (including fish, reptiles, birds, and mammals) consume entire fruits and disperse seeds that are adapted to survive passage through the digestive tract (Crawley, 1989; de Souza-Stevaux *et al.*, 1994; Horn, 1997; Sallabanks and Courtney, 1992; Temple, 1977). **Seed predators** include a number of insect, bird, and rodent species that consume seeds where found. Some seed predators eat the entire seed (e.g., vertebrates and ants), but others penetrate the seed coat and consume only the endosperm (e.g., seed bugs and weevils) or develop and feed within the seed (e.g., seed wasps and seed maggots) (Brown *et al.*, 1979; Crawley, 1989; Louda *et al.*, 1990b; Schowalter, 1993;, Turgeon *et al.*, 1994). **Seed cachers** eat some seeds and move others from their original location to storage locations. Although ants and rodents are best known for caching seeds (Brown *et al.*, 1979), at least one carabid beetle, *Synuchus impunctatous* caches seeds of *Melampyrum* in hiding places after consuming the caruncle at the end of the seed (Manley, 1971). **Seed vectors** include primarily vertebrates that carry seeds adapted to stick to fur or feathers. Insects generally are too small to transport seeds in this way but can transmit spores or microorganisms adapted to adhere to insect exoskeletons.

These functional groups can be subdivided on the basis of pre- or postdispersal seed predation, seed size, etc. Predispersal frugivores and seed predators feed on the concentrated fruits and seeds developing on the parent plant, whereas postdispersal frugivores and seed predators must locate scattered fruits and seeds that have fallen to the ground. Rodents and birds typically exploit larger seeds than do insects, and species within taxonomic groups also partition seeds

---

large (2.2 ha) and small (1 ha) fragments) and by time of day in Argentina. Vertical lines represent standard errors; bars under the same letter do not differ at $P < 0.05$. From Aizen and Feinsinger (1994) by permission from the Ecological Society of America.

on the basis of size (e.g., Brown *et al.*, 1979; Davidson *et al.*, 1984; Whitford, 1978). Vertebrates are more likely to disperse seeds from consumed fruits than are insects, which (due to their small size) typically feed on portions of fruits and on or in seeds. Insects, especially ants, are more likely to disperse small seeds, particularly of plant species adapted for dispersal by ants (myrmecochory).

B. Efficiency of Seed Production and Dispersal

A number of factors influence the efficiency of seed production and dispersal. The extent of seed mortality, mechanism of seed transport, distance moved from the parent plant, attraction of particular dispersal agents, and thermodynamic constraints determine the probability that seeds will survive and be moved to suitable or distant locations.

Seed predators are capable of consuming or destroying virtually the entire production of viable seed of a given plant species in some years (Ehrlén, 1996; Robertson *et al.*, 1990; Schowalter, 1993; Turgeon *et al.*, 1994). The intensity of seed predation depends to a large extent on seed availability. Seed predators focus on the largest or most concentrated seed resources (Ehrlén, 1996). During years of poor seed production, most or all seeds may be consumed, whereas during years of abundant seed production, predator satiation enables many seeds to survive (Schowalter, 1993; Turgeon *et al.*, 1994). Long-lived plant species need produce few offspring over time to balance mortality. Hence, many tree species produce abundant seed only once every several years. Years of abundant seed production are known as mast years. Poor seed production during intervening years reduces seed predator populations and increases efficiency of seed production during mast years (Fig. 13.5).

Insects generally are more important predispersal seed predators than are

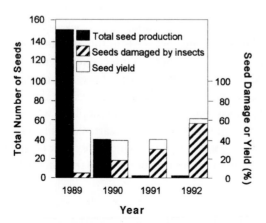

FIG. 13.5   Relationship between total seed produced, seed loss to insects, and seed yield in a Douglas-fir, *Pseudotsuga menziesii*, seed orchard in western Oregon. Data from Schowalter (1993).

vertebrates, but vertebrates are more important postdispersal seed predators (Crawley, 1989; Davidson *et al.*, 1984; Louda *et al.*, 1990b; Schupp, 1988). Predispersal seed predators greatly reduce seed production efficiency and reduce the number of seeds available for postdispersal seed predators and dispersal. Christensen and Whitham (1991) reported that seed-dispersing birds avoided foraging in piñon pine trees in which the stem- and cone-boring moth, *Dioryctria albovitella*, had inhibited cone development and increased cone mortality. Sallabanks and Courtney (1992) suggested that seed predators and dispersers often may exert opposing selection pressures on temporal and spatial patterns of fruit and seed production.

Seed dispersal is an important mechanism for plant colonization of new sites. However, dispersal also may increase seed and seedling survival. Schupp (1988) reported that vertebrate seed predators limited seed survival under the parent tree to 15%, but that dispersal distances of only 5 m significantly increased seed survival to nearly 40% over a 7-month period (Fig. 13.6). The efficiency with which seeds reach favorable sites is critical to plant population dynamics.

Seeds transported by wind or water often have low dispersal efficiency, for which plants must compensate by producing large numbers of seeds. Animals are presumed to be more efficient dispersal agents but this may not be accurate. Seeds drop from animal vectors with no more likelihood of landing on suitable germination sites than do seeds deposited by wind or water, unless animal dens or habitats provide suitable germination sites. However, the direction of animal movement is more variable than that of wind or water. Birds, in particular, quickly cover large areas, but local seed redistribution by ants also can significantly affect plant demographies (O'Dowd and Hay, 1980). In any case, a number of plant species are specifically adapted to seed dispersal by animals, and

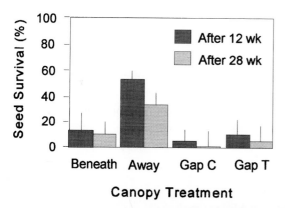

FIG. 13.6 Survival of *Faramea occidentalis* seeds beneath fruiting parent trees (Beneath), away from parent trees (Away; 5 m from crown perimeter of nearest fruiting adult), and within the canopy (Gap C) and trunk (Gap T) zones of treefall gaps on Barro Colorado, Panama. Survival of seed was significantly ($P < 0.05$) higher 5 m from parent trees than beneath parent trees or in treefall gaps. Data from Schupp (1988).

some species with large seeds or thick seed coats may show reduced dispersal or germination ability where movement by animals or seed scarification are prevented (Culver and Beattie, 1980; Temple, 1977).

Seed storage underground by ants and rodents may move seeds to sites of better soil conditions or reduce vulnerability to further predation. A number of studies have demonstrated that seedlings germinating from ant nests are larger and have higher survival rates than do seedlings emerging elsewhere (Andersen, 1988; Bennett and Krebs, 1987; Culver and Beattie, 1980; Rissing, 1986; Wagner, 1997). Ant nests may or may not enrich surrounding soils (see Chapter 14). Wagner (1997) demonstrated that soil from ant nests had significantly higher concentrations of nitrate, ammonium, phosphorus, and water and significantly higher nitrogen mineralization rates than did soil away from nests. On the other hand, Hughes (1990) and Rice and Westoby (1986) found that myrmecochorous plants do not necessarily show distribution patterns associated with soil fertility. However, Hughes (1990) reported that, after 1 year, 5–40% of the ground surface had been within 10 cm of a nest entrance, as a result of changes in nest structure. Such rapid influence of ant nests on surrounding soil could homogenize soil fertility.

Plants also may benefit from seed removal to ant nests where they are deposited at suitable depths for germination or protected from intense predation by vertebrates (Cowling *et al.*, 1994). Shea *et al.* (1979) found that germination of serotinous seeds of several legume species in Western Australia was enhanced by seed redistribution by ants to depths that were heated sufficiently, but protected from higher surface temperatures, during high intensity autumn fires. O'Dowd and Hay (1980) reported that transport of diaspores of *Datura discolor* by ants, to nests averaging only 2.3 m from the nearest plant, reduced seed predation by desert rodents from 25–43% of seeds in dishes under parent plants to <1% of seeds in dishes near ant nests. Heithaus (1981) found that when seed dispersal by ants was experimentally prevented, rodents removed 70–84% of *Asarum canadense* and *Sanguinaria canadensis* seeds, compared to 13–43% of seeds lost when ants were present. Furthermore, laboratory experiments demonstrated that rodents located buried seeds less frequently than seeds on the surface and consumed buried seeds less often when elaiosomes were removed, as done by ants. Hughes (1990) reported that changes in nest structure, indicated by relocation of nest entrances, may provide refuges for seeds remaining in abandoned portions of nests and reduce seedling competition by preventing long-term concentration of seeds in localized sections of nests.

C. Patterns of Seed Mortality and Dispersal among Ecosystems

Few studies have compared seed predation and dispersal among ecosystems. Different agents dominate these processes in different ecosystems (Moll and

McKenzie, 1994). For example, dominant plant species in temperate, especially arid, ecosystems frequently have wind-dispersed seed, whereas plant species on oceanic islands often are water-dispersed (Howe and Smallwood, 1982). Howe and Smallwood (1982) concluded that consistently windy ecosystems promote wind-driven dispersal, whereas more mesic conditions promote animal-driven dispersal. Old World deserts have relatively few (<5%) animal-dispersed plant species (Howe and Smallwood 1982). More than 60% of temperate and tropical forest plant species are dispersed by animals (Howe and Smallwood, 1982). Fruits and seeds in seasonally flooded tropical forests often are dispersed by fish during periods of inundation (de Souza-Stevaux *et al.*, 1994; Horn, 1997; Howe and Smallwood, 1982). Bats and primates are more important frugivores and seed dispersers in tropical forests than in temperate ecosystems. Insects are ubiquitous frugivores and seed predators, but may be more important dispersers in grassland and desert ecosystems, where transport to ant nests may be critical to protection of seeds from vertebrate seed predators and from competition (e.g., Louda *et al.*, 1990b; Rice and Westoby, 1986).

Rice and Westoby (1986), Rissing (1986), and Westoby *et al.* (1991) discussed a number of potential factors affecting differences in the incidence of ant-dispersed seeds among biogeographic regions. Myrmecochory appears to be more prevalent in Australia and South Africa than in other regions. One hypothesis is that smaller plants (characteristic of arid biomes) generally are more likely to be ant-dispersed than are larger plants. A second hypothesis is that the relatively infertile soils of Australia and South Africa preclude nutrient allocation to fruit production, forcing plants to adapt to seed dispersal by ants rather than vertebrates. Finally, Australia and South Africa lack the large seed-harvester ants, e.g., *Pogonomyrmex* spp., *Messor* spp., and *Veromessor* spp., common in arid regions of North America and Eurasia. These ants consume relatively large seeds, limiting the value of an elaiosome as a food reward for seed dispersal.

IV. EFFECTS OF SEED PREDATION AND DISPERSAL

Seed predators and dispersers influence plant population dynamics and community structure by affecting both seed survival and seedling recruitment. Robertson *et al.* (1990) reported that predispersal seed predation rates varied widely among mangrove species at study sites in northeastern Australia. Three species (*Ceriops australis*, *C. tagal*, and *Rhizophora apiculata*) had fewer than 10% of seeds damaged by insects, whereas six species (*Avicennia marina*, *Bruguiera gymnorrhiza*, *B. parviflora*, *Heritiera littoralis*, *Xylocarpus australasicus*, and *X. granatum*) consistently had > 40% of seeds damaged. These mangrove species also showed variation in survival and growth rates (height and diameter) of seedlings from insect-damaged seeds. Ehrlén (1996) reported a significant positive correlation between the change in population growth rate induced by seed predation and the reproductive value of seeds and seedlings, indicating that survival of seeds and seedlings is the important aspect of seed predator effects on plant population growth.

Postdispersal seed predators similarly affect the survival and growth of seeds and seedlings. Seeds selected for storage in ant nests or refuse piles often show increased survival and seedling growth, relative to seeds in control sites (Andersen, 1988; Culver and Beattie, 1980; Hughes, 1990; Rissing, 1986). Enhanced seedling growth on ant nests may reflect nutrient enrichment (but see Rice and Westoby, 1986) or greater water-holding capacity (Jonkman, 1978; Wagner, 1997).

The composition of the granivore community affects plant community development. Inouye *et al.* (1980) reported that exclusion of granivorous rodents or ants altered densities and community composition of annual plant species (Table 13.1). Rodents preyed selectively on large-seeded species (e.g., *Erodium* spp. and *Lotus humistratus*). These species increased to dominate vegetative biomass and replace small-seeded plant species, especially *Euphorbia polycarpa*, on plots from which rodents were excluded. Ants preyed most intensively on the most abundant plant species (*Filago californica*). When ants were excluded, this small-seeded composite became numerically dominant and reduced species diversity.

Many plant species have become dependent on animal mutualists for seed-dispersal. Seed and seedling survival for some species depend on distance from parent plants, where seed predation may be concentrated (O'Dowd and Hay, 1980; Schupp, 1988). As found by Powell and Powell (1988) and Steffan-Dewenter and Tscharntke (1997) for pollinators (as previously discussed), decline in abundance of seed dispersal agents may threaten some plant species. The hard seed coats of plant species adapted for dispersal by vertebrates often require scarification during passage through the digestive systems before germination is possible. Temple (1977) noted the coincidence between the age (300–400 years) of the last 13 naturally regenerated tambalacoque trees, *Sideroxylon sessiliflorum* (= *Calvaria major*) and the disappearance of the dodo in 1680 on the South Pacific island of Mauritanius. When *S. sessiliflorum* seeds were force-fed to turkeys (approximately the size of the dodo), the seed coats were sufficiently abraded during gut passage to permit germination, demonstrating a potential role of the dodo in dispersal and survival of this once dominant tree.

V. SUMMARY

Insects are the major agents of pollination, seed predation, or seed dispersal in many ecosystems. Although few studies have evaluated the effects of pollinators, seed predators, and seed dispersers on ecosystem processes, these organisms have important effects on seedling recruitment and vegetation dynamics.

Pollination is an important means of increasing genetic heterogeneity and improving plant fitness. Pollination can be accomplished by abiotic (wind) or

TABLE 13.1 Effects of Removal of Ants, Rodents, or Both on Densities of Certain Annual Plant Species, All Plants, Plant Biomass and Two Measures of Species Diversity

	+ Rodents + Ants	+ Rodents - Ants	- Rodents + Ants	- Rodents - Ants	Effects of removal of	
					Rodents	Ants
Initial Census 29 January 1977						
1. Large plants	1.00 (35.8)	0.98	2.08	2.35	Increase[b]	NS
2. Small plants	1.00 (292.5)	3.30	3.32	3.17	NS	Increase[b]
Final Census 2 April 1977						
3. *Erodium cicutarium* (seed mass = 1.6 mg)	1.00 (1.8)	1.83	7.03	16.11	Increase[b]	NS
4. *E. texanum* (seed mass = 1.6 mg)	1.00 (0.6)	0.88	2.07	0.78	Increase[a]	NS
5. *Euphorbia polycarpa* (seed mass = 0.2 mg)	1.00 (0.6)	2.00	0.14	0.29	Decrease[a]	NS
6. *Filago californica* (seed mass = 0.04 mg)	1.00 (142.1)	1.90	1.43	2.59	NS	Increase[a]
7. *Lotus humistratus* (seed mass = 1.5 mg)	1.00 (11.4)	1.14	2.43	5.22	Increase[b]	NS
8. All plants	1.00 (209.6)	1.35	1.34	1.94	Increase[a]	Increase[b]
9. Dry mass (all species)	1.00 (5.8)	1.07	2.09	2.17	Increase[b]	NS
10. Species diversity (H′)	1.00 (2.78)	0.73	0.99	0.89	NS	Decrease[a]
11. Species evenness (E)	1.00 (0.53)	0.77	1.99	1.04	NS	Decrease[a]

[a]Significant at $P < 0.05$.
[b]Significant at $P < 0.01$.
Values given are ratios of treatment to control (+ Rodents + Ants) means. Numbers in parentheses are mean values for unthinned plots except for plant biomass and the two measures of diversity, which are for control plots. Statistical analysis was by ANOVA; NS, not significant. Reproduced from Inouye *et al.* (1980) by permission from the Ecological Society of America.

biotic (insects, birds, and bats) agents. Wind pollination is inefficient but sufficiently effective for species that dominate temperate ecosystems. However, animal agents increase pollination efficiency for more isolated plants and are critical to survival of many plant species that typically occur as widely scattered individuals. Pollinator functional groups can be distinguished on the basis of their degree of specialization on particular floral resources, their seasonality, or habitat preferences.

Seed predators often consume the entire reproductive effort of host plants. Predispersal seed predators typically focus on concentrated seed resources on the parent plants whereas postdispersal seed predators must locate more scattered seed resources on the ground. Insects are more important predispersal seed predators but vertebrates are more important postdispersal seed predators in most ecosystems.

Seed dispersal is critical to plant species survival both because new habitats can be colonized and because seed relocation often improves seed and seedling survival. Seeds can be dispersed by abiotic (wind and water) or biotic (insect and vertebrate) agents. Animals can increase dispersal efficiency by moving seeds to more suitable germination sites, especially if seeds are buried. Ants, in particular, can increase seed survival and seedling growth by relocating seeds to nests, where seeds are protected from further predation, from suboptimal surface conditions, and from competition with parent plants. Ant nests also may provide more suitable soil conditions for germination and growth. Some seeds require scarification of hard seed coats and must pass through vertebrate digestive systems before germination can occur.

Both pollination and seed dispersal affect plant population and community dynamics. Differential pollination, seed predation and seed dispersal efficiencies among plant species affect seedling recruitment and growth. Survival of some plant species depends on sufficient abundance of pollinators or seed dispersers.

Decomposition and Pedogenesis

DECOMPOSITION IS THE BREAKDOWN OF DEAD ORGANIC MATTER THAT eventually results in release of CO_2, other organic trace gases, water, mineral nutrients, and energy. Pedogenesis (soil development) largely reflects the activities of decomposers that mix organic matter with mineral soil. These two processes contribute greatly to the capacity of a site to support primary production. Accumulated organic litter represents a major pool of energy and nutrients in many ecosystems. Carbon and other nutrients released through decomposition can be acquired by plants or microbes or returned to abiotic pools (see Chapter 11). Incorporation of decay-resistant organic matter and nutrients into soil increases fertility, aeration, and water-holding capacity. Release of CO_2, CH_4, and other trace gases affects atmospheric conditions and global climate.

Decomposition can be categorized into four component processes: **photo-oxidation**, abiotic catabolism resulting from exposure to solar radiation; **leaching**, the loss of soluble materials as a result of percolation of water through material; **comminution**, the fragmentation of organic litter, largely as a result of detritivory; and **mineralization**, the catabolism of organic molecules by saprophytic microorganisms. Vossbrink *et al.* (1979) found that when arthropods and microbes were excluded, detritus lost 5% mass, due entirely to leaching or photo-oxidation; further decomposition was negligible. A variety of macro-, meso-, and microarthropods are the primary detritivores in most ecosystems.

The feeding and burrowing activities of many animals, including ants, termites, and other arthropods, redistribute and mix soil and organic material. Burrowing also increases soil porosity, thereby increasing aeration and water-holding capacity.

The roles of arthropod detritivores and burrowers in decomposition and soil development are the most widely studied effects of arthropods on ecosystem processes (e.g., Ausmus, 1977; Crossley, 1977; Eldridge, 1993, 1994; Seastedt, 1984; Swift, 1977; Swift et al., 1979; Whitford, 2000; Wotton et al., 1998). Arthropod burrowers and detritivores are relatively accessible and often can be manipulated for experimental purposes. Their key contributions to decomposition and mineralization of litter (both fine or suspended organic matter and coarse woody debris) and pedogenesis has been demonstrated in virtually all ecosystems. Indeed, some aquatic and glacial ecosystems consist of arthropod detritivores and associated microorganisms feeding entirely on allochthonous detritus (Edwards and Sugg, 1990; Oertli, 1993; Wallace et al., 1992). Effects of detritivorous and fossorial species on decomposition and soil mixing depend on the size of the organism, its food source, type and rate of detritivory, volume of displaced litter or soil, and type of saprophytic microorganisms inoculated into litter. Although most studies have addressed the effects of detritivores and burrowers on soil processes, their contributions to soil development and biogeochemical cycling affect primary production, as well.

I. TYPES AND PATTERNS OF DETRITIVORY AND BURROWING

A. Detritivore and Burrower Functional Groups

Functional groups of detritivorous and burrowing arthropods have been distinguished on the basis of principal food source, mode of feeding, and microhabitat preferences (e.g., Moore et al., 1988; Wallace et al., 1992). For example, functional groups can be distinguished on the basis of seasonal occurrence, habitats, and substrates (e.g., terrestrial versus aquatic, animal versus plant, foliage versus wood, arboreal versus fossorial) or particular stages in the decomposition process (Anderson et al., 1984; Hawkins and MacMahon, 1989; Schowalter and Sabin, 1991; Schowalter et al., 1998; Seastedt, 1984; Siepel and de Ruiter-Dijkman, 1993; Tantawi et al., 1996; Tullis and Goff, 1987; Wallace et al., 1992; Winchester, 1997; Zhong and Schowalter, 1989).

General functional groupings for detritivores are based on their effect on decomposition processes. **Coarse** and **fine comminuters** are instrumental in the fragmentation of litter material. Major taxa in terrestrial ecosystems include millipedes, earthworms, termites, beetles, and wasps (coarse) and mites, collembolans, and various other small arthropods (fine). Many species are primarily **fungivores** or **bacteriovores**, capable of fragmenting substrates while feeding on the surface microflora. A number of species, including dung beetles, millipedes, and termites, are **coprophages**, either feeding on feces of larger species or reingesting their own feces following microbial decay and enrichment (Coe, 1977; Dangerfield, 1994; Holter, 1979; McBrayer, 1975). In aquatic ecosystems **scrapers,** that graze or scrape microflora from mineral and organic substrates, and **shredders,** that chew or gouge large pieces of decomposing material, represent coarse comminuters; **gatherers,** that feed on fine particles of decomposing organic material deposited in streams, and **filterers,** that have specialized structures for sieving fine suspended organic material, represent fine comminuters (Cummins, 1973; Wallace and Webster, 1996; Wallace *et al.*, 1992). **Xylophages** are a specialized group of detritivores that excavate and fragment woody litter. Major species include bark beetles, ambrosia beetles, wood-boring beetles, wasps, carpenter ants, and termites (Fig. 14.1). Most of these species either feed on fungal-colonized wood or support mutualistic internal or external fungi or bacteria that digest cellulose and enhance the nutritional quality of wood (e.g., Breznak and Brune, 1994; Siepel and de Ruiter-Dijkman, 1993; see Chapter 8). Many fungivores and bacteriovores, including nematodes and protozoa, as well as arthropods, feed exclusively on microflora and affect the abundance and distribution of these decomposers (e.g., Santos *et al.*, 1981).

An important consequence of fragmentation is increased surface area for microbial colonization and decomposition. Microbes also are carried, either passively through transport of microbes acquired during feeding or dispersal or

FIG. 14.1 *Melanophila* sp. (Coleoptera: Buprestidae) larva in mine in phloem of recently killed Douglas-fir, *Pseudotsuga menziesii*, tree in western Oregon. The entire phloem volume of this tree has been fragmented and converted to frass packed behind mining larvae of this species, demonstrating detritivore capacity to reduce detrital biomass.

actively through inoculation of mutualistic associates, to fresh surfaces during feeding.

Many detritivores redistribute large amounts of soil or detritus during foraging or feeding activities. However, nondetritivores also contribute to mixing of soil and organic matter. Fossorial functional groups can be distinguished on the basis of their food source and mechanism and volume of soil/detrital mixing. **Subterranean nesters** burrow primarily for shelter. Vertebrates, e.g., squirrels, woodrats and coyotes, and many invertebrates excavate tunnels of various sizes, typically depositing soil on the surface and introducing some organic detritus into nests. **Gatherers**, primarily social insects, actively concentrate organic substrates in colonies. Ants and termites redistribute large amounts of soil and organic matter during construction of extensive subterranean, surficial, or arboreal nests (Anderson, 1988; Haines, 1978). Subterranean species concentrate organic matter in nests excavated in soil, but many species bring fine soil particles to the surface that are mixed with organic matter in arboreal nests or foraging tunnels. These insects can affect a large volume of substrate (up to 10^3 m^3), especially as a result of restructuring and lateral movement of the colony (Hughes, 1990; Moser, 1963; Whitford *et al.*, 1976). **Fossorial feeders**, such as earthworms, gophers, moles, mole crickets (*Gryllotal pidae*), and benthic invertebrates, feed on subsurface resources (plant, animal, or detrital substrates) as they burrow, constantly mixing mineral substrate and organic material in their wake.

B. Measurement of Detritivory, Burrowing, and Decomposition Rates

Evaluation of the effects of detritivory and burrowing on decomposition and soil mixing requires appropriate methods for measuring rates of these processes. Detritivory and burrowing can be measured by extrapolating from individual consumption rates, by presenting decomposer communities with experimental substrates, or by manipulating abundances of detritivores or burrowers. Several methods have been used to measure rates of decomposition and soil mixing.

Detritivory can be measured by providing experimental substrates and measuring colonization and consumption rates. Johnson and Whitford (1975) measured the rate of termite feeding on an artificial carbohydrate source and natural substrates in a desert ecosystem. Edmonds and Eglitis (1989) and Zhong and Schowalter (1989) measured the rate of wood-borer colonization and excavation in freshly cut tree boles.

Detritivory often has been estimated by multiplying the per capita feeding rate for each functional group by its abundance (Anderson *et al.*, 1984; Crossley *et al.*, 1995; Dangerfield, 1994). Laboratory conditions, however, might not represent the choices of substrates available under field conditions. For example, Dangerfield (1994) noted that laboratory studies might encourage coprophagy by millipedes by restricting the variety of available substrates, thereby overrepresenting this aspect of consumption. Mankowski *et al.* (1998) used both forced-feeding and choice tests to measure wood consumption by termites when a variety of substrate types is available or restricted.

Abundances of detritivore functional groups can be manipulated to some extent by use of microcosms (Setälä and Huhta, 1991; Setälä *et al.*, 1996) or selective biocides or other exclusion techniques (Crossley and Witkamp, 1964; Edwards and Heath, 1963; Ingham *et al.*, 1986; Macauley, 1975; Santos and Whitford, 1981; Schowalter *et al.*, 1992; Seastedt and Crossley, 1983; Wallace *et al.*, 1991). Naphthalene and chlordane in terrestrial studies (Crossley and Witkamp, 1964; Santos and Whitford, 1981; Seastedt and Crossley, 1983; Whitford, 1986) and methoxychlor in aquatic studies (Wallace *et al.*, 1991) have been used to exclude arthropods. However, Ingham (1985) reviewed the use of selective biocides and concluded that none had effects limited to a particular target group, limiting their utility for evaluating effects of particular functional groups. Furthermore, Seastedt (1984) noted that biocides provide a carbon and, in some cases, nitrogen source that may alter the activity or composition of microflora. Mesh sizes of litterbags (see following) can be manipulated to exclude detritivores larger than particular sizes, but this technique often alters litter environment and may reduce fragmentation regardless of faunal presence (Seastedt, 1984).

The most widely used techniques for measuring decomposition rates in terrestrial and aquatic ecosystems involve measurement of respiration rate, comparison of litterfall and litter standing crop, and measurement of mass loss (Anderson and Swift, 1983; Bernhard-Reversat, 1982; Seastedt, 1984; Witkamp, 1971; Woods and Raison, 1982). These techniques oversimplify representation of the decomposition process and consequently yield biased estimates of decay rate. Radioisotope movement from litter also provided early data on decomposition rate (Witkamp, 1971). Stable isotopes (e.g., ^{13}C, ^{14}C, and ^{15}N) are becoming widely used to measure fluxes of particular organic fractions (Ågren *et al.*, 1996; Andreux *et al.*, 1990; Horwath *et al.*, 1996; Mayer *et al.*, 1995; Spain and Le Feuvre, 1997; Wedin *et al.*, 1995).

Soil respiration represents the entire heterotrophic community, as well as living roots. Most commonly, a chamber containing sodalime or a solution of NaOH is sealed over litter for a 24-hour period, and CO_2 efflux is measured as the weight gain of sodalime or volume of acid neutralized by NaOH (Edwards, 1982). Comparison of respiration rates in plots with litter present to those with litter removed provides a more accurate estimate of respiration rates from decomposing litter, but separation of litter from soil is difficult and often arbitrary (Anderson and Swift, 1983; Woods and Raison, 1982). More recently, gas chromatography and infrared gas analysis (IRGA) have been used to measure CO_2 efflux (Nakadai *et al.*, 1993; Parkinson, 1981; Raich *et al.*, 1990).

The ratio of litterfall mass to litter standing crop provides an estimate of the decay constant, k, when litter stand crop is constant (Olson, 1963). Decay rate can be calculated if the rate of change in litter standing crop is known

(Woods and Raison, 1982). This technique also is limited by the difficulty of separating litter from underlying soil for mass measurement (Anderson and Swift, 1983; Spain and Le Feuvre, 1987; Woods and Raison, 1982).

Weight loss of fine litter has been measured using tethered litter, litterbags, and litter boxes. Tethering allows litter to take a natural position in the litterbed and does not restrict detritivore activity or alter microclimate, but is subject to loss of fragmented material and difficulty in separating late stages from surrounding litter and soil (Anderson *et al.*, 1984; Birk, 1979; Witkamp and Olson, 1963; Woods and Raison, 1982).

Litterbags provide a convenient means for studying litter decomposition (Crossley and Hoglund, 1962; Edwards and Heath, 1963). Litterbags retain selected litter material, and mesh size can be used to selectively restrict entry by larger functional groups (e.g., Edwards and Heath, 1963; Wise and Schaefer, 1994). However, litterbags may alter litter microclimate and restrict detritivore activity, depending on litter conformation and mesh size. Moisture retention between flattened leaves apparently is independent of mesh size. Exclusion of larger detritivores by small mesh sizes has little effect, at least until litter has been preconditioned by microbial colonization (Anderson and Swift, 1983; Macauley, 1975; O'Connell and Menagé, 1983; Spain and Le Feuvre, 1987; Woods and Raison, 1982). Large woody litter (e.g., tree boles) also can be enclosed in mesh cages for experimental restriction of colonization by wood-boring insects. The potential interference with decomposition by small mesh sizes has been addressed in some studies by minimizing leaf overlap (and prolonged moisture retention) in larger litterbags, using small mesh on the bottom to retain litter fragments and large mesh on the top to maximize exchange of moisture and detritivores, and measuring decomposition over several years to account for differences due to changing environmental conditions (Anderson *et al.*, 1983; Cromack and Monk, 1975; Woods and Raison, 1982, 1983). Despite limitations, litterbags have been the simplest and most widely used method for measuring decomposition rates and probably provide reasonably accurate estimates (Seastedt, 1984; Spain and Le Feuvre, 1987; Woods and Raison, 1982).

More recently, litter boxes have been designed to solve problems associated with litterbags. Litter boxes can be inserted into the litter, with the open top providing unrestricted exchange of moisture and detritivores (Seastedt and Crossley, 1983), or used as laboratory microcosms to study effects of decomposers (Haimi and Huhta, 1990; Huhta *et al.*, 1991).

Measurement of wood decomposition presents special problems, including the long time-frame of wood decomposition, the logistical difficulties of experimental placement, and manipulation of large, heavy material. Decomposition of large woody debris represents one of the longest ecological processes, often spanning centuries (Harmon *et al.*, 1986). This process traditionally was studied by comparing mass of wood of estimated age to the mass expected for the estimated original volume, based on particular tree species. The poor resolution resulting from this approach permitted modeling only as a single exponential decay model, although decomposition of some wood components begins only after lag times of up to several years, and differences in chemistry and volume between bark and wood components affect overall decay rates (Harmon et al., 1986; Schowalter et al., 1998).

Few experimental studies have compared effects of manipulated abundances of boring insects on wood decomposition (Schowalter et al., 1992). Some studies have compared species or functional group abundances in wood of estimated age or decay class, but such comparison ignores the effect of initial conditions on subsequent community development and decomposition rate. Prevailing weather conditions, the physical and chemical condition of the wood at the time of plant death, and prior colonization determine the species pools and establishment of potential colonists. Penetration of the bark and transmission by wood-boring insects generally facilitates microbial colonization of subcortical tissues (Ausmus, 1977; Dowding, 1984; Swift, 1977). Käärik (1974) reported that wood previously colonized by mold fungi (Ascomyctina and Fungi Imperfecti) was less suitable for establishment by decay fungi (Basidiomycotina) than was uncolonized wood. Mankowski et al. (1998) reported that wood consumption by termites was affected by wood species and fungal preconditioning. Hence, experiments should be designed to evaluate species or functional group effects on decomposition over long time periods using wood of standard size, composition, and condition.

Assessing rates of burrowing and mixing of soil and litter is even more problematic. A few studies have provided limited data on the volume of soil affected through excavation of ant nests (Moser, 1963; Whitford et al., 1976). However, the difficulty of separating litter from soil limits measurement of mixing. Tunneling through woody litter presents similar problems. Zhong and Schowalter (1989) dissected decomposing tree boles to assess volume of wood excavated and/or mixed among bark, wood, and fecal substrates.

C. Spatial and Temporal Patterns in Processing of Organic Matter

All, or most, dead organic matter eventually is catabolized to CO_2, water, and energy, reversing the process by which energy and matter were fixed in primary production. Some materials are decomposed more readily than are others, some processes release carbon primarily as methane, and some enter long-term storage as humus, peat, coal, or oil. Moisture, litter quality (especially lignin and nitrogen content), and oxygen supply are extremely important to the decomposition process (Birk, 1979; Fogel and Cromack, 1977; Meentemeyer, 1978; Seastedt, 1984; Whitford et al., 1981). For example, animal carrion is readily digestible by many organisms and decomposes rapidly (Payne, 1965), whereas plant materials, especially those composed largely of lignin and cellulose, can be decomposed only by relatively few species of fungi, bacteria, or protozoa and may require long time periods for complete decomposition (Harmon *et al.*, 1986). Conifer litter tends to decompose more slowly than does angiosperm litter because of low nitrogen content and high lignin content. Low soil or litter pH inhibits decomposition. Rapid burial or saturation with water inhibits decomposition of litter, because of limited oxygen availability. Submerged litter is degraded primarily by aquatic gougers and scrapers that slowly fragment and digest consumed organic matter from the surface inward (Anderson *et al.*, 1984).

Decomposition processes differ among ecosystem types. Physical factors may predominate in xeric ecosystems where decomposition of exposed litter reflects catabolic effects of ultraviolet light. Decomposition resulting from biological processes is favored by warm, moist conditions. Decomposition is most rapid in wet tropical ecosystems, where litter disappears quickly, and slowest in desert, tundra, and boreal ecosystems because of dry or cold conditions. Nevertheless, decomposition may continue underground, or under snow in tundra and boreal regions, if temperature and moisture are adequate (e.g., Santos *et al.*, 1981). Termites may be particularly important to decomposition processes in forest and grassland ecosystems (Lee and Butler, 1977; Whitford, 1986). Jones (1989, 1990) reported that termites in dry tropical ecosystems in Africa so thoroughly decompose organic matter that little or no carbon is incorporated into the soil. As noted above, decomposition rates may be lower in aquatic ecosystems as a result of saturation and limited oxygen supply. Low decomposition rates generally result in the accumulation of large standing crops of woody and fine litter.

Different groups of detritivores and decomposers dominate different ecosystems. For example, shredders and gatherers were more abundant in headwater streams with substantial inputs of largely unfragmented organic matter, whereas filter-feeders were more abundant in a seventh-order stream (the Little Tennessee River) with highly fragmented, suspended organic matter (Fig. 14.2). Fungi and associated fungivores are more prevalent in forests, whereas bacteria, bacteriovores, and earthworms are more prevalent in grasslands. Termites are the most important detritivores in arid and semiarid ecosystems. Wood-boring insects occur only in ecosystems with woody litter accumulation but are the predominant detritivore group in woody litter. Dung feeders are important in ecosystems where vertebrate herbivores are abundant (Coe, 1977; Holter, 1979). Oribatid mites and Collembola generally dominate microarthropod assemblages in forests, whereas prostigmatid mites and Collembola dominate microarthropod assemblages in deserts and grasslands (Seastedt, 2000).

The relative contributions of physical and biological factors to pedogenesis vary among ecosystems. Erosion and earth movements (e.g., soil creep and landslides) mix soil and litter in ecosystems with steep topography or high wind or raindrop impact on surface material. Burrowing animals are common in ecosystems with loose substrates suitable for excavation. Grasslands and forests on sandy or loamy soils support the highest diversity and abundances of burrowers. Ants often excavate nests through rocky or other substrates which

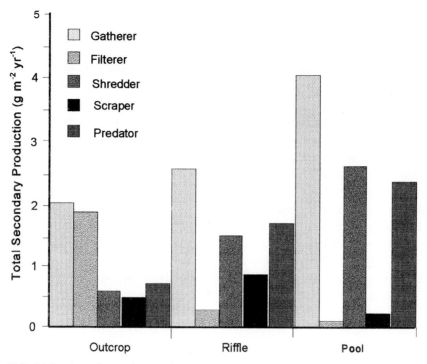

FIG. 14.2 Annual secondary production for aquatic functional groups in bedrock outcrop, riffle, and pool habitats of upper Ball Creek, North Carolina, during July 1983–June 1984. Data from Huryn and Wallace (1987).

would preclude burrowing by larger or softer animals and are the dominant burrowers in many ecosystems.

Distinct temporal patterns in decomposition rates often reflect preconditioning requirements for further degradation or inhibition or facilitation of new colonizers by established groups. For example, leaching of toxic chemicals may be necessary before many groups are able to colonize litter (Barz and Weltring, 1985). Hulme and Shields (1970) and Käärik (1974) reported that wood decay is inhibited by competition for labile carbohydrates, necessary for early growth of decay fungi, by nondecay fungi. On the other hand, Blanchette and Shaw (1978) found that decay fungus growth in wood with bacteria and yeasts was twice that in wood without bacteria and yeasts, presumably because bacteria and yeasts provide fixed nitrogen, vitamins, and other nutrients while exploiting carbohydrates from lignocellulose degradation. Microbes typically require bark penetration, and often inoculation, by insects in order to colonize woody litter. Many saprophagic arthropods require some preconditioning of litter by bacteria, fungi, or other arthropods prior to feeding. Small comminuters typically feed on fragments or feces left by larger comminuters (O'Connell and Menagé, 1983). Shredders in streams convert coarse particulate organic matter (CPOM) to fine particulate organic matter (FPOM) that can be acquired by filterers (Wallace and Webster, 1996; Wallace *et al.*, 1991). Santos and Whitford (1981) reported that a consistent succession of microarthropods was related to the percentage of organic matter lost.

Decomposition often begins long before detritus reaches the soil. Considerable detrital accumulation occurs in forest canopies (Coxson and Nadkarni, 1995; Paoletti *et al.*, 1991). Processes of decomposition and pedogenesis in these suspended sediments are poorly known, but Paoletti *et al.* (1991) reported that suspended soils associated with bromeliads in a Venezuelan cloud forest had higher concentrations of organic matter, nitrogen, calcium, and magnesium and higher densities (based on bulk density of soil) of macro- and microinvertebrates than did forest floor soils. However, rates of litter decomposition as measured in litterbags were similar in the canopy and forest floor. Oribatid mites and Collembola are the most abundant detritivores in temperate and tropical forest canopies (Paoletti *et al.*, 1991; Schowalter and Ganio, 1998; Walter and O'Dowd, 1995; Winchester, 1997).

Decomposition is an easily modeled process. Typically, an initial period of leaching or microbial oxidation of simple organic molecules results in a short-term, rapid loss of mass, followed by a longer-term, slower decay of recalcitrant compounds. Decomposition of foliage litter has been expressed as a single- or double-component negative exponential model (Olson, 1963)

$$N_t = S_0 e^{-kt} + L_0 e^{-kt} \qquad (14.1)$$

where N_t is mass at time t, S_0, and L_0 are masses in short- and long-term components, respectively, and ks are the respective decay constants. The short-term rate of decay reflects the mass of labile organic molecules and the long-term rate of decay reflects lignin content and actual evapotranspiration (AET) rate, based on temperature and moisture conditions (Meentemeyer, 1978; Seastedt, 1984). Long-term decay constants for foliage litter range from -0.14 yr^{-1} to -1.4 yr^{-1}, depending on nutritional value for decomposers (Table 14.1) (Laskowski *et al.*, 1995; Seastedt, 1984; Schowalter *et al.*, 1991). Decay constants for wood range from -0.004 yr^{-1} to -0.5 yr^{-1} (Harmon *et al.*, 1986). Schowalter *et al.* (1998) monitored decomposition of freshly cut oak, *Quercus* spp., logs over a 5-year period and found that a 3-component exponential model was necessary to account for differential decay rates among bark and wood tissues. An initial decay rate of -0.12 yr^{-1} during the first year reflected primarily the rapid loss of the nutritious inner bark (phloem), which largely disappeared by the end of the second year as a result of rapid exploitation by insects and fungi. An intermediate decay rate of -0.06 yr^{-1} for years 2–5 reflected the slower decay rate for sapwood and outer bark, and a long-term decay rate of -0.012 yr^{-1} was predicted, based on the slow decomposition of heartwood.

Decomposition often is not constant but shows seasonal peaks and annual variation that reflect periods of suitable temperature and moisture for decomposers. Patterns of nutrient mineralization from litter reflect periods of storage and loss, depending on activities of various functional groups. For example,

TABLE 14.1   Annual Decay Rates of Various Litter Types with Microarthropods Present and Experimentally Excluded

Litter type	Decay constant (year⁻¹)			Faunal effect (%)	Reference
	Without fauna	With fauna	Faunal component		
Dogwood foliage[a] (Cornus florida)	-0.69	-0.82	-0.13	16	Cromack (unpublished data), Seastedt and Crossley (1980, 1983)
Chestnut oak foliage[a] (Quercus prinus)	-0.48	-0.50	-0.02	4	Cromack (unpubulished data), Seastedt and Crossley (1980, 1983)
White oak foliage (Quercus alba)	-0.60	-0.92	-0.32	35	Witkamp and Crossley (1966)
Beech foliage[a] (Fagus sylvatica)	-0.41	-0.50	-0.09	18	Anderson (1973)
Chestnut foliage[a] (Castanea sativa)	-0.27	-0.28	-0.01	4	Anderson (1973)
Mixed hardwood foliage	-0.40	-0.70	-0.30	43	Cromack (1973)
Eucalypt foliage[b] (Eucalyptus pauciflora)	-0.45	-0.73	-0.28	38	Madge (1969)
Eucalypt foliage[c] (Eucalyptus pauciflora)	-0.69	-0.73	-0.04	8	Madge (1969)
Shinnery oak foliage (Quercus harvardii)	-0.22	-0.43	-0.21	49	Elkins and Whitford (1982)
Broomsedge (Andropogon virginicus)	-0.30	-0.36	-0.06	17	Williams and Wiegert (1971)
Blue grama grass (Bouteloua gracilis)	-0.14	-0.45	-0.31	69	Vossbrinck et al. (1979)
Mixed pasture grasses					
Surface	-1.15	-1.24	-0.09	7	Curry (1969)
Buried	-1.55	-1.34	+0.21	-16	Curry (1969)
Mixed tundra grasses[a]	-0.22	-0.32	-0.10	31	Douce and Crossley (1982)

[a]Mean values for experiments replicated over sites (Anderson, 1973; Douce and Crossley, 1982) or years (Cromack unpublished data; Seastedt and Crossley, 1980, 1983).

[b]Control versus insecticide comparison.

[c]Medium mesh (1 mm) versus fine mesh (0.5 mm) comparison. Fine mesh bags probably did not exclude all microarthropods.

From Seastedt (1984) by permission from the *Annual Review of Entomology*, Vol. 29, © 1984 by Annual Reviews.

Schowalter and Sabin (1991) reported that nitrogen and calcium content of decomposing Douglas-fir needle litter, in litterbags, in western Oregon peaked in spring each year, when microarthropod abundances were lowest, and declined during winter, when microarthropod abundances were highest. High rates of comminution by microarthropods and decay by microorganisms during the wet winters likely contributed to release of nutrients from litter, whereas reduced comminution and decay during dry springs and summers led to nutrient immobilization in microbial biomass. Similarly, fluctuating concentrations of nutrients in decomposing oak wood over time probably reflect patterns of colonization and mobilization (Schowalter *et al.* 1998).

II. EFFECTS OF DETRITIVORY AND BURROWING

Arthropod detritivores and burrowers directly and indirectly affect decomposition, carbon flux, biogeochemical cycling, pedogenesis, and primary production. The best known effects are on decomposition and mineralization (Seastedt, 1984). Detritivorous and fossorial arthropods are capable of significantly affecting global carbon budgets and ecosystem capacity to store and release nutrients and pollutants.

A. Decomposition and Mineralization

An extensive literature has addressed the effects of detritivores on decomposition and mineralization rates. Generally, the effect of arthropods on the decay rate of litter can be calculated by subtracting the decay rate when arthropods are excluded from the decay rate when arthropods are present (Table 14.1). Detritivores affect decomposition and mineralization processes, including fluxes of carbon as CO_2 or CH_4, by fragmenting litter and by affecting rates of microbial catabolism of organic molecules. The magnitude of these effects depends on the degree to which feeding increases the surface area of litter and inoculates or reduces microbial biomass.

1. Comminution

Large comminuters are responsible for the fragmentation of large detrital materials into finer particles that can be processed by fine comminuters and saprophytic microorganisms. Cuffney *et al.* (1990) and Wallace *et al.* (1991) reported that exclusion of shredders from a small headwater stream in North Carolina (USA) reduced leaf litter decay rates by 25–28% and export of fine particulate organic matter by 56%. Wise and Schaefer (1994) found that excluding macroarthropods and earthworms from leaf litter of selected plant species in a beech forest reduced decay rates 36–50% for all litter types except fresh beech litter. When all detritivores were excluded, comparable reduction in decay rate was 36–93%, indicating the prominent role of large comminuters in decomposition. Tian *et al.* (1995) manipulated abundances of millipedes and earthworms in tropical agricultural ecosystems. They found that millipedes alone significantly accounted for 10–65% of total decay over a 10-week period. Earthworms did not affect decay significantly by themselves, but earthworms and millipedes combined significantly accounted for 11–72% of total decay. Haimi and Huhta (1990) demonstrated that earthworms significantly increased mass loss of litter by 13–41%. Anderson *et al.* (1984) noted that aquatic xylophagous tipulid larvae fragmented >90% of decayed alder, *Alnus rubra*, wood in a 1-year period.

Termites have received considerable attention because of their substantial ecological and economic importance in forest, grassland, and desert ecosystems. Based on laboratory feeding rates, Lee and Butler (1977) estimated wood consumption by termites in dry sclerophyll forest in South Australia. They reported that wood consumption by termites was equivalent to about 25% of annual woody litter increment and 5% of total annual litterfall. Based on termite exclusion plots, Whitford *et al.* (1982) reported that termites consumed up to 40% of surficial leaf litter in a warm desert ecosystem in the southwestern U.S. (Fig. 14.3) Overall, termites in this ecosystem consume at least 50% of estimated annual litterfall (Johnson and Whitford, 1975; Silva *et al.*, 1985). Collins (1981) reported that termites in tropical savannas in west Africa consume 60% of annual wood-fall and 3% of annual leaf fall (24% of total litter production), but fire removes 0.2% of annual wood-fall and 49% of annual leaf fall (31% of total litter production). In that study, fungus-feeding Macrotermitinae were responsible for 95% of the litter removed by termites. Termites apparently consume virtually all litter in tropical savannas in east Africa (Jones, 1989, 1990). Termites consume a lower proportion of annual litter inputs in more mesic ecosystems. Collins (1983) reported that termites consumed about 16% of annual litter production in a Malaysian rain forest receiving 2000 mm precipitation yr^{-1} and 1–3% of annual litter production in a Malaysian rain forest receiving 5000 mm precipitation yr^{-1}.

Accumulation of dung from domestic mammalian grazers has become a serious problem in many arid and semiarid ecosystems. Termites remove as much as 100% of cattle dung over 3 months in Kenya (Coe, 1977), 80–85% over 5–9 months in tropical pastures in Costa Rica (Herrick and Lal, 1996) and 47% over 4 months in the Chihuahuan Desert in the southwestern U.S. (Whitford *et al.*, 1982). In the absence of termites, dung would require 25–30 years to disappear (Whitford, 1986). Dung beetles (Scarabaeidae) and earthworms also are important consumers of dung in many tropical and subtropical ecosystems (e.g., Coe, 1977; Holter, 1979).

Relatively few studies have provided estimates of wood consumption by bark- and wood-boring insects, despite their recognized importance to wood decomposition. Zhong and Schowalter (1989) reported that bark beetles consumed 0.1–7.6% of inner bark and wood-boring beetles consumed an additional 0.05–2.3% during the first year of decomposition, depending on conifer tree species. Ambrosia beetles consumed 0–0.2% of the sapwood during the

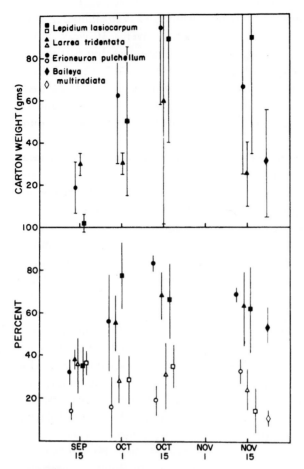

FIG. 14.3 Rate of gallery carton deposition (top) and mass loss (bottom) of pepperweed, *Lepidium lasiocarpum*, creosote bush, *Larrea tridentata*, fluff grass, *Erioneuron pulchellum*, and desert marigold, *Baileya multiradiata*, foliage when subterranean termites, *Gnathamitermes tubiformans*, were present (black symbols) or absent (white symbols) in experimental plots in southern New Mexico. Litter (10 g) was placed in aluminum mesh cylinders on the soil surface on 15 August 1979. Vertical lines represent standard errors. From Whitford *et al.* (1982) by permission from Springer-Verlag.

first year. Schowalter *et al.* (1998) found that virtually the entire inner bark of oak logs was consumed by beetles during the first two years of decomposition, facilitating separation of the outer bark and exposing the sapwood surface to generalized saprophytic microorganisms. Edmonds and Eglitis (1989) used exclusion techniques to demonstrate that, over a 10-year period, bark beetles and wood borers increased decay rates of large Douglas-fir logs (42 cm diameter at breast height) by 12% and of small logs (26 cm diameter at breast height) by 70%.

Payne (1965) explored the effects of carrion feeders on carrion decay during the summer in South Carolina, U.S. He placed baby pig carcasses under replicated treatment cages, open at the bottom, that either permitted or restricted access to insects. Carcasses were weighed carefully at intervals. Carcasses exposed to insects lost 90% of their mass in six days, whereas carcasses protected from insects lost only 30% of their mass in this period, followed by a gradual loss of mass, with 20% mass remaining in mummified pigs after 100 days.

Not all studies indicate significant effects of litter fragmentation by macroarthropods. Setälä *et al.* (1996) reported that manipulation of micro-, meso-, and macroarthropods in litter baskets resulted in slower decay rates in the presence of macroarthropods. Most litter in baskets with macroarthropods (millipedes and earthworms) was converted into large fecal pellets that decayed slowly.

A number of studies have demonstrated that microarthropods are responsible for 4–69% of the total decay rate, depending on litter quality and ecosystem (Table 14.1, Fig 14.4) (Seastedt, 1984). Seastedt (1984) suggested that an apparent, but nonsignificant, inverse relationship between decay rate due to microarthropods and total decay rate indicated a greater contribution of arthropods to decomposition of recalcitrant litter fractions compared to more labile fractions. Tian *et al.* (1995) subsequently reported that millipedes and earthworms contributed more to the decomposition of plant residues with high C/N, lignin, and polyphenol contents than to high quality plant residues.

2. Microbial Respiration

Microbial decomposers are responsible for about 95% of total heterotrophic respiration in soil. Arthropods generally increase microbial respiration rates and carbon flux but may reduce respiration rates if they overgraze microbial resources (Huhta *et al.*, 1991; Seastedt, 1984). Several studies have documented increased microbial respiration as a result of increased arthropod access to detrital substrate and stimulation of microbial production.

Litter fragmentation greatly increases the surface area exposed for micro-

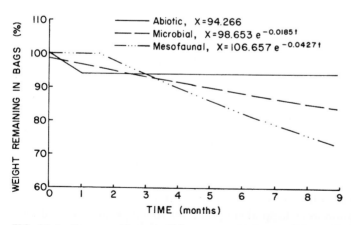

FIG. 14.4 Decomposition rate of blue grama grass in litterbags treated to permit decomposition by abiotic factors alone, abiotic factors + microbes, and abiotic factors + microbes + mesofauna (microarthropods). Decomposition in the abiotic treatment was insignificant after the first month; decomposition showed a 2-month time lag in the treatment including mesofauna. From Vossbrinck *et al.* (1979) by permission from the Ecological Society of America.

bial colonization. Zhong and Schowalter (1989) reported that ambrosia beetle densities averaged 300 m^{-2} bark surface in Douglas-fir and western hemlock, *Tsuga heterophylla*, logs, and their galleries extended 9–14 cm in 4–9 cm thick sapwood, indicating that considerable sapwood volume was made accessible to microbes colonizing gallery walls. The entire sapwood volume of these logs was colonized by various fungi within one year (Schowalter *et al.*, 1992). Mixing of organic material and microbes during passage through detritivore guts ensures infusion of consumed litter with decomposers and may alter litter quality in ways that stimulate microbial production (Maraun and Scheu, 1996). Gut mixing is especially important for species such as termites and other wood borers that require microbial digestion of cellulose and lignin into labile carbohydrates (Breznak and Brune, 1994).

Many arthropods directly transport and inoculate saprophytic microorganisms into organic residues. For example, Schowalter *et al.* (1992) documented transport of a large number of fungal genera by wood-boring insects. Some of these fungi are mutualists that colonize wood in advance of insects and degrade cellulose to labile carbohydrates that subsequently are used by insects (Bridges and Perry 1985, French and Roeper, 1972; Morgan, 1968). Others may be acquired accidentally by insects during feeding or movement through colonized material (Schowalter *et al.*, 1992). Behan and Hill (1978) documented transmission of fungal spores by oribatid mites.

Fungivorous and bacteriophagous arthropods stimulate microbial activity by maximizing microbial production. As discussed for herbivores in Chapter 12, low to moderate levels of grazing often stimulate productivity of the microflora by alleviating competition, altering microbial species composition, and gouging new detrital surfaces for microbial colonization. Microarthropods also can stimulate microbial respiration by preying on bacteriophagous and mycophagous nematodes (Seastedt, 1984; Setälä *et al.*, 1996). Higher levels of grazing may depress microbial biomass and reduce respiration rates (Huhta *et al.*, 1991; Seastedt, 1984).

Seastedt (1984) suggested a way to evaluate the importance of three pathways of microbial enhancement by arthropods, based on the tendency of microbes to immobilize nitrogen in detritus until C:N ratio approaches 10–20:1. Where arthropods affect decomposition primarily through comminution, nitrogen content of litter should be similar with or without fauna. Alternatively, where arthropods stimulate microbial growth and respiration rates, the C:N ratio of litter with fauna should be less than the ratio without fauna. Finally, where arthropods graze microbial tissues as fast as they are produced, C:N ratio of litter should be constant and mass should decrease.

Seasonal variation in arthropod effects on microbial production and biomass may explain variable results and conclusions from earlier studies. Maraun and Scheu (1996) reported that fragmentation and digestion of beech leaf litter by the millipede, *Glomeris marginata*, increased microbial biomass and respiration in February and May but reduced microbial biomass and respiration in August and November. They concluded that millipede feeding generally increased nutrient (nitrogen and phosphorus) availability, but that these nutrients were only used for microbial growth when carbon resources were adequate, as occurred early in the year. Depletion of carbon resources relative to nutrient availability in detritus limited microbial growth later in the year.

Although CO_2 is the major product of litter decomposition, incomplete oxidation of organic compounds occurs in some ecosystems, resulting in evolution of other trace gases, especially methane (Khalil *et al.*, 1990). Zimmerman *et al.* (1982) first suggested that termites could contribute up to 35% of global emissions of methane. A number of arthropod species, including most tropical representatives of millipedes, cockroaches, termites, and scarab beetles, are important hosts for methanogenic bacteria and are relatively important sources of biogenic global methane emissions (Hackstein and Stumm, 1994).

Termites have received the greatest attention as sources of methane because their relatively sealed colonies are warm and humid, with relatively low oxygen concentrations that favor fermentation processes and emission of methane or acetate (Brauman *et al.*, 1992; Wheeler *et al.*, 1996). Thirty of 36 temperate and tropical termite species assayed by Brauman *et al.* (1992), Hackstein and Stumm (1994), and Wheeler *et al.* (1996) produce methane and/or acetate. Generally, acetogenic bacteria outproduce methanogenic bacteria in wood- and grass-feeding termites, but methanogenic bacteria are much more important in fungus-growing and soil-feeding termites (Brauman *et al.*, 1992).

Zimmerman *et al.* (1982) suggested that tropical deforestation and conversion to pasture and agricultural land could increase the biomass and methane emissions of fungus-growing and soil-feeding termites, but Martius *et al.* (1996) concluded that methane emissions from termites in deforested areas in Amazonia would not contribute significantly to global methane fluxes. Khalil *et al.* (1990), Martius *et al.* (1993), and Sanderson (1996) calculated CO_2 and methane fluxes based on global distribution of termite biomass and concluded that termites contribute approximately 2% of the total global flux of CO_2 (3500 tg yr^{-1}) and 4–5% of the global flux of methane ($\leq$20 tg yr^{-1}) (Fig. 14.5). However, emissions of CO_2 by termites are 25–50% of annual emissions from fossil fuel combustion (Khalil *et al.*, 1990). Contributions to atmospheric composition by this ancient insect group may have been more substantial prior to accumulation of anthropogenic sources of CO_2, methane, and other trace gases.

3. Mineralization

Measurements of changes in elemental concentrations represent net mineralization rates. Net mineralization includes loss of elements due to mineralization and accumulation by microflora of elements entering as microparticulates, precipitation and leachate, or transferred (e.g., via hyphae) from other organic material (Schowalter *et al.*, 1998; Seastedt, 1984). Although microbial biomass typically is a negligible component of litter mass, microbes often represent a large proportion of the total nutrient content of decomposing detritus and significantly affect the nutrient content of the litter–microbial complex

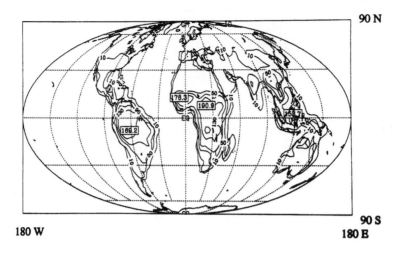

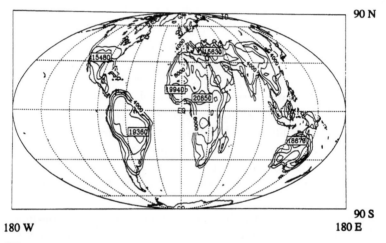

FIG. 14.5 Geographic distribution of emissions of methane (top) and carbon dioxide (bottom) by termites. Units are 10^6 kg yr^{-1}. From Sanderson (1996) courtesy of the American Geophysical Union.

(e.g., Seastedt, 1984). Arthropods affect net mineralization in two measurable ways: through mass loss and assimilation of consumed nutrients and through effects on nutrient content of the litter–microbe system. Seastedt (1984) proposed the following equation to indicate the relative effect of arthropods on mineralization

$$Y = [\% \text{ mass}_i/\% \text{ mass}_x) \times (\text{concentration}_i/\text{concentration}_x) \quad (14.2)$$

where Y is the relative arthropod effect, % mass$_i$ is the percentage of initial mass remaining that has been accessible to arthropods, % mass$_x$ is the percentage of initial mass remaining that has been unavailable to arthropods, and concentration$_i$ and concentration$_x$ are the respective concentrations of a given element. Net immobilization of an element is indicated by $Y > 1$, and net loss is indicated by $Y < 1$. Temporal changes in nutrient content depend on the structural position of the element within organic molecules, microbial use of the element, and the form and amounts of the element entering the detritus from other sources.

Nitrogen generally is considered to be the element most likely to limit growth of plants and animals, and its release from decomposing litter often is correlated with plant productivity (Vitousek, 1982). As noted above, saprophytic microbes typically immobilize nitrogen until sufficient carbon has been respired to make carbon or some other element more limiting than nitrogen (Maraun and Scheu, 1996; Schowalter *et al.*, 1998; Seastedt, 1984). Thereafter, the amount of nitrogen released should equal the amount of carbon oxidized. Microbes have considerable capacity to absorb nitrogen from precipitation, canopy leachate, and animal excrement (Fig. 12.14) (Lovett and Ruesink, 1995; Seastedt and Crossley, 1983; Stadler and Müller, 1996), permitting nitrogen mineralization and immobilization even at high C:N ratios. Generally, exclusion of microarthropods decreases the concentration of nitrogen in litter, but the absolute amounts of nitrogen in litter are decreased or unaffected by microarthropod feeding activities (Seastedt, 1984).

Phosphorus concentrations often show initial decline due to leaching but subsequently reach an asymptote determined by microbial biomass (Schowalter and Sabin, 1991; Schowalter *et al.*, 1998; Seastedt, 1984). Microarthropods can increase or decrease rates of phosphorus mineralization, presumably as a result of their effect on microbial biomass (Seastedt, 1984).

Calcium dynamics are highly variable. This element often is bound in organic acids (e.g., calcium oxalate) as well as in elemental and inorganic forms in detritus. Some fungi accumulate high concentrations of this element (Cromack *et al.*, 1975, 1977; Schowalter *et al.*, 1998), and some litter arthropods, especially millipedes and oribatid mites, have highly calcified exoskeletons (Norton and Behan-Pelletier, 1991; Reichle *et al.*, 1969). Nevertheless, calcium content in arthropod tissues is low compared to annual inputs in litter. No consistent arthropod effects on calcium mineralization have been apparent (Seastedt, 1984).

Potassium and sodium are highly soluble elements, and their initial losses (via leaching) from decomposing litter invariably exceed mass losses (Schowalter and Sabin, 1991; Schowalter *et al.*, 1998; Seastedt, 1984). Amounts of these elements entering the litter in precipitation or throughfall approach or exceed amounts entering as litterfall. In addition, these elements are not bound in organic molecules, so their supply in elemental form is adequate to meet the needs of microflora. Arthropods have been shown to affect mineralization of [134]Cs or ^{137}Cs, used as analogs of potassium (Crossley and Witkamp, 1964; Witkamp and Crossley, 1966) but not mineralization of potassium (Seastedt, 1984). Sodium content often increases in decomposing litter, especially decomposing wood

(Cromack *et al.,* 1977; Schowalter *et al.,* 1998). Sollins *et al.,* (1987) suggested that this increase represented accumulation of arthropod tissues and products, which typically contain high concentrations of sodium (e.g., Reichle *et al.,* 1969). However, Schowalter et al. (1998) reported increased concentrations of sodium during early stages of wood decomposition, prior to sufficient accumulation of arthropod tissues. They suggested that increased sodium concentrations in wood reflected accumulation by decay fungi, which contained high concentrations of sodium in fruiting structures. Fungi and bacteria have no known physiological requirement for sodium (Cromack *et al.,* 1977). Accumulation of sodium, and other nutrients, in decomposing wood may represent a mechanism for attracting sodium-limited animals that transport fungi to new wood resources.

The generally insignificant effects of arthropods on net mineralization rates, compared to their substantial effects on mass loss, can be attributed to the compensatory effects of arthropods on microbial biomass. The stimulation by arthropods of microbial respiration and immobilization of nutrients results in loss of litter mass but not of the standing crops of elements within litter (Seastedt, 1984). Other aspects of fragmentation also may contribute to nutrient retention, rather than loss. Aquatic comminuters generally fragment detritus into finer particles more amenable to downstream transport (Wallace and Webster, 1996). However, some filter-feeders concentrate fine detrital material into larger fecal pellets that are more likely to remain in the aquatic ecosystem (e.g., Wotton *et al.,* 1998). Some shredders deposit feces in burrows, thereby incorporating the nutrients into the substrate (Wagner, 1991). Furthermore, Seastedt (2000) noted that most studies of terrestrial detritivore effects have been relatively short-term. Accumulating data (e.g., Setälä *et al.,* 1996) suggest that mixing of recalcitrant organic matter and mineral soil in the guts of some arthropods may produce stable soil aggregates that reduce the decay rate of organic material.

B. Soil Structure and Infiltration

Fossorial arthropods alter soil structure by redistributing soil and organic material and increasing soil porosity (Anderson, 1988). Porosity determines the depth to which air and water penetrate the substrate. A variety of substrate-nesting vertebrates, colonial arthropods, and detritivorous arthropods and earthworms affect substrate structure and infiltration. Defecation by a larval caddisfly, *Sericostoma personatum,* that feeds nocturnally on surficial detritus and burrows in the stream bed during the day, increases subsurface organic content by 75–185% (Wagner, 1991).

Ants and termites are particularly important modifiers of soils. Colonies of these insects often occur at high densities and introduce cavities into large volumes of substrate. Eldridge (1993) reported that densities of funnel ant, *Aphaenogaster barbigula,* nest entrances could reach 37 m^{-2}, equivalent to 9% of the surface area over portions of the landscape. Nests of leaf-cutting ants,

Atta vollenweideri, reach depths of >3 m in pastures in western Paraguay (Jonkman, 1978). Whitford *et al.* (1976) excavated nests of desert harvester ants, *Pogonomyrmex* spp., in New Mexico and mapped the 3-dimensional structure of interconnected chambers radiating from a central tunnel (Fig. 14.6). They reported colony densities of 21–23 ha^{-1} at four sites. Each colony consisted of 12–15 interconnected galleries (each about 0.035 m^3) within a 1.1 m^3 volume (1.5 m diameter $\times$ 2 m deep) of soil, equivalent to about 10 m^3 ha^{-1} cavity space (Fig. 14.6). These colonies frequently penetrated the hardpan (caliche) layer 1.7–1.8 m below the surface. Moser (1963) partially excavated a leaf-cutting ant, *Atta texana*, nest in central Louisiana, U.S. He found 93 fungus-garden chambers, 12 dormancy chambers, and 5 detritus chambers (for disposal of depleted foliage substrate) in a volume measuring 12 $\times$ 17 m on the surface by at least 4 m deep (the bottom of the colony could not be reached).

The infusion of large soil volumes with galleries and tunnels greatly alters soil chemistry and structure. Jones (1990) and Salick *et al.* (1983) noted that termites concentrate organic material in their nests. As a consequence, soils outside nest zones become relatively depleted of organic matter and nutrients. Parker *et al.* (1982) reported that experimental exclusion of termites for 4 years

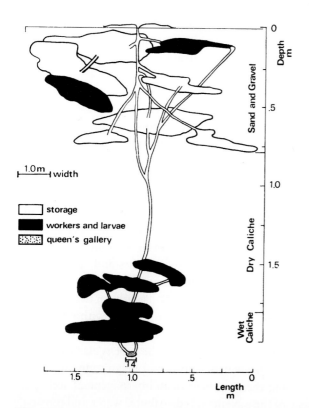

FIG. 14.6 Vertical structure of a harvester ant, *Pogonomyrmex rugosus*, nest in southern New Mexico. From Whitford, W. G., P. Johnson, and J. Ramirez. 1976. "Comparative ecology of the harvester ants *Pogonomyrmex barbatus* (F. Smith) and *Pogonomyrmex rugosus* (Emery). *Insectes Sociaux* **23**: 117–132 by permission of Birkhäuser Verlag.

increased soil nitrogen concentration 11%. Similarly, ant nests represent sites of concentrated organic matter and nutrients (Anderson, 1988; Wagner, 1997; Wagner et al., 1997), but such concentration may be limited to the galleries within the nest (Jonkman, 1978; Westoby et al., 1991; see Chapter 13). Jonkman (1978) noted that soil within leaf-cutter ant, *Atta* spp., nests tended to have higher pH than did soil outside the nest. However, Wagner et al. (1997) measured significantly lower pH (6.1) in nests of harvester ants, *Pogonomyrmex barbatus*, than in reference soil (6.4).

Conversely, termites transport large amounts of soil from lower horizons to the surface and above for construction of nests (Fig. 14.7), gallery tunnels, and "carton," the soil deposited around litter material to retain moisture during termite feeding in arid and semiarid ecosystems (Fig. 14.8) (Whitford, 1986). Whitford et al. (1982) reported that termites brought $10–27$ g m^{-2} of fine-textured soil material (35% coarse sand, 45% medium fine sand, and 21% very fine sand, clay, and silt) to the surface and deposited $6–20$ g of soil carton per gram of litter removed (Fig. 14.3). Herrick and Lal (1996) found that termites deposited an average of 2.0 g of soil at the surface for every gram of dung removed.

A number of studies have demonstrated effects of soil animals on soil moisture (Fig. 14.9). Litter reduction or removal increases soil temperature and evaporation but reduces infiltration of water. Burrowing and redistribution of soil and litter increase soil porosity, water infiltration, and stability of soil aggregates.

Ant and termite nests have particularly important effects on soil moisture because of the large substrate surface areas and volumes affected. Wagner (1997) reported that soil near ant nests had higher moisture content than did more distant soil. Elkins et al. (1986) compared runoff and water infiltration in plots with termites present or excluded during the previous 4 years. Infiltration rates in plots with low plant cover ($<10\%$) were greater with termites present (88 mm hr^{-1}) than with termites absent (51 mm hr^{-1}); runoff volumes were twice as high in the termite-free plots with low plant cover (40 mm) as in untreated plots (20 mm). Infiltration and runoff volumes did not differ between shrub-dominated plots (higher vegetation cover) with or without termites.

Eldridge (1993, 1994) measured effects of funnel ants and subterranean harvester termites, *Drepanotermes* spp., on infiltration of water in semiarid eastern Australia. He found that infiltration rates in soils with ant nest entrances were 4- to 10-fold higher (1030–1380 mm hr^{-1}) than in soils without nest entrances (120–340 mm hr^{-1}). Infiltration rate was correlated positively with nest entrance diameter. However, infiltration rate on the subcircular pavements covering the surface over termite nests was an order of magnitude lower than in the annular zone surrounding the pavement or in interpavement soils (Fig. 14.10). The cemented surface of the pavement redistributes water and nutrients from the pavement to the surrounding annular zone. Ant and termite control of infiltration creates moist patches in moisture-limited environments.

FIG. 14.7 Termite castle in northern Australian woodland. Dimensions are approximately 3 m height and 1.5 m diameter.

C. Primary Production and Vegetation Dynamics

Through control of decomposition, mineralization, and pedogenesis, detritivorous and fossorial arthropods have the capacity to control nutrient availability for, and perhaps uptake by, plants (Crossley, 1977). In particular, release of ni-

FIG. 14.8   Termite gallery carton on stems of dead creosote bush. Soil particles are cemented together to provide protection and moisture control during termite feeding on detrital material.

trogen from decaying organic matter often is correlated with plant productivity (Vitousek, 1982). However, relatively few studies have measured the effect of detritivores on plant growth or vegetation dynamics.

Edwards and Lofty (1978) compared seedling emergence and shoot and root growth of barley between pots of intact, sterilized soil (from fields in which

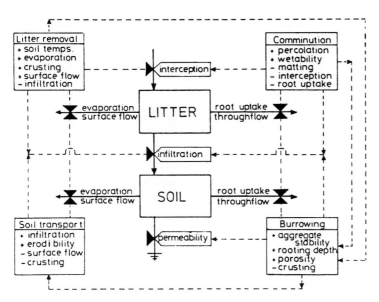

FIG. 14.9 Effects of soil invertebrates on soil water balance. Reprinted from *Agriculture, Ecosystems, and Environment* 24, Anderson, J. M., "Invertebrate-mediated transport processes in soils," pp. 5–19. Copyright 1988, with permission from Elsevier Science.

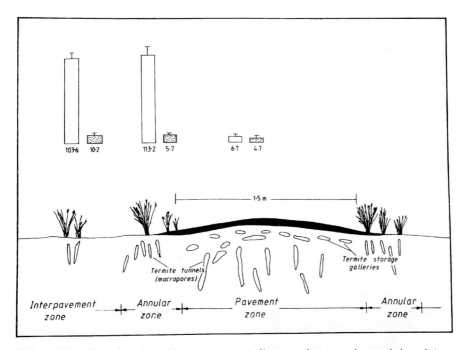

FIG. 14.10 Effect of termite colony structure on infiltration of water under ponded conditions (clear) and under tension (hatched). Vertical lines indicate 1 standard error of the mean. From Eldridge (1994) by permission from Gustav Fischer Verlag.

seed had been either drilled into the soil or planted during ploughing) with microarthropods or earthworms absent or reintroduced. Percent seedling emergence, plant height, and root weight were higher in ploughed soil and direct-drilled soil with animals, compared to sterile direct-drilled soil, suggesting important effects of soil animals on soil porosity and infiltration.

Ingham *et al.* (1985) inoculated microcosms of blue grama grass, *Bouteloua gracilis*, in sandy loam soil, low in inorganic nitrogen, with bacteria or fungi; half of each microflora treatment was inoculated with microbivorous nematodes. Plants growing in soil with bacteria and bacteriophagous nematodes grew faster and acquired more nitrogen initially than did plants in soil with bacteria only. Addition of mycophagous nematodes did not increase plant growth. These differences in plant growth resulted from greater nitrogen mineralization by bacteria (compared to fungi), excretion of NH_4^+–N by bacteriophagous (but not mycophagous) nematodes, and rapid uptake of available nitrogen by plants. Mycophagous nematodes did not increase plant growth or nitrogen uptake over fungi alone, because these nematodes excreted less NH_4^+–N, and the fungus alone mineralized sufficient nitrogen for plant growth.

Setälä and Huhta (1991) created laboratory microcosms with birch seedlings, *Betula pendula*, planted in partially sterilized soil reinoculated with soil microorganisms only or with soil microorganisms and a diverse soil fauna. During two growing periods the presence of soil fauna increased birch leaf, stem, and root biomass by 70, 53, and 38%, respectively, and increased foliar nitrogen and phosphorus contents 3-fold and 1.5-fold, respectively, compared to controls with microorganisms only (Fig. 14.11).

In addition to direct effects on nutrient availability, soil arthropods can influence plant growth indirectly by affecting mycorrhizal fungi. Grazing on mycorrhizal fungi by fungivorous arthropods could inhibit plant growth by interfering with nutrient uptake. Conversely, many fungivorous arthropods disperse mycorrhizal spores or hyphae to new hosts. Rabatin and Stinner (1988) reported that 28–97% of soil animals contained mycorrhizal spores or hyphae in their guts.

By affecting plant growth differentially, soil animals also influence vegetation dynamics. In one study (Guo, 1998), diversity of annual and perennial plants was highest on ant mounds and under shrubs, compared to kangaroo rat mound, half-shrub, and open area microsites; biomass of these plants was highest under shrubs, followed by kangaroo rat mounds and ant mounds, indicating that ant mounds are important determinants of vegetation structure.

Jonkman (1978) reported that abandoned nests of leaf-cutter ants served as sites of accelerated succession in Paraguayan pastures. Collapse of the nest chamber formed a depression that held water and facilitated development of woody vegetation. At high nest densities, these oases coalesced, greatly increasing forest area and reducing area suitable for leaf-cutter ant nests.

Parker *et al.* (1982) reported that termite exclusion significantly reduced biomass of four annual plant species and significantly increased biomass of one annual plant species. They observed an overall trend toward increased biomass of annual plants in plots with termites excluded. These results likely reflected

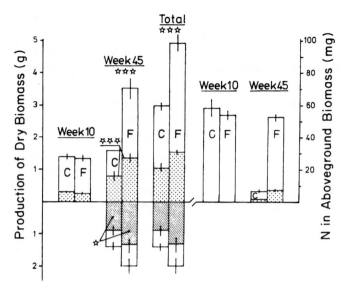

FIG. 14.11 Biomass production (left of break in horizontal axis) and nitrogen accumulation (right of break in horizontal axis) of birch, *Betula pendula*, seedlings. Bars above the horizontal axis are stems (stippled) and leaves (clear); bars below the horizontal axis are roots in humus (hatched) and roots in mineral soil (clear). C, fauna removed; F, refaunated. Vertical lines represent 1 standard deviation for all data (except nitrogen at week 45, where vertical lines represent minimum and maximum values). For C versus F, * = $P < 0.05$; *** = $P < 0.001$. Stem nitrogen was not measured week 10. From Setälä and Huhta (1991) by permission from the Ecological Society of America.

increased nitrogen availability in termite exclusion plots, compared to plots with unmanipulated termite abundance.

III. SUMMARY

Decomposition and pedogenesis are major ecosystem processes that affect biogeochemical cycling, trace gas fluxes, soil fertility, and primary production. Decomposition of organic matter involves four component processes: photo-oxidation, leaching, comminution, and mineralization. Arthropods are key factors influencing comminution and mineralization.

Functional groups involved in comminution involve coarse comminuters that fragment large materials and fine comminuters that fragment smaller materials, often those produced by large comminuters. In aquatic ecosystems, scrapers and shredders represent coarse comminuters, whereas gatherers and filterers represent fine comminuters. Xylophages represent a specialized group of comminuters that fragment woody litter; coprophages feed on animal excrement. Fungivores and bacteriovores fragment detrital material while grazing on microflora. Fossorial functional groups include subterranean nesters that excavate simple burrows, gatherers that return detrital or other organic materials to nesting areas, and fossorial feeders that consume organic material and soil and mix these biotic and abiotic materials in their wake.

Evaluation of detritivore and burrower effects on decomposition and pedo-genesis requires appropriate methods for measuring animal abundances and process rates. Abundances of detritivores or burrowers can be manipulated using exclusion and microcosm techniques, and detritivory can be measured as the product of detritivore abundance and individual consumption rate or as the rate of disappearance of substrate. Decomposition most commonly is measured as respiration rate, as the ratio of litter input to litter standing crop, or as the rate of litter disappearance. Isotopic tracers also provide data on decomposition rate.

Decomposition rate typically is higher in more mesic ecosystems. Different functional groups dominate different ecosystems, depending on availability and quality of detrital resources. For example, shredders dominate headwater streams where coarse detrital inputs are the primary resource, whereas filterers dominate larger streams with greater availability of suspended fine organic material. Xylophages occur only in ecosystems with woody residues. Decomposition generally can be modeled as a multiple negative exponential decay function over time, with decay constants proportional to the quality of litter components. Typically, an initial large decay constant represents rapid loss of labile materials and successively smaller decay constants represent slower losses of recalcitrant materials, e.g., lignin and cellulose. Most studies have been relatively short term. Recent long-term studies suggest that mixing of recalcitrant materials and soil in arthropod guts may create stable aggregates that decay very slowly.

Detritivores affect decomposition in three ways: through comminution, effects on microbial biomass, and effects on mineralization. Comminution increases detrital surface area and facilitates colonization and decay by microflora. Low-to-moderate levels of grazing on microflora stimulate microbial productivity and biomass, maximizing microbial activity and respiration. High levels of grazing may reduce microbial biomass and decomposition. Grazers also disperse fungi and bacteria to new substrates. Not all organic material is converted to CO_2. The low oxygen concentrations characterizing warm, humid termite colonies favor reduction of organic molecules to methane and other trace gases. Arthropod detritivores affect mineralization in different ways, depending on the chemical characteristics and biological use of the element. Detritivores often increase mineralization of nitrogen, but nitrogen released from detritus may be immobilized quickly by microbes.

Burrowers affect soil development by redistributing soil and organic matter. Ants and termites, in particular, excavate large volumes of soil and accumulate organic material in their centralized nests, mixing soil with organic material and influencing the distribution of soil nutrients and organic matter. Surrounding soils may become depleted in soil carbon and nutrients.

Detritivore and burrower effects on mineralization and soil composition can affect primary production and vegetation dynamics. However, despite documented effects of detritivores and burrowers on nutrient availability, relatively few studies have demonstrated increased plant growth and altered vegetation as a result of detritivory or burrowing activities.

Insects as Regulators of Ecosystem Processes

INSECTS AND OTHER ORGANISMS INEVITABLY AFFECT THEIR ENVIRON-
ment through spatial and temporal patterns of resource acquisition and redistribution. Insects respond to environmental changes in ways that dramatically
alter ecosystem conditions, as discussed in Chapters 12–14. These effects of organisms do not necessarily provide cybernetic (stabilizing) regulation. However, the hypothesis that insects stabilize ecosystem properties through feedback
regulation is one of the most important and revolutionary concepts to emerge
from research on insect ecology.

The concept of self-regulation is a key aspect of ecosystem ecology. Vegetation has a documented role in ameliorating variation in climate and biogeochemical cycling (Chapter 11), and vegetative succession facilitates recovery of
ecosystem functions following disturbances. However, the concept of self-regulating ecosystems has seemed to be inconsistent with evolutionary theory (emphasizing selection of "selfish" attributes) (e.g., Pianka, 1974), with variable
successional trends following disturbance (e.g., Horn, 1981) and with the lack
of obvious mechanisms for maintaining homeostasis (e.g., Engelberg and Boyarsky, 1979).

The debate over the self-regulating capacity of ecosystems, and especially
the role of insects, is somewhat reminiscent of debate on the now-recognized
importance of density-dependent feedback regulation of population size (Chapter 6) and is a useful example of how science develops. The outcome of this debate has significant consequences for how we manage ecosystems and their bi-

otic resources. Although controversial, this concept is an important aspect of insect ecology, and its major issues are the subject of this chapter.

I. DEVELOPMENT OF THE CONCEPT

The intellectual roots of ecosystem self-regulation lie in Darwin's (1859) recognition that some adaptations apparently benefit a group of organisms more than the individual, leading to selection for population stability. The concept of altruism and selection for homeostasis at supraorganismal levels has remained an important issue, despite recurring challenges and alternative models (e.g., Axelrod and Hamilton, 1981; Schowalter, 1981; Wilson, 1973).

Behavioral ecologists have been challenged to explain the evolution of altruistic behaviors that are fundamental to social organization. Even sexual reproduction could be considered a form of self-restraint, because individuals contribute only half the genotype of their progeny through sexual reproduction, compared to the entire genotype of their progeny through asexual reproduction (Pianka, 1974). Cooperative interactions, such as mutualism, and self-sacrificing behavior, such as suppression of reproduction and suicidal defense by workers of social insects, have been more difficult to explain in terms of individual selection. Haldane (1932) proposed a model in which altruism would have a selective advantage if the starting gene frequency were high enough and the benefits to the group outweighed individual disadvantage. This model raised obvious questions about the origin of altruist genes and the relative advantages and disadvantages that would be necessary for increased frequency of altruist genes.

Group selection theory was advanced during the early 1960s by Wynne-Edwards (1963, 1965), who proposed that social behavior arose as individuals evolved to curtail their own individual fitnesses to enhance survival of the group. Populations that do not restrain combat among their members or that overexploit their resources have a higher probability of extinction than do populations that regulate their use of resources. Selection thus should favor demes with traits to regulate their densities, i.e., maintain homeostasis in group size. Behaviors such as territoriality, restraint in conflict, and suppressed reproduction by subordinate individuals (including workers in social insect colonies) thereby reflect selection (feedback) for traits that prevent destructive interactions or oscillations in group size.

This hypothesis was challenged for lack of explicit evolutionary models or experimental tests that could explain the progressive evolution of homeostasis at the group level, i.e., demonstration of an individual advantage to altruistic individuals over selfish individuals. Furthermore, Wynne-Edwards' proposed devices by which individuals curtail their individual fitnesses and communicate their density and the degree to which each individual should decrease its individual fitness were inconsistent with available evidence or could be explained better by models of individual fitness (Wilson, 1973). Nevertheless, the concept of group selection was recognized as an important aspect of social evolution (Wilson, 1973). Hamilton (1964) and Smith (1964) developed an evolutionary model, based on **kin selection**, whereby individual fitness is increased by behaviors that favor survival of relatives with similar genomes. They introduced a new term, **inclusive fitness**, to describe the contributions of both personal reproduction and reproduction by near kin to individual fitness. For example, care for offspring of one's siblings increases an individual's fitness to the extent that it contributes to the survival of related genomes. Failure to provide sufficient care for offspring of siblings reduces survival of family members.

This concept explained evolution of altruistic behaviors, such as maternal care, shared rearing of offspring among related individuals, alarm calls (that may draw attention of predators to the caller), and voluntary suppression of reproduction and suicidal defense by workers in colonies of social insects, which typically benefit close relatives. For social insects, in particular, Hamilton (1964) noted that males are produced from unfertilized eggs and have unpaired chromosomes. Therefore, all the daughters in the colony inherit only one type of gamete from their father and thereby share 50% of their genes through this source. In addition, they share another 25%, on average, of their genes in common from their mother. Overall, the daughters share 75% of their genes with each other, compared to only 50% of their genes with their mother. Accordingly, workers maximize their fitness by helping to rear siblings rather than by having their own offspring.

Levins (1970) and Boorman and Levitt (1972) proposed **interdemic selection** models to account for differential extinction rates among demes of metapopulations that differ in altruistic traits. Both the Levins and Boorman-Levitt models are based on differential extinction among demes founded by colonists differing in altruist traits (Wilson, 1973). In the Levins model, colonists from small populations found other small populations in habitable sites. Increasing frequency of altruist genes decreases the probability of extinction of these small populations, i.e., cooperation elevates and maintains each deme above the extinction threshold (see Chapters 6 and 7). In the Boorman-Levitt model, colonists from a large, stable population found small, marginal populations in satellite habitats. Altruist genes do not influence extinction rates until marginal populations reach demographic carrying capacity, i.e., altruism prevents destructive population increase above carrying capacity (see Chapters 6 and 7). Both models require restrictive conditions for evolution of altruist genes. Matthews and Matthews (1978) noted that group selection requires that an allele become established by selection at the individual level. Thereafter, selection could favor demes with altruist genes that reduce extinction rates, relative to demes without these genes. Interdemic selection has become a central theme in developing concepts of metapopulation dynamics (Chapter 7).

Meanwhile, the concept of group selection was implicit in early models of ecological succession and community development. The facilitation model of succession proposed by Clements (1916) and elaborated by Odum (1953,

1969) emphasized the apparently progressive development of a stable, "climax," ecosystem through succession. Each successional stage altered conditions in ways that benefitted the replacing species more than itself. However, such facilitation contradicted individual self-interest that was fundamental to the theory of natural selection. Furthermore, identification of alternative models of succession, including the inhibition model (Chapter 10), made succession appear to be more consistent with evolutionary theory.

Wilson (1976) developed a model that specifically applied the concept of group selection to the community level. Wilson recognized that individuals and species affect themselves, hence their fitness, through their effects on their environment, including effects on the fitness of other individuals. For example, earthworm effects on soil development stimulate plant growth (see Chapter 14) and thereby increase the detrital resources exploited by the worms, a positive feedback. Furthermore, spatial heterogeneity, from large geographic to microsite scales, in population distribution results in intrademic variation in effects of organisms on their community. Given sufficient iterations of Wilson's model, every effect of a species on its community eventually affects that species, positively or negatively, through all possible feedback pathways. Intrademic variation in effects on the environment is subject to selection for adaptive traits of individuals.

The models described help explain the increased frequency of altruist genes, but what selective factors can maintain altruist genes in the face of evolutionary pressure to "cheat" among nonrelated individuals? Trivers (1971) and Axelrod and Hamilton (1981) developed a model of **reciprocal altruism** based on the Prisoner's Dilemma (Fig. 15.1), in which each of two players can cooperate or defect. Each player can choose to cooperate or defect if the other player chooses to cooperate or defect. If the first player cooperates, the benefit for cooperation by the second player (reward for mutual cooperation) is less than that for defection (temptation for the first player to defect in the future); if the first player defects, the benefit for cooperation by the second player (sucker's payoff) is less than that for defection (punishment for mutual defection). Therefore, if the interaction occurs only once, defection (noncooperation) is always the optimal strategy, despite both individuals doing worse than they would if they both cooperate.

However, Axelrod and Hamilton (1981) recognized the probability of repeated interaction between pairs of unrelated individuals and addressed the initial viability (as well as final stability) of cooperative strategies in environments dominated by (a) noncooperating individuals or (b) more heterogeneous environments composed of other individuals using a variety of strategies. After numerous computer simulations with a variety of strategies, they concluded that the most robust strategy in an environment of multiple strategies also was the simplest, Tit-for-Tat. This strategy is cooperation based on reciprocity and a memory extending only one move back, i.e., never being the first to defect but retaliating after a defection by the other and forgiving after just one act of re-

	Player B	
Player A	C Cooperation	D Defection
C Cooperation	R = 3 Reward for mutual cooperation	S = 0 Sucker's payoff
D Defection	T = 5 Temptation to defect	P = 1 Punishment for mutual defection

FIG. 15.1 Prisoner's Dilemma, defined by T > R > P > S and R > (S + T)/2, with payoff to player A shown using illustrative values. From Axelrod and Hamilton (1981) by permission from the American Association for the Advancement of Science. Reprinted with permission from Axelrod, R., and W. D. Hamilton. "The evolution of cooperation," *Science* **211**: pp. 1390–1396. Copyright 1981 American Association for the Advancement of Science.

taliation. They also found that once Tit-for-Tat was established, it resisted invasion by possible mutant strategies as long as the interacting individuals had a sufficiently large probability of meeting again.

Axelrod and Hamilton emphasized that Tit-for-Tat is not the only strategy that can be evolutionarily stable. The Always Defect strategy also is evolutionarily stable, no matter what the probability of future interaction. They postulated that altruism could appear between close relatives when each individual has part interest in the partner's gain (i.e., rewards in terms of inclusive fitness), whether or not the partner cooperated. Once the altruist gene exists, selection would favor strategies that base cooperative behavior on recognition of cues, such as relatedness or previous reciprocal cooperation. Therefore, individuals in stable environments are more likely to experience repeated interaction and selection for reciprocal cooperation than are individuals in unstable environments that provide low probabilities of future interaction.

These models demonstrate that selection at supraorganismal levels must be viewed as contributing to the inclusive fitness of individuals. Cooperating individuals have demonstrated greater ability in finding or exploiting uncommon or aggregated resources, defending shared resources, and mutual protection (Hamilton, 1964). Cooperating predators, e.g., wolves and ants, have higher capture efficiency and can acquire larger prey, compared to solitary predators. The mass attack behavior of bark beetles is critical to successful colonization of living trees. Animals in groups are more difficult for predators to attack.

Reciprocal cooperation reflects selection via feedback (selection) from individual effects on their environment. The strength of individual effects on the environmental is greatest among directly interacting individuals and declines from the population to community levels (Fig. 1.2) (e.g., Lewinsohn and Price, 1996). Reciprocal cooperation can explain the evolution of sexual reproduction and social behavior as the net result of trade-offs between maximizing the contribution of an individual's own genes to its progeny and maximizing the contribution of genes represented in the individual to progeny of its relatives.

Population distribution in time and space (i.e., metapopulation dynamics; see Chapter 7) is a major factor affecting interaction strengths. Individuals dispersed in a regular pattern (Chapter 5) over an area will affect a large proportion of the total habitat and interact widely with cooccurring populations, whereas the same total number of individuals dispersed in an aggregated pattern will affect a smaller proportion of the total habitat, but may have a higher frequency of interactions with cooccurring populations in areas of local abundance. Consistency of population dispersion through time affects the long-term frequency of interactions and reinforcement of selection from generation to generation. Metapopulation dynamics interacting with disturbance dynamics provide the template for selection of species assemblages best adapted to local environmental variation.

II. PROPERTIES OF CYBERNETIC SYSTEMS

The cybernetic nature of ecosystems, from patch to global scales, has been a central theme of ecosystem ecology. Lovelock (1988) suggested that autotroph–heterotroph interactions have been responsible for the development and regulation of atmospheric composition and climate that are suitable for the persistence of life. The ability of ecosystems to minimize variability in climate and rates of energy and nutrient fluxes is an important issue and could affect responses to anthropogenic changes in global conditions.

Cybernetic systems generally are characterized by (1) information systems that integrate system components, (2) low-energy feedback regulators that have high-energy effects, and (3) goal-directed stabilization of high-energy processes. Mechanisms that sense deviation (perturbation) in system condition communicate with mechanisms that function to reduce the amplitude and period of deviation. Negative feedback is the most commonly recognized method for stabilizing outputs. A thermostat represents a simple example of a negative feedback mechanism. The thermostat senses a departure in room temperature from a set level and communicates with a temperature control system that interacts with the thermostat to readjust temperature back to the set level. The room system is maintained at temperatures within a narrow equilibrial range.

Organisms are recognized as cybernetic systems with neurological networks for communicating physiological conditions and various feedback loops for maintaining homeostasis of biological functions. Cybernetic function is perhaps best developed among endotherms. These organisms are capable of self-regulating internal temperature through physiological mechanisms that sense change in body temperature and trigger changes in metabolic rate, blood flow, and sweat that increase or decrease temperature, as necessary. Ectotherms also have physiological and behavioral mechanisms for adjusting body temperature within a somewhat wider range (see Chapter 2). Regardless of mechanism, the result is sufficient stability of metabolic processes for survival.

III. ECOSYSTEMS AS CYBERNETIC SYSTEMS

Although self-adjusting mechanical systems and organisms are the best-recognized examples of cybernetic systems, the properties of self-regulating systems have analogues at supraorganismal levels (Patten and Odum, 1981; Schowalter, 1985, 2000). Human families and societies express goals in terms of survival, economic growth, improved living conditions, etc., and accomplish these goals culturally through governing bodies, communication networks, and balances between reciprocal cooperation (e.g., trade agreements, treaties) and negative feedback (e.g., economic regulations, warfare).

Odum (1969) presented a number of testable hypotheses concerning ecosystem capacity to develop and maintain homeostasis, in terms of energy flow and biogeochemical cycling, during succession. Although subsequent research has shown that many of the predicted trends are not observed, at least in some ecosystems, Odum's hypotheses focused debate on ecosystems as cybernetic systems. Engleberg and Boyarsky (1979) argued that ecosystems do not possess the critical goal-directed communication and low-cost/large-effect feedback systems required of cybernetic systems. Ecosystems can be shown to possess these properties of cybernetic ecosystems, as will be described, but this debate cannot be resolved until ecosystem ecologists reach consensus on a definition and measurable criteria of stability and demonstrate that potential homeostatic mechanisms, such as biodiversity and insects (see following), function to reduce variability in ecosystem conditions.

A. Ecosystem Homeostasis

Although discussion of ecosystem goals appears to be teleological, nonteleological goals can be identified, e.g., maximizing distance from thermodynamic ground (see Patten, 1995), a requisite for all life. Stabilizing ecosystem conditions obviously would reduce exposure of individuals and populations to extreme, and potentially lethal, departures from normal conditions. Furthermore, stable population sizes would prevent extreme fluctuations in abundances that would jeopardize stability of other variables. Hence, environmental heterogeneity might select for individual traits that contribute to stability of the ecosystem.

The argument that ecosystems do not possess centralized mechanisms for communicating departure in system condition and initiating responses (e.g., Engleberg and Boyarsky, 1979) ignores the pervasive communication network in ecosystems (see Chapters 2, 3, and 8). However, the importance of volatile chemicals for communicating resource conditions among species has been recognized relatively recently (Baldwin and Schultz, 1983; Rhoades, 1983; Sticher et al., 1997; Turlings et al., 1990; Zeringue, 1987). The airstream carries a blend of volatile chemicals, produced by the various members of the community, that advertises the abundance, distribution, and condition of various organisms within the community. Changes in the chemical composition of the local atmosphere indicate changes in the relative abundance and suitability of hosts or the presence and proximity of competitors and predators. Sensitivity among organisms to the chemical composition of the atmosphere or water column may provide a global information network that communicates conditions for a variety of populations and initiates feedback responses.

Feedback loops are the mechanisms for maintaining ecosystem stability, regulating abundances and interaction strengths (de Ruiter et al., 1995; Patten and Odum, 1981). The combination of bottom-up (resource availability) and top-down (predation) and lateral (competitive) interactions generally represents negative feedback, stabilizing food webs by reducing the probability that populations increase to levels that threaten their resources (and, thereby, other species requiring those resources). Mutualistic interactions and other positive feedbacks reduce the probability of population decline to extinction thresholds. Although positive feedback often is viewed as destabilizing, such feedback may be most important when populations are small and likely is limited by negative feedbacks as populations grow beyond threshold sizes (Ulanowicz, 1995). Such compensatory interactions may maintain ecosystem properties within relatively narrow ranges, despite spatial and temporal variation in abiotic conditions (Kratz et al., 1995; Ulanowicz, 1995). Ecological succession represents one mechanism for recovery of ecosystem properties following disturbance-induced departures from nominal conditions.

The concept of self-regulation does not require efficient feedback by all ecosystems or ecosystem components. Just as some organisms (recognized as cybernetic systems) have greater homeostatic ability than do others (e.g., endotherms versus ectotherms), some ecosystems demonstrate greater homeostatic ability than do others (Webster et al., 1975). Frequently disturbed ecosystems may be reestablished by relatively random assemblages of opportunistic colonists and select genes for rapid exploitation and dispersal. Their short duration provides little opportunity for repeated interaction that could lead to stabilizing cooperation (cf. Axelrod and Hamilton, 1981). Some species increase variability or promote disturbance, e.g., brittle or flammable species (e.g., easily toppled Cecropia and flammable Eucalyptus). Insect outbreaks increase variation in some ecosystem parameters (Romme et al., 1986), often in ways that promote regeneration of resources (e.g., Schowalter et al., 1981a). On the other hand, relatively stable environments, such as tropical rain forests, also might not select for stabilizing interactions. However, stable environmental conditions should favor consistent species interactions and the evolution of reciprocal cooperation, such as demonstrated by a diversity of mutualistic interactions in tropical forests. Selection for stabilizing interactions should be greatest in ecosystems characterized by intermediate levels of environmental variation. Interactions that reduce such variation would contribute to individual fitnesses.

B. Definition of Stability

Patten and Odum (1981) proposed that a number of time-invariant or regularly oscillating ecosystem parameters represent potential goals for stabilization. These included total system production (P) and respiration (R), P:R ratio, total chlorophyll, total biomass, nutrient pool sizes, species diversity, and population sizes. However, the degree of spatial and temporal variability of these parameters remains poorly known for most, even intensively studied, ecosystems (Kratz et al., 1995).

Kratz et al. (1995) compiled data on the variability of climatic, edaphic, plant, and animal variables from 12 Long-Term Ecological Research (LTER) sites, representing forest, grassland, desert, lotic, and lacustrine ecosystems, in the U.S. Unfortunately, given the common long-term goals of these projects, comparison was limited because different variables and measurement techniques were represented among these sites. Nevertheless, Kratz et al. offered several important conclusions concerning variability.

First, the level of species combination (e.g., species, family, guild, total plants or animals) had a greater effect on observed variability in community structure than did spatial or temporal extent of data. For plant parameters, species- and guild-level data were more variable than were data for total plants; for animal parameters, species-level data were more variable than guild-level data, and both were more variable than total animal data. As discussed for food web properties in Chapter 9, the tendency to ignore diversity, especially of insects (albeit for logistic reasons), clearly affects our perception of variability. Detection of long-term trends or spatial patterns depends on data collection for parameters sufficiently sensitive to show significant differences but not so sensitive that their variability hinders detection of differences.

Second, spatial variability exceeded temporal variability. This result indicates that individual sites are inadequate to describe the range of variation among ecosystems within a landscape. Variability must be examined over larger spatial scales. Edaphic data were more variable than climatic data, indicating high spatial variation in substrate properties, whereas common weather across landscapes homogenizes microclimatic conditions. This result also could be explained as the result of greater biotic modification of climatic variables, compared to substrate variables (see following).

Third, biotic data were more variable than climatic or edaphic data. Organisms can exhibit exponential responses to incremental changes in abiotic conditions (see Chapter 6). The ability of animals to move and alter their spatial distribution quickly in response to environmental changes is reflected in greater variation in animal data, compared to plant data. However, animals also have greater ability to hide or escape sampling devices.

Finally, two sites, a desert and a lake, provided a sufficiently complete array of biotic and abiotic variables to permit comparison. These two ecosystem types represent contrasting properties. Deserts are exposed to highly variable and harsh abiotic conditions but are interconnected within landscapes, whereas lakes exhibit relatively constant abiotic conditions (buffered from thermal change by mass and latent heat capacity of water, from pH change by bicarbonates, and from biological invasions by their isolation) but isolated by land barriers. Comparison of variability between these contrasting ecosystems supported the hypothesis that deserts are more variable than lakes among years, but lakes are more variable than deserts among locations.

These studies provide important data on variation in a variety of ecosystem parameters among ecosystem types. However, important questions remain. Which parameters are most important for stability? How much deviation can be tolerated? What temporal and spatial scales are relevant to ecosystem stability?

Among the parameters that could be stabilized as a result of species interactions, net primary production and biomass structure (living and dead) may be particularly important. Many other parameters, including energy, water and nutrient fluxes, trophic interactions, species diversity, population sizes, climate, and soil development, are directly or indirectly determined by net primary production or biomass structure (see Chapter 11). In particular, the ability of ecosystems to modify internal microclimate, protect and modify soils, and provide stable resource bases for primary and secondary producers depends on NPP and biomass structure. Therefore, natural selection over long periods of coevolution should favor individuals whose interactions stabilize these ecosystem parameters. NPP may be stabilized over long time periods as a result of compensatory community dynamics and biological interactions, such as those resulting from biodiversity and herbivory (see following).

No studies have addressed the limits of deviation, for any parameter, within which ecosystems can be regarded as qualitatively stable. Traditional views of stability have emphasized consistent species composition, at the local scale, but shifts in species composition may be a mechanism for maintaining stability in other ecosystem parameters at the landscape or watershed scale. This obviously is an important issue for evaluating stability and predicting effects of global environmental changes. However, given the variety of ecosystem parameters and their integration at the global scale, this issue will be difficult to resolve.

The range of parameter values within which ecosystems are conditionally stable may be related to characteristic fluctuations in environmental conditions. For example, biomass accumulation increases ecosystem storage capacity and ability to resist variation in resource availability (Webster *et al.*, 1975), but also increases ecosystem vulnerability to some disturbances, including fire and storms. Complex ecosystems with high storage capacity (i.e., forests) are the most buffered ecosystems, in terms of regulation of internal climate, soil conditions, and resource supply, but also fuel the most catastrophic fires under drought conditions and suffer the greatest damage during hurricanes or typhoons. Hence, ecosystems with lower biomass may be more stable under some environmental conditions. Species interactions that periodically reduce biomass (e.g., herbivore outbreaks) traditionally have been viewed as evidence of instability but may contribute to stability of ecosystems in which biomass accumulation is destabilizing.

No studies have addressed the appropriate temporal and spatial scales over which stability should be evaluated or whether these scales should be the same for all ecosystems. Most studies of ecosystem processes represent periods of < 5 years, although some ecosystem studies now span 40 years. The long time scales representing processes such as succession exceed the scale of human lifetimes and have required substitution of temporal variation by spatial variation (e.g., chronosequences within a landscape). Data from such studies have limited utility because individual patches have unique conditions and are influenced by the conditions of surrounding patches (Woodwell, 1993). Therefore, temporal changes at the patch scale often follow different successional trajectories. Although individual patches may change dramatically over time, or recover to variable endpoints, the dynamic mosaic of ecosystem-types (e.g., successional stages or community types) at the landscape or watershed scale may stabilize the proportional area represented by each ecosystem type (see Chapter 10). Changing land use practices have disrupted this conditionally stable heterogeneity of patch types at the landscape scale.

Finally, the time frame of stability must be considered within the context of the ecosystem. For example, forests appear to be less stable than grasslands because of the long time period required for recovery of forests to predisturbance conditions, compared to rapid refoliation of grasses from surviving underground rhizomes. However, forests typically are disturbed less frequently. NPP typically recovers to predisturbance levels within 2–3 years, although biomass requires longer periods to reach predisturbance levels (e.g., Boring et al., 1988; Scatena et al., 1996; Zimmerman et al., 1996).

C. Regulation of NPP by Biodiversity

The extent to which biodiversity contributes to ecosystem stability has been highly controversial (see Chapter 10). Some species have been shown to control aspects of ecosystem function, e.g., production, decomposition, and nutrient fluxes, demonstrating that biodiversity in its broadest sense affects ecosystem function (Beare et al., 1995; Vitousek and Hooper, 1993; Waide et al., 1999; Woodwell, 1993). The presence or absence of individual species affects biotic, atmospheric, hydrospheric, and soil conditions. However, relatively few species have been studied sufficiently, under different conditions, to evaluate their effects on ecosystem functions. The debate depends, to a large extent, on definitions and measures of stability (as has been discussed) and diversity (see Chapter 9).

Vitousek and Hooper (1993) suggested that the relationship between biodiversity and ecosystem function could take several forms. Their Type 1 relationship implies that each species has the same effect on ecosystem function. Therefore, the effect of adding species to the ecosystem is incremental, producing a line with constant slope. The Type 2 relationship represents a decreasing and eventually disappearing effect of additional species, producing a curve that approaches an asymptote. The Type 3 relationship indicates no further effect of additional species.

Communities are not random assemblages of species but, rather, functionally linked groups of species. Therefore, the Type 2 relationship probably represents most ecosystems, with additional species contributing incrementally to ecosystem function and stability until all functional groups are represented (Vitousek and Hooper, 1993). Further additions have progressively smaller effects, as species packing within functional groups simply redistributes the overall contribution among species. Hence, ecosystem function is not linearly related to diversity (Waide *et al.*, 1999).

Within-group diversity could affect the persistence or sustainability of a given function, more than its rate or regulation, and thereby increase the reliability of that function (Fig. 15.2) (Naeem, 1998). Tilman *et al.* (1997) reported that both plant species diversity and functional diversity significantly influenced six ecosystem response variables, including primary productivity and nitrogen pools in plants and soil, when analyzed in separate univariate regressions, but that only functional diversity significantly affected these variables in a multiple regression. Hooper and Vitousek (1997) reported that variability in

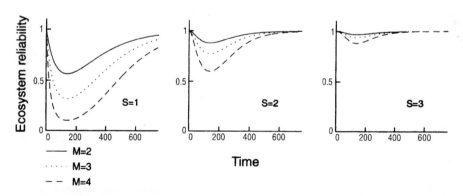

FIG. 15.2 Ecosystem reliability over time as a function of the number of functional groups (M) and number of species per functional group (S), for a probability of species colonization over time of 0.005 and a probability of species presence over time of 0.005. From Naeem (1998), reprinted by permission of Blackwell Science, Inc.

ecosystem parameters was significantly related to the composition of functional groups, rather than the number of functional groups, further supporting the concept of complementarity among species or functional groups.

Dominant organisms in any ecosystem are adapted to survive environmental changes or disturbances that recur regularly with respect to generation time. Therefore, adaptation to prevailing conditions (evolution) constitutes a feedback that reduces ecosystem deviation from nominal conditions. For example, many grassland and pine forest species are adapted to survive low-intensity fires and drought (e.g., underground rhizomes and insulating bark, respectively) that characterize these ecosystems, thereby stabilizing vegetation structure and primary production.

However, all ecosystems are subject to periodic catastrophic disturbances and subsequent community recovery through species replacement (succession). Ecosystem diversity at large spatial or temporal scales provides for reestablishment of key species from neighboring patches or seed banks. The rapid development of early successional communities limits loss of ecosystem assets, especially soil and limiting nutrients. Hence, succession represents a mechanism for reducing deviation in ecosystem parameters, but some early or mid-successional stages are capable of inhibiting further succession. Herbivores may be instrumental in facilitating replacement of inhibitive successional stages under suitable conditions (Chapter 10).

Few studies have measured the effect of biodiversity on stability of ecosystem parameters. Most are based on selection of plots that differ in plant species diversity and, therefore, potentially are confounded by other factors that could vary among plots.

McNaughton (1985, 1993b) studied the effects of plant species diversity on the persistence and productivity of biomass in grazed grasslands in the Serengeti Plain in east Africa. Portions of areas differing in plant diversity were fenced to exclude ungulate grazers. Stability was measured as both resistance (change in productivity resulting from grazing) and resilience (recovery to fenced control condition following cessation of grazing). Grazing reduced diversity 27% in the more diverse community but had no effect on the less diverse community. The percentage biomass eaten was 67 and 76% in the more and less diverse communities, respectively, a nonsignificant difference. By 4 weeks after cessation of grazing, the more diverse community had recovered to 89% of control productivity, but the less diverse community recovered to only 31% of control productivity, a significant difference.

McNaughton (1977, 1993b) also compared resistance of adjacent grasslands of differing diversities to environmental fluctuation. Stability, measured as resistance to deviation in photosynthetic biomass, increased with diversity, as a result of compensation between species with rapid growth following rain but rapid drying between showers and species with slower growth after showers but slower drying between showers. Eight of 10 tests demonstrated a positive relationship between diversity and stability (McNaughton, 1993b).

Frank and McNaughton (1991) similarly compared effects of drought on plant species composition among communities of differing diversities in Yellowstone National Park in the western U.S. Stability of species composition to this environmental change was strongly correlated to diversity (Fig. 15.3).

Ewel (1986) and Ewel *et al.* (1991) evaluated effects of experimental manipulation of plant diversity on biogeochemical processes in a tropical rain forest in Costa Rica. This study included five treatments: a diverse natural succession; a modified succession with the same number and growth form of successional species but no species in common with natural succession; an enriched species diversity with species added to a natural succession; a crop monoculture (replicates of three different crop species); and bare ground (vegetation-free). After 5 years, this design yielded plots with no plants (vegetation-free), single species (monoculture), >100 species (natural and modified succession), and 25% more species (enriched succession). Elemental pool sizes always were significantly larger in the more diverse plots, reflecting a greater variety of mechanisms for retention of nutrients and maintenance of soil processes favorable for plant production. The results suggested a Type 2 relationship between biodiversity and stability, with most change occurring at low species diversity. However, the absence of intermediate levels of diversity, be-

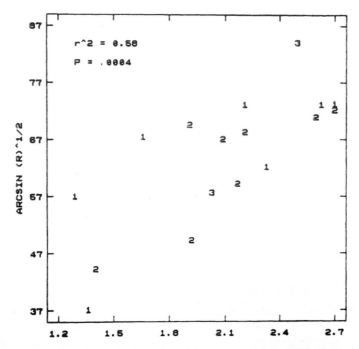

FIG. 15.3 Relationship between stability (measured as resistance (R) to change in species abundances, in degrees,) and diversity (H′) in grasslands subject to grazing and drought at Yellowstone National Park, Wyoming. 1, early season, ungrazed; 2, peak season, grazed; 3, peak season, ungrazed. From Frank and McNaughton (1991) by permission from *Oikos*.

tween the monoculture and > 100 species treatments, limited interpolation of results.

Tilman and Downing (1994) established replicated plots, in 1982, in which the number of plant species was altered through different rates of nitrogen addition. These plots subsequently (1987–1988) were subjected to a record drought. During the drought, plots with > 9 species averaged about half of their predrought biomass, but plots with < 5 species averaged only about 12% of their predrought biomass (Fig. 15.4). Hence, the more diverse plots were better buffered against this disturbance because they were more likely to include drought-tolerant species, compared to less diverse plots. More diverse plots also recovered biomass more quickly following the drought. When biomass was measured in 1992, plots with ≥ 6 species had biomass equivalent to predrought levels, but plots with ≤ 5 species had significantly lower biomass, with deviations ranging from 8–40% (Fig. 15.5). Tilman and Dowing (1994) and Tilman *et al.* (1997) concluded that more diverse ecosystems represent a greater variety of ecological strategies that confer both greater resistance and greater re-

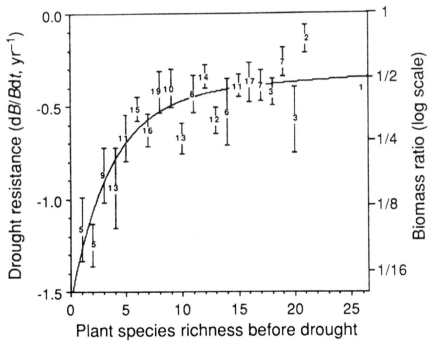

FIG. 15.4 Relationship between plant species diversity prior to drought and drought resistance in experimental grassland plots planted with different species diversities. Mean, standard error, and number of plots with given species richness are shown. 1 dB/Bdt $(yr^{-1}) = 0.5 \ln(1988$ biomass/ 1986 biomass), where 1988 was the peak drought year and 1986 was the year preceding drought. The biomass 1988/1986 ratio (right-hand scale) indicates the proportional decrease in plant biomass associated with dB/Bdt values. From Tilman and Downing (1994) by permission from *Nature*, © 1994 Macmillan Magazines, Ltd.

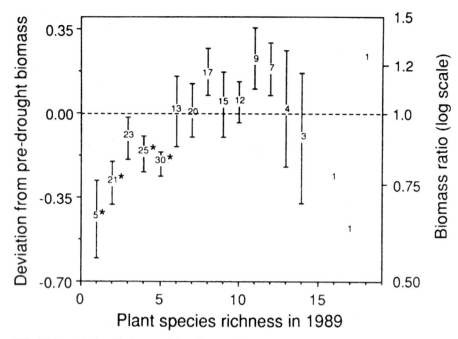

FIG. 15.5 Relationship between plant species diversity and deviation in 1992 biomass (following drought) from mean (1982–1986) predrought biomass in experimental grassland plots planted with different species diversities. Mean, standard error, and number of plots with given species richness are shown. Negative values indicate 1992 biomass lower than predrought mean. Biomass ratio is biomass 1992/predrought. Plots with 1, 2, 4, or 5 species (but not plots with > 5 species) differed significantly from predrought means. From Tilman and Downing (1994) by permission from *Nature*, © 1994 Macmillan Magazines, Ltd.

silience to environmental variation. However, the contribution of diversity to ecosystem stability may be related to environmental heterogeneity, i.e., diversity does not necessarily increase stability in more homogeneous environments.

A number of studies have demonstrated that ecosystem resistance to elevated herbivory is positively correlated to vegetation diversity (e.g., McNaughton, 1985; Schowalter and Lowman, 1999; Schowalter and Turchin, 1993; Chapters 6 and 7). As vegetation diversity increases, the ability of any particular herbivore species to find and exploit its particular hosts decreases, leading to increasing stability of herbivore–plant interactions.

Experimental studies relating ecosystem stability to diversity generally have been limited to manipulation of plant species diversity. However, diversity typically increases from lower to higher trophic levels. Insects represent the bulk of diversity in virtually all ecosystems (e.g., Table 9.1) and are capable of controlling a variety of ecosystem conditions (Chapters 12–14). A few studies have addressed the significance of diversity at higher trophic levels to ecosystem processes but not to ecosystem stability (Lewinsohn and Price, 1996).

Klein (1989) found that diversity of dung beetles (*Scarabaeidae*) and the rate of dung decomposition were positively correlated to the size of forest fragments in central Amazonia. However, abiotic conditions that also affect decomposition likely differed among fragment sizes, as well.

Coûteaux et al. (1991) manipulated diversity of decomposer communities in microcosms with ambient or elevated concentrations of CO_2. They found that decomposition and respiration rates were significantly related to decomposer diversity, as affected by species shifts following CO_2 treatment. This study demonstrated an effect of biodiversity on rates of a key ecosystem process, but did not address long-term stability of this process.

Herbivore and predator diversities have not been experimentally manipulated to evaluate the effect of diversity at these levels on processes at lower trophic levels, except for biological control purposes, which may not represent interactions in natural ecosystems. For example, McEvoy et al. (1993) examined the effects of two insect species with complementary feeding strategies (cinnabar moth, *Tyria jacobaeae*, a foliage and inflorescence feeder, and ragwort flea beetle, *Longitarsus jacobaeae*, a root feeder) introduced to control the exotic ragwort, *Senecio jacobaea*, in coastal Oregon, U.S. They manipulated the presence of these two herbivores in 0.25 m^2 cages. Their results indicated that increasing diversity (from no herbivores to one herbivore to both herbivores) decreased local stability of the herbivore–plant interaction, as increasing herbivory drove the host to local extinction, at the plot scale. However, this plant species persists at low densities over the landscape, suggesting that the interaction is stable at larger spatial scales. Croft and Slone (1997) reported that European red mite *Panonychus ulmi*, abundances in apple orchards were maintained at lower, equilibrial, levels by three predaceous mite species than by any single predaceous species.

Ultimately, the capacity of ecosystems to endure or modify the range of environmental conditions is the primary measure of stability (McNaughton, 1993b) (Fig. 15.2). In this regard, Boucot (1990) noted that characteristic species assemblages (hence, ecosystems) often have persisted for many thousands of years over large areas.

D. Regulation of NPP by Insects

During the 1960s, a number of studies, including Crossley and Howden (1961), Crossley and Witkamp (1964), Edwards and Heath (1963), and Zlotin and Khodashova (1980), indicated that arthropods potentially control energy and nutrient fluxes in ecosystems. Clearly, phytophages could affect, without regulating, ecosystem properties. However, phytophages respond to changes in vegetation density or physiological condition in ways that provide both positive and negative feedback, depending on the direction of deviation in primary production from nominal levels (Figs. 12.4 and 15.6).

Mattson and Addy (1975) introduced the hypothesis that phytophagous insects regulate primary production, based on observations that low intensities

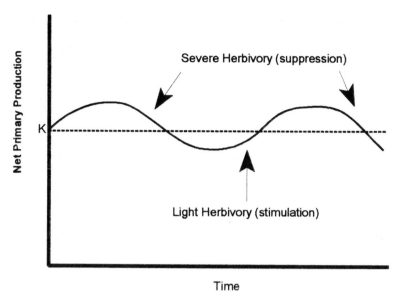

FIG. 15.6 Stimulation of primary production at NPP < K and suppression of primary production at NPP > K by phytophages (see Fig. 12.4) could stabilize primary production.

of herbivory on healthy plants often stimulate primary production but high intensities of herbivory on stressed or dense plants suppress primary production. Furthermore, productivity by surviving plants often is greater following herbivore outbreaks (see Chapter 12). Schowalter (1981) proposed that phytophage outbreaks, triggered by host stress and density as resources become limiting, function to advance succession from communities with high demands for resources to communities with lower demands for resources. Davidson (1993) and Schowalter and Lowman (1999) refined this hypothesis by noting that herbivores and granivores can advance, retard, or reverse succession, depending on environmental conditions.

Despite the obvious influence of animals on key ecosystem processes, their regulatory role has remained controversial and largely untested. Insects possess the characteristics of cybernetic regulators (i.e., low maintenance cost and rapidly amplified effects, sensitivity to deviation in ecosystem parameters, and capacity to dramatically alter primary production through positive and negative feedback) and appear, in many cases, to stabilize net primary production. For example, inconsequential biomass of phytophagous insects, even at outbreak densities, is capable of removing virtually all foliage and altering plant species composition. Virtually undetectable biomass of termites accounts for substantial decomposition, soil redistribution, and gas fluxes that could affect global climate. The following model for insect effect on ecosystem stability focuses on herbivores, but detritivores, pollinators, and seed dispersers also are capable of modifying ecosystem conditions in ways that might promote stability.

Primary production often peaks at low to moderate intensities of pruning and thinning (Fig. 12.4), supporting the grazing optimization hypothesis. Herbivores apparently stimulate primary production at low levels of herbivory, when host density is low or condition good, and reduce primary production at high levels, when host density is high or condition poor (Fig. 15.6), potentially stabilizing primary production at intermediate levels. Furthermore, primary production often is higher following herbivore outbreaks than during the pre-outbreak period (e.g., Alfaro and Shepherd, 1991; Romme *et al.*, 1986), suggesting alleviation of stressful conditions that could lead to instability. By stabilizing primary production, herbivores also stabilize internal climate and soil conditions, biogeochemical fluxes, etc., that affect survival and reproduction of associated organisms. Romme *et al.* (1986) reported that mountain pine beetle, *Dendroctonus ponderosae*, outbreaks appeared to increase variation (destabilization) of some ecosystem properties. However, these outbreaks represent a response to a deviation in primary production. No data are available to indicate whether or not long-term variation is reduced by such outbreaks. Annual wood production following mountain pine beetle outbreaks equaled or exceeded preoutbreak levels within 10 years, suggesting relatively rapid recovery of primary production (Romme *et al.* 1986).

Outbreaks of phytophagous insects are most likely to occur under two interrelated conditions, both of which represent responses to departure from nominal ecosystem conditions, often resulting from anthropogenic alteration (Schowalter, 1985; Schowalter and Lowman, 1999). First, adverse environmental conditions, such as inadequate water or nutrient availability, changing climate, and atmospheric pollution, cause changes in plant physiological conditions that increase suitability for phytophages. High intensities of herbivory under these conditions generally reduce biomass and improve water or nutrient balance or, in extreme cases, reduce biomass of the most stressed plants, regardless of their abundance, and promote replacement by better adapted plants (e.g., Ritchie *et al.*, 1998; Schowalter and Lowman, 1999). Second, high densities of particular plant species, as a result of artificial planting or of inhibitive successional stages, enhance host availability for associated phytophages. High intensities of herbivory represent a major mechanism for reversing site dominance by such plant species, by facilitating their replacement and increasing diversity.

If communities evolve to minimize environmental variation, then herbivore interactions with disturbances are particularly important. Although outbreaks of herbivores traditionally have been viewed as disturbances (together with events such as fire, storm damage, and drought), their response to host density or stress appears to reduce the severity of abiotic disturbances. Herbivore outbreaks commonly co-occur with drought conditions (Mattson and Haack, 1987; White, 1969, 1976, 1984), suggesting that plant moisture stress may be a particularly important trigger for feedback responses that reduce transpiration and improve water balance (Webb, 1978). Fuel accumulation, as a result of herbivore-induced fluxes of material from living to dead biomass, often predisposes ecosystems to fire in arid environments. Whether such predisposition is stabilizing or destabilizing depends on the degree to which outbreaks modify the severity and temporal or spatial scale of such disturbances. Schowalter (1985) and Schowalter et al. (1981a) suggested that herbivore-induced disturbances might occur more regularly with respect to host generation times or stages of ecosystem development, as a result of specific plant–herbivore interactions, and thereby facilitate rapid adaptation to disturbance conditions. Although such interaction would appear to increase variation in the short term, accelerated adaptation would contribute to stability over longer temporal scales. Furthermore, increased likelihood of disturbance during particular seres should maintain that sere on the landscape, contributing to stability over larger spatial scales. The following example demonstrates the potential stabilization of ecosystem properties over the large spatial scales of western North America.

Conifer forests dominate much of the montane and high latitude region of western North America. The large, contiguous, lower elevation zone is characterized by relatively arid conditions and frequent droughts that historically maintained a sparse woodland dominated by drought- and fire-tolerant (but shade-intolerant) pine trees and a ground cover of grasses and shrubs, with little understory (Fig. 15.7). Low intensity fires occurred frequently, at intervals of 15–100 years, and covered large areas (Agee, 1993), minimizing drought-intolerant vegetation and litter accumulation. The relatively isolated higher elevation and riparian zones were more mesic and supported shade-tolerant (but fire- and drought-intolerant) fir and spruce forests. Fire was less frequent (every 150–1000 years) but more catastrophic, as a result of the greater tree densities and understory development that facilitated fire access to tree canopies (Agee, 1993; Veblen et al., 1994).

As a result of fire suppression during the past century, much of the lower elevation zone has undergone succession from pine forest to later successional fir forest (Fig. 15.7), a conspicuous deviation from historic conditions. Outbreaks of a variety of folivore and bark beetle species have become more frequent in these altered forests. During mesic periods and in more mesic locations, e.g., riparian corridors and higher elevations, the mountain pine beetle has advanced succession by facilitating the replacement of competitively stressed pines by more competitive firs. However, during inevitable drought periods, such as occurred during the 1980s, moisture limitation increases the vulnerability of these firs to several folivores and bark beetles specific to fir species (Fig. 15.8). Insect-induced mortality of the firs reversed succession by favoring the remaining drought- and fire-tolerant pines. Tree mortality also increases the severity and scale of catastrophic fires that historically were rare in these forests. However, this altered fire regime represents a continuation of the anthropogenic effect of fire suppression in these forests and likely will be mitigated in ecological time by eventual reestablishment of the pine sere following catastrophic fire.

FIG. 15.7 The relatively arid interior forest region of North America was characterized by open-canopied forests dominated by drought- and fire-tolerant pines and by sparse understories, prior to fire suppression beginning in the late 1800s (left). Fire suppression has transformed forests into dense, multistoried ecosystems stressed by competition for water and nutrients (right). From Goyer *et al.* (1998) by permission from the Society of American Foresters. Reprinted from the *Journal of Forestry* (Vol. 96, no. 12, p. 29) published by the Society of American Foresters, 5400 Grosvenor Lane, Bethesda, MD 20814-2198. Not for further reproduction.

FIG. 15.8   Phytophage modification of succession in central Sierran mixed conifer ecosystems during 1998. Understory white fir (*Abies concolor*), the late successional dominant, is increasingly stressed by competition for water in this arid forest type. An outbreak of the Douglas-fir tussock moth, *Orgyia pseudotsugata* (top) has completely defoliated the white fir (grey trees, bottom), restoring the ecosystem to a more stable condition dominated by earlier successional, drought- and fire-tolerant sequoias and pines (dark, foliated trees).

A similar situation has been inferred from insect demography in pine–hardwood forests of the southern U.S. (Fig. 10.5).

To what extent do insects contribute to stability and "health" of various ecosystems? Until recently, insect outbreaks and disturbances have been viewed as destructive forces. The increased productivity of ecosystems in which fire and insect outbreaks were prevented supported a view that resource production could be freed from limitations imposed by these disturbances. However, fire now is recognized as an important tool for restoring sustainable (stable) ecosystem conditions and characteristic communities. Accumulating evidence also suggests that outbreaks of native insects represent feedback that maintains ecosystem production within sustainable ranges. Regulation of primary production by phytophagous insects could stabilize other ecosystem variables, as well. Clearly, experimental studies should address the long-term effects of phytophagous insects on variability of ecosystem parameters. Our management of ecosystem resources and, in particular, our approach to managing phytophagous insects requires that we understand the extent to which phytophages contribute to ecosystem stability.

IV. SUMMARY

The hypothesis that phytophagous insects regulate ecosystem processes is one of the most important and controversial concepts to emerge from research on insect ecology. The extent to which ecosystems are random assemblages of species that simply affect ecosystem processes or are tightly coevolved groups of species that stabilize ecosystem function has important implications for management of ecosystem resources and "pests." Although this hypothesis is not contingent on selection at the supraorganismal level, concepts of group selection have developed from and contributed to this hypothesis.

Debate on the issue of group selection has solidified consensus on the dominance of direct selection on individual attributes. However, individual attributes affect other organisms and generate feedback on individual fitness. Such feedback selection contributes to the inclusive fitness of an individual. The intensity of this feedback is proportional to the relatedness of interacting individuals. The greatest feedback selection is between near kin (kin selection). The frequency of interaction and the intensity of feedback selection declines as interacting individuals become less related. However, frequent interspecies interactions can lead to negative feedback interactions (e.g., competition and predation) and reciprocal cooperation (mutualism), based on the trade-off between gain and loss to each individual from such interaction.

Homeostasis at supraorganismal levels depends only in part on selection for attributes that benefit assemblages of organisms (i.e., group selection). The critical issue is the trade-off required to balance individual sacrifice, if any, and inclusive fitness accruing from traits that benefit the group. Stabilization of environmental conditions through species interactions favor survival and reproduction of the constituent individuals. Therefore, feedback selection over evolutionary time scales should select for species interactions that contribute to ecosystem stability and mutually assured survival.

Major challenges for ecologists include defining stability, i.e., which ecosystem properties are stabilized, what range of deviation is tolerated, what temporal and spatial scales are appropriate levels for measurement of stability, and what ecosystem mechanisms contribute to stability. Traditionally, stability has been viewed as constancy or recovery of species composition over narrow ranges of time and space. Alternative views include net primary production and biomass structure that affect the stability of internal climate and soil conditions and biogeochemical pools and fluxes over larger ranges of time and space. Stability may be achieved, not so much at the patch scale, as at the landscape scale where conditional stability is achieved through relatively constant proportions of various ecosystem types.

The relationship of stability to diversity has been a major topic of debate. Some species are known to control ecosystem properties and their loss or gain can severely affect ecosystem structure or function. Furthermore, effects of different species often are complementary, such that diverse assemblages should be better buffered against changes in ecosystem properties in heterogenous environments. A few experimental manipulations of plant species diversity have shown that more diverse communities can have lower variability in primary production than do less diverse communities.

Phytophagous insects have been identified as potentially important regulators of primary production, and hence, of ecosystem properties dependent on primary production. Phytophagous insects possess the key criteria of cybernetic regulators, i.e., small biomass, rapid amplification of effect at the ecosystem level, sensitivity to airborne or waterborne cues indicating ecosystem conditions, and stabilizing feedback on primary production and other processes. Low intensity of herbivory, under conditions of low densities or optimal condition of hosts, tends to stimulate primary production, whereas higher intensities, under conditions of high density or stressed condition of hosts, tend to reduce primary production. Clearly, this aspect of insect ecology has significant implications.

V

SYNTHESIS

THE PREVIOUS FOUR SECTIONS HAVE ADDRESSED THE integration of insect ecology at the individual, population, community, and ecosystem levels of organization. Resource acquisition and allocation by individuals (Section I) can be seen to depend on population (Section II), community (Section III), and ecosystem conditions that the individual also influences, as described in Section IV. Insects are involved in a particularly rich variety of feedbacks between individual, population, community, and ecosystem levels as a consequence of their dominance and diversity in terrestrial and freshwater ecosystems and their sensitivity and dramatic responses to environmental changes. The hypothesis that insects are major regulatory mechanisms in homeostatic ecosystems has important ecological and management implications and warrants critical testing.

The importance of temporal and spatial scales is evident at each level of the ecological hierarchy. Individuals have a period and range of occurrence, populations are characterized by temporal dynamics and dispersion patterns, communities and ecosystems are represented over temporal and spatial scales. In particular, ecosystem stability and its effect on component individuals traditionally has been evaluated at relatively small scales, in time and space, but larger scales are more appropriate. The dynamic mosaic of ecosystem types at the landscape or biome level is conditionally stable in its proportional representation of ecosystem types.

This concluding chapter summarizes and synthesizes the study of insect ecology. The focus will be on important aspects of insect ecology and intriguing questions for future study.

Synthesis

THE STUDY OF INSECT ECOLOGY HAS TRADITIONALLY ADDRESSED INSECT adaptations to their environment, including interactions with other organisms. Insects represent the full scope of heterotrophic strategies, from sessile species whose ecological strategies resemble those of plants to social insects whose range of behavioral attributes is more like that of advanced vertebrates. The variety of their interactions with other species spans the range of complexity and often brings them to the attention of natural resource managers as crop pests or biological control agents. Three of the four sections in this book emphasize this traditional approach to the study of insect ecology.

However, this traditional focus on species adaptations and community interactions does not portray the full scope of insect ecology. Whereas the evolutionary perspective emphasizes insect responses to environmental conditions, as demonstrated by adaptive physiology, behavior, and interspecific interactions, the ecosystem perspective emphasizes feedbacks between organisms and their environment. Insects, as well as other organisms, influence their environment in complex ways. The foraging path of an organism affects its interactions with other organisms and the resulting distribution of resources. Natural selection represents a major form of feedback between ecosystem conditions and individual attributes that affect ecosystem parameters. Other feedback mechanisms between individuals, populations, and communities can stabilize or destabilize ecosystem, landscape, and global processes. Understanding these feedbacks is critical to prediction of ecosystem responses to environmental changes. Phytophages dramatically alter the structure of landscapes and potentially stabilize primary production and other processes affecting global climate and biogeochemistry (Chapter 12). Termites account for substantial portions of carbon flux in some ecosystems (Chapter 14). Section IV, dealing with feedbacks between insects and ecosystem properties, is the major contribution of this book. This synthesis summarizes key ecological issues, synthesizes key integrating variables, and identifies critical issues for future study.

I. SUMMARY

The hierarchical organization of this text emphasizes linkages and feedbacks among the levels of ecological organization. Linkages and feedbacks are strongest between neighboring levels but are significant even between individual and ecosystem levels of the hierarchy. Physiological and behavioral responses to environmental variation determine individual fitness, but also affect the rate and geographic pattern of resource acquisition and allocation that control climate and energy and biogeochemical fluxes at the ecosystem level. These feedbacks are an important and largely neglected aspect of insect ecology that affect ecosystem stability and global processes.

The geographic distribution of individual species generally reflects the environmental template established by continental history, latitude, mountain ranges, and global atmospheric and oceanic circulation patterns. The great diversity of insects reflects their rapid adaptation, conferred by small size, short life spans, and rapid reproductive rates, to environmental variation. These attributes have facilitated speciation at multiple scales: among geographic regions, habitats, resources, and at microscales on or within resources (e.g., individual leaves). However, within the potential geographic range of a species, the spatial and temporal patterns of abundance reflect disturbance dynamics, resource distribution, and interactions with other species that affect individual fitnesses and enhance or limit colonization and population growth.

Energy and resource balances are key aspects of individual fitness, population persistence, and community interactions. All organisms require energy to accumulate resources, necessary for growth and reproduction, against resource concentration gradients and thereby maintain the thermodynamic disequilibrium characteristic of life. Where resources are more concentrated, relative to individual needs, less energy is required for acquisition. Although many small insects live within their resources and are relatively sedentary, e.g., scale insects on large plants, most species must seek resources. All species have mobile stages adapted to find new resources before current resources are depleted. The early evolution of flight among insects greatly facilitated foraging, escape from unsuitable environmental or resource conditions, and discovery of more optimal conditions. Individuals or populations that fail to acquire sufficient energy and nutrients to grow and reproduce do not survive.

Adaptations for detecting and acquiring resources are highly developed among insects. Many insects can detect the presence and location of resources from chemical cues carried at low concentrations on wind or water currents. The diversity of strategies among insect species for acquiring resources has perhaps drawn the most ecological attention. These strategies range from ambush to active foraging, often demonstrate considerable learning ability (especially among social insects), and involve insects in all types of interactions with other organisms, including competition (e.g., for food, shelter, and oviposition site resources), predation and parasitism (on plant, invertebrate, and vertebrate prey or hosts), and mutualism (e.g., for protection, pollination, and seed dispersal).

Spatial and temporal variation in population and community structure reflects net effects of environmental conditions. Changes in population and community structure also constrain survival and reproduction of associated species. Population density and competitive, predatory, and mutualistic interactions affect foraging behavior and energy and nutrient balances of individuals. Individuals forced to move constantly to avoid intra- or interspecific competitors or predators will be unable to forage sufficiently for energy and nutrient resources. On the other hand, energy and nutrient balances can be improved through mutualistic interactions that enhance the efficiency of resource acquisition. The relative contributions of intra- and interspecific interactions to individual survival and reproduction remain a central theme of ecology, but have been poorly integrated with ecosystem conditions. Debate over the importance of bottom-up versus top-down controls of populations perhaps reflects investigator biases as well as spatial and temporal variation in the importance of various regulatory mechanisms. However, the abundance of any particular species determines a variety of ecological processes, e.g., NPP, succession, pollination, seed dispersal, decomposition, fluxes of water, carbon and other nutrients, climate, and disturbance dynamics.

Ecosystems represent the level at which complex feedbacks among abiotic and biotic processes are integrated. Ecosystems can be viewed as dynamic energy- and nutrient-processing engines that modify global energy and nutrient fluxes. Cycling and storage processes controlled by organisms reduce variation in abiotic conditions and resource availability. Although ecosystem properties are largely determined by vegetation structure and composition, insects and other animals modify ecosystem conditions, often dramatically, through effects on primary production, decomposition and mineralization, and pedogenesis. Insect herbivore effects on vegetation structure affect albedo, evapotranspiration, and wind abatement. Changes in decomposition processes affect fluxes of carbon and trace gases, as well as soil structure and fertility. Insect roles as ecosystem engineers mitigate or exacerbate environmental changes resulting from anthropogenic activities. Resolution of environmental issues requires attention to these roles of insects, as well as to their responses to environmental changes.

II. SYNTHESIS

Insect ecology addresses an astounding variety of interactions between insects and their environment. However, key aspects of insect ecology involve feed-

backs between insect responses to changes in abiotic conditions, especially resource supply, and their capacity to modify, and potentially stabilize, energy
and nutrient fluxes. Feedback integration among hierarchical levels occurs primarily through responses to, and modification of, variation in environmental
conditions (see Fig. 1.3). Evolution represents feedback from ecosystem conditions to individual attributes that affect ecosystem conditions.

The importance of environmental change and disturbance as a central
theme in insect ecology has been recognized only recently. Disturbance, in particular, provides a context for individual adaptations, population strategies, organization and succession of community types, and rates and regulation of
ecosystem processes. Environmental changes or disturbances kill individuals or
affect their activity and reproduction. Some populations are reduced to local
extinction, but others exploit the altered conditions. Population strategies and
interactions with other species also affect ecosystem properties in ways that increase the probability of disturbance (or other changes) or that mitigate environmental changes and favor persistence of species less tolerant to change.
Insects contribute greatly to feedback between ecosystem properties and environmental variation. This aspect of insect ecology has important consequences
for ecosystem responses to global changes resulting from anthropogenic activities.

Energy and biogeochemical fluxes integrate individuals, populations, and
communities with their abiotic environment. Energy flow and biogeochemical
cycling processes determine rates and spatial patterns of resource availability.
Many, perhaps most, species attributes can be shown to represent trade-offs between maximizing resource acquisition and optimizing resource allocation
among metabolic pathways, e.g., foraging activity, defensive strategies, or
growth and reproduction. The patterns of energy and nutrient acquisition and
allocation by individuals determine the patterns of storage and fluxes among
populations, fluxes among species at the community level, and storage and flux
at the ecosystem level that, in turn, determine resource availability for individuals, populations and communities. Resource availability is fundamental to
ecosystem productivity and diversity. Resource limitation, including reduced
availability resulting from inhibition of water and nutrient fluxes, is a key factor affecting species interactions. Herbivore and predator populations grow
when host or prey populations are incapable of escape or defense because of insufficient resource acquisition or poor food quality.

Regulatory mechanisms emerge at all levels of the ecological hierarchy.
Negative feedback and reciprocal cooperation are apparent at population, community, and ecosystem levels. Cooperation benefits individuals by improving
ability to acquire limiting resources. This positive feedback balances the negative feedbacks that limit population density, growth, and ecological processes.
At the population level, positive and negative feedbacks maintain density within narrower ranges than occur when populations are released from regulatory
mechanisms. The responsiveness of insect herbivores to changes in plant den-

sity and condition, especially resulting from crop management, introduction into new habitats, and land use, bring some species into conflict with human interests. However, insect outbreaks in natural ecosystems appear to be restricted in time and space and function to (1) maintain net primary production within relatively narrow ranges imposed by the carrying capacity of the ecosystem and (2) facilitate replacement of plant species that are poorly adapted to current conditions by species that are better adapted to these conditions. Regulatory capacity appears to reflect selection for recognition of cues that signal changes in host density or condition that affect long-term carrying capacity of the ecosystem.

The issue of ecosystem self-regulation is a key concept that significantly broadens the scope of insect ecology. Although this idea remains controversial, accumulating evidence supports a view that insect outbreaks function to reduce long-term deviation in net primary production, at least in some ecosystems. Although outbreaks appear to increase short-term variation in some ecosystem parameters, reversal of potentially destructive increases in NPP could reduce long-term variation in ecosystem conditions.

Models of group selection predict that stabilizing interactions are most likely in ecosystems where pairs of organisms interact consistently. Hence, selection for stabilizing selection might be least likely in ecosystems where such interactions are inconsistent, such as in harsh or frequently disturbed environments. However, selection for stabilizing interactions also should be unlikely in productive ecosystems with little variation in abiotic conditions or resource availability, such as tropical rain forest ecosystems. Stabilizing interactions are most likely in ecosystems where selection would favor interactions that reduce moderate levels of variation in abiotic conditions or resource availability.

Insects play key roles in regulation of primary and secondary production. Their large numbers, rapid reproduction and mobility may maximize their interactions with other organisms and the rate at which they evolve reciprocal cooperation.

III. CRITICAL ISSUES

Perhaps the most important goal for insect ecologists is resolution of the debate over the regulatory roles of insects in natural ecosystems. Whether we view insects as disturbances that destabilize ecosystems or as regulators that contribute to stability determines not only our approach to managing insects in natural or engineered ecosystems, but also our approaches to managing our ecosystem resources and to responding to global changes.

Clearly, exotic species freed from both bottom-up and top-down regulation function in the same way as pollutants or other abiotic disturbances, i.e., with little ecosystem control over their effects. By contrast, population size and effects of native species are regulated by a variety of bottom-up, top-down, and lateral factors. However, native species are adapted to respond to natural or anthropogenic alteration of vegetation and landscape structure, with effects that often are contrary to management goals but perhaps conducive to ecological balances. If native insects function as regulators that contribute to ecosystem stability, then traditional management approaches that emphasize suppression may interfere with this natural feedback mechanism and maintain anthropogenic imbalances, at least in some ecosystems. In any case, insect outbreaks are responses to high density or stress of host plants, making outbreaks a symptom of imbalances in ecosystem conditions, rather than a pest problem. The solution, therefore, requires that the management priority be remedies for the imbalances, rather than suppression of the outbreak.

Predicting and alleviating effects of anthropogenic changes requires understanding of insect roles and how these roles affect ecosystem responses to anthropogenic changes. Anthropogenic changes will continue to trigger insect outbreaks whether as destructive events or regulatory responses. Land use, in particular, affects patch structure and interactions among demes, greatly altering the spatial and temporal patterns of insect abundances. Ruderal plant species, valued for crop production but also adapted for rapid colonization of new habitats, are increasingly likely to dominate fragmented landscapes. The rapid growth and poor competitive ability of these species in crowded ecosystems make them targets of their associated insects. Such ecosystems will require constant human intervention. Protection or restoration of natural ecosystems will require attention to interactions necessary to maintain key species, including pollinators, seed dispersers, and decomposers.

Accomplishment of this primary goal requires broadening of research approaches to address the breadth of insect effects on ecosystem structure and function. This, in turn, requires changes in research approaches and integration of population and ecosystem models.

Testing of ecosystem-level hypotheses involves different approaches than does testing of population- and community-level hypotheses. At least three considerations are particularly important.

First, experimental design requires attention to statistical independence of samples. Whereas individuals within populations can serve as replicates for population and community properties, data must be pooled at the site (ecosystem) level for comparison of ecosystem properties. Ecosystem studies often have provided inconclusive data because a single site representing each of several ecosystem types or experimental treatments provides no error degrees of freedom for statistical analysis. The many samples collected within each site are not statistically independent (Hurlbert, 1984). Therefore, the experimental design must incorporate multiple, geographically interspersed sites representing each ecosystem type or treatment. A greater number of replicate sites provides a greater range of inference than do multiple samples within sites, permitting a trade-off between sampling effort within sites and between sites.

Second, research should address a greater range of ecosystem variables than has been common in studies of insect ecology. Insects respond to multiple factors simultaneously, not just to one or a few factors subject to experimental manipulation, and their responses reflect trade-offs that might not be reflected in studies that control some of these factors. A greater breadth of parameters can be addressed through multidisciplinary research, with experts in different aspects of ecosystems contributing to a common goal. Involvement of insect ecologists in established multidisciplinary projects, such as the Long-Term Ecological Research (LTER) sites, can facilitate integration of insect ecology and ecosystem ecology.

Third, spatial and temporal scales of research and perspectives must be broadened. Most ecosystem studies address processes at relatively small spatial and temporal scales. However, population dynamics and capacity to influence ecosystem and global properties span landscape and watershed scales, at least. Feedbacks often may be delayed, or operate over long time periods, especially in ecosystems with substantial buffering capacity, requiring long-term institutional and financial commitments for adequate study. Linkage of population and ecosystem variables using GIS will become an increasingly important aspect of insect ecology. Nevertheless, ecosystems with large biomass or high complexity may require use of simplified field mesocosms to test some hypotheses.

The complexity of ecosystem interactions and information linkages has limited incorporation of detail, such as population dynamics, in ecosystem models. Modeling methodology for ecosystem description is necessarily simplified, relative to that for population models. However, population models have largely ignored feedbacks between population and ecosystem processes. Hierarchical structure in ecosystem models facilitates integration of more detailed insect population (and other) submodels, and their linkages and feedbacks with other levels, as data become available.

Several ecosystem components should be given special attention. Subterranean and forest canopy subsystems represent two ecological frontiers. Logistical difficulties in gaining nondestructive or nonintrusive access to these two subsystems have limited data available for insect effects on canopy–atmosphere and canopy–rhizosphere–soil interactions that control climate and energy and matter fluxes. Improved canopy access methods, such as construction cranes for ecological use (Schowalter and Ganio, 1998; Shaw, 1998), and rhizotron technology (Sackville *et al.*, 1991; Sword, 1998) offer opportunities for scientific advances in the structure and function of these components.

Finally, principles of insect ecology must be applied to improved management of insect populations and ecosystem resources. Ecosystem engineering can make crop systems less conducive to insect population irruption. Alternative cropping systems include protection of soil systems to enhance energy and matter availability and polyculture cropping and landscape patterns of crop patches to restrict herbivore dispersal among hosts (Kogan, 1998; Lowrance *et al.*, 1984; Rickson and Rickson, 1998; Risch, 1980, 1981; Schowalter, 1996). These cropping systems also enhance conditions for predators that control potentially irruptive insect species. Promotion of interactions that tend to stabilize populations of irruptive species is more effective in the long term than is reliance on pesticides or genetically engineered crops. Examples include provision or retention of hedgerows, ant-attracting plants, or other refuges within agricultural landscapes that maintain predator populations (Kruess and Tscharntke, 1994; Rickson and Rickson, 1998; Schowalter, 1996).

IV. CONCLUSIONS

Insects are involved in virtually all aspects of terrestrial and freshwater ecosystems. Environmental issues directly or indirectly involve insects, in either their capacity to respond to environmental changes or their capacity to alter environmental conditions. Therefore, insect ecology is fundamental to our ability to understand ecosystem structure and function and to solve environmental problems.

The hierarchical ecosystem approach to insect ecology emphasizes linkages and feedbacks among individual, population, community, and ecosystems levels and clarifies the basis and consequences of insect adaptive strategies. This approach also indicates which level best addresses environmental problems. For example, if the issue is factors controlling plant susceptibility to herbivores, then individual responses to environmental cues are the appropriate focus. If the issue is spread of exotic species or restoration of native species, then metapopulation dynamics and regulatory interactions within communities are the levels of focus. If the issue is factors affecting global carbon flux, then ecosystem structure and function is the appropriate focus.

Our most significant scientific advances in the next decades will be in demonstrating the degree to which ecosystems modify environmental conditions and persist in the face of changing global conditions. Insects are major contributors to the ways in which ecosystems modify local and global conditions. Natural selection can be viewed as a major form of feedback between ecosystem conditions and individual adaptations that modify or stabilize ecosystem parameters. The degree to which insects regulate ecosystem parameters remains a key issue and one that significantly broadens the scope and value of insect ecology.

BIBLIOGRAPHY

Adler, P. H., and J. W. McCreadie. 1997. The hidden ecology of black flies: Sibling species and ecological scale. *American Entomologist* **43**: 153–161.

Agee, J. K. 1993. *Fire Ecology of Pacific Northwest Forests.* Island Press, Washington, DC.

Ågren, G. I., E. Bosatta, and J. Balesdent. 1996. Isotope discrimination during decomposition of organic matter: A theoretical analysis. *Soil Science Society of America Journal* **60**: 1121–1126.

Aide, T. M. 1992. Dry season leaf production: An escape from herbivory. *Biotropica* **24**: 532–537.

Aide, T. M. 1993. Patterns of leaf development and herbivory in a tropical understory community. *Ecology* **74**: 455–466.

Aide, T. M., and J. K. Zimmerman. 1990. Patterns of insect herbivory, growth, and survivorship in juveniles of a neotropical liana. *Ecology* **71**: 1412–1421.

Aizen, M. A., and P. Feinsinger. 1994. Habitat fragmentation, native insect pollinators, and feral honey bees in Argentine "Chaco Serano." *Ecological Applications* **4**: 378–392.

Akiyama, T., S. Takahashi, M. Shiyomi, and T. Okubo. 1984. Energy flow at the producer level: The energy dynamics of grazed grassland 1. *Oikos* **42**: 129–137.

Alfaro, R. I., and R. F. Shepherd. 1991. Tree-ring growth of interior Douglas-fir after one year's defoliation by Douglas-fir tussock moth. *Forest Science* **37**: 959–964.

Allee, W. C. 1931. *Animal Aggregations: A Study in General Sociology.* University of Chicago Press, Chicago, IL.

Allen, E. B., and M. F. Allen. 1990. The mediation of competition by mycorrhizae in successional and patchy environments. In *Perspectives on Plant Competition* (J.B. Grace and D. Tilman, Eds.), pp. 367–389. Academic Press, San Diego.

Alstad, D. N., and D. A. Andow. 1995. Managing the evolution of insect resistance to transgenic plants. *Science* **268**: 1894–1896.

Alstad, D. N., G. F. Edmunds, Jr., and L. H. Weinstein. 1982. Effects of air pollutants on insect populations. *Annual Review of Entomology* **27**: 369–384.

Amman, G. D., M. D. McGregor, R. F. Schmitz, and R. D. Oakes. 1988. Susceptibility of lodgepole pine to infestation by mountain pine beetles following partial cutting of stands. *Canadian Journal of Forest Research* **18**: 688–695.

Andersen, A. N. 1988. Soil of the nest-mound of the seed-dispersing ant, *Aphaenogaster longiceps,* enhances seedling growth. *Australian Journal of Ecology* **13**: 469–471.

Andersen, A. N., and W. M. Lonsdale. 1990. Herbivory by insects in Australian tropical savannas: A review. *Journal of Biogeography* **17**: 433–444.

Andersen, D. C., and J. A. MacMahon. 1985. Plant succession following the Mount St. Helens volcanic eruption: Facilitation by a burrowing rodent, *Thomomys talpoides. American Midland Naturalist* **114**: 63–69.

Andersen, P. C., B. V. Brodbeck, and R. F. Mizell, III. 1992. Feeding by the leafhopper, *Homalodisca coagulata,* in relation to xylem fluid chemistry and tension. *Journal of Insect Physiology* **38**: 611–622.

Anderson, J. M. 1973. The breakdown and decomposition of sweet chestnut (*Castanea sativa* Mill.) and beech (*Fagus sylvatica* L.) leaf litter in two deciduous woodland soils. I. Breakdown, leaching and decomposition. *Oecologia* **12**: 251–274.

Anderson, J. M. 1988. Invertebrate-mediated transport processes in soils. *Agriculture, Ecosystems and Environment* 24: 5–19.

Anderson, J. M., and M. J. Swift. 1983. Decomposition in tropical forests. In *Tropical Rain Forest: Ecology and Management* (S. L Sutton, T. C. Whitmore, and A. C. Chadwick, Eds.), pp. 287–309. Blackwell, London.

Anderson, J. M., J. Proctor, and H. W. Vallack. 1983. Ecological studies in four contrasting lowland rain forests in Gunung Mulu National Park, Sarawak. *Journal of Ecology* 71: 503–527.

Anderson, N. H., R. J. Steedman, and T. Dudley. 1984. Patterns of exploitation by stream invertebrates of wood debris (xylophagy). *Verhandlungen der Internationalen Vereinigung für Theoretische und Angewandte Limnologie* 22: 1847–1852.

Anderson, V. J., and D. D. Briske. 1995. Herbivore-induced species replacement in grasslands: Is it driven by herbivory tolerance or avoidance? *Ecological Applications* 5: 1014–1024.

Andreux, F., C. Cerri, P. B. Vose, and V. A. Vitorello. 1990. Potential of stable isotope, ^{15}N and ^{13}C, methods for determining input and turnover in soils. In *Nutrient Cycling in Terrestrial Ecosystems: Field Methods, Application and Interpretation* (A. F. Harrison, P. Ineson, and O. W. Heal, Eds.), pp. 259–275. Elsevier, London.

Andrewartha, H. G., and L. C. Birch. 1954. *The Distribution and Abundance of Animals.* Univ. of Chicago, Chicago, IL.

Appanah, S. 1990. Plant-pollinator interactions in Malaysian rain forests. In *Reproductive Ecology of Tropical Forest Plants* (K. Bawa and M. Hadley, Eds.), pp. 85–100. UNESCO/Parthenon, Paris.

Archer, S., and D. A. Pyke. 1991. Plant-animal interactions affecting plant establishment and persistence on revegetated rangeland. *Journal of Range Management* 44: 558–565.

Arnone, J. A., III, J. G. Zaller, C. Ziegler, H. Zandt, and C. Körner. 1995. Leaf quality and insect herbivory in model tropical plant communities after long-term exposure to elevated atmospheric CO_2. *Oecologia* 104: 72–78.

Auclair, J. L. 1958. Honeydew excretion in the pea aphid *Acyrthosiphum pisum* (Harr.) (Homoptera: Aphididae). *Journal of Insect Physiology* 2: 330–337.

Auclair, J. L. 1959. Feeding and excretion of the pea aphid, *Acyrthosiphum pisum* (Harr.), reared on different varieties of peas. *Entomologia Experimentalis et Applicata* 2: 279–286

Auclair, J. L. 1965. Feeding and nutrition of the pea aphid, *Acyrthosiphum pisum* (Harr.) (Homoptera: Aphididae), on chemically defined diets of various pH and nutrient levels. *Annals of the Entomological Society of America* 58: 855–875.

Ausmus, B. S. 1977. Regulation of wood decomposi-

tion rates by arthropod and annelid populations. *Ecological Bulletin (Stockholm)* 25: 180–192.

Axelrod, R.. and W. D. Hamilton. 1981. The evolution of cooperation. *Science* 211: 1390–1396.

Bach, C. E. 1990. Plant successional stage and insect herbivory:Flea beetles on sand-dune willow. *Ecology* 71: 598–609.

Baker, R. R. 1972. Territorial behaviour of the nymphalid butterflies, *Aglais urticae* (L.) and *Inachis io* (L.). *Journal of Animal Ecology* 41: 453–469.

Baldwin, I. T. 1990. Herbivory simulations in ecological research. *Trends in Ecology and Evolution* 5: 91–93.

Baldwin, I. T., and J. C. Schultz. 1983. Rapid changes in tree leaf chemistry induced by damage: evidence for communication between plants. *Science* 221: 277–279.

Banks, C. J., and E. D. M. Macaulay. 1964. The feeding, growth and reproduction of *Aphis fabae* Scop. on *Vicia faba* under experimental conditions. *Annals of Applied Biology* 53: 229- 242.

Banks, C. J., and H. L. Nixon. 1958. Effects of the ant, *Lasius niger* (L.) on the feeding and excretion of the bean aphid, *Aphis fabae* Scop. *Journal of Experimental Biology* 35: 703- 711.

Banks, C. J., and H. L. Nixon. 1959. The feeding and excretion rates of *Aphis fabae* Scop. on *Vicia faba* L. *Entomologia Experimentalis et Applicata* 2: 77–81.

Barbosa, P., and M. R. Wagner. 1989. *Introduction to Forest and Shade Tree Insects.* Academic Press, San Diego, CA.

Bardgett, R. D., D. K. Leemans, R. Cook, and P. J. Hobbs. 1997. Seasonality of the soil biota of grazed and ungrazed hill grasslands. *Soil Biology and Biochemistry* 29: 1285–1294.

Bardgett, R. D., D. A. Wardle, and G. W. Yeates. 1998. Linking above-ground and below-ground interactions: How plant responses to foliar herbivory influence soil organisms. *Soil Biology and Biochemistry* 30: 1867–1878.

Barras, S. J. 1970. Antagonism between *Dendroctonus frontalis* and the fungus *Ceratocystis minor. Annals of the Entomological Society of America* 63: 1187–1190.

Barz, W., and K. Weltring. 1985. Biodegradation of aromatic extractives of wood. In *Biosynthesis and Biodegradation of Wood Components* (T. Higuchi, Ed.), pp. 607–666. Academic Press, New York.

Baskerville, G. L., and P. Emin. 1969. Rapid estimation of heat accumulation from maximum and minimum temperatures. *Ecology* 50: 514–522.

Basset, Y. 1996. Local communities of arboreal herbivores in Papua New Guinea: Predictors of insect variables. *Ecology* 77: 1906–1919,

Batra, L. R. 1966. Ambrosia fungi: Extent of specificity to ambrosia beetles. *Science* 153: 193–195.

Bawa, K. S. 1990. Plant–pollinator interactions in tropical rain forests. *Annual Review of Ecology and Systematics* 21: 399–422.

Bayliss-Smith, T. P. 1990. The integrated analysis of seasonal energy deficits: Problems and prospects. *European Journal of Clinical Nutrition* 44 (supplement 1): 113–121.

Bazykin, A. D., F. S. Berezovskaya, A. S. Isaev, and R. G. Khlebopros. 1997. Dynamics of forest insect density: Bifurcation approach. *Journal of Theoretical Biology* 186: 267–278.

Bazzaz, F. A. 1975. Plant species diversity in old-field successional ecosystems in southern Illinois. *Ecology* 56: 485–488.

Bazzaz, F. A. 1990. The response of natural ecosystems to the rising global CO_2 levels. *Annual Review of Ecology and Systematics* 21: 167–196

Beare, M. H., D. C. Coleman, D. A. Crossley, Jr., P. F. Hendrix, and E. P. Odum. 1995. A hierarchical approach to evaluating the significance of soil biodiversity to biogeochemical cycling. *Plant and Soil* 170: 5–22.

Becerra, J. X. 1994. Squirt-gun defense in *Bursera* and the chrysomelid counterploy. *Ecology* 75: 1991–1996.

Becerra, J. X. 1997. Insects on plants: Macroevolutionary chemical trends in host use. *Science* 276: 253–256.

Begon, M., and M. Mortimer. 1981. *Population Ecology: A Unified Study of Animals and Plants*. Blackwell Scientific, Oxford, U. K.

Behan, V. M., and S. B. Hill. 1978. Feeding habits and spore dispersal of oribatid mites in the North American Arctic. *Revue d'Ecologie et Biologie du Sol* 15: 497–516.

Bell, W. J. 1990. Searching behavior patterns in insects. *Annual Review of Entomology* 35: 447–467.

Belnap, J., and D. A. Gillette. 1998. Vulnerability of desert biological soil crusts to wind erosion: The influences of crust development, soil texture, and disturbance. *Journal of Arid Environments* 39: 133–142.

Belsky, A. J. 1986. Does herbivory benefit plants? A review of the evidence. *American Naturalist* 127: 870–892.

Benedict, F. 1976. *Herbivory Rates and Leaf Properties in Four Forests in Puerto Rico and Florida*. Ph. D. Dissertation, University of Florida, Gainesville, FL.

Benke, A. C., and J. B. Wallace. 1997. Trophic basis of production among riverine caddisflies: Implications for food web analysis. *Ecology* 78: 1132–1145.

Bennett, A., and J. Krebs. 1987. Seed dispersal by ants. *Trends in Ecology and Evolution* 2:291–292.

Bernays, E. A. 1989. Insect–Plant Interactions. CRC Press, Boca Raton, FL.

Bernays, E. A., and R. F. Chapman. 1994. *Host-plant Selection by Phytophagous Insects*. Chapman and Hall, New York.

Bernays, E. A., and S. Woodhead. 1982. Plant phenols utilized as nutrients by a phytophagous insect. *Science* 216: 201–203.

Bernhard-Reversat, F. 1982. Measuring litter decomposition in a tropical forest ecosystem: Comparison of some methods. *International Journal of Ecology and Environmental Science* 8: 63–71.

Berryman, A. A. 1981. *Population Systems: A General Introduction*. Plenum Press, New York.

Berryman, A. A. 1996. What causes population cycles of forest Lepidoptera? *Trends in Ecology and Evolution* 11: 28–32.

Berryman, A. A. 1997. On the principles of population dynamics and theoretical models. *American Entomologist* 43: 147–151.

Berryman, A. A., N. C. Stenseth, and A. S. Isaev. 1987. Natural regulation of herbivorous forest insect populations. *Oecologia* 71: 174–184.

Birk, E. M. 1979. Disappearance of overstorey and understorey litter in an open eucalypt forest. *Australian Journal of Ecology* 4: 207–222.

Birks, H. J. B. 1980. British trees and insects: A test of the time hypothesis over the last 13,000 years. *American Naturalist* 115: 600–605.

Blanchette, R. A., and C. G. Shaw. 1978. Associations among bacteria, yeasts, and basiodiomycetes during wood decay. *Phytopathology* 68: 631–637.

Blanton, C. M. 1990. Canopy arthropod sampling: A comparison of collapsible bag and fogging methods. *Journal of Agricultural Entomology* 7: 41–50.

Blum, M. S. 1980. Arthropods and ecomones: Better fitness through ecological chemistry. In *Animals and Environmental Fitness* (R. Gilles, Ed.), pp. 207–222. Pergamon Press, Oxford, UK.

Blum, M. S. 1981. *Chemical Defenses of Arthropods*. Academic Press, New York.

Blum, M. S. 1992. Ingested allelochemicals in insect wonderland: A menu of remarkable functions. *American Entomologist* 38: 222–234.

Blumer, P., and M. Diemer. 1996. The occurrence and consequences of grasshopper herbivory in an alpine grassland, Swiss central Alps. *Arctic and Alpine Research* 28: 435–440.

Boecklen, W. J. 1991. The conservation status of insects: mass extinction, scientific interest, and statutory protection. In *Entomology Serving Society: Emerging Technologies and Challenges* (S. B. Vinson and R. L. Metcalf, Eds.), pp. 40–57. Entomological Society of America, Lanham, MD.

Boethel, D. J., and R. D. Eikenbary, Eds. 1986. *Interactions of Plant Resistance and Parasitoids and Predators of Insects*. Ellis Horwood Ltd., Chichester, UK.

Bond, W. J. 1993. Keystone species. In *Biodiversity and Ecosystem Function* (E. D. Schulze and H. A. Mooney, Eds.), pp. 237–253. Springer-Verlag, Berlin, Germany.

Boorman, S. A., and P. R. Levitt. 1972. Group selection on the boundary of a stable population. *Proceedings of the National Academy of Sciences, USA* **69**: 2711–2713.

Boring, L. R., W. T. Swank, and C. D. Monk. 1988. Dynamics of early successional forest structure and processes in the Coweeta Basin. In *Forest Hydrology and Ecology at Coweeta* (W. T. Swank and D. A. Crossley, Jr., Eds.), pp. 161–179. Springer-Verlag, New York.

Bormann, F. H., and G. E. Likens. 1979. *Pattern and Process in a Forested Ecosystem.* Springer-Verlag, New York.

Botkin, D. B. 1981. Causality and succession. In *Forest Succession: Concepts and Application* (D. C. West, H. H. Shugart and D. B. Botkin, Eds.), pp. 36–55. Springer-Verlag, New York.

Boucot, A. J. 1990. *Evolutionary Paleobiology of Behavior and Coevolution.* Elsevier, Amsterdam.

Bowers, M. D., and G. M. Puttick. 1988. Response of generalist and specialist insects to qualitative allelochemical variation. *Journal of Chemical Ecology* **14**: 319–334.

Boyce, M. S. 1984. Restitution of r- and K-selection as a model of density-dependent natural selection. *Annual Review of Ecology and Systematics* **15**: 427–447.

Bozer, S. F., M. S. Traugott, and N. E. Stamp. 1996. Combined effects of allelochemical-fed and scarce prey of the generalist insect predator *Podisus maculiventris.* *Ecological Entomology* **21**: 328–334.

Bradshaw, J. W. S., and P. E. Howse. 1984. Sociochemicals of ants. In *Chemical Ecology of Insects* (W. J. Bell and R. T. Cardé, Eds.), pp. 429–473. Chapman and Hall, London, UK.

Braithwaite, R. W., and J. A. Estbergs. 1985. Fire patterns and woody litter vegetation trends in the Alligator Rivers region of northern Australia. In *Ecology and Management of the World's Savannahs* (J. C. Tothill and J. J. Mott, Eds.), pp. 359–364. Australian Academy of Science, Canberra, Australia.

Brauman, A., M. D. Kane, M. Labat, and J. A. Breznak. 1992. Genesis of acetate and methane by gut bacteria of nutritionally diverse termites. *Science* **257**: 1384–1387.

Breznak, J. A., and A. Brune. 1994. Role of microorganisms in the digestion of lignocellulose by termites. *Annual Review of Entomology* **39**: 453–487.

Briand, F., and J. E. Cohen. 1984. Community food webs have scale-invariant structure. *Nature* **307**: 264–267.

Bridges, J. R. 1983. Mycangial fungi of *Dendroctonus frontalis* (Coleoptera: Scolytidae) and their relationship to beetle population trends. *Environmental Entomology* **12**: 858–861.

Bridges, J. R., and J. C. Moser. 1983. Role of two phoretic mites in transmission of bluestain fungus, *Ceratocystis minor. Ecological Entomology* **8**: 9–12.

Bridges, J. R., and J. C. Moser. 1986. Relationship of phoretic mites (Acari: Tarsonemidae) to the bluestaining fungus, *Ceratocystis minor,* in trees infested by southern pine beetle (Coleoptera: Scolytidae). *Environmental Entomology* **15**: 951–953.

Bridges, J. R., and T. J. Perry. 1985. Effects of mycangial fungi on gallery construction and distribution of bluestain in southern pine beetle-infested pine bolts. *Journal of Entomological Science* **20**: 271–275.

Bridges, J. R., W. A. Nettleton, and M. D. Connor. 1985. Southern pine beetle (Coleoptera: Scolytidae) infestations without the bluestain fungus, *Ceratocystis minor. Journal of Economic Entomology* **78**: 325–327.

Bristow, C. M. 1991. Why are so few aphids ant-tended? In *Ant–Plant Interactions* (C. R. Huxley and D. F. Cutler, Eds.), pp. 104–119. Oxford Univ. Press, Oxford, UK.

Brokaw, N. V. L. 1985. Treefalls, regrowth, and community structure in tropical forests. In *The Ecology of Natural Disturbance and Patch Dynamics* (S. T. A. Pickett and P. S. White, Eds.), pp. 53–69. Academic Press, Orlando, FL.

Bronstein, J. L. 1998. The contribution of ant–plant protection studies to our understanding of mutualism. *Biotropica* **30**: 150–161.

Brower, A. V. Z. 1996. Parallel race formation and the evolution of mimicry in *Heliconius* butterflies: A phylogenetic hypothesis from mitochondrial DNA sequences. *Evolution* **50**: 195–221.

Brower, L. P., J. V. Z. Brower, and F. P. Cranston. 1965. Courtship behavior of the queen butterfly, *Danaus gilippus berenice* (Cramer). *Zoologica* **50**:1–39.

Brower, L. P., W. N. Ryerson, L. L. Coppinger, and S. C. Glazier. 1968. Ecological chemistry and the palatability spectrum. *Science* **161**: 1349–1351.

Brown, B. J., and J. J. Ewel. 1987. Herbivory in complex and simple tropical successional ecosystems. *Ecology* **68**: 108–116.

Brown, J. H., O. J. Reichman, and D. W. Davidson. 1979. Granivory in desert ecosystems. *Annual Review of Ecology and Systematics* **10**: 201–227.

Brown, K. S., and J. R. Trigo. 1995. The ecological activity of alkaloids. In *The Alkaloids: Chemistry and Pharmacology* (G. A. Cordell, Ed.), pp. 227–354. Academic Press, San Diego.

Brown, M. V., T. E. Nebeker, and C. R. Honea. 1987. Thinning increases loblolly pine vigor and resistance to bark beetles. *Southern Journal of Applied Forestry* **11**: 28–31.

Brown, S., and A. E. Lugo. 1982. Storage and production of organic matter in tropical forests and their role in the global carbon cycle. *Biotropica* **14**: 161–187.

Brown, V. C. 1995. Insect herbivores and gaseous air

pollutants—Current knowledge and predictions. In *Insects in a Changing Environment* (R. Harrington and N. E. Stork, Eds.), pp. 219–249. Academic Press, London, UK.

Brown, V. K. 1982. Size and shape as ecological discriminants in successional communities of Heteroptera. *Biological Journal of the Linnean Society* **18**: 279–290.

Brown, V. K. 1984. Secondary succession: Insect–plant relationships. *BioScience* **34**: 710–716.

Brown, V. K. 1986. Life cycle strategies and plant succession. In *The Evolution of Insect Life Cycles* (F. Taylor and R. Karban, Eds.), pp. 105–124. Springer-Verlag, New York.

Brown, V. K., and A. C. Gange. 1989. Differential effects of above- and below-ground insect herbivory during early plant succession. *Oikos* **54**: 67–76.

Brown, V. K., and A. C. Gange. 1991. Effects of root herbivory on vegetation dynamics. In *Plant Root Growth: An Ecological Perspective* (D. Atkinson, Ed.), pp. 453–470. Blackwell Scientific, Oxford, UK.

Brown, V. K., and P. S. Hyman. 1986. Successional communities of plants and phytophagous Coleoptera. *Journal of Ecology* **74**: 963–975.

Brown, V. K., and M. Llewellyn. 1985. Variation in aphid weight and reproductive potential in relation to plant growth form. *Journal of Animal Ecology* **54**: 651–661.

Brown, V. K., and T. R. E. Southwood. 1983. Trophic diversity, niche breadth and generation times of exopterygote insects in a secondary succession. *Oecologia* **56**: 220–225.

Brown, V. K., A. C. Gange, I. M. Evans, and A. L. Storr. 1987. The effect of insect herbivory on the growth and reproduction of two annual *Vicia* species at different stages in plant succession. *Journal of Ecology* **75**: 1173–1189.

Brown, V. K., M. Jepson, and C. W. D. Gibson. 1988. Insect herbivory: Effects on early old field succession demonstrated by chemical exclusion methods. *Oikos* **52**: 293–302.

Bullock, S. H. 1991. Herbivory and the demography of the chaparral shrub *Ceanothus greggii* (Rhamnaceae). *Madroňo* **38**: 63–72.

Byers, J. A., and D. L. Wood. 1981 Antibiotic-induced inhibition of pheromone synthesis in a bark beetle. *Science* **213**: 763–764.

Camilo, G. R., and M. R. Willig. 1995. Dynamics of a food chain model from an arthropod-dominated lotic community. *Ecological Modelling* **79**: 121–129.

Capinera, J. L. 1987. Population ecology of rangeland grasshoppers. In *Integrated Pest Management on Rangeland: A Shortgrass Prairie Perspective* (J. L. Capinera, Ed.), pp. 162–182. Westview Press, Boulder, CO.

Cappuccino, N. 1992. The nature of population sta-

bility in *Eurosta solidaginis*, a nonoutbreaking herbivore of goldenrod. *Ecology* **73**: 1792–1801.

Cardé, R. T. 1996. Odour plumes and odour-mediated flight in insects. In *Olfaction in Mosquito–Host Interactions* (Ciba Foundation Symposium 200), pp. 54–70. John Wiley & Sons, Chichester, UK.

Cardé, R. T. and T. C. Baker. 1984. Sexual communication with pheromones. In *Chemical Ecology of Insects* (W. J. Bell and R. T. Cardé, Eds.), pp. 355–383. Chapman and Hall, London, UK.

Carpenter, S. R., and J. F. Kitchell. 1984. Plankton community structure and limnetic primary production. *American Naturalist* **124**: 159–172.

Carpenter, S. R., and J. F. Kitchell. 1987. The temporal scale of variance in lake productivity. *American Naturalist* **129**: 417–433.

Carpenter, S. R., and J. F. Kitchell. 1988. Consumer control of lake productivity. *BioScience* **38**: 764–769.

Carpenter, S. R., J. F. Kitchell, and J. R. Hodgson. 1985. Cascading trophic interactions and lake productivity. *BioScience* **35**: 634–639.

Carroll, C. R., and C. A. Hoffman. 1980. Chemical feeding deterrent mobilized in response to insect herbivory and counter adaptation by *Epilachna trededimnotata*. *Science* **209**: 414–416.

Carroll, G. 1988. Fungal endophytes in stems and leaves: From latent pathogen to mutualistic symbiont. *Ecology* **69**: 2–9.

Carter, J. L., S. V. Fend, and S. S. Kennelly. 1996. The relationships among three habitat scales and stream benthic invertebrate community structure. *Freshwater Biology* **35**: 109–124.

Casey, T. M. 1988. Thermoregulation and heat exchange. *Advances in Insect Physiology* **20**: 119–146.

Cates, R. G. 1980. Feeding patterns of monophagous, oligophagous, and polyphagous insect herbivores: the effect of resource abundance and plant chemistry. *Oecologia* **46**: 22–31.

Cavalieri, L. F., and H. Koçak. 1994. Chaos in biological control systems. *Journal of Theoretical Biology* **169**: 179–187.

Cebrián, J., and C. M., Duarte. 1994. The dependence of herbivory on growth rate in natural plant communities. *Functional Ecology* **8**: 518–525.

Chabot, B. F., and D. J. Hicks. 1982. The ecology of leaf life spans. *Annual Review of Ecology and Systematics* **13**: 229–259.

Chapin, F. S., III, A. J. Bloom, C. B. Field, and R. H. Waring. 1987. Plant responses to multiple environmental factors. *BioScience* **37**: 49–57.

Chapman, R. F. 1982. *The Insects: Structure and Function*, 3rd ed. Harvard Univ. Press, Cambridge, MA.

Chapman, R. F., and E. A. Bernays. 1989. Insect behavior at the leaf surface and learning aspects of host plant selection. *Experientia* **45**: 215–222.

Chase, J. M. 1996. Abiotic controls of trophic cascades in a simple grassland food chain. *Oikos* **77**: 495–506.

Chase, T. N., R. A. Pielke, T. G. F. Kittel , R. Nemani, and S. W. Running. 1996. Sensitivity of a general circulation model to global changes in leaf area index. *Journal of Geophysical Research* **101**: 7393–7408.

Chazdon, R., and N. Fetcher 1984. Photosynthetic light environments in a lowland tropical rainforest in Costa Rica. *Journal of Ecology* **72**: 553–564.

Chen, J., J. F. Franklin, and T. A. Spies. 1995. Growing-season microclimatic gradients from clearcut edges into old-growth Douglas-fir forests. *Ecological Applications* **5**: 74–86

Chittka, L., and R. Menzel. 1992. The evolutionary adaptation of flower colours and the insect pollinators' colour vision. *Journal of Comparative Physiology A.* **171**: 171–181.

Christensen, K. M., and T. G. Whitham. 1991. Indirect herbivore mediation of avian seed dispersal in pinyon pine. *Ecology* **72**: 534–542.

Clark, D. B., and D. A. Clark. 1985. Seedling dynamics of a tropical tree: impacts of herbivory and meristem damage. *Ecology* **66**: 1884–1892.

Clark, L. R., P. W. Geier, R. D. Hughes, and R. F. Morris. 1967. *The Ecology of Insect Populations in Theory and Practice*. Methuen, London, UK.

Clark, W. C. 1979. Spatial structure relationship in a forest insect system: Simulation models and analysis. *Mittelungen der Schweizerischen Entomologischen Gesellschaft* **52**: 235–257.

Clarke, C. M., and R. L. Kitching. 1995. Swimming ants and pitcher plants: A unique ant–plant interaction from Borneo. *Journal of Tropical Ecology* **11**: 589–602.

Clay, K. 1990. Fungal endophytes of grasses. *Annual Review of Ecology and Systematics* **21**: 275–297.

Clay, K., S. Marks, and G. P. Cheplick. 1993. Effects of insect herbivory and fungal endophyte infection on competitive interactions among grasses. *Ecology* **74**: 1767–1777.

Clements, F. E. 1916. *Plant Succession: An Analysis of the Development of Vegetation*. Carnegie Institute of Washington Publication 242, Washington, DC.

Codella, S. G., Jr., and K. F. Raffa. 1993. Defense strategies of folivorous sawflies. In *Sawfly Life History Adaptations to Woody Plants*. (M. R. Wagner and K. F. Raffa, Eds.), pp. 261–294. Academic Press, San Diego, CA.

Coe, M. 1977. The role of termites in the removal of elephant dung in the Tsavo (East) National Park Kenya. *East African Wildlife Journal* **15**: 49–55.

Cohen, J. E., and Z. J. Palka. 1990. A stochastic theory of community food webs. V. Intervality and triangulation in the trophic-niche overlap graph. *American Naturalist* **135**: 435–463.

Cohen, J. E., F. Briand, and C. M. Newman. 1990. *Community Food Webs: Data and Theory*. Springer-Verlag, Berlin, Germany.

Colbert, J. J., and R. W. Campbell. 1978. The inte-grated model. In *The Douglas-fir Tussock Moth: A Synthesis*. (M. H. Brookes, R. W. Stark and R. W. Campbell, Eds.), pp. 216–230. USDA Forest Service, Tech. Bull. 1585, USDA Forest Service, Washington, DC.

Colegrave, N. 1997. Can a patchy population structure affect the evolution of competition strategies? *Evolution* **51**: 483–492.

Coley, P. D. 1980. Effects of leaf age and plant life history patterns on herbivory. *Nature* **284**: 545–546.

Coley, P. D. 1982. Rates of herbivory on different tropical trees. In *The Ecology of a Tropical Forest: Seasonal Rhythms and Long-term Changes*. (E. G. Leigh, Jr., A. S. Rand, and D. M. Windsor, Eds.), pp. 123–132. Smithsonian Institution Press, Washington, DC.

Coley, P. D. 1983. Herbivory and defensive characteristics of tree species in a lowland tropical forest. *Ecological Monographs* **53**: 209–233.

Coley, P. D. 1986. Costs and benefits of defense by tannins in a neotropical tree. *Oecologia* **70**: 238–241.

Coley, P. D., and T. M. Aide. 1991. Comparison of herbivory and plant defenses in temperate and tropical broad-leaved forests. In *Plant-Animal Interactions: Evolutionary Ecology in Tropical and Temperate Regions*. (P. W. Price, T. M. Lewinsohn, G. W. Fernandes, and W. W. Benson, Eds.), pp. 25–49. John Wiley & Sons, Inc., New York.

Coley, P. D., and J. A. Barone. 1996. Herbivory and plant defenses in tropical forests. *Annual Review of Ecology and Systematics* **27**: 305–335.

Coley, P. D., J. P. Bryant, and F. S. Chapin III. 1985. Resource availability and plant antiherbivore defense. *Science* **230**: 895–899.

Collins, N. C., R. Mitchell, and R. G. Wiegert. 1976. Functional analysis of a thermal spring ecosystem, with an evaluation of the role of consumers. *Ecology* **57**: 1221–1232.

Collins, N. M. 1981. The role of termites in the decomposition of wood and leaf litter in the southern Guinea savanna of Nigeria. *Oecologia* **51**: 389–399.

Collins, N. M. 1983. Termite populations and their role in litter removal in Malaysian rain forests. In *Tropical Rain Forest: Ecology and Management* (S. L. Sutton, T. C. Whitmore and A. C. Chadwick, Eds.), pp. 311–325. Blackwell, London.

Connell, J. H. 1978. Diversity in tropical rain forests and coral reefs. *Science* **199**: 1302–1310.

Connell, J. H. 1980. Diversity and the coevolution of competitors, or the ghost of competition past. *Oikos* **35**: 131–138.

Connell, J. H. 1983. On the prevalence and relative importance of interspecific competition: Evidence from field experiments. *American Naturalist* **122**: 661–696.

Connell, J. H., and R. O. Slatyer. 1977. Mechanisms of succession in natural communities and their role in community stability and organization. *American Naturalist* **111**: 1119–1144.

Constantino, R. F., R. A. Desharnais, J. M. Cushing, and B. Dennis. 1997. Chaotic dynamics in an insect population. *Science* **275**: 389–391.

Corbet, P. S. 1962. *A Biology of Dragonflies*. H. F. & G. Witherby, London, UK.

Corbet, S. A. 1997. Role of pollinators in species preservation, conservation, ecosystem stability and genetic diversity. In *Pollination: From Theory to Practise* (K. W. Richards, Ed.), pp. 219–229. Proc. 7th International Symposium on Pollination. Acta Horticulturae #437.

Costanza, R., R. d'Arge, R. de Groot, S. Farger, M. Grasso, B. Hannon, K. Limburg, S. Naeem, R. V. O'Neill, J. Paruelo, R. G. Raskin, P. Sutton, and M. van den Belt. 1997. The value of the world's ecosystem services and natural capital. *Nature* **387**: 253–260.

Coulson, R. N. 1979. Population dynamics of bark beetles. *Annual Review of Entomology* 24:417–447.

Coulson, R. N., and D. A. Crossley, Jr. 1987. What is insect ecology? A commentary. *Bulletin of the Entomological Society of America* **33**: 64–68.

Coulson, R. N., D. N. Pope, J. A. Gagne, W. S. Fargo, P. E. Pulley, L. J. Edson, and T. L. Wagner. 1980. Impact of foraging by *Monochamus titillator* (Col.: Cerambycidae) on within-tree populations of *Dendroctonus frontalis* (Col.: Scolytidae). *Entomophaga* **25**: 155–170.

Coulson, R. N., P. B. Hennier, R. O. Flamm, E. J. Rykiel, L. C. Hu, and T. L. Payne. 1983. The role of lightning in the epidemiology of the southern pine beetle. *Zeitschrift für angewandte Entomologie* **96**: 182–193.

Coulson, R. N., R. O. Flamm, P. E. Pulley, T. L. Payne, E. J. Rykiel, and T. L. Wagner. 1986. Response of the southern pine bark beetle guild (Coleoptera: Scolytidae) to host disturbance. *Environmental Entomology* **15**: 850–858.

Coulson, R. N., J. W. Fitzgerald, B. A. McFadden, P. E. Pulley, C. N. Lovelady, and J. R. Giardino. 1996. Functional heterogeneity of forest landscapes: How host defenses influence epidemiology of the southern pine beetle. In *Dynamics of Forest Herbivory: Quest for Pattern and Principle*. (W. J. Mattson, P. Niemela, and M. Rousi, Eds.), USDA Forest Serv. Gen. Tech. Rep. NC-183. USDA Forest Serv., North Central Forest Exp. Stn., St. Paul, MN.

Courtney, S. P. 1985. Apparency in coevolving relationships. *Oikos* **44**: 91–98.

Courtney, S. P. 1986. The ecology of pierid butterflies: Dynamics and interactions. *Advances in Ecological Research* **15**: 51–131.

Coûteaux, M. M., M. Mousseau, M. L. Célérier, and P. Bottner. 1991. Increased atmospheric CO_2 and litter quality: Decomposition of sweet chestnut leaf litter with animal food webs of different complexities. *Oikos* **61**: 54–64.

Cowles, H. C. 1911. The causes of vegetative cycles. *Botanical Gazette* **51**: 161–183.

Cowling, R. M., S. M. Pierce, W. D. Stock, and M. Cocks. 1994. Why are there so many myrmecochorous species in the Cape fynbos? In *Plant–Animal Interactions in Mediterranean-type Ecosystems*. (M. Arianoutsou and R. H. Graves, Eds.), pp. 159–168. Kluwer, Dordrecht, Netherlands.

Coxson, D. S., and N. M. Nadkarni. 1995. Ecological roles of epiphytes in nutrient cycles of forest canopies. In *Forest Canopies* (M. D. Lowman and N. M. Nadkarni, Eds.), pp. 495–543. Academic Press, San Diego, CA.

Crawford, C. S. 1978. Seasonal water balance in *Orthoporus ornatus*, a desert millipede. *Ecology* **59**: 996–1004.

Crawford, C. S. 1986. The role of invertebrates in desert ecosystems. In *Pattern and Process in Desert Ecosystems*. (W. G. Whitford, Ed.), pp. 73–91. University of New Mexico Press, Albuquerque, NM.

Crawley, M. J. 1983. *Herbivory: The Dynamics of Animal-plant Interactions*. Univ. of California Press, Berkeley.

Crawley, M. J. 1989. Insect herbivores and plant population dynamics. *Annual Review of Entomology* **34**: 531–564.

Croft, B. A. 1990. *Arthropod Biological Control Agents and Pesticides*. Wiley, New York.

Croft, B. A., and A. P. Gutierrez. 1991. Systems analysis role in modeling and decision-making. In *Entomology Serving Society: Emerging Technologies and Challenges* (S. B. Vinson and R. L. Metcalf, Eds.), pp. 298–319. Entomological Society of America, Lanham, MD.

Croft, B. A., and D. H. Slone. 1997. Equilibrium densities of European red mite (Acari: Tetranychidae) after exposure to three levels of predaceous mite diversity on apple. *Environmental Entomology* **26**: 391–399.

Cromack, K., Jr. 1973. *Litter Production and Decomposition in a Mixed Hardwood Watershed at Coweeta Hydrologic Station, North Carolina*. Ph. D. Dissertation, Univ. of Georgia, Athens.

Cromack, K., Jr., and C. D. Monk. 1975. Litter production, decomposition, and nutrient cycling in a mixed-hardwood watershed and a white pine watershed. In *Mineral Cycling in Southeastern Ecosystems* (F. G. Howell, J. B. Gentry, and M. H. Smith, Eds.), pp. 609–624. U.S. Energy Research and Development Administration. Technical Information Center, Washington, DC.

Cromack, K., Jr., R. L. Todd, and C. D. Monk. 1975. Patterns of basidiomycete nutrient accumulation in conifer and deciduous forest litter. *Soil Biology and Biochemistry* **7**: 265–268.

Cromack, K., Jr., P. Sollins, R. L. Todd, D. A. Crossley, Jr., W. M. Fender, R. Fogel, and A. W. Todd. 1977. Soil microorganism–arthropod interactions: Fungi as major calcium and sodium sources. In *The Role of Arthropods in Forest Ecosystems* (W. J. Mattson, Ed.), pp. 78–84. Springer-Verlag, New York.

Cromartie, W. J., Jr. 1975. The effect of stand size and vegetational background on the colonization of cruciferous plants by herbivorous insects. *Journal of Applied Ecology* **12**: 517–533.

Crossley, D. A., Jr. 1977. The roles of terrestrial saprophagous arthropods in forest soils: Current status of concepts. In *The Role of Arthropods in Forest Ecosystems* (W. J. Mattson, Ed.), pp. 49–56. Springer-Verlag, New York.

Crossley, D. A., Jr., and M. P. Hoglund. 1962. A litter-bag method for the study of microarthropods inhabiting leaf litter. *Ecology* **43**: 571–573.

Crossley, D. A., Jr., and H. F. Howden. 1961. Insect–vegetation relationships in an area contaminated by radioactive wastes. *Ecology* **42**: 302–317.

Crossley, D. A., Jr. and M. Witkamp. 1964. Effects of pesticide on biota and breakdown of forest litter. In *8th International Congress of Soil Science*, Bucharest, Romania, pp. 887–892. Publishing House of the Academy of the Socialist Republic of Romania.

Crossley, D. A., Jr., E. R. Blood, P. F. Hendrix, and T. R. Seastedt. 1995. Turnover of cobalt-60 by earthworms (*Eisenia foetida*) (Lumbricidae, Oligochaeta). *Applied Soil Ecology* **2**: 71–75.

Cuffney, T. F., J. B. Wallace, and G. J. Lugthart. 1990. Experimental evidence quantifying the role of benthic invertebrates in organic matter dynamics in headwater streams. *Freshwater Biology* **23**: 281–299.

Culver, D. C., and A. J. Beattie. 1980. The fate of *Viola* seeds dispersed by ants. *American Journal of Botany* **67**: 710–714.

Cummins, K. W. 1973. Trophic relations of aquatic insects. *Annual Review of Entomology* **18**: 183–206.

Curry, J. P. 1969. The decomposition of organic matter in the soil. I. The role of fauna in decaying grassland herbage. *Soil Biology and Biochemistry* **1**: 253–258.

Curry, J. P. 1994. *Grassland Invertebrates*. Chapman & Hall, London, UK.

Cushman, J. H., and J. F. Addicott. 1991. Conditional interactions in ant–plant–herbivore mutualisms. In *Ant–Plant Interactions* (C. R. Huxley and D. F. Cutler, Eds.), pp. 92–103. Oxford Univ. Press, Oxford, UK.

Daily, G. C., Ed. 1997. *Nature's Services: Societal Dependence on Natural Ecosystems*. Island Press, Washington, DC.

Dambacher, J. M., H. W. Li, J. O. Wolff, and P. A. Rossignol. 1999. Parsimonious interpretation of the impact of vegetation, food, and predation on snowshoe hare. *Oikos* **84**: 530–532.

Dangerfield, J. M. 1994. Ingestion of leaf litter by millipedes: The accuracy of laboratory estimates for predicting litter turnover in the field. *Pedobiologia* **38**: 262–265.

Darlington, P. J. 1943. Carabidae of mountains and islands: Data on the evolution of isolated faunas, and on atrophy of wings. *Ecological Monographs* **13**: 37–61.

Darwin, C. 1859. *The Origin of Species by Means of Natural Selection or the Preservation of Favored Races in the Struggle for Life*. Murray, London, UK.

Davidson, D. W. 1993. The effects of herbivory and granivory on terrestrial plant succession. *Oikos* **68**: 23–35.

Davidson, D. W., and B. L. Fisher. 1991. Symbiosis of ants with *Cecropia* as a function of light regime. In *Ant–Plant Interactions* (C. R. Huxley and D. F. Cutler, Eds.), pp. 289–309. Oxford Univ. Press, Oxford, UK.

Davidson, D. W., R. S. Inouye, and J. H. Brown. 1984. Granivory in a desert ecosystem: Experimental evidence for indirect facilitation of ants by rodents. *Ecology* **65**: 1780–1786.

Davison, E. A. 1987. Respiration and energy flow in two Australian species of desert harvester ants, *Chelaner rothsteini* and *Chelaner whitei*. *Journal of Arid Environments* **12**: 61–82.

Day, M. F., and H. Irzykiewicz. 1953. Feeding behavior of the aphids *Myzus persicae* and *Brevicoryne brassicae*, studied with radiophosphorus. *Australian Journal of Biological Science* **6**: 98–108.

Day, M. F., and A. McKinnon. 1951. A study of some aspects of the feeding of the jassid *Orosius*. *Australian Journal of Scientific Research (B)* **4**: 125–135.

Dean, A. M. 1983. A simple model of mutualism. *American Naturalist* **121**: 409–417.

de Carvalho, E. L., and M. Kogan. 1991. Order Strepsiptera. In *Immature Insects. Vol. 2* (F. Stehr, Ed.), pp. 659–673. Kendell/Hunt, Dubuque, IA.

de la Cruz, M., and R. Dirzo. 1987. A survey of the standing levels of herbivory in seedlings from a Mexican rain forest. *Biotropica* **19**: 98–106.

Deevy, E. S. 1947. Life tables for natural populations of animals. *Quarterly Review of Biology* **22**: 283–314.

Delphin, F. 1965. The histology and possible functions of neurosecretory cells in the ventral ganglia of *Schistocerca gregaria* Forskal (Orthoptera: Acrididae). *Transactions of the Royal Entomological Society London* **117**: 167–214.

DeMers, M. N. 1993. Roadside ditches as corridors for range expansion of the western harvester ant (*Pogonomyrmex occidentalis* Cresson). *Landscape Ecology* **8**: 93–102.

Denno, R. F., M. S. McClure, and J. R. Ott. 1995. Interspecific interactions in phytophagous insects: Competition reexamined and resurrected. *Annual Review of Entomology* 40: 297–331.

Denslow, J. S. 1985. Disturbance-mediated coexistence of species. In *Ecology of Natural Disturbance and Patch Dynamics* (S. T. A. Pickett and P. S. White, Eds.), pp. 307–323. Academic Press, Orlando, FL.

Denslow, J. S. 1995. Disturbance and diversity in tropical rain forests: The density effect. *Ecological Applications* 5: 962–968.

de Ruiter, P. C., A. M. Neutel, and J. C. Moore. 1995. Energetics, patterns of interaction strengths, and stability in real ecosystems. *Science* 269: 1257–1260.

de Souza-Stevaux, M. C., R. R. B. Negrelle, and V. Citadini-Zanette. 1994. Seed dispersal by the fish *Pterodoras granulosus* in the Paraná River Basin, Brazil. *Journal of Tropical Ecology* 10: 621–626.

Detling, J. K. 1987. Grass response to herbivory. In *Integrated Pest Management on Rangeland: A Shortgrass Prairie Perspective* (J. L. Capinera, Ed.), pp. 56–68. Westview Press, Boulder, CO.

Detling, J. K. 1988. Grasslands and savannas: regulation of energy flow and nutrient cycling by herbivores. In *Ecosystems: Analysis and Synthesis* (L. R. Pomeroy and J. A. Alberts, Eds.), pp. 131–148. Springer-Verlag, New York.

Detling, J. K., M. I. Dyer, C. Procter-Gregg, and D. T. Winn. 1980. Plant-herbivore interactions: examination of potential effects of bison saliva on regrowth of *Bouteloua gracilis* (H. B. K.) Lag. *Oecologia* 45: 26–31.

Dévai, G., and J. Moldován. 1983. An attempt to trace eutrophication in a shallow lake (Balaton, Hungary) using chironomids. *Hydrobiologia* 103: 169–175.

Dial, R., and J. Roughgarden. 1995. Experimental removal of insectivores from rain forest canopy: Direct and indirect effects. *Ecology* 76: 1821–1834.

Dial, R., and J. Roughgarden. 1996. Natural history observations of *Anolisomyia rufianalis* (Diptera: Sarcophagidae) infesting *Anolis* lizards in a rain forest canopy. *Environmental Entomology* 25: 1325–1328.

Diamond, J. M., and R. M. May. 1981. Island biogeography and the design of natural reserves. In *Theoretical Ecology: Principles and Applications* (R. M. May, Ed.), pp. 228–252. Blackwell, Oxford, UK.

Didham, R. K., J. Ghazoul, N. E. Stork, and A. J. Davis. 1996. Insects in fragmented forests: a functional approach. *Trends in Ecology and Evolution* 11: 255–260.

Dik, A. J., and J. A. van Pelt. 1993. Interaction between phyllosphere yeasts, aphid honeydew and fungicide effectiveness in wheat under field conditions. *Plant Pathology* 41: 661–675.

Dirzo, R. 1984. Herbivory: A phytocentric overview. In *Perspectives on Plant Population Ecology* (R. Dirzo and J. Sarukhán, Eds.), pp. 141–165. Sinauer Assoc., Inc., Sunderland, MA.

Dixon, A. F. G. 1985. *Aphid Ecology*. Blackie & Son Ltd., Glasgow, UK.

Douce, G. K., and D. A. Crossley, Jr. 1982. The effect of soil fauna on litter mass loss and nutrient loss dynamics in arctic tundra at Barrow, Alaska. *Ecology* 63: 523–537.

Dowding, P. 1984. The evolution of insect–fungus relationships in the primary invasion of forest timber. In *Invertebrate Microbial Interactions* (J. M. Anderson, A. D. M. Rayner, and D. W. H. Walton, Eds.), pp. 135–153. British Mycological Society Symposium 6, Cambridge Univ. Press, Cambridge, UK.

Downes, J. A. 1970. The feeding and mating behaviour of the specialized Empididae (Diptera): Observations on four species of *Rhamphomyia* in the high Arctic and a general discussion. *Canadian Entomologist* 102: 769–791.

Doyle, T. W. 1981. The role of disturbance in the gap dynamics of a montane rain forest: an application of a tropical forest succession model. In *Forest Succession: Concepts and Application* (D. C. West, H. H. Shugart, and D. B. Botkin, Eds.), pp. 56–73. Springer-Verlag, New York.

Dreisig, H. 1988. Foraging rate of ants collecting honeydew or extrafloral nectar and some possible constraints. *Ecological Entomology* 13: 143–154.

Drury, W. H., and I. C. T. Nisbet. 1973. Succession. *Journal of the Arnold Arboretum* 54: 331–368.

Dudt, J. F., and D. J. Shure. 1994. The influence of light and nutrients on foliar phenolics and insect herbivory. *Ecology* 75: 86–98.

Dyer, L. A. 1995. Tasty generalists and nasty specialists? Antipredator mechanisms in tropical lepidopteran larvae. *Ecology* 76: 1483–1496.

Dyer, M. I., M. A. Acra, G. M. Wang, D. C. Coleman, D. W. Freckman, S. J. McNaughton, and B. R. Strain. 1991. Source-sink carbon relations in two *Panicum coloratum* ecotypes in response to herbivory. *Ecology* 72: 1472–1483.

Dyer, M. I., C. L. Turner, and T. R. Seastedt. 1993. Herbivory and its consequences. *Ecological Applications* 3: 10–16.

Dyer, M. I., A. M. Moon, M. R. Brown, and D. A. Crossley, Jr. 1995. Grasshopper crop and midgut extract effects on plants: An example of reward feedback. *Proceedings of the National Academy of Sciences USA* 92: 5475–5478.

Edmonds, R. L., and A. Eglitis. 1989. The role of the Douglas-fir beetle and wood borers in the decomposition of and nutrient release from Douglas-fir logs. *Canadian Journal of Forest Research* 19: 853–859.

Edmunds, G. F., Jr., and D. N. Alstad. 1978. Coevolution in insect herbivores and conifers. *Science* 199: 941–945.

Edmunds, G. F., Jr., and D. N. Alstad. 1985. Malathion-induced sex ratio changes in black pine-leaf scale (Hemiptera: Diaspididae). *Annals of the Entomological Society of America* 78: 403–405.

Edney, E. B. 1974. Desert arthropods. In *Desert Biology* (G. W. Brown, Ed.), pp. 311–384. Academic Press, New York.

Edson, K. M., S. B. Vinson, D. B. Stoltz, and M. D. Summers. 1981. Virus in a parasitoid wasp: Suppression of the cellular immune response in the parasitoid's host. *Science* 211: 582–583.

Edwards, C. A., and G. W. Heath. 1963. The role of soil animals in breakdown of leaf material. In *Soil Organisms* (J. Doeksen and J. van der Drift, Eds.), pp. 76–84. North-Holland, Amsterdam, Netherlands.

Edwards, C. A., and J. R. Lofty. 1978. The influence of arthropods and earthworms upon root growth of direct drilled cereals. *Journal of Applied Ecology* 15: 789–795.

Edwards, E. P. 1982. Hummingbirds feeding on an excretion produced by scale insects. *Condor* 84: 122.

Edwards, J. S., and P. Sugg. 1990. Arthropod fallout as a resource in the recolonization of Mt. St. Helens. *Ecology* 74: 954–958.

Edwards, N. T. 1982. The use of soda-lime for measuring respiration rates in terrestrial systems. *Pedobiologia* 23: 321–330.

Edwards, P. J. 1989. Insect herbivory and plant defence theory. In *Toward a More Exact Ecology* (P. J. Grubb and J. B. Whittaker, Eds.), pp. 275–297. Blackwell Scientific, Oxford, UK.

Egler, F. E. 1954. Vegetation science concepts. I. Initial floristic composition, a factor in old-field vegetation development. *Vegetatio* 4: 412–417.

Ehrlén, J. 1996. Spatiotemporal variation in predispersal seed predation intensity. *Oecologia* 108: 708–713.

Eisner, T., R. E. Silberglied, D. Aneshansley, J. E. Carrel, and H. C. Howland. 1969. Ultraviolet video-viewing: The television camera as an insect eye. *Science* 166: 1172–1174.

Eldridge, D. J. 1993. Effect of ants on sandy soils in semi-arid eastern Australia: Local distribution of nest entrances and their effect on infiltration of water. *Australian Journal of Soil Research* 31: 509–518.

Eldridge, D. J. 1994. Nests of ants and termites influence infiltration in a semi-arid woodland. *Pedobiologia* 38: 481–492.

Elkins, N. Z., and W. G. Whitford. 1982. The role of microarthropods and nematodes in decomposition in a semi-arid ecosystem. *Oecologia* 55: 303–310.

Elkins, N. Z., G. V. Sabol, T. J. Ward, and W. G. Whitford. 1986. The influence of subterranean termites on the hydrological characteristics of a Chihuahuan Desert ecosystem. *Oecologia* 68: 521–528.

Elliott, E. T., D. C. Coleman, R. E. Ingham, and J. A. Trofymow. 1984. Carbon and energy flow through microflora and microfauna in the soil subsystem of terrestrial ecosystems. In *Current Perspectives in Microbial Ecology* (M. J. Klug and C. A. Reddy, Eds.), pp. 424–433. American Society for Microbiology, Washington, DC.

Elliott, N. C., G. A. Simmons, and F. J. Sapio. 1987. Honeydew and wildflowers as food for the parasites *Glypta fumiferanae* (Hymenoptera: Ichneumonidae) and *Apanteles fumiferanae* (Hymenoptera: Braconidae). *Journal of the Kansas Entomological Society* 60: 25–29.

Elton, C. 1939. *Animal Ecology*. Macmillan, New York.

Engleberg, J., and L. L. Boyarsky. 1979. The noncybernetic nature of ecosystems. *American Naturalist* 114: 317–324.

Ernsting, G., and D. C. van der Werf. 1988. Hunger, partial consumption of prey and prey size preference in a carabid beetle. *Ecological Entomology* 13: 155–164.

Erwin, T. L. 1995. Measuring arthropod diversity in the tropical forest canopy. In *Forest Canopies* (M. D. Lowman and N. M. Nadkarni, Eds.), pp. 109–127. Academic Press, San Diego, CA.

Ewel, J. J. 1986. Designing agricultural ecosystems for the humid tropics. *Annual Review of Ecology and Systematics* 17: 245–271.

Ewel, J. J., M. J. Mazzarino, and C. W. Berish. 1991. Tropical soil fertility changes under monocultures and successional communities of different structure. *Ecological Applications* 1: 289–302.

Faeth, S. H., E. F. Connor, and D. Simberloff. 1981. Early leaf abscission: A neglected source of mortality for folivores. *American Naturalist* 117: 409–415.

Fajer, E. D., M. D. Bowers, and F. A. Bazzaz. 1989. The effects of enriched carbon dioxide atmospheres on plant–insect herbivore interactions. *Science* 243: 1198–1200.

Fares, Y., P. J. H. Sharpe, and C. E. Magnusen. 1980. Pheromone dispersion in forests. *Journal of Theoretical Biology* 84: 335–359.

Fargo, W. S., T. L. Wagner, R. N. Coulson, J. D. Cover, T. McAudle, and T. D. Schowalter. 1982. Probability functions for components of the *Dendroctonus frontalis*-host tree population system and their potential use with population models. *Researches in Population Ecology* 24: 123–131.

Feener, D. H., Jr. 1981. Competition between ant species: Outcome controlled by parasitic flies. *Science* 214: 815–817.

Feeny, P. P. 1969. Inhibitory effect of oak leaf tannins on the hydrolysis of proteins by trypsin. *Phytochemistry* 8: 2119–2126.

Feeny, P. P. 1970. Seasonal changes in oak leaf tannins and nutrients as a cause of spring feeding by winter moth caterpillars. *Ecology* 51: 565–581.

Feinsinger, P. 1983. Coevolution and pollination. In *Coevolution* (D. J. Futuyma and M. Slatkin, Eds.), pp. 282–310. Sinauer Associates Inc., Sunderland, MA.

Fernandez, D. S., and N. Fetcher. 1991. Changes in light availability following Hurricane Hugo in a subtropical montane forest in Puerto Rico. *Biotropica* 23: 393–399.

Fewell, J. H., J. F. Harrison, J. R. B. Lighton, and M. D. Breed. 1996. Foraging energetics of the ant, *Paraponera clavata*. *Oecologia* 105: 419–427.

Fielden, L. J., F. D. Duncan, Y. Rechav, and R. M. Crewe. 1994. Respiratory gas exchange in the tick *Amblyomma hebraeum* (Acari: Ixodidae). *Journal of Medical Entomology* 31: 30–35.

Fielding, D. J., and M. A. Brusven. 1993. Grasshopper (Orthoptera: Acrididae) community composition and ecological disturbances on southern Idaho rangeland. *Environmental Entomology* 22: 71–81.

Fielding, D. J., and M. A. Brusven. 1995. Ecological correlates between rangeland grasshopper (Orthoptera: Acrididae) and plant communities of southern Idaho. *Environmental Entomology* 24: 1432–1441.

Filip, V., R. Dirzo, J. M. Maass, and J. Sarukhán. 1995. Within- and among-year variation in the levels of herbivory on the foliage of trees from a Mexican tropical deciduous forest. *Biotropica* 27: 78–86.

Finch, V. C., and G. T. Trewartha. 1949. *Elements of Geography: Physical and Cultural*. McGraw-Hill, New York.

Fitzgerald, T. D. 1995. *The Tent Caterpillars*. Cornell Univ. Press, Ithaca, NY.

Flamm, R. O., P. E. Pulley, and R. N. Coulson. 1993. Colonization of disturbed trees by the southern pine beetle guild (Coleoptera: Scolytidae). *Environmental Entomology* 22: 62–70.

Flinn, P. W., D. W. Hagstrum, W. E. Muir, and K. Sudayappa. 1992. Spatial model for simulating changes in temperature and insect population dynamics in stored grain. *Environmental Entomology* 21: 1351–1356.

Florence, L. Z., P. C. Johnson, and J. E. Coster. 1982. Behavioral and genetic diversity during dispersal: Analysis of a polymorphic esterase locus in southern pine beetle, *Dendroctonus frontalis*. *Environmental Entomology* 11: 1014–1018.

Fogel, R., and K. Cromack, Jr. 1977. Effect of habitat and substrate quality on Douglas-fir litter decomposition in western Oregon. *Canadian Journal of Botany* 55: 1632–1640.

Foley, P. 1997. Extinction models for local populations. In *Metapopulation Biology: Ecology, Genetics, and Evolution* (I. A. Hanski and M. E. Gilpin, Eds.), pp. 215–246. Academic Press, San Diego, CA.

Fonseca. C. R. 1994. Herbivory and the long-lived leaves of an Amazonian ant-tree. *Journal of Ecology* 82: 833–842.

Fowler, S. V., and J. H. Lawton. 1985. Rapidly induced defenses and talking trees: The devil's advocate position. *American Naturalist* 126: 181–195.

Fox, L. R. 1975a. Cannibalism in natural populations. *Annual Review of Ecology and Systematics* 6: 87–106.

Fox, L. R., 1975b. Some demographic consequences of food shortage for the predator, *Notonecta hoffmanni*. *Ecology* 56: 868–880.

Fox, L. R., and B. J. Macauley. 1977. Insect grazing on Eucalyptus in response to variation in leaf tannins and nitrogen. *Oecologia* 29: 145–162.

Fox, L. R., and P. A. Morrow. 1981. Specialization: Species property or local phenomenon? *Science* 211: 887–893.

Fox, L. R., and P. A. Morrow. 1983. Estimates of damage by herbivorous insects on *Eucalyptus* trees. *Australian Journal of Ecology* 8: 139–147.

Fox, L. R., and P. A. Morrow. 1992. Eucalypt responses to fertilization and reduced herbivory. *Oecologia* 89: 214–222.

Fox, L. R., and W. W. Murdoch. 1978. Effects of feeding history on short-term and long-term functional responses in *Notonecta hoffmanni*. *Journal of Animal Ecology* 47: 945–959.

Fox, L. R., S. P. Ribeiro, V. K. Brown, G. J. Masters, and I. P. Clarke. 1999. Direct and indirect effects of climate change on St. John's Wort, *Hypericum perforatum* L. (Hypericaceae). *Oecologia* 120: 113–122.

Fraenkel, G., and M. Blewett. 1946. Linoleic acid, vitamin E and other fat-soluble substances in the diet of certain insects, *Ephestria kuehniella*, *E. elutella*, *E. cautella*, and *Plodia interpunctella* (Lepidoptera). *Journal of Experimental Biology* 22: 172–190.

Frank, D. A., and S. J. McNaughton. 1991. Stability increases with diversity in plant communities: Empirical evidence from the 1988 Yellowstone drought. *Oikos* 62: 360–362.

Franklin, J. F., F. J. Swanson, M. E. Harmon, D. A. Perry, T. A. Spies, V. H. Dale, A. McKee, W. K. Ferrell, J. E. Means, S. V. Gregory, J. D. Lattin, T. D. Schowalter, and D. Larsen. 1992. Effects of global climatic change on forests in northwestern North America. In *Global Warming and Biological Diversity* (R. L. Peters and T. E. Lovejoy, Eds.), pp. 244–257. Yale Univ. Press, New Haven, CT.

French, J. R. J., and R. A. Roeper. 1972. Interactions of the ambrosia beetle, *Xyleborus dispar*, with its symbiotic fungus *Ambrosiella hartigii* (Fungi Imperfecti). *Canadian Entomologist* 104: 1635–1641.

Fritz, R. S. 1983. Ant protection of a host plant's defoliator: Consequences of an ant–membracid mutualism. *Ecology* 64: 789–797.

Furniss, R. L., and V. M. Carolin. 1977. *Western Forest Insects*. USDA Forest Service Misc. Publ. 1339. USDA Forest Service, Washington, DC.

Futuyma, D. J., and S. S. Wasserman. 1980. Resource concentration and herbivory in oak forests. *Science* 210: 920–922.

Gandar, M. V. 1982. The dynamics and trophic ecology of grasshoppers (Acridoidea) in a South African savanna. *Oecologia* 54: 370–378.

Gange, A. C., and V. K. Brown. 1989. Insect herbivory affects size variability in plant populations. *Oikos* 56: 351–356.

Gara, R. I., D. R. Geiszler, and W. R. Littke. 1984. Primary attraction of the mountain pine beetle to lodgepole pine in Oregon. *Annals of the Entomological Society of America* 77: 333–334.

Gardner, K. T., and D. C. Thompson. 1998. Influence of avian predation on a grasshopper (Orthoptera: Acrididae) assemblage that feeds on threadleaf snakeweed. *Environmental Entomology* 27: 110–116.

Gear, A. J., and B. Huntley. 1991. Rapid changes in the range limits of Scots pine 4000 years ago. *Science* 251: 544–547.

Gehring, C. A., and T. G. Whitham. 1991. Herbivore-driven mycorrhizal mutualism in insect-susceptible pinyon pine. *Nature* 353: 556–557.

Gehring, C. A., and T. G. Whitham. 1995. Duration of herbivore removal and environmental stress affect the ectomycorrhizae of pinyon pine. *Ecology* 76: 2118–2123.

Gibson, D. J., C. C. Freeman, and L. C. Hulbert. 1990. Effects of small mammal and invertebrate herbivory on plant species richness and abundance in tallgrass prairie. *Oecologia* 84: 169–175.

Gist, C. S., and D. A. Crossley, Jr. 1975. The litter arthropod community in a southern Appalachian hardwood forest: Numbers, biomass and mineral element content. *American Midland Naturalist* 93: 107–122.

Gleason, H. A. 1917. The structure and development of the plant association. *Bulletin of the Torrey Botanical Club* 44: 463–481.

Gleason, H. A. 1926. The individualistic concept of the plant association. *Bulletin of the Torrey Botanical Club* 53: 7–26.

Gleason, H. A. 1927. Further views on the succession-concept. *Ecology* 8: 299–326.

Glenn-Lewin, D. C., R. K. Peet, and T. T. Veblen. 1992. *Plant Succession: Theory and Prediction*. Chapman and Hall, New York.

Godfray, H. C. J. 1994. *Parasitoids: Behavioral and Evolutionary Ecology*. Princeton Univ. Press, Princeton, NJ.

Goh, B. S. 1979. Stability of models of mutualism. *American Naturalist* 113: 261–275.

Goldwasser, L., and J. Roughgarden 1997. Sampling effects and the estimation of food-web properties. *Ecology* 78: 41–54.

Golley, F. B. 1968. Secondary productivity in terrestrial communities. *American Zoologist* 8: 53–59.

Golley, F. B. 1977. Insects as regulators of forest nutrient cycling. *Tropical Ecology* 18: 116–123.

Golley, F. B. 1993. *A History of the Ecosystem Concept in Ecology*. Yale Univ. Press, New Haven, CT.

Gordon, D. M., and A. W. Kulig. 1996. Founding, foraging, and fighting: Colony size and the spatial distribution of harvester ant nests. *Ecology* 77: 2393–2409.

Gould, J. L. 1985. How bees remember flower shapes. *Science* 227: 1492–1494.

Gould, J. L. 1986. The locale map of honey bees: Do insects have cognitive maps? *Science* 232: 861–863.

Gould, J. L., and W. F. Towne. 1988. Honey bee learning. *Advances in Insect Physiology* 20: 55–86.

Goyer, R. A., M. R. Wagner, and T. D. Schowalter. 1998. Current and proposed technologies for bark beetle management. *Journal of Forestry* 96(12): 29–33.

Grant, G. G., and W. E. Miller. 1995. Larval images on lepidopteran wings—An unrecognized defense mechanism? *American Entomologist* 41: 44–48.

Greenbank, D. O. 1957. The role of climate and dispersal in the initiation of outbreaks of the spruce budworm in New Brunswick. I. The role of climate. *Canadian Journal of Zoology* 34: 453–476.

Greenbank, D. O. 1963. The development of the outbreak. In (R. F. Morris, Ed.), *The Dynamics of Epidemic Spruce Budworm Populations. Memoirs of the Entomological Society of Canada* 31: 19–23.

Gressitt, J. L., J. Sedlacek, and J. J. H. Szent-Ivany. 1965. Flora and fauna on backs of large Papuan moss-forest weevils. *Science* 150: 1833–1835.

Gressitt, J. L., G. A. Samuelson, and D. H. Vitt. 1968. Moss growing on living Papuan moss-forest weevils. *Nature* 217: 765–767.

Gribko, L. S., A. M. Liebhold, and M. E. Hohn. 1995. Model to predict gypsy moth (Lepidoptera: Lymantriidae) defoliation using kriging and logistic regression. *Environmental Entomology* 24: 529–537.

Grier, C. C., and D. J. Vogt. 1990. Effects of aphid honeydew on soil nitrogen availability and net primary production in an *Alnus rubra* plantation in western Washington. *Oikos* 57: 114–118.

Grilli, M. P., and D. E. Gorla. 1997. The spatio-temporal pattern of *Delphacodes kuscheli* (Homoptera: Delphacidae) abundance in central Argentina. *Bulletin of Entomological Research* 87: 45–53.

Grime, J. P. 1977. Evidence for the existence of three primary strategies in plants and its relevance to ecological and evolutionary theory. *American Naturalist* 111: 1169–1194.

Grime, J. P. 1997. Biodiversity and ecosystem function: The debate deepens. *Science* 277: 1260–1261.

Grime, J. P., J. H. C. Cornelissen, K. Thompson, and J. G. Hodgson. 1996. Evidence of a causal connection between anti-herbivore defense and the decomposition rate of leaves. *Oikos* 77: 489–494.

Guo, Q. 1998. Microhabitat differentiation in Chihuahuan Desert plant communities. *Plant Ecology* **139**: 71–80.

Gutierrez, A. P. 1986. Analysis of the interactions of host plant resistance, phytophagous and entomophagous species. In *Interactions of Plant Resistance and Parasitoids and Predators of Insects* (D. J. Boethel and R. D. Eikenbary, Eds.), pp. 198–215. Ellis Horwood Ltd., Chichester, UK.

Gutierrez, A. P. 1996. *Applied Population Ecology: A Supply–Demand Approach.* John Wiley & Sons, Inc., New York.

Hackstein, J. H. P., and C. K. Stumm. 1994. Methane production in terrestrial arthropods. *Proceedings of the National Academy of Sciences, USA* **91**: 5441–5445.

Hadley, K. S., and T. T. Veblen. 1993. Stand response to western spruce budworm and Douglas-fir bark beetle outbreaks, Colorado Front Range. *Canadian Journal of Forest Research* **23**: 479–491.

Haglund, B. M. 1980. Proline and valine—Cues which stimulate grasshopper herbivory during drought stress? *Nature* **288**: 697–698.

Haimi, J., and V. Huhta. 1990. Effects of earthworms on decomposition processes in raw humus forest soil: A microcosm study. *Biology and Fertility of Soils* **10**: 178–183.

Hain, F. P. 1980. Sampling and predicting population trends. In *The Southern Pine Beetle* (R. C. Thatcher, J. L. Searcy, J. E. Coster, and G. D. Hertel, Eds.), pp. 107–135. USDA Forest Service Tech. Bull. 1631. USDA Forest Service, Washington, DC.

Haines, B. L. 1978. Element and energy flows through colonies of the leaf-cutting ant, *Atta columbica*, in Panama. *Biotropica* **10**: 270–277.

Hairston, N. G., F. E. Smith, and L. B. Slobodkin. 1960. Community structure, population control, and competition. *American Naturalist* **94**: 421–425.

Hajek, A. E., and R. J. St. Leger. 1994. Interactions between fungal pathogens and insect hosts. *Annual Review of Entomology* **39**: 293–322.

Halaj, J., D. W. Ross, and A. R. Moldenke. 1997. Negative effects of ant foraging on spiders in Douglas-fir canopies. *Oecologia* **109**: 313–322.

Haldane, J. B. S. 1932. *The Causes of Evolution.* Harper and Brothers, New York.

Hamilton, W. D. 1964. The genetic evolution of social behavior. I. and II. *Journal of Theoretical Biology* **7**: 1–52.

Hanski, I. 1989. Metapopulation dynamics: Does it help to have more of the same? *Trends in Ecology and Evolution* **4**: 113–114.

Hanski, I. 1997. Metapopulation dynamics: From concepts and observations to predictive models. In *Metapopulation Biology: Ecology, Genetics and Evolution* (I. A. Hanski and M. E. Gilpin, Eds.), pp. 69–91. Academic Press, San Diego, CA.

Hanski, I. A., and M. E. Gilpin (Eds.). 1997. *Metapopulation Biology: Ecology, Genetics and Evolution.* Academic Press, San Diego, CA.

Hanski, I., and D. Simberloff. 1997. The metapopulation approach, its history, conceptual domain, and application to conservation. In *Metapopulation Biology: Ecology, Genetics and Evolution* (I. A. Hanski and M. E. Gilpin, Eds.), pp. 5–26. Academic Press, San Diego, CA.

Harborne, J. B. 1994. *Introduction to Ecological Biochemistry*, 4th ed. Academic Press, London, UK.

Hargrove, W. W. 1988. A photographic technique for tracking herbivory on individual leaves through time. *Ecological Entomology* **13**: 359–363.

Harmon, M. E., J. F. Franklin, F. J. Swanson, P. Sollins, S. V. Gregory, J. D. Lattin, N. H. Anderson, S. P. Cline, N. G. Aumen, J. R. Sedell, G. W. Lienkaemper, K. Cromack, Jr., and K. W. Cummins. 1986. Ecology of coarse woody debris in temperate ecosystems. *Advances in Ecological Research* **15**: 133–302.

Harris, L. D. 1984. *The Fragmented Forest.* Univ. of Chicago Press, Chicago, IL.

Harrison, S. 1994. Resources and dispersal as factors limiting a population of the tussock moth (*Orgyia vetusta*), a flightless defoliator. *Oecologia* **99**: 27–34.

Harrison, S., and N. Cappuccino. 1995. Using density-manipulation experiments to study population regulation. In *Population Dynamics: New Approaches and Synthesis* (N. Cappuccino and P. W. Price, Eds.), pp. 131–147. Academic Press, San Diego, CA.

Harrison, S., and R. Karban. 1986. Effects of an early-season folivorous moth on the success of a later-season species, mediated by a change in the quality of the shared host, *Lupinus arboreus* Sims. *Oecologia* **69**: 354–359.

Harrison, S. and A. D. Taylor. 1997. Empirical evidence for metapopulation dynamics. In *Metapopulation Biology: Ecology, Genetics and Evolution* (I. A. Hanski and M. E. Gilpin, Eds.), pp. 27–42. Academic Press, San Diego, CA.

Hart, D. D. 1992. Community organization in streams: The importance of species interactions, physical factors, and chance. *Oecologia* **91**: 220–228.

Hassell, M. P., and G. C. Varley. 1969. New inductive population model for insect parasites and its bearing on biological control. *Nature* **223**: 1133–1136.

Hassell, M. P., H. N. Comins, and R. M. May. 1991. Spatial structure and chaos in insect population dynamics. *Nature* **353**: 255–258.

Hatcher, P. E., P. G. Ayres, and N. D. Paul. 1995. The effect of natural and simulated insect herbivory, and leaf age, on the process of infection of *Rumex crispus* L. and *R. obtusifolius* L. by *Uromyces rumicis* (Schum.) *Wint. New Phytologist* **130**: 239–249.

Haukioja, E. 1990. Induction of defenses in trees. *Annual Review of Entomology* **36**: 25–42.

Havens, K. 1992. Scale and structure in natural food webs. *Science* **257**: 1107–1109.

Hawkins, C. P., and J. A. MacMahon. 1989. Guilds: the multiple meanings of a concept. *Annual Review of Entomology* **34**: 423–451.

Hayes, J. L., B. L. Strom, L. M. Roton, and L. L. Ingram, Jr. 1994. Repellent properties of the host compound 4-allylanisole to the southern pine beetle. *Journal of Chemical Ecology* **20**: 1595–1615.

Hazlett, D. L. 1998. *Vascular Plant Species of the Pawnee National Grasslands*. USDA Forest Service Gen. Tech. Rpt. RM-GTR-17, USDA Forest Service, Rocky Mountain Exp. Stn., Ft. Collins, CO.

He, F., and R. I. Alfaro. 1997. White pine weevil (Coleoptera: Curculionidae) attack on white spruce: Spatial and temporal patterns. *Environmental Entomology* **26**: 888–895.

Hedin, P. A. (Ed.). 1983. *Plant Resistance to Insects*. American Chemical Society, Symposium Series 208, Washington, DC.

Hedrick, P. W., and M. E. Gilpin. 1997. Genetic effective size of a metapopulation. In *Metapopulation Biology: Ecology, Genetics and Evolution* (I. A. Hanski and M. E. Gilpin, Eds.), pp. 165–181. Academic Press, San Diego, CA.

Heinrich, B. 1974. Thermoregulation in endothermic insects. *Science* **185**: 747–756.

Heinrich, B. 1979. *Bumble bee Economics*. Harvard Univ. Press, Cambridge, MA.

Heinrich, B. 1981. *Insect Thermoregulation*. Wiley, New York.

Heinrich, B. 1993. *The Hot-blooded Insects: Strategies and Mechanisms of Thermoregulation*. Harvard Univ. Press, Cambridge, MA.

Heithaus, E. R. 1979. Flower-feeding specialization in wild bee and wasp communities in seasonal neotropical habitats. *Oecologia* **42**: 179–194.

Heithaus, E. R. 1981. Seed predation by rodents on three ant-dispersed plants. *Ecology* **62**: 136–145.

Heliövaara, K. 1986. Occurrence of *Petrova resinella* (Lepidoptera: Tortricidae) in a gradient of industrial air pollutants. *Silva Fennica* **20**: 83–90.

Heliövaara, K., and R. Väisänen. 1986. Industrial air pollution and the pine bark bug, *Aradus cinnamomeus* Panz. (Het., Aradidae). *Zeitschrift für angewandte Entomologie* **101**: 469–478.

Heliövaara, K., and R. Väisänen. 1993. *Insects and Pollution*. CRC Press, Boca Raton, FL.

Hendrix, P. F., D. A. Crossley, Jr., J. M. Blair, and D. C. Coleman. 1990. Soil biota as components of sustainable agroecosystems. In *Sustainable Agricultural Systems* (C. A. Edwards, R. Lal, P. Madden, R. H. Miller, and G. House, Eds.), pp. 637–654. Soil and Water Conservation Society, Ankeny, IA.

Herms, D. A., and W. J. Mattson. 1992. The dilemma of plants: To grow or defend. *Quarterly Review of Biology* **67**: 283–335.

Herrick, J. E., and R. Lal. 1996. Dung decomposition and pedoturbation in a seasonally dry tropical pasture. *Biology and Fertility of Soils* **23**: 177–181.

Hik, D. S., and R. L. Jefferies. 1990. Increases in the net above-ground primary production of a salt-marsh forage grass: A test of the predictions of the herbivore-optimization model. *Journal of Ecology* **78**: 180–195.

Hirai, H., W. S. Procunier, J. O. Ochoa, and K. Uemoto. 1994. A cytogenetic analysis of the *Simulium ochraceum* species complex (Diptera: Simuliidae) in Central America. *Genome* **37**: 36–53.

Hirschel, G., C. Körner, and J. A. Arnone III. 1997. Will rising atmospheric CO_2 affect leaf litter quality and in situ decomposition rates in native plant communities? *Oecologia* **110**: 387–392.

Hochberg, M. E. 1989. The potential role of pathogens in biological control. *Nature* **337**: 262–265.

Hodges, J. D., S. J. Barras, and J. K. Mauldin. 1968. Free and protein-bound amino acids in inner bark of loblolly pine. *Forest Science* **14**: 330–333.

Hohn, M. E., A. M. Liebhold, and L. S. Gribko. 1993. Geostatistical model for forecasting spatial dynamics of defoliation caused by the gypsy moth (Lepidoptera: Lymantriidae). *Environmental Entomology* **22**: 1066–1075.

Holland, J. N. 1995. Effects of above-ground herbivory on soil microbial biomass in conventional and no-tillage agroecosystems. *Applied Soil Ecology* **2**: 275–279.

Holland, J. N., W. Cheng, and D. A. Crossley, Jr. 1996. Herbivore-induced changes in plant carbon allocation: Assessment of below-ground C fluxes using carbon-14. *Oecologia* **107**: 87–94.

Hölldobler, B. 1995. The chemistry of social regulation: multicomponent signals in ant societies. *Proceedings of the National Academy of Sciences USA* **92**: 19–22.

Holling, C. S. 1959. Some characteristics of simple types of predation and parasitism. *Canadian Entomologist* **91**: 385–398.

Holling, C. S. 1965. The functional response of predators to prey density and its role in mimicry and population regulation. *Memoirs of the Entomological Society of Canada* **45**: 1–60.

Holling, C. S. 1966. The functional response of invertebrate predators to prey density. *Memoirs of the Entomological Society of Canada* **48**: 1–86.

Holling, C. S. 1973. Resilience and stability of ecological systems. *Annual Review of Ecology and Systematics* **4**: 1–23.

Holling, C. S. 1992. Cross-scale morphology, geometry, and dynamics of ecosystems. *Ecological Monographs* **62**: 447–502.

Hollinger, D. Y. 1986. Herbivory and the cycling of nitrogen and phosphorus in isolated California oak trees. *Oecologia* 70: 291–297.

Holter, P. 1979. Effect of dung-beetles (*Aphodius* spp.) and earthworms on the disappearance of cattle dung. *Oikos* 32: 393–402.

Honkanen, T., E. Haukioja, and J. Suomela. 1994. Effects of simulated defoliation and debudding on needle and shoot growth in Scots pine (*Pinus sylvestris*): implications of plant source/sink relationships for plant–herbivore studies. *Functional Ecology* 8: 631–639.

Hooker, J. D. 1847. *The Botany of the Antarctic Voyage of H. M. Discovery Ships Erebus and Terror, in the Years 1839–1843: Vol. 1. Flora Antarctica.* Reeve, Brothers, London, UK.

Hooker, J. D. 1853. *The Botany of the Antarctic Voyage of H. M. Discovery Ships Erebus and Terror, in the Years 1839–1843: Vol. 2. Flora Novae-Zelandiae.* Reeve, Brothers, London, UK.

Hooker, J. D. 1860. *The Botany of the Antarctic Voyage of H. M. Discovery Ships Erebus and Terror, in the Years 1839–1843: Vol. 3. Flora Tasmaniae.* Reeve, Brothers, London, UK.

Hooper, D. U., and P. M. Vitousek. 1997. The effects of plant composition and diversity on ecosystem processes. *Science* 277: 1302–1305.

Horn, H. S. 1981. Some causes of variety in patterns of secondary succession. In *Forest Succession: Concepts and Application* (D. C. West, H. H. Shugart, and D. B. Botkin, Eds.), pp. 24–35. Springer-Verlag, New York.

Horn, M. H. 1997. Evidence for dispersal of fig seeds by the fruit-eating characid fish *Brycon guatemalensis* Regan in a Costa Rican tropical rain forest. *Oecologia* 109: 259–264.

Horvitz, C. C., and D. W. Schemske. 1986. Ant-nest soil and seedling growth in a neotropical ant-dispersed herb. *Oecologia* 70: 318–320.

Horwath, W. R., E. A. Paul, D. Harris, J. Norton, L. Jagger, and K. A. Horton. 1996. Defining a realistic control for the chloroform fumigation–incubation method using microscopic counting and ^{14}C-substrates. *Canadian Journal of Soil Science* 76: 459–467.

Howe, H. F., and J. Smallwood. 1982. Ecology of seed dispersal. *Annual Review of Ecology and Systematics* 13: 201–228.

Huettl, R. F., and D. Mueller-Dombois (Eds.). 1993. *Forest Declines in the Atlantic and Pacific Regions.* Springer-Verlag, Berlin, Germany.

Hughes, L. 1990. The relocation of ant nest entrances: Potential consequences for ant-dispersed seeds. *Australian Journal of Ecology* 16: 207–214.

Hughes, L., and F. A. Bazzaz. 1997. Effect of elevated CO_2 on interactions between the western flower thrips, *Frankliniella occidentalis* (Thysanoptera: Thripidae) and the common milkweed, *Asclepias syriaca. Oecologia* 109: 286–290.

Huhta, V., J. Haimi, and H. Setälä. 1991. Role of the fauna in soil processes: Techniques using simulated forest floor. *Agriculture, Ecosystems and Environment* 34: 223–229.

Hulme, M. A., and J. K. Shields. 1970. Biological control of decay fungi in wood by competition for nonstructural carbohydrates. *Nature* 227: 300–301.

Hulme, P. E. 1994. Seedling herbivory in grassland: Relative impact of vertebrate and invertebrate herbivores. *Journal of Ecology* 82: 873–880.

Hung, C. F., C. H. Kao, C. C. Liu, J. G. Lin, and C. N. Sun. 1990. Detoxifying enzymes of selected insect species with chewing and sucking habits. *Journal of Economic Entomology* 83: 361–365.

Hunt, H. W., D. C. Coleman, E. R. Ingham, R. E. Ingham, E. T. Elliott, J. C. Moore, S. L. Rose, C. P. P. Reid, and C. R. Morley. 1987. The detrital food web in a shortgrass prairie. *Biology and Fertility of Soils* 3: 57–68.

Hunter, A. F., and L. W. Arssen. 1988. Plants helping plants. *BioScience* 38: 34–40.

Hunter, M. D. 1987. Opposing effects of spring defoliation on late season oak caterpillars. *Ecological Entomology* 12: 373–382.

Hunter, M. D. 1992. A variable insect–plant interaction: The relationship between tree budburst phenology and population levels of insect herbivores among trees. *Ecological Entomology* 17: 91–95.

Hunter, M. D., and P. W. Price. 1992. Playing chutes and ladders: Heterogeneity and the relative roles of bottom-up and top-down forces in natural communities. *Ecology* 73: 724–732.

Hunter, M. D., and J. C. Schultz. 1993. Induced plant defenses breached? Phytochemical induction protects an herbivore from disease. *Oecologia* 94: 195–203.

Hunter, M. D., and J. C. Schultz. 1995. Fertilization mitigates chemical induction and herbivore responses within damaged oak trees. *Ecology* 76: 1226–1232.

Huntly, N. 1991. Herbivores and the dynamics of communities and ecosystems. *Annual Review of Ecology and Systematics* 22: 477–503.

Hurlbert, S. H. 1984. Pseudoreplication and the design of ecological field experiments. *Ecological Monographs* 54: 187–211.

Huryn, A. D., and J. B. Wallace. 1987. Local geomorphology as a determinant of macrofaunal production in a mountain stream. *Ecology* 68: 1932–1942.

Huston, M. 1979. A general hypothesis of species diversity. *American Naturalist* 113: 81–101.

Huxley, C. R., and D. F. Cutler. 1991. Ant–plant Interactions. Oxford Univ. Press, Oxford, UK.

Ingham, E. R. 1985. Review of the effects of 12 selected biocides on target and non-target soil organisms. *Crop Protection* 4: 3–32.

Ingham, E. R., C. Cambardella, and D. C. Coleman.

1986. Manipulation of bacteria, fungi and protozoa by biocides in lodgepole pine forest soil microcosms: Effects on organism interactions and nitrogen mineralization. *Canadian Journal of Soil Science* **66**: 261–272.

Ingham, R. E., J. A. Trofymow, E. R. Ingham, and D. C. Coleman. 1985. Interactions of bacteria, fungi, and their nematode grazers: Effects on nutrient cycling and plant growth. *Ecological Monographs* **55**: 119–140.

Inouye, D. W. 1982. The consequences of herbivory: A mixed blessing for *Jurinea mollis* (Asteracea). *Oikos* **39**: 269–272.

Inouye, R. S., G. S. Byers, and J. H. Brown. 1980. Effects of predation and competition on survivorship, fecundity, and community structure of desert annuals. *Ecology* **61**: 1344–1351.

Isaev, A. S., and R. G. Khlebopros. 1979. Inertial and noninertial factors regulating forest insect population density. In *Pest Management. Proc. Internat. Conf. Oct. 25–29, 1976* (G. A. Norton and C. S. Holling, Eds.), pp. 317–339. Pergamon Press, Oxford, UK.

Istock, C. A. 1973. Population characteristics of a species ensemble of water-boatmen (Corixidae). *Ecology* **54**: 535–544

Istock, C. A. 1977. Logistic interaction of natural populations of two species of waterboatmen. *American Naturalist* **111**: 279–287.

Istock, C. A. 1981. Natural selection and life history variation: theory plus lessons from a mosquito. In *Insect Life History Patterns: Habitat and Geographic Variation* (R. F. Denno and H. Dingle, Eds.), pp. 113–127. Springer-Verlag, New York.

Iwasaki, T. 1990. Predatory behavior of the praying mantis, *Tenodera aridifolia*. I. Effect of prey size on prey recognition. *Journal of Ethology* **8**: 75–79.

Iwasaki, T. 1991. Predatory behavior of the praying mantis, *Tenodera aridifolia*. II. Combined effect of prey size and predator size on the prey recognition. *Journal of Ethology* **9**: 77–81.

Jackson, R. V., J. Kollmann, and R. J. Grubb. 1999. Insect herbivory on five European tall-shrub species related to leaf ontogeny and quality, and the need to distinguish pre-expansion and expanding leaves. *Oikos* (in press).

Janzen, D. H. 1966. Coevolution of mutualism between ants and acacias in Central America. *Evolution* **20**: 249–275.

Janzen, D. H. 1977. What are dandelions and aphids? *American Naturalist* **111**: 586–589.

Janzen, D. H. 1981. Patterns of herbivory in a tropical deciduous forest. *Biotropica* **13**: 271–282.

Jepson, P. C., and J. R. M. Thacker. 1990. Analysis of the spatial component of pesticide side-effects on non-target invertebrate populations and its relevance to hazard analysis. *Functional Ecology* **4**: 349–355.

Joel, D. M., B. E. Juniper, and A. Dafni. 1985. Ultraviolet patterns in the traps of carnivorous plants. *New Phytologist* **101**: 585–593.

Johnson, K. A., and W. G. Whitford. 1975. Foraging ecology and relative importance of subterranean termites in Chihuahuan Desert ecosystems. *Environmental Entomology* **4**: 66–70.

Johnson, M. P., and D. S. Simberloff. 1974. Environmental determinants of island species numbers in the British Isles. *Journal of Biogeography* **1**: 149–154.

Johnson, S. D., and W. J. Bond. 1994. Red flowers and butterfly pollination in the fynbos of South Africa. In *Plant-animal Interactions in Mediterranean-type Ecosystems* (M. Arianoutsou and R. H. Graves, Eds.), pp. 137–148. Kluwer, Dordrecht, Netherlands.

Jolivet, P. 1996. *Ants and Plants: An Example of Coevolution.* Backhuys Publishers, Leiden, The Netherlands.

Jones, C. G. 1984. Microorganisms as mediators of plant resource exploitation by insect herbivores. In *A New Ecology: Novel Approraches to Interactive Systems* (P. W. Price, W. S. Gaud, and C. N. Slobodchikoff, Eds.), pp. 53–100. John Wiley, New York.

Jones, J. A. 1989. Environmental influences on soil chemistry in central semiarid Tanzania. *Soil Science Society of America Journal* **53**: 1748–1758.

Jones, J. A. 1990. Termites, soil fertility and carbon cycling in dry tropical Africa: A hypothesis. *Journal of Tropical Ecology* **6**: 291–305.

Jonkman, J. C. M. 1978. Nests of the leaf-cutting ant *Atta vollenweideri* as accelerators of succession in pastures. *Zeitschrift für angewandte Entomologie* **86**: 25–34.

Juniper, B. E., R. J. Robins, and D. M. Joel. 1989. *The Carnivorous Plants.* Academic Press, London, UK.

Käärik, A. A. 1974. Decomposition of wood. In *Biology of Plant Litter Decomposition* (C. H. Dickinson and G. J. F. Pugh, Eds.), pp. 129–174. Academic Press, London, UK.

Kaczmarek, M.. and A. Wasilewski. 1977. Dynamics of numbers of the leaf-eating insects and its effect on foliage production in the "Grabowy" Reserve in the Kampinos National Park. *Ekologia Polska* **25**: 653–673.

Kamil, A. C., J. R. Krebs, and H. R. Pulliam (Eds.). 1987. *Foraging Behavior.* Plenum Press, New York.

Karban, R., and C. Niiho. 1995. Induced resistance and susceptibility to herbivory: Plant memory and altered plant development. *Ecology* **76**: 1220–1225.

Kareiva, P. 1983. Influence of vegetation texture on herbivore populations: resource concentration and herbivore movement. In *Variable Plants and Herbivores in Natural and Managed Systems* (R. F. Denno and M. S. McClure, Eds.), pp. 259–289. Academic Press, New York.

Keeling, C. D., T. P. Whorf, M. Wahlen, and J. van der

Pilcht. 1995. Interannual extremes in the rate of rise of atmospheric carbon dioxide since 1980. *Science* **375**: 666–670.

Kempton, R. A. 1979. The structure of species abundance and measurement of diversity. *Biometrics* **35**: 307–321.

Kennedy, J. S. 1975. Insect dispersal. In *Insects, Science, and Society* (D. Pimentel, Ed.), pp. 103–119. Academic Press, New York.

Kettlewell, H. B. D. 1956. Further selection experiments on industrial melanism in the Lepidoptera. *Heredity* **10**: 287–301.

Key, K. H. L., and M. F. Day. 1954. A temperature-controlled physiological colour response in the grasshopper *Kosciuscola tristis* Sjost. (Orthoptera: Acrididae). *Australian Journal of Zoology* **2**: 309–339.

Khalil, M. A. K., R. A. Rasmussen, J. R. J. French, and J. A. Holt. 1990. The influence of termites on atmospheric trace gases: CH_4, CO_2, $CHCl_3$, N_2O, CO, H_2, and light hydrocarbons. *Journal of Geophysical Research* **95**: 3619–3634.

Kharboutli, M. S., and T. P. Mack. 1993. Tolerance of the striped earwig (Dermaptera: Labiduridae) to hot and dry conditions. *Environmental Entomology* **22**: 663–668.

Kiilsgaard, C. W., S. E. Greene, and S. G. Stafford. 1987. Nutrient concentration in litterfall from some western conifers with special reference to calcium. *Plant and Soil* **102**: 223–227.

Kim, Y., and N. Kim. 1997. Cold hardiness in *Spodoptera exigua* (Lepidoptera: Noctuidae). *Environmental Entomology* **26**: 1117–1123.

Kimmins, J. P., 1972. Relative contributions of leaching, litterfall, and defoliation by *Neodiprion sertifer* (Hymenoptera) to the removal of cesium-134 from red pine. *Oikos* **23**: 226–234.

Kinn, D. N. 1980. Mutualism between *Dendrolaelaps neodisetus* and *Dendroctonus frontalis*. *Environmental Entomology* **9**: 756–758.

Kinney, K. K., R. L. Lindroth, S. M. Jung, and E. V. Nordheim. 1997. Effects of CO_2 and NO_3^- availability on deciduous trees: Phytochemistry and insect performance. *Ecology* **78**: 215–230.

Kinyamario, J. I., and S. K. Imbamba. 1992. Savanna at Nairobi National Park, Nairobi. In *Primary Productivity of Grass Ecosystems of the Tropics and Sub-tropics* (S. P. Long, M. B. Jones and M. J. Roberts, Eds.), pp. 25–69. Chapman and Hall, London, UK.

Kitchell, J. F., R. V. O'Neill, D. Webb, G. W. Gallepp, S. M. Bartell, J. F. Koonce, and B. S. Ausmus. 1979. Consumer regulation of nutrient cycling. *BioScience* **29**: 28–34.

Klein, B. C. 1989. Effects of forest fragmentation on dung and carrion beetle communities in central Amazonia. *Ecology* **70**: 1715–1725.

Klock, G. O., and B. E. Wickman. 1978. Ecosystem effects. In *The Douglas-fir Tussock Moth: A Synthesis* (M. H. Brookes, R. W. Stark, and R. W. Campbell, Eds.), pp. 90–95. USDA Forest Service, Tech. Bull. 1585, USDA Forest Service, Washington, DC.

Knapp, A. K., and T. R. Seastedt. 1986. Detritus accumulation limits productivity of tallgrass prairie. *BioScience* **36**: 662–668.

Kogan, M. 1975. Plant resistance in pest management. In *Introduction to Insect Pest Management* (R. L. Metcalf and W. H. Luckmann, Eds.), pp. 103–146. Wiley, New York.

Kogan, M. 1981. Dynamics of insect adaptations to soybean: impact of integrated pest management. *Environmental Entomology* **10**: 363–371.

Kogan, M. 1998. Integrated pest management: Historical perspectives and contemporary developments. *Annual Review of Entomology* **43**: 243–270.

Kogan, M., and J. Paxton. 1983. Natural inducers of plant resistance to insects. In *Plant Resistance to Insects* (P. A. Hedin, Ed.), pp. 153–170. ACS Symposium Series 208, American Chemical Society, Washington, DC.

Koricheva, J., S. Larsson, and E. Haukioja. 1998. Insect performance on experimentally stressed woody plants: A meta-analysis. *Annual Review of Entomology* **43**: 195–216.

Körner, C. 1993. Scaling from species to vegetation: the usefulness of functional groups. In *Biodiversity and Ecosystem Function* (E. D. Schulze and H. A. Mooney, Eds.), pp. 117–140. Springer-Verlag, Berlin, Germany.

Kozár, F. 1991. Recent changes in the distribution of insects and the global warming. In *Proc. 4th European Congress of Entomology* (pp. 406–413). Gödöllö, Hungary.

Kozár, F. 1992a. Organization of arthropod communities in agroecosystems. *Acta Phytopathologica et Entomologica Hungarica* **27**: 365–373.

Kozár, F. 1992b. Resource partitioning of host plants by insects on a geographic scale. In *Proc. 8th International Symposium on Insect–Plant Relationships* (S. B. J. Menken, J. H. Visser and P. Harrewijn, Eds.), pp. 46–48. Kluwer Academic Publ., Dordrecht, Netherlands.

Krafft, C. C., and S. N. Handel. 1991. The role of carnivory in the growth and reproduction of *Drosera filiformis* and *D. rotundifolia*. *Bulletin of the Torrey Botanical Club* **118**: 12–19.

Krantz, G. W. 1978. *A Manual of Acarology*. Oregon State University Book Stores, Corvallis, OR.

Krantz, G. W., and J. L. Mellott. 1972. Studies on phoretic specificity in *Macrocheles mycotrupetes* and *M. peltotrupetes* Krantz and Mellott (Acari: Macrochelidae), associates of geotrupine Scarabaeidae. *Acarologia* **14**: 317–344.

Kratz, T. K., J. J. Magnuson, P. Bayley, B. J. Benson,

C. W. Berish, C. S. Bledsoe, E. R. Blood, C. J. Bowser, S. R. Carpenter, G. L. Cunningham, R. A. Dahlgren, T. M. Frost, J. C. Halfpenny, J. D. Hansen, D. Heisey, R. S. Inouye, D. W. Kaufman, A. McKee, and J. Yarie. 1995. Temporal and spatial variability as neglected ecosystem properties: lessons learned from 12 North American ecosytems. In *Evaluating and Monitoring the Health of Large-scale Ecosystems* (D. J. Rapport, C. L. Gaudet, and P. Calow, Eds.), pp. 359–383. NATO ASI Series, Vol. 128. Springer-Verlag, Berlin, Germany.

Kruess, A., and T. Tscharntke. 1994. Habitat fragmentation, species loss, and biological control. *Science* **264**: 1581–1584.

Labandeira, C. C., and J. J. Sepkoski, Jr. 1993. Insect diversity in the fossil record. *Science* **261**: 310–315.

Landsberg, J. 1989. A comparison of methods for assessing defoliation, tested on eucalypt trees. *Australian Journal of Ecology* **14**: 423–440.

Landsberg, J., and C. Ohmart. 1989. Levels of insect defoliation in forests: patterns and concepts. *Trends in Ecology and Evolution* **4**: 96–100

Larsson, S., C. Bjorkman, and N. A. C. Kidd. 1993. Outbreaks in diprionid sawflies: Why some species and not others? In *Sawfly Life History Adaptations to Woody Plants* (M. R. Wagner and K. F. Raffa, Eds.), pp. 453–483. Academic Press, San Diego, CA.

Laskowski, R., M. Niklińska, and M. Maryański. 1995. Then dynamics of chemical elements in forest litter. *Ecology* **76**: 1393–1406.

Lavigne, R., R. Kumar, and J. A. Scott. 1991. Additions to the Pawnee National Grasslands insect checklist. *Entomological News* **102**: 150–164.

Law, J. M., and F. E. Regnier. 1971. Pheromones. *Annual Review of Biochemistry* **40**: 533–548.

Lawrence, R. K., W. J. Mattson, and R. A. Haack. 1997. White spruce and the spruce budworm: Defining the phenological window of susceptibility. *Canadian Entomologist* **129**: 291–318.

Lawton, J. H. 1982. Vacant niches and unsaturated communities: A comparison of bracken herbivores at sites on two continents. *Journal of Animal Ecology* **51**: 573–595.

Lawton, J. H. 1983. Plant architecture and the diversity of phytophagous insects. *Annual Review of Entomology* **28**: 23–39.

Lawton, J. H. 1995. Response of insects to environmental change. In *Insects in a Changing Environment* (R. Harrington and N. E. Stork, Eds.), pp. 5–26. Academic Press, London, UK.

Lawton, J. H., and V. K. Brown. 1993. Redundancy in ecosystems. In *Biodiversity and Ecosystem Function* (E. D. Schulze and H. A. Mooney, Eds.), pp. 255–270. Springer-Verlag, Berlin, Germany.

Lawton, J. H., and D. R. Strong. 1981. Community patterns and competition in folivorous insects. *American Naturalist* **118**: 317–338.

Lee, K. E., and J. H. A. Butler. 1977. Termites, soil organic matter decomposition and nutrient cycling. *Ecological Bulletin* (Stockholm) **25**: 544–548.

Leigh, E. G., and N. Smythe. 1978. Leaf production, leaf consumption and the regulation of folivory on Barro Colorado Island. In *The Ecology of Arboreal Folivores* (E. G. Leigh, A. S. Rand, and D. M. Windsor Eds.), pp. 33–50. Smithsonian Institution Press, Washington, DC.

Leigh, E. G., and D. M. Windsor. 1982. Forest production and regulation of primary consumers on Barro Colorado Island. In *The Ecology of Arboreal Folivores* (E. G. Leigh, A. S. Rand, and D. M. Windsor, Eds.), pp. 109–123. Smithsonian Institution Press, Washington, DC.

Leonard, D. E. 1970. Intrinsic factors causing qualitative changes in populations of *Porthetria dispar* (Lepidoptera: Lymantriidae). *Canadian Entomologist* **102**: 239–249.

Letourneau, D. K., and L. A. Dyer. 1998. Density patterns of *Piper* ant-plants and associated arthropods: Top-predator trophic cascades in a terrestrial system? *Biotropica* **30**: 162–169.

Leuschner, W. A. 1980. Impacts of the southern pine beetle. In *The Southern Pine Beetle* (R. C. Thatcher, J. L. Searcy, J. E. Coster, and G. D. Hertel, Eds.), pp. 137–151. USDA Forest Service Tech. Bull. 1631. USDA Forest Service, Washington, DC.

Levins, R. 1970. Extinction. *Lectures on Mathematics in the Life Sciences* **2**: 77–107.

Lewinsohn, T. M., and P. W. Price. 1996. Diversity of herbivorous insects and ecosystem processes. In *Biodiversity and Savanna Ecosystem Processes* (O. T. Solbrig, E. Medina, and J. F. Silva, Eds.), pp. 143–157. Springer-Verlag, Berlin, Germany.

Lewis, A. C. 1979. Feeding preference for diseased and wilted sunflower in the grasshopper, *Melanoplus differentialis*. *Entomologia Experimentalis et Applicata*. **26**: 202–207.

Lewis, A. C. 1986. Memory constraints and flower choice in *Pieris rapae*. *Science* **232**: 863–865.

Lewis, T. 1998. The effect of deforestation on ground surface temperatures. *Global and Planetary Change* **18**: 1–13.

Lewis, W. J., and J. H. Tumlinson. 1988. Host detection by chemically mediated associative learning in a parasitic wasp. *Nature* **331**: 257–259.

Liebhold, A. M., and J. S. Elkinton. 1989. Characterizing spatial patterns of gypsy moth regional defoliation. *Forest Science* **35**: 557–568.

Liebhold, A. M., R. E. Rossi, and W. P. Kemp. 1993. Geostatistics and geographic information systems in applied insect ecology. *Annual Review of Entomology* **38**: 303–327.

Lincoln, D. E., E. D. Fajer, and R. H. Johnson. 1993. Plant–insect herbivore interactions in elevated CO_2

environments. *Trends in Ecology and Evolution* 8: 64–68.

Lindeman, R. L. 1942. The trophic-dynamic aspect of ecology. *Ecology* 23: 399–418.

Linley, J. R. 1966. The ovarian cycle of *Culicoides barbosai* Wirth & Blanton and *C. furens* (Poey) (Ceratopogonidae). *Bulletin of Entomological Research* 57: 1–17.

Llewellyn, M. 1972. The effects of the lime aphid, *Eucallipterus tiliae* L. (Aphididae) on the growth of the lime, *Tilia* x *vulgaris* Hayne. *Journal of Applied Ecology* 9: 261–282.

Lloyd, J. E. 1983. Bioluminescence and communication in insects. *Annual Review of Entomology* 28: 131–160

Lockwood, J. A., and L. D. DeBrey. 1990. A solution for the sudden and unexplained extinction of the Rocky Mountain grasshopper (Orthoptera: Acrididae). *Environmental Entomology* 19: 1194–1205.

Logan, J. A., and J. C. Allen. 1992. Nonlinear dynamics and chaos in insect populations. *Annual Review of Entomology* 37: 455–477.

Lorio, P. L., Jr. 1993. Environmental stress and whole-tree physiology. In *Beetle–pathogen Interactions in Conifer Forests* (T. D. Schowalter and G. M. Filip, Eds.), pp. 81–101. Academic Press, London, UK.

Lotka, A. J. 1925. *Elements of Physical Biology.* Williams and Wilkins, Baltimore, MD.

Louda, S. M. 1982. Inflorescence spiders: a cost/benefit analysis for the host plant, *Haplopappus venetus* Blake (Asteraceae). *Oecologia* 55: 185–191.

Louda, S. M., and J. E. Rodman. 1996. Insect herbivory as a major factor in the shade distribution of a native crucifer (*Cardamine cordifolia* A. Gray, bittercress). *Journal of Ecology* 84: 229–237.

Louda, S. M., K. H. Keeler, and R. D. Holt. 1990a. Herbivore influences on plant performance and competitive interactions. In *Perspectives on Plant Competition* (J. B. Grace and D. Tilman, Eds.), pp. 413–444. Academic Press, San Diego, CA.

Louda, S. M., M. A. Potvin, and S. K. Collinge. 1990b. Predispersal seed predation, postdispersal seed predation and competition in the recruitment of seedlings of a native thistle in sandhills prairie. *American Midland Naturalist* 124: 105–113.

Lovelock, J. 1988. *The Ages of Gaia.* W. W. Norton, New York.

Lovett, G. M., and A. E. Ruesink. 1995. Carbon and nitrogen mineralization from decomposing gypsy moth frass. *Oecologia* 104: 133–138.

Lovett, G., and P. Tobiessen. 1993. Carbon and nitrogen assimilation in red oaks (*Quercus rubra* L.) subject to defoliation and nitrogen stress. *Tree Physiology* 12: 259–269.

Lovett, G. M., S. S. Nolan, C. T. Driscoll, and T. J. Fahey. 1996. Factors regulating throughfall flux in a New Hampshire forested landscape. *Canadian Journal of Forest Research* 26: 2134–2144.

Lowman, M. D. 1982. The effects of different rates and methods of leaf area removal on coachwood (*Ceratopetalum apetalum*). *Australian Journal of Botany* 30: 477–483.

Lowman, M. D. 1984. An assessment of techniques for measuring herbivory: Is rainforest defoliation more intense than we thought? *Biotropica* 16: 264–268.

Lowman, M. D. 1985. Spatial and temporal variability in herbivory of Australian rain forest canopies. *Australian Journal of Ecology* 10: 7–14.

Lowman, M. D. 1992. Leaf growth dynamics and herbivory in five species of Australian rain forest canopy trees. *Journal of Ecology* 80: 433–447.

Lowman, M. D. 1995. Herbivory as a canopy process in rain forest trees. In *Forest Canopies* (M. D. Lowman and N. M. Nadkarni, Eds.), pp. 431–455. Academic Press, San Diego, CA.

Lowman, M. D., and J. H. Box. 1983. Variation in leaf toughness and phenolic content among 5 species of Australia rain forest trees. *Australian Journal of Ecology* 8: 17–25.

Lowman, M. D., and H. H. Heatwole. 1992. Spatial and temporal variability in defoliation of Australian eucalypts. *Ecology* 73: 129–142.

Lowman, M. D., M. Moffett, and H. B. Rinker. 1993. A technique for taxonomic and ecological sampling in rain forest canopies. *Selbyana* 14: 75–79.

Lowrance, R., B. R. Stinner, and G. J. House, Eds. 1984. *Agricultural Ecosystems: Unifying Concepts.* Wiley, New York.

Lubchenco, J. 1978. Plant species diversity in a marine intertidal community: Importance of herbivore food preference and algal competitive abilities. *American Naturalist* 112: 23–39.

Lunderstädt, J. 1981. The role of food as a density-determining factor for phytophagous insects with reference to the relationship between Norway spruce (*Picea abies* Karst) and *Gilpinia hercyniae* Htg. (Hymenoptera, Diprionidae). *Forest Ecology and Management* 3: 335–353.

Lundheim, R., and K. E. Zachariassen. 1993. Water balance of over-wintering beetles in relation to strategies for cold tolerance. *Journal of Comparative Physiology B* 163: 1–4.

Lüscher, M. 1961. Air-conditioned termite nests. *Scientific American* 205: 138–145 (July).

MacArthur, R. H., and E. O. Wilson. 1967. *The Theory of Island Biogeography.* Princeton Univ. Press, Princeton, NJ.

Macauley, B. J. 1975. Biodegradation of litter in *Eucalyptus pauciflora* communities. I: Techniques for comparing the effects of fungi and insects. *Soil Biology and Biochemistry* 7: 341–344.

MacMahon, J. A. 1981. Successional processes: comparisons among biomes with special reference to probable roles of and influences on animals. In *Forest Succession: Concepts and Application* (D. C. West, H. H. Shugart, and D. B. Botkin, Eds.), pp. 277–304. Springer-Verlag, New York.

Maddrell, S. H. P. 1962. A diuretic hormone in *Rhodnius prolixus* Stal. *Nature* 194: 605–606.

Madge, D. S. 1969. Litter disappearance in forest and savanna. *Pedobiologia* 9: 288–299.

Mafra-Neto, A., and R. T. Cardé. 1995. Influence of plume structure and pheromone concentration on upwind flight by *Cadra cautella* males. *Physiological Entomology* 20: 117–133.

Magurran, A. E. 1988. *Ecological Diversity and its Measurement.* Princeton Univ. Press, Princeton, NJ.

Mahunka, S. (Ed.) 1981 and 1983. *The Fauna of the Hortobágy National Park, Vols. 1 and 2.* Akadémiai Kiadó, Budapest, Hungary.

Mahunka, S. (Ed.) 1986 and 1987. *The Fauna of the Kiskunság National Park, Vols. 4 and 5.* Akadémiai Kiadó, Budapest, Hungary.

Mahunka, S. (Ed.) 1991. *Bátorliget Nature Reserve After Forty Years, Vols. 1 and 2.* Hungarian Natural History Museum, Budapest, Hungary.

Majer, J. D., and H. F. Recher. 1988. Invertebrate communities on Western Australian eucalypts—A comparison of branch clipping and chemical knockdown procedures. *Australian Journal of Ecology* 13: 269–278.

Malcolm, S. B. 1992. Prey defense and predator foraging. In *Natural Enemies: The Population Biology of Predators, Parasites and Diseases* (M. J. Crawley, Ed.), pp. 458–475. Blackwell Scientific, London, UK.

Malthus, T. R. 1789. *An Essay on the Principle of Population as it Affects the Future Improvement of Society.* Johnson, London, UK.

Mankowski, M. E., T. D. Schowalter, J. J. Morrell, and B. Lyons. 1998. Feeding habits and gut fauna of *Zootermopsis angusticollis* (Isoptera: Termopsidae) in response to wood species and fungal associates. *Environmental Entomology* 27: 1315–1322.

Manley, G. V. 1971. A seed-cacheing carabid (Coleoptera). *Annals of the Entomological Society of America* 64: 1474–1475.

Maraun, M., and S. Scheu. 1996. Changes in microbial biomass, respiration and nutrient status of beech (*Fagus silvatica*) leaf litter processed by millipedes (*Glomeris marginata*). *Oecologia* 107: 131–140.

Mark, S., and J. M. Olesen. 1996. Importance of elaiosome size to removal of ant-dispersed seeds. *Oecologia* 107: 95–101.

Marks, S., and Lincoln, D. E. 1996. Antiherbivore defense mutualism under elevated carbon dioxide levels: A fungal endophyte and grass. *Environmental Entomology* 25: 618–623.

Marquis, R. J., 1984. Leaf herbivores decrease fitness of a tropical plant. *Science* 226: 537–539.

Marquis, R. J., and C. J. Whelan. 1994. Insectivorous birds increase growth of white oak through consumption of leaf-chewing insects. *Ecology* 75: 2007–2014.

Martinez, N. D. 1992. Constant connectance in community food webs. *American Naturalist* 139: 1208–1218.

Martius, C., R. Wassmann, U. Thein, A. Bandeira, H. Rennenberg, W. Junk, and W. Seiler. 1993. Methane emission from wood-feeding termites in Amazonia. *Chemosphere* 26: 623–632.

Martius, C., P. M. Fearnside, A. G. Bandeira, and R. Wassmann. 1996. Deforestation and methane release from termites in Amazonia. *Chemosphere* 33: 517–536.

Maschinski, J., and T. G. Whitham. 1989. The continuum of plant responses to herbivory: The influence of plant association, nutrient availability, and timing. *American Naturalist* 134: 1–19.

Mason, R. R. 1996. Dynamic behavior of Douglas-fir tussock moth populations in the Pacific Northwest. *Forest Science* 42: 182–191.

Mason, R. R., and R. F. Luck. 1978. Population growth and regulation. In *The Douglas-fir Tussock Moth: A Synthesis* (M. H. Brookes, R. W. Stark, and R. W. Campbell, Eds.), pp. 41–47. USDA Forest Service Technical Bulletin. 1585. USDA Forest Service, Washington, DC.

Masters, G. J., V. K. Brown, and A. C. Gange. 1993. Plant mediated interactions between above- and below-ground insect herbivores. *Oikos* 66: 148–151.

Matis, J. H., T. R. Kiffe, and G. W. Otis. 1994. Use of birth–death–migration processes for describing the spread of insect populations. *Environmental Entomology* 23: 18–28.

Matthews, R. W., and J. R. Matthews. 1978. *Insect Behavior.* John Wiley & Sons, New York.

Mattson, W. J. 1980. Herbivory in relation to plant nitrogen content. *Annual Review of Ecology and Systematics* 11: 119–161.

Mattson, W. J., and N. D. Addy. 1975. Phytophagous insects as regulators of forest primary production. *Science* 190: 515–522.

Mattson, W. J., and R. A. Haack. 1987. The role of drought in outbreaks of plant-eating insects. *BioScience* 37: 110–118.

May, R. M. 1973. Qualitative stability in model ecosystems. *Ecology* 54: 638–641.

May, R. M. 1981. Models for two interacting populations. In *Theoretical Ecology: Principles and Applications* (R. M. May, Ed.), pp. 78–104. Blackwell Scientific, Oxford, UK.

May, R. M. 1983. The structure of food webs. *Nature* 301: 566–568.

May, R. M. 1988. How many species are there on Earth? *Science* 241: 1441–1449.

Mayer, B., K. H. Feger, A. Giesemann, and H. J. Jäger.

1995. Interpretation of sulfur cycling in two catchments in the Black Forest (Germany) using stable sulfur and oxygen isotope data. *Biogeochemistry* 30: 31–58.

McBrayer, J. F. 1975. Exploitation of deciduous leaf litter by *Apheloria montana* (Diplopoda: Eurydesmidae). *Pedobiologia* 13: 90–98.

McClure, M. S. 1991. Density-dependent feedback and population cycles in *Adelges tsugae* (Homoptera: Adelgidae) on *Tsuga canadensis*. *Environmental Entomology* 20: 258–264.

McCreadie, J. W., and M. H. Colbo. 1993. Larval and pupal microhabitat selection by *Simulium truncatum, S. rostratum* and *S. verecundum* AA (Diptera: Simuliidae). *Canadian Journal of Zoology* 71: 358–367.

McCullough, D. G., and M. R. Wagner. 1993. Defusing host defenses: Ovipositional adaptations of sawflies to plant resins. In *Sawfly Life History Adaptations to Woody Plants* (M. R. Wagner and K. F. Raffa, Eds.), pp. 157–172. Academic Press, San Diego, CA.

McCullough, D. G., R. A. Werner, and D. Neumann. 1998. Fire and insects in northern and boreal forest ecosystems of North America. *Annual Review of Entomology* 43: 107–127.

McEvoy, P. B., C. Cox, and E. Coombs. 1991. Successful biological control of ragwort, *Senecio jacobaea*, by introduced insects in Oregon. *Ecological Applications* 1: 430–442.

McEvoy, P. B., N. T. Rudd, C. S. Cox, and M. Huso. 1993. Disturbance, competition, and herbivory effects on ragwort *Senecio jacobaea* populations. *Ecological Monographs* 63: 55–75.

McIntosh, R. P. 1981. Succession and ecological theory. In *Forest Succession: Concepts and Application* (D. C. West, H. H. Shugart, and D. B. Botkin, Eds.), pp. 10–23. Springer-Verlag, New York.

McKey, D. 1979. The distribution of secondary compounds within plants. In *Herbivores: Their Interactions with Secondary Plant Metabolites* (G. A. Rosenthal and D. H. Janzen, Eds.), pp. 55–133. Academic Press, New York.

McNaughton, S. J. 1977. Diversity and stability of ecological communities: A comment on the role of empiricism in ecology. *American Naturalist* 111: 515–525.

McNaughton, S. J. 1979. Grazing as an optimization process: Grass–ungulate relationships in the Serengeti. *American Naturalist* 113: 691–703.

McNaughton, S. J. 1985. Ecology of a grazing system: The Serengeti. *Ecological Monographs* 55: 259–294.

McNaughton, S. J. 1986. On plants and herbivores. *American Naturalist* 128: 765–770.

McNaughton, S. J. 1993a. Grasses and grazers, science and management. *Ecological Applications* 3: 17–20.

McNaughton, S. J. 1993b. Biodiversity and function of grazing ecosystems. In *Biodiversity and Ecosystem Function* (E. D. Schulze and H. A. Mooney, Eds.), pp. 361–383. Springer-Verlag, Berlin, Germany.

McNeill, S., and J. H. Lawton. 1970. Annual production and respiration in animal populations. *Nature* 225: 472–474.

Meentemeyer, V. 1978. Macroclimate and lignin control of litter decomposition rates. *Ecology* 59: 465–472.

Meher-Homji, V. M. 1991. Probable impact of deforestation on hydrological processes. *Climate Change* 19: 163–173.

Meinwald, J., and T. Eisner. 1995. The chemistry of phyletic dominance. *Proceedings of the National Academy of Sciences USA* 92: 14–18.

Michener, C. D. 1969. Comparative and social behavior of bees. *Annual Review of Entomology* 14: 299–334.

Miles, P. W. 1972. The saliva of Hemiptera. *Advances in Insect Physiology* 9: 183–255.

Miller, J. C., and P. E. Hanson. 1989. *Laboratory Feeding Tests on the Development of Gypsy Moth Larvae with Reference to Plant Taxa and Allelochemicals*. Bulletin 674, Oregon Agricultural Exp. Stn., Oregon State University, Corvallis, OR.

Miller, K. K., and M. R. Wagner. 1984. Factors influencing pupal distribution of the pandora moth (Lepidoptera: Saturniidae) and their relationship to prescribed burning. *Environmental Entomology* 13: 430–431.

Misra, R. 1968. Energy transfer along terrestrial food chain. *Tropical Ecology* 9: 105–118.

Mitchell, R. 1970. An analysis of dispersal in mites. *American Naturalist* 104: 425–431.

Mitchell, R. 1975. The evolution of oviposition tactics in the bean weevil, *Callosobruchus maculatus* (F.). *Ecology* 56: 696–702.

Mitchell, R. G., and R. E. Martin. 1980. Fire and insects in pine culture of the Pacific Northwest. *Proceedings of the Conference on Fire and Forest Meteorology* 6: 182–190.

Mitchell, R. G., and H. Preisler. 1992. Analysis of spatial patterns of lodgepole pine attacked by outbreak populations of mountain pine beetle. *Forest Science* 29: 204–211.

Mittler, T. E. 1958. The excretion of honeydew by *Tuberolachnus salignus* (Gmelin) (Homoptera: Aphididae). *Proceedings of the Royal Entomological Society of London (A)* 33: 49–55.

Mittler, T. E. 1970. Uptake rates of plant sap and synthetic diet by the aphid *Myzus persicae*. *Annals of the Entomological Society of America* 63: 1701–1705.

Mittler, T. E., and E. S. Sylvester. 1961. A comparison of the injury to alfalfa by the aphids, *Therioaphis maculata* and *Microsiphum pisi*. *Journal of Economic Entomology* 54: 615–622.

Moldenke, A. R. 1976. California pollination ecology and vegetation types. *Phytologia* 34: 305–361.

Moldenke, A. R. 1979. Pollination ecology as an assay for ecosystemic organization: Convergent evolution in Chile and California. *Phytologia* 42: 415–454.

Moll, E. J., and B. McKenzie. 1994. Modes of dispersal of seeds in the Cape fynbos. In *Plant–animal Interactions in Mediterranean–type Ecosystems* (M. Arianoutsou and R. H. Graves, Eds.), pp. 151–157. Kluwer, Dordrecht, Netherlands.

Momose, K., T. Nagamitsu, and T. Inoue. 1998. Thrips cross-pollination of *Popowia pisocarpa* (Annonaceae) in a lowland dipterocarp forest in Sarawak. *Biotropica* 30: 444–448.

Monteith, J. L. 1973. *Principles of Environmental Physics*. American Elsevier, New York.

Moore, J. C., and H. W. Hunt. 1988. Resource compartmentation and the stability of real ecosystems. *Nature* 333: 261–263.

Moore, J. C., D. E. Walter, and H. W. Hunt. 1988. Arthropod regulation of micro- and mesobiota in below-ground detrital food webs. *Annual Review of Entomology* 33: 419–439.

Moore, R., and B. J. Francis. 1991. Factors influencing herbivory by insects on oak trees in pure stands and paired mixtures. *Journal of Applied Ecology* 28: 305–317.

Moore, R., S. Warrington, and J. B. Whittaker 1991. Herbivory by insects on oak trees in pure stands compared with paired mixtures. *Journal of Applied Ecology* 28: 290–304.

Mopper, S. 1996. Adaptive genetic structure in phytophagous insect populations. *Trends in Ecology and Evolution* 11: 235–238.

Mopper, S., and S. Y. Strauss. 1998. *Genetic Structure and Local Adaptation in Natural Insect Populations: Effects of Ecology, Life History, and Behavior*. Chapman and Hall, New York.

Moran, N. A., and T. G. Whitham. 1990. Differential colonization of resistant and susceptible host plants: *Pemphigus* and *Populus*. *Ecology* 71: 1059–1067.

Moran, P. A. P. 1953. The statistical analysis of the Canadian lynx cycle. II. Synchronization and meteorology. *Australian Journal of Zoology* 1: 291–298.

Moran, V. C., and T. R. E. Southwood. 1982. The guild composition of arthropod communities in trees. *Journal of Animal Ecology* 51: 289–306.

Morgan, F. D. 1968. Bionomics of siricidae. *Annual Review of Entomology* 13: 239–256.

Morón-Ríos, A., R. Dirzo, and V. J. Jaramillo. 1997. Defoliation and below-ground herbivory in the grass *Muhlenbergia quadridentata*: Effects on plant performance and on the root-feeder *Phyllophaga* sp. (Coleoptera: Melolonthidae). *Oecologia* 110: 237–242.

Morris, R. F. 1969. Approaches to the study of population dynamics. In *Forest Insect Population Dynamics* (W. E. Waters, Ed.), pp. 9–28. U.S.D.A. Forest Service, Northeast Forest Exp. Stn., Hamden, CT.

Morrow, P. A., and V. C. LaMarche, Jr. 1978. Tree ring evidence for chronic insect suppression of productivity in subalpine *Eucalyptus*. *Science* 201: 1244–1246.

Moser, J. C. 1963. Contents and structure of *Atta texana* nest in summer. *Annals of the Entomological Society of America* 56: 286–291.

Moser, J. C. 1985. Use of sporothecae by phoretic *Tarsonemus* mites to transport ascospores of coniferous bluestain fungi. *Transactions of the British Mycological Society* 84: 750–753.

Murlis, J., J. S. Elkinton, and R. T. Cardé. 1992. Odor plumes and how insects use them. *Annual Review of Entomology* 37: 505–532.

Mustaparta, H. 1984. Olfaction. In *Chemical Ecology of Insects* (W. J. Bell and R. T. Carde, Eds.), pp. 37–70. Chapman and Hall, London, UK.

Myers, J. H. 1988. Can a general hypothesis explain population cycles of forest Lepidoptera? *Advances in Ecological Research* 18: 179–242.

Myers, N. 1996. Environmental services of biodiversity. *Proceedings of the National Academy of Sciences, USA* 93: 2764–2769.

Naeem, S. 1998. Species redundancy and ecosystem reliability. *Conservation Biology* 12: 39–45.

Nakadai, T., H. Koizumi, Y. Usami, M. Satoh, and T. Oikawa. 1993. Examination of the method for measuring soil respiration in cultivated land: Effect of carbon dioxide concentration on soil respiration. *Ecological Research* 8: 65–71.

Nault, L. R., and E. D. Ammar. 1989. Leafhopper and planthopper transmission of plant viruses. *Annual Review of Entomology* 34: 503–529.

Nebeker, T. E., J. D. Hodges, and C. A. Blanche. 1993. Host response to bark beetle and pathogen colonization. In *Beetle–pathogen Interactions in Conifer Forests* (T. D. Schowalter and G. M. Filip, Eds.), pp. 157–173. Academic Press, London, UK.

Newman, R. M. 1990. Herbivory and detritivory on freshwater macrophytes by invertebrates: A review. *Journal of the North American Benthological Society* 10: 89–114.

Nicholson, A. J. 1933. The balance of animal populations. *Journal of Animal Ecology* 2(suppl.): 131–178.

Nicholson, A. J. 1954a. Compensatory reactions of populations to stress, and their evolutionary significance. *Australian Journal of Zoology* 2: 1–8

Nicholson, A. J. 1954b. An outline of the dynamics of animal populations. *Australian Journal of Zoology* 2: 9–65.

Nicholson, A. J. 1958. Dynamics of insect populations. *Annual Review of Entomology* 3: 107–136.

Nicholson, A. J., and V. A. Bailey. 1935. The balance

of animal populations. Part I. *Proceedings of the Zoological Society of London*, pp. 551–598.

Nielsen, B. O. 1978. Above ground food resources and herbivory in a beech forest ecosystem. *Oikos* **31**: 273–279.

Niesenbaum, R. A. 1992. The effects of light environment on herbivory and growth in the dioecious shrub *Lindera benzoin* (Lauraceae). *American Midland Naturalist* **128**: 270–275.

Norton, R. A., and V. M. Behan-Pelletier. 1991. Calcium carbonate and calcium oxalate as cuticular hardening agents in oribatid mites (Acari: Oribatida). *Canadian Journal of Zoology* **69**: 1504–1511.

Nothnagle, P. J., and J. C. Schultz. 1987. What is a forest pest? In *Insect Outbreaks* (P. Barbosa and J. C. Schultz, Eds.), pp. 59–80. Academic Press, San Diego, CA.

O'Connell, A. M., and P. Menagé. 1983. Decomposition of litter from three major plant species of jarrah (*Eucalyptus marginata* Donn ex Sm.) forest in relation to site fire history and soil type. *Australian Journal of Ecology* **8**: 277–286.

O'Dowd, D. J., and M. E. Hay. 1980. Mutualism between harvester ants and a desert ephemeral: seed escape from rodents. *Ecology* **61**: 531–540.

O'Dowd, D. J., and M. F. Willson. 1991. Associations between mites and leaf domatia. *Trends in Ecology and Evolution* **6**: 179–182.

Odum, E. P. 1953. *Fundamentals of Ecology*. W. B. Saunders, Philadelphia, PA.

Odum, E. P. 1969. The strategy of ecosystem development. *Science* **164**: 262–270.

Odum, E. P. 1971. *Fundamentals of Ecology*, 3rd Ed. W. B. Saunders, Philadelphia, PA.

Odum, E. P., and A. E. Smalley. 1959. Comparison of population energy flow of a herbivorous and a deposit-feeding invertebrate in a salt marsh ecosystem. *Proceedings of the National Academy of Sciences, USA* **45**: 617–622.

Odum, H. T. 1957. Trophic structure and productivity of Silver Springs, Florida. *Ecological Monographs* **27**: 55–112.

Odum, H. T. 1996. *Environmental Accounting: Emergy and Environmental Decision Making*. John Wiley, New York.

Odum, H. T., and R. C. Pinkerton. 1955. Time's speed regulator: The optimum efficiency for maximum power output in physical and biological systems. *American Scientist* **43**: 331–343.

Odum, H. T., and J. Ruiz-Reyes. 1970. Holes in leaves and the grazing control mechanism. In *A Tropical Rain Forest* (H. T. Odum and R. F. Pigeon, Eds.), pp. I-69–I-80. U.S. Atomic Energy Commission, Oak Ridge, TN.

Oertli, B. 1993. Leaf litter processing and energy flow through macroinvertebrates in a woodland pond (Switzerland). *Oecologia* **96**: 466–477.

Oesterheld, M., and S. J. McNaughton. 1988. Intraspecific variation in the response of *Themeda triandra* to defoliation: The effect of time of recovery and growth rates on compensatory growth. *Oecologia*. **77**: 181–186.

Oesterheld, M., and S. J. McNaughton. 1991. Effect of stress and time for recovery on the amount of compensatory growth after grazing. *Oecologia* **85**: 305–313.

Oesterheld, M., O. E. Sala, and S. J. McNaughton. 1992. Effect of animal husbandry on herbivore-carrying capacity at a regional scale. *Nature* **356**: 234–236.

Ohgushi, T. 1995. Adaptive behavior produces stability of herbivorous lady beetle populations. In *Population Dynamics: New Approaches and Synthesis* (N. Cappuccino and P. W. Price, Eds.), pp. 303–319. Academic Press, San Diego, CA.

Ohgushi, T., and H. Sawada. 1985. Population equilibrium with respect to available food resource and its behavioural basis in an herbivorous lady beetle *Henosepilachna niponica*. *Journal of Animal Ecology* **54**: 781–796.

Ohkawara, K., S. Higashi, and M. Ohara. 1996. Effects of ants, ground beetles and the seed-fall patterns on myrmecochory of *Erythronium japonicum* Decne. (Liliaceae). *Oecologia* **106**: 500–506.

Ohmart, C. P., L. G. Stewart, and J. R. Thomas. 1983. Phytophagous insect communities in the canopies of three *Eucalyptus* forest types in south-eastern Australia. *Australian Journal of Ecology* **8**: 395–403.

Oksanen, L. 1983. Trophic exploitation and arctic phytomass patterns. *American Naturalist* **122**: 45–52.

Oliveira, P. S., and C. R. F. Brandâo. 1991. The ant community associated with extrafloral nectaries in the Brazilian cerrados. In *Ant–plant Interactions* (C. R. Huxley and D. F. Cutler, Eds.), pp. 198–212. Oxford Univ. Press, Oxford, UK.

Olson, J. S. 1963. Energy storage and the balance of producers and decomposers in ecological systems. *Ecology* **44**: 322–331.

Oman, P. 1987. *The Leafhopper Genus Errhomus (Homoptera: Cicadellidae: Cicadelliniae) Systematics and Biogeography*. Occasional Publication No. 1, Department of Entomology, Oregon State University, Corvallis, OR.

O'Neill, R. V., D. L. DeAngelis, J. B. Waide, and T. F. H. Allen. 1986. *A Hierarchical Concept of Ecosystems*. Princeton Univ. Press, Princeton, NJ.

Ostfeld, R. S., R. H. Manson, and C. D. Canham. 1997. Effects of rodents on survival of tree seeds and seedlings invading old fields. *Ecology* **78**: 1531–1542.

Ostrom, P. H., M. Colunga-Garcia, and S. H. Gage. 1997. Establishing pathways of energy flow for insect predators using stable isotope ratios: Field and laboratory evidence. *Oecologia* **109**: 108–113.

Otte, D., and A. Joern. 1975. Insect territoriality and its evolution: Population studies of desert grasshoppers on creosote bushes. *Journal of Animal Ecology* 44: 29–54.

Owen, D. F. 1978. Why do aphids synthesize melezitose? *Oikos* 31: 264–267.

Owen, D. F., and Wiegert, R. G. 1976. Do consumers maximize plant fitness? *Oikos* 27: 488–492.

Ozanne, C. M. P., C. Hambler, A. Foggo, and M. R. Speight. 1997. The significance of edge effects in the management of forests for invertebrate diversity. In *Canopy Arthropods* (N. E. Stork, J. Adis, and R. K. Didham, Eds.), pp. 534–550. Chapman and Hall, London, UK.

Paige, K. N., and T. G. Whitham. 1987. Overcompensation in response to mammalian herbivory: The advantage of being eaten. *American Naturalist* 129: 407–416.

Paine, R. T. 1966. Food web complexity and species diversity. *American Naturalist* 100: 65–75.

Paine, R. T. 1969a. The *Pisaster-Tegula* interaction: Prey patches, predator food preference, and intertidal community structure. *Ecology* 50: 950–961.

Paine, R. T. 1969b. A note on trophic complexity and community stability. *American Naturalist* 103: 91–93.

Paine, T. D., and F. A. Baker. 1993. Abiotic and biotic predisposition. In *Beetle–pathogen Interactions in Conifer Forests* (T. D. Schowalter and G. M. Filip, Eds.), pp. 61–79. Academic Press, London.

Painter, E. L., and A. J. Belsky. 1993. Application of herbivore optimization theory to rangelands of the western United States. *Ecological Applications* 3: 2–9.

Palmisano, S., and L. R. Fox. 1997. Effects of mammal and insect herbivory on population dynamics of a native Californian thistle, *Cirsium occidentale*. *Oecologia* 111: 413–421.

Paoletti, M. G., R. A. J. Taylor, B. R. Stinner, D. H. Stinner, and D. H. Benzing. 1991. Diversity of soil fauna in the canopy and forest floor of a Venezuelan cloud forest. *Journal of Tropical Ecology* 7: 373–383.

Papaj, D. R., and R. J. Prokopy. 1989. Ecological and evolutionary aspects of learning in phytophagous insects. *Annual Review of Entomology* 34: 315–350.

Park, T. 1948. Experimental studies of interspecies competition. I. Competition between populations of the flour beetles, *Tribolium confusum* Duval and *Tribolium castaneum* Herbst. *Ecological Monographs* 18: 265–308.

Park, T. 1954. Experimental studies of interspecies competition. II. Temperature, humidity and competition in two species of *Tribolium*. *Physiological Zoology* 27: 177–238.

Parker, G. G. 1983. Throughfall and stemflow in the forest nutrient cycle. *Advances in Ecological Research* 13: 57–133.

Parker, G. G. 1995. Structure and microclimate of forest canopies. In *Forest Canopies* (M. D. Lowman, and N. M. Nadkarni, Eds.), pp. 73–106. Academic Press, San Diego, CA.

Parker, L. W., H. G. Fowler, G. Ettershank, and W. G. Whitford. 1982. The effects of subterranean termite removal on desert soil nitrogen and ephemeral flora. *Journal of Arid Environments* 5: 53–59.

Parker, M. A. 1985. Size dependent herbivore attack and the demography of an arid grassland shrub. *Ecology* 66: 850–860.

Parkinson, K. J. 1981. An improved method for measuring soil respiration in the field. *Journal of Applied Ecology* 18: 221–228.

Parks, C. G. 1993. *The Influence of Induced Host Moisture Stress on the Growth and Development of Western Spruce Budworm and* Armillaria ostoyae *on Grand Fir Seedlings*. Ph. D. Dissertation, Oregon State University, Corvallis, OR.

Parry, D., J. R. Spence, and W. J. A. Volney. 1997. Responses of natural enemies to experimentally increased populations of the forest tent caterpillar, *Malacosoma disstria*. *Ecological Entomology* 22: 97–108.

Parsons, G. L., G. Cassis, A. R. Moldenke, J. D. Lattin, N. H. Anderson, J. C. Miller, P. Hammond, and T. D. Schowalter. 1991. *Invertebrates of the H. J. Andrews Experimental Forest, Western Cascade Range, Oregon. V: An Annotated List of Insects and Other Arthropods*. Gen. Tech. Rpt. PNW-GTR-290. USDA Forest Service, Pacific Northwest Research Station, Portland, OR.

Parsons, K. A., and A. A. de la Cruz. 1980. Energy flow and grazing behavior of conocephaline grasshoppers in a *Juncus roemerianus* marsh. *Ecology* 61: 1045–1050.

Parton, W. J., J. M. O. Scurlock, D. S. Ojima, T. G. Gilmanov, R. J. Scholes, D. S. Schimel, T. Kirchner, J-C. Menaut, T. Seastedt, E. G. Moya, A. Kamnalrut, and J. I. Kinyamario. 1993. Observations and modeling of biomass and soil organic matter dynamics for the grassland biome worldwide. *Global Biogeochemical Cycles* 7: 785–809.

Patnaik, S., and P. S. Ramakrishnan. 1989. Comparative study of energy flow through village ecosystems of two co-existing communities (the Khasis and the Nepalis) of Meghalaya in north-east India. *Agricultural Systems* 30: 245–267.

Patten, B. C. 1995. Network integration of ecological extremal principles: Exergy, emergy, power, ascendency, and indirect effects. *Ecological Modelling* 79: 75–84.

Patten, B. C., and E. P. Odum. 1981. The cybernetic nature of ecosystems. *American Naturalist* 118: 886–895.

Patten, D. T. 1993. Herbivore optimization and overcompensation: Does native herbivory on western rangelands support these theories? *Ecological Applications* 3: 35–36.

Payne, J. A. 1965. A summer carrion study of the baby pig *Sus scrofa* Linnaeus. *Ecology* 46: 592- 602.

Peakall, R., A. J. Beattie, and S. H. James. 1987. Pseudocopulation of an orchid by male ants: A test of two hypotheses accounting for the rarity of ant pollination. *Oecologia* 73: 522–524.

Pearl, R. 1928. *The Rate of Living*. Knopf, New York.

Pearl, R., and L. J. Reed. 1920. On the rate of growth of the population of the United States since 1790 and its mathematical representation. *Proceedings of the National Academy of Sciences, USA* 6: 275–288.

Peet, R. K., and N. L. Christensen. 1980. Succession: A population process. *Vegetatio* 43: 131–140.

Perry, D. R. 1984. The canopy of the tropical rain forest. *Scientific American* 251(5): 138–147.

Petelle, M. 1980. Aphids and melezitose: A test of Owen's 1978 hypothesis. *Oikos* 35: 127–128.

Peterson, J. R., and D. J. Merrell. 1983. Rare male mating disadvantage in *Drosophila melanogaster*. *Evolution* 37: 1306–1316.

Petrusewicz, K. (Ed.). 1967. *Secondary Productivity of Terrestrial Ecosystems: Principles and Methods*. Państwowe Wydawnictwo Naukowe, Warszawa, Poland.

Phillipson, J. 1981. Bioenergetic options and phylogeny. In *Physiological Ecology: an Evolutionary Approach to Resource Use* (C. R. Townsend and P. Calow, Eds.), pp. 20–45. Blackwell Scientific, Oxford, UK.

Pianka, E. R. 1974. *Evolutionary Ecology*. Harper & Row, New York.

Pianka, E. R. 1981. Competition and niche theory. In *Theoretical Ecology: Principles and Applications* (R. M. May, Ed.), pp. 167–196. Blackwell Scientific, Oxford, UK.

Pickett, S. T. A., and P. S. White. 1985. Patch dynamics: a synthesis. In *The Ecology of Natural Disturbance and Patch Dynamics* (S. T. A. Pickett and P. S. White, Ed.), pp. 371–384. Academic Press, Orlando, FL.

Pielke, R. A., and P. L. Vidale. 1995. The boreal forest and the polar front. *Journal of Geophysical Research* 100: 25755–25758.

Pimm, S. L. 1980. Properties of food webs. *Ecology* 61: 219–225.

Pimm, S. L. 1982. *Food Webs*. Chapman and Hall, London, UK.

Pimm, S. L., and R. L. Kitching. 1987. The determinants of food chain length. *Oikos* 50: 302–307.

Pimm, S. L., and J. H. Lawton. 1977. Number of trophic levels in ecological communities. *Nature* 268: 329–331.

Pimm, S. L., and J. H. Lawton. 1980. Are food webs divided into compartments? *Journal of Animal Ecology* 49: 879–898.

Pimm, S. L., and J. C. Rice. 1987. The dynamics of multispecies, multi-life-stage models of aquatic food webs. *Theoretical Population Biology* 32: 303–325.

Pimm, S. L., J. H. Lawton, and J. E. Cohen. 1991. Food web patterns and their consequences. *Nature* 350: 669–674.

Pinder, L. C. V., and D. J. Morley. 1995. Chironomidae as indicators of water quality—with a comparison of the chironomid faunas of a series of contrasting Cumbrian tarns. In *Insects in a Changing Environment* (R. Harrington and N. E. Stork, Eds.), pp. 271–293. Academic Press, London, UK.

Poinar, G. O., Jr. 1993. Insects in amber. *Annual Review of Entomology* 38: 145–159.

Polis, G. A. 1991a. Desert communities: an overview of patterns and processes. In *The Ecology of Desert Communities* (G. A. Polis, Ed.), pp. 1–26. Univ. of Arizona Press, Tucson, AZ.

Polis, G. A. 1991b. Food webs in desert communities: complexity via diversity and omnivory. In *The Ecology of Desert Communities* (G. A. Polis, Ed.), pp. 383–429. Univ. of Arizona Press, Tucson, AZ.

Polis, G. A., and D. R. Strong. 1996. Food web complexity and community dynamics. *American Naturalist* 147: 813–846.

Polis, G. A., C. Myers, and M. Quinlan. 1986. Burrowing biology and spatial distribution of desert scorpions. *Journal of Arid Environments* 10: 137–145.

Polis, G. A., W. B. Anderson, and R. D. Holt. 1997. Toward an integration of landscape and food web ecology: The dynamics of spatially subsidized food webs. *Annual Review of Ecology and Systematics* 28: 289–316.

Ponyi, J. E., I. Tátrai, and A. Frankó. 1983. Quantitative studies on Chironomidae and Oligochaeta in the benthos of Lake Balaton. *Archiv für Hydrobiologie* 97: 196–207.

Pope, D. N., R. N. Coulson, W. S. Fargo, J. A. Gagne, and C. W. Kelly. 1980. The allocation process and between-tree survival probabilities in *Dendroctonus frontalis* infestations. *Researches in Population Ecology* 22: 197–210.

Porter, E. E., and R. A. Redak. 1996. Short-term recovery of grasshopper communities (Orthoptera: Acrididae) of a California native grassland after prescribed burning. *Environmental Entomology* 25: 987–992.

Powell, A. H., and G. V. N. Powell. 1987. Population dynamics of male euglossine bees in Amazonian forest fragments. *Biotropica* 19: 176–179.

Power, M. E., D. Tilman, J. A. Estes, B. A. Menge, W. J. Bond, L. S. Mills, G. Daily, J. C. Castilla, J. Lubchenco, and R. T. Paine. 1996. Challenges in the quest for keystones. *BioScience* 46: 609–620.

Prange, H. D., and B. Pinshow. 1994. Thermoregulation of an unusual grasshopper in a desert environment: The importance of food source and body size. *Journal of Thermal Biology* **19**: 75–78.

Price, P. W. 1980. Evolutionary biology of parasites. *Monographs in Population Biology* **15**. Princeton Univ. Press, Princeton, NJ.

Price, P. W. 1986. Ecological aspects of host plant resistance and biological control: interactions among three trophic levels. In *Interactions of Plant Resistance and Parasitoids and Predators of Insects* (D. J. Boethel and R. D. Eikenbary, Eds.), pp. 11–30. Ellis Horwood Ltd., Chichester, UK.

Price, P. W. 1991. The plant vigor hypothesis and herbivore attack. *Oikos* **62**: 244–251.

Price, P. W. 1997. *Insect Ecology* (3rd ed.). John Wiley & Sons, New York.

Price, P. W., C. E. Bouton, P. Gross, B. A. McPheron, J. N. Thompson, and A. E. Weis. 1980. Interactions among three trophic levels: Influence of plants on interactions between insect herbivores and natural enemies. *Annual Review of Ecology and Systematics* **11**: 41–65.

Pringle, C. M. 1997. Exploring how disturbance is transmitted upstream: Going against the flow. *Journal of the North American Benthological Society* **16**: 425–438.

Pritchard, I. M., and R. James. 1984a. Leaf mines: Their effect on leaf longevity. *Oecologia* **64**: 132–140.

Pritchard, I. M., and R. James. 1984b. Leaf fall as a source of leaf miner mortality. *Oecologia* **64**: 140–142.

Punttila, P., Y. Haila, N. Niemelä, and T. Pajunen. 1994. Ant communities in fragments of old-growth taiga and managed surroundings. *Annales Zoologici Fennici* **31**: 131–144.

Rabatin, S. C., and B. R. Stinner. 1988. Indirect effects of interactions between VAM fungi and soil-inhabiting invertebrates on plant processes. *Agriculture, Ecosystems and Environment* **24**: 135–146.

Rácz, V., and I. Bernath. 1993. Dominance conditions and population dynamics of *Lygus* (Het., Miridae) species in Hungarian maize stands (1976–1985), as functions of climatic conditions. *Journal of Applied Entomology* **115**: 511–518.

Raffa, K. F., T. W. Phillips, and S. M. Salom. 1993. Strategies and mechanisms of host colonization by bark beetles. In *Beetle–pathogen Interactions in Conifer Forests* (T. D. Schowalter and G. M. Filip, Eds.), pp. 103–128. Academic Press, London, UK.

Raich, J. W., R. D. Bowden, and P. A. Steudler. 1990. Comparison of two static chamber techniques for determining carbon dioxide efflux from forest soils. *Soil Science Society of America Journal* **54**: 1754–1757.

Rankin, M. A., and J. C. A. Burchsted. 1992. The cost of migration in insects. *Annual Review of Entomology* **37**: 533–559.

Rasmussen, E. M., and J. M. Wallace. 1983. Meteorological aspects of the El Niño/southern oscillation. *Science* **222**: 1195–1202.

Rastetter, E. B., M. G. Ryan, G. R. Shaver, J. M. Melillo, K. J. Nadelhoffer, J. E. Hobbie, and J. D. Aber. 1991. A general biogeochemical model describing the responses of the C and N cycles in terrestrial ecosystems to changes in CO_2, climate and N deposition. *Tree Physiology* **9**: 101–126.

Raven, J. A. 1983. Phytophages of xylem and phloem: A comparison of animal and plant sap-feeders. *Advances in Ecological Research* **13**: 136–204.

Reagan, D. P., G. R. Camilo, and R. B. Waide. 1996. The community food web: Major properties and patterns of organization. In *The Food Web of a Tropical Rain Forest* (D. P. Reagan and R. B. Waide, Eds.), pp. 462–488. Univ. of Chicago Press, Chicago, IL.

Regal, R. J. 1982. Pollination by wind and animals: Ecology of geographic patterns. *Annual Review of Ecology and Systematics*. **13**: 497–424.

Reice, S. R. 1985. Experimental disturbance and the maintenance of species diversity in a stream community. *Oecologia* **67**: 90–97.

Reichle, D. E., and D. A. Crossley, Jr. 1967. Investigation on heterotrophic productivity in forest insect communities. In *Secondary Productivity of Terrestrial Ecosystems: Principles and Methods* (K. Petrusewicz, Ed.), pp. 563–587. Państwowe Wydawnictwo Naukowe, Warszawa, Poland.

Reichle, D. E., M. H. Shanks, and D. A. Crossley, Jr. 1969. Calcium, potassium, and sodium content of forest floor arthropods. *Annals of the Entomological Society of America* **62**: 57–62.

Reichle, D. E., R. A. Goldstein, R. I. Van Hook, and G. J. Dodson. 1973. Analysis of insect consumption in a forest canopy. *Ecology* **54**: 1076–1084.

Rhoades, D. F. 1983. Responses of alder and willow to attack by tent caterpillars and webworms: evidence for pheromonal sensitivity of willows. In *Plant Resistance to Insects* (P. A. Hedin, Ed.), pp. 55–68. ACS Symposium Series 208, American Chemical Society, Washington, DC.

Ribeiro, S. P., H. R. Pimenta, and G. W. Fernandes. 1994. Herbivory by chewing and sucking insects on *Tabebuia ochracea*. *Biotropica* **26**: 302–307.

Rice, B., and M. Westoby 1986. Evidence against the hypothesis that ant-dispersed seeds reach nutrient-enriched microsites. *Ecology* **67**: 1270–1274.

Richerson, J. V., and P. E. Boldt. 1995. Phytophagous insect fauna of *Flourensia cernua* (Asteraceae: Heliantheae) in trans-Pecos Texas and Arizona. *Environmental Entomology*, **24**: 588–594.

Richter, M. R. 1990. Hunting social wasp interactions: Influence of prey size, arrival order, and wasp species. *Ecology* **71**: 1018–1030.

Rickson, F. R. 1971. Glycogen plastids in Müllerian body cells of *Cecropia peltata*—a higher green plant. *Science* **173**: 344–347.

Rickson, F. R. 1977. Progressive loss of ant-related traits of *Cecropia peltata* on selected Caribbean islands. *American Journal of Botany* **64**: 585–592.

Rickson, F. R., and M. M. Rickson. 1998. The cashew nut, *Anacardium occidentale* (Anacardiaceae), and its perennial association with ants: Extrafloral nectary location and the potential for ant defense. *American Journal of Botany* **85**: 835–849.

Risch, S. 1980. The population dynamics of several herbivorous beetles in a tropical agroecosystem: The effect of intercropping corn, beans and squash in Costa Rica. *Journal of Applied Ecology* **17**: 593–612.

Risch, S. J. 1981. Insect herbivore abundance in tropical monocultures and polycultures: An experimental test of two hypotheses. *Ecology* **62**: 1325–1340.

Risley, L. S., and D. A. Crossley, Jr. 1993. Contribution of herbivore-caused greenfall to litterfall nitrogen flux in several southern Appalachian forested watersheds. *American Midland Naturalist* **129**: 67–74.

Rissing, S. W. 1986. Indirect effects of granivory by harvester ants: Plant species composition and reproductive increase near ant nests. *Oecologia* **68**: 231–234.

Ritchie, M. E., D. Tilman, and J. M. H. Knops. 1998. Herbivore effects on plant and nitrogen dynamics in oak savanna. *Ecology* **79**: 165–177.

Ritland, D. B., and L. P. Brower. 1991. The viceroy butterfly is not a batesian mimic. *Nature* **350**: 497–498.

Ritter, H., Jr. 1964. Defense of mate and mating chamber in a wood roach. *Science* **143**: 1459–1460.

Robertson, A. I., R. Giddins, and T. J. Smith. 1990. Seed predation by insects in tropical mangrove forests: Extent and effects on seed viability and the growth of seedlings. *Oecologia* **83**: 213–219.

Robinson, M. H., and B. C. Robinson. 1974. Adaptive complexity: The thermoregulatory postures of the golden-web spider, *Nephila clavipes*, at low latitudes. *American Midland Naturalist* **92**: 386–396.

Rodgers, H. L., M. P. Brakke, and J. J. Ewel. 1995. Shoot damage effects on starch reserves of *Cedrela odorata*. *Biotropica* **27**: 71–77.

Roelofs, W. L. 1995. Chemistry of sex attraction. *Proceedings of the National Academy of Sciences USA* **92**: 44–49.

Roland, J. 1993. Large-scale forest fragmentation increases the duration of tent caterpillar outbreak. *Oecologia* **93**: 25–30.

Roland, J., and W. J. Kaupp. 1995. Reduced transmission of forest tent caterpillar (Lepidoptera: Lasiocampidae) nuclear polyhedrosis virus at the forest edge. *Environmental Entomology* **24**: 1175–1178.

Roland, J., and P. D. Taylor. 1997. Insect parasitoid species respond to forest structure at different spatial scales. *Nature* **386**: 710–713.

Romme, W. H., D. H. Knight, and J. B. Yavitt. 1986. Mountain pine beetle outbreaks in the Rocky Mountains: Regulators of primary productivity? *American Naturalist* **127**: 484–494.

Romoser, W. S., and J. G. Stoffolano, Jr. 1998. *The Science of Entomology* (4th ed.). McGraw-Hill, Boston, MA.

Root, R. B. 1967. The niche exploitation pattern of the blue-gray gnatcatcher. *Ecological Monographs* **37**: 317–350.

Root, R. B. 1973. Organization of a plant–arthropod association in simple and diverse habitats: The fauna of collards (*Brassica oleracea*). *Ecological Monographs* **43**: 95–124.

Rosenthal, G. A., and M. R. Berenbaum (Eds.). 1991. *Herbivores: Their Interactions with Secondary Plant Metabolites. Vol. 1. The Chemical Participants* (2nd ed.). Academic Press, San Diego, CA.

Rosenthal, G. A., and M. R. Berenbaum (Eds.). 1992. *Herbivores: Their Interactions with Secondary Plant Metabolites. Vol. 2. Ecological and Evolutionary Processes* (2nd ed.). Academic Press, San Diego, CA.

Rosenthal, G. A., and D. H. Janzen, (Eds.). 1979. *Herbivores: Their Interactions with Secondary Plant Metabolites*. Academic Press, New York.

Rosenzweig, M. L., and Z. Abramsky. 1993. How are diversity and productivity related? In *Species Diversity in Ecological Communities: Historical and Geographic Perspectives* (R. E. Ricklefs and D. Schluter, Eds.), pp. 52–65. Univ. of Chicago Press, Chicago, IL.

Roth, S. K., and R. L. Lindroth. 1994. Effects of CO_2-mediated changes in paper birch and white pine chemistry on gypsy moth performance. *Oecologia* **98**: 133–138.

Roubik, D. W. 1989. *Ecology and Natural History of Tropical Bees*. Cambridge Univ. Press, Cambridge, UK.

Roubik, D. W. 1993. Tropical pollinators in the canopy and understory: Field data and theory for stratum "preferences." *Journal of Insect Behavior* **6**: 659–673.

Royama, T. 1984. Population dynamics of the spruce budworm *Choristoneura fumiferana*. *Ecological Monographs* **54**: 429–462.

Royama, T. 1992. *Analytical Population Dynamics*. Chapman & Hall, London, UK.

Ruangpanit, N. 1985. Percent crown cover related to water and soil losses in mountainous forest in Thailand. In *Soil Erosion and Conservation* (S. A. El-Swaify, W. C. Moldenhauer, and A. Lo, Eds.), pp. 462–471. Soil Conservation Society of America, Ankeny, IA.

Rubenstein, D. I. 1992. The greenhouse effect and changes in animal behavior: Effects on social structure and life-history strategies. In *Global Warming and Biological Diversity* (R. L. Peters and T. E. Lovejoy, Eds.), pp. 180–192. Yale Univ. Press, New Haven, CT.

Rudd, W. G., and R. W. Gandour. 1985. Diffusion model for insect dispersal. *Journal of Economic Entomology* **78**: 295–301.

Rudinsky, J. A., and L. C. Ryker. 1977. Olfactory and auditory signals mediating behavioral patterns of bark beetles. In *Coll. Internat., Centre National de al Recherche Scientifique* pp. 195–207, No. 265, Paris, France.

Running, S. W., and S. T. Gower. 1991. FOREST-BGC, a general model of forest ecosystem processes for regional applications. II. Dynamic carbon allocation and nitrogen budgets. *Tree Physiology* 9: 147–160.

Rykiel, E. J., M. C. Saunders, T. L. Wagner, D. K. Loh, R. H. Turnbow, L. C. Hu, P. E. Pulley, and R. N. Coulson. 1984. Computer-aided decision making and information accessing in pest management systems, with emphasis on the southern pine beetle (Coleoptera: Scolytidae). *Journal of Economic Entomology* 77: 1073–1082.

Rykken, J. J., D. E. Capen, and S. P. Mahabir. 1997. Ground beetles as indicators of land type diversity in the Green Mountains of Vermont. *Conservation Biology* 11: 522–530.

Sackville Hamilton, C. A. G., J. M. Cherrett, J. B. Ford, G. R. Sagar, and R. Whitbread. 1991. A modular rhizotron for studying soil organisms: construction and establishment. In *Plant Root Growth: An Ecological Perspective* (D. Atkinson, Ed.), pp. 49–59. Blackwell Scientific, London, UK.

Salick, J., R. Herrera. and C. F. Jordan. 1983. Termitaria: Nutrient patchiness in nutrient-deficient rain forests. *Biotropica* 15: 1–7.

Sallabanks, R., and S. P. Courtney. 1992. Frugivory, seed predation, and insect–vertebrate interactions. *Annual Review of Entomology* 37: 377–400.

Salati, E. 1987. The forest and the hydrologic cycle. In *The Geophysiology of Amazonia: Vegetation and Climate Interactions* (R. E. Dickinson, Ed.), pp. 273–296. John Wiley & Sons, New York.

Salt, D. T., P. Fenwick, and J. B. Whittaker. 1996. Interspecific herbivore interactions in a high CO_2 environment: Root and shoot aphids feeding on *Cardamine*. *Oikos* 77: 326–330.

Samways, M. J. 1995. Southern hemisphere insects: Their variety and the environmental pressures upon them. In *Insects in a Changing Environment* (R. Harrington and N. E. Stork, Eds.), pp. 297–320. Academic Press, London, UK.

Samways, M. J., P. M. Caldwell, and R. Osborn. 1996. Ground-living invertebrate assemblages in native, planted and invasive vegetation in South Africa. *Agriculture, Ecosystems and Environment* 59: 19–32.

Sanderson, M. G. 1996. Biomass of termites and their emissions of methane and carbon dioxide: A global database. *Global Biogeochemical Cycles* 10: 543–557.

Sandlin, E. A., and M. R. Willig. 1993. Effects of age, sex, prior experience, and intraspecific food variation on diet composition of a tropical folivore (Phasmatodea: Phasmatidae). *Environmental Entomology* 22: 625–633.

Santos, P. F., and W. G. Whitford 1981. The effects of microarthropods on litter decomposition in a Chihuahuan Desert ecosystem. *Ecology* 62: 654–663.

Santos, P. F., J. Phillips, and W. G. Whitford. 1981. The role of mites and nematodes in early stages of buried litter decomposition in a desert. *Ecology* 62: 664–669.

Sarmiento, J. L., and Le Quéré, C. 1996. Oceanic carbon dioxide uptake in a model of century-scale global warming. *Science* 274: 1346–1350.

Sartwell, C., and R. E. Stevens. 1975. Mountain pine beetle in ponderosa pine: Prospects for silvicultural control in second-growth stands. *Journal of Forestry* 73: 136–140.

Savely, H. E., Jr. 1939. Ecological relations of certain animals in dead pine and oak logs. *Ecological Monographs* 9: 321–385.

Scatena, F. N., S. Moya, C. Estrada, and J. D. Chinea. 1996. The first five years in the reorganization of aboveground biomass and nutrient use following Hurricane Hugo in the Bisley Experimental Watersheds, Luquillo Experimental Forest, Puerto Rico. *Biotropica* 28: 424–440.

Schell, S. P., and J. A. Lockwood. 1997. Spatial characteristics of rangeland grasshopper (Orthoptera: Acrididae) population dynamics in Wyoming: Implications for pest management. *Environmental Entomology* 26: 1056–1065.

Schlesinger, W. H., J. F. Reynolds, G. L. Cunningham, L. F. Huenneke, W. M. Jarrell, R. A. Virginia, and W. G. Whitford. 1990. Biological feedbacks in global desertification. *Science* 247: 1043–1048.

Schmidt, J. O. 1982. Biochemistry of insect venoms. *Annual Review of Entomology* 27: 339–368.

Schnierla, T. C. 1953. Modifiability in insect behavior. In *Insect Physiology* (K. D. Roeder, Ed.), pp. 723–747. John Wiley & Sons, New York.

Schoener, T. W. 1982. The controversy over interspecific competition. *American Scientist* 70:586–595.

Schöpf, R., C. Mignat, and P. Hedden. 1982. As to the food quality of spruce needles for forest damaging insects: 18: Resorption of secondary plant metabolites by the sawfly, *Gilpinia hercyniae* Htg. (Hym., Diprionidae). *Zeitschrift für angewandte Entomologie* 93: 244–257.

Schowalter, T. D. 1981. Insect herbivore relationship to the state of the host plant: Biotic regulation of ecosystem nutrient cycling through ecological succession. *Oikos* 37: 126–130.

Schowalter, T. D. 1985. Adaptations of insects to disturbance. In *The Ecology of Natural Disturbance and Patch Dynamics* (S. T. A. Pickett and P. S. White, Eds.), pp. 235–252. Academic Press, Orlando, FL.

Schowalter, T. D. 1986. Overwintering aggregation of *Boisea rubrolineatus* (Heteroptera: Rhopalidae) in western Oregon. *Environmental Entomology* 15: 1055–1056.

Schowalter, T. D. 1989. Canopy arthropod community structure and herbivory in old-growth and regenerating forests in western Oregon. *Canadian Journal of Forest Research* **19**: 318–322.

Schowalter, T. D. 1991. Forest ecology. In *1992 Yearbook of Science and Technology* pp. 169–172. McGraw-Hill, New York.

Schowalter, T. D. 1993. Cone and seed insect phenology in a Douglas-fir seed orchard during three years in western Oregon. *Journal of Economic Entomology* **87**: 758–765.

Schowalter, T. D. 1994a. Invertebrate community structure and herbivory in a tropical rainforest canopy in Puerto Rico following Hurricane Hugo. *Biotropica* **26**:312–319.

Schowalter, T. D. 1994b. An ecosystem-centered view of insect and disease effects on forest health. In *Sustainable Ecological Systems: Implementing an Ecological Approach to Land Management* (W. W. Covington and L. F. DeBano, Eds), pp. 189–195. USDA Forest Service Gen. Tech. Rpt. RM-247, USDA Forest Service, Rocky Mountain Forest and Range Exp. Stn., Fort Collins, CO.

Schowalter, T. D. 1995a. Canopy arthropod communities in relation to forest age and alternative harvest practices in western Oregon. *Forest Ecology and Management* **78**: 115–125.

Schowalter, T. D. 1995b. Canopy invertebrate community response to disturbance and consequences of herbivory in temperate and tropical forests. *Selbyana* **16**: 41–48.

Schowalter, T. D. 1996. Stand and landscape diversity as a mechanism of forest resistance to insects. In *Dynamics of Forest Herbivory: Quest for Pattern and Principle* (W. J. Mattson, P. Niemela, and M. Rousi, Eds.), pp. 21–27. USDA Forest Serv. Gen. Tech. Rep. NC-183. USDA Forest Serv., North Central Forest Exp. Stn., St. Paul, MN.

Schowalter, T. D. 2000. Insects as regulators of ecosystem development. In *Invertebrates as Webmasters in Ecosystems* (D. C. Coleman and P. Hendrix, Eds.), CAB International, Wallingford, Oxon, UK (in press).

Schowalter, T. D. and D. A. Crossley, Jr. 1982. Bioelimination of ^{51}Cr and ^{85}Sr by cockroaches, *Gromphadorhina portentosa* (Orthoptera: Blaberidae), as affected by mites, *Gromphadorholaelaps schaeferi* (Parasitiformes: Laelapidae). *Annals of the Entomological Society of America* **75**: 158–160.

Schowalter, T. D., and D. A. Crossley, Jr. 1983. Forest canopy arthropods as sodium, potassium, magnesium and calcium pools in forests. *Forest Ecology and Management* **7**: 143–148.

Schowalter, T. D., and D. A. Crossley, Jr. 1988. Canopy arthropods and their response to forest disturbance. In *Forest Hydrology and Ecology at Coweeta* (W. T. Swank and D. A. Crossley, Jr., Eds.), pp. 207–218. Springer-Verlag, New York.

Schowalter, T. D., and L. M. Ganio. 1998. Vertical and seasonal variation in canopy arthropod communities in an old-growth conifer forest in southwestern Washington, USA. *Bulletin of Entomological Research* **88**: 633–640.

Schowalter, T. D., and L. M. Ganio. 1999. Invertebrate communities in a tropical rain forest canopy in Puerto Rico following Hurricane Hugo. *Ecological Entomology* **24**: 1–11.

Schowalter, T. D., and M. D. Lowman. 1999. Forest herbivory by insects, In *Ecosystems of the World: Ecosystems of Disturbed Ground* (L. R. Walker, Ed.), pp. 269–285. Elsevier, Amsterdam, Netherlands (in press).

Schowalter, T. D., and J. E. Means. 1989. Pests link site productivity to the landscape. In *Maintaining the Long-term Productivity of Pacific Northwest Forest Ecosystems* (D. A. Perry, R. Meurisse, B. Thomas, R. Miller, J. Boyle, J. Means, C. R. Perry, and R. F. Powers, Eds.), pp. 248–250. Timber Press, Portland, OR.

Schowalter, T. D., and T. E. Sabin. 1991. Litter microarthropod responses to canopy herbivory, season and decomposition in litterbags in a regenerating conifer ecosystem in western Oregon. *Biology and Fertility of Soils* **11**: 93–96.

Schowalter, T. D., and P. Turchin. 1993. Southern pine beetle infestation development: Interaction between pine and hardwood basal areas. *Forest Science* **39**: 201–210.

Schowalter, T. D., and W. G. Whitford. 1979. Territorial behavior of *Bootettix argentatus* Bruner (Orthoptera: Acrididae). *American Midland Naturalist* **102**: 182–184.

Schowalter, T. D., W. G. Whitford, and R. B. Turner. 1977. Bioenergetics of the range caterpillar, *Hemileuca oliviae* (Ckll.). *Oecologia* **28**: 153–161.

Schowalter, T. D., R. N. Coulson, and D. A. Crossley, Jr. 1981a. Role of southern pine beetle and fire in maintenance of structure and function of the southeastern coniferous forest. *Environmental Entomology* **10**: 821–825.

Schowalter, T. D., D. N. Pope, R. N. Coulson, and W. S. Fargo. 1981b. Patterns of southern pine beetle (*Dendroctonus frontalis* Zimm.) infestation enlargement. *Forest Science* **27**: 837- 849.

Schowalter, T. D., J. W. Webb, and D. A. Crossley, Jr. 1981c. Community structure and nutrient content of canopy arthropods in clearcut and uncut forest ecosystems. *Ecology* **62**:1010–1019.

Schowalter, T. D., R. N. Coulson, R. H. Turnbow, and W. S. Fargo. 1982. Accuracy and precision of procedures for estimating populations of the southern pine beetle (Coleoptera: Scolytidae) by using host tree correlates. *Journal of Economic Entomology* **75**: 1009–1016.

Schowalter, T. D., W. W. Hargrove, and D. A. Crossley, Jr. 1986. Herbivory in forested ecosystems. *Annual Review of Entomology* **31**: 177–196.

Schowalter, T. D., T. E. Sabin, S. G. Stafford, and J. M. Sexton. 1991. Phytophage effects on primary production, nutrient turnover, and litter decomposition of young Douglas-fir in western Oregon. *Forest Ecology and Management* **42**: 229–243.

Schowalter, T. D., B. A. Caldwell, S. E. Carpenter, R. P. Griffiths, M. E. Harmon, E. R. Ingham, R. G. Kelsey, J. D. Lattin, and A. R. Moldenke. 1992. Decomposition of fallen trees: Effects of initial conditions and heterotroph colonization rates. In *Tropical Ecosystems: Ecology and Management* (K. P. Singh and J. S. Singh, Eds.), pp. 373–383. Wiley Eastern Ltd., New Delhi, India.

Schowalter, T. D., Y. L. Zhang, and T. E. Sabin. 1998. Decomposition and nutrient dynamics of oak *Quercus* spp. logs after five years of decomposition. *Ecography* **21**: 3–10.

Schowalter, T. D., D. C. Lightfoot, and W. G. Whitford. 1999. Diversity of arthropod responses to host-plant water stress in a desert ecosystem in southern New Mexico. *American Midland Naturalist* (in press).

Schroll, H. 1994. Energy-flow and ecological sustainability in Danish agriculture. *Agriculture, Ecosystems and Environment* **51**: 301–310.

Schultz, J. C. 1983. Habitat selection and foraging tactics of caterpillars in heterogeneous trees. In *Variable Plants and Herbivores in Natural and Managed Systems* (R. F. Denno and M. S. McClure, Eds.), pp. 61–90. Academic Press, New York.

Schultz, J. C. 1988. Many factors influence the evolution of herbivore diets, but plant chemistry is central. *Ecology* **69**: 896–897.

Schulze, E. D., and H. A. Mooney, Eds. 1993. *Biodiversity and Ecosystem Function.* Springer-Verlag, Berlin, Germany.

Schupp, E. W. 1988. Seed and early seedling predation in the forest understory and in treefall gaps. *Oikos* **51**: 71–78.

Schupp, E. W., and D. H. Feener, Jr. 1991. Phylogeny, lifeform, and habitat dependence of ant-defended plants in a Panamanian forest. In *Ant–Plant Interactions.* (C. R. Huxley and D. F. Cutler, Eds.), pp. 175–197. Oxford Univ. Press, Oxford, UK.

Scott, A. C., and T. N. Taylor. 1983. Plant/animal interactions during the Upper Carboniferous. *The Botanical Review* **49**: 259–307.

Scriber, J. M., and F. Slansky, Jr. 1981. The nutritional ecology of immature insects. *Annual Review of Entomology* **26**: 183–211.

Seastedt, T. R. 1984. The role of microarthropods in decomposition and mineralization processes. *Annual Review of Entomology* **29**: 25–46.

Seastedt, T. R. 1985. Maximization of primary and secondary productivity by grazers. *American Naturalist* **126**: 559–564.

Seastedt, T. R. 2000. Soil fauna and controls of carbon dynamics: Comparisons of rangelands and forests across latitudinal gradients. In *Invertebrates as Webmasters in Ecosystems* (D.C. Coleman and P. Hendrix, Eds.), CAB International, Wallingford, Oxon, UK.

Seastedt, T. R., and D. A. Crossley, Jr. 1980. Effects of microarthropods on the seasonal dynamics of nutrients in forest litter. *Soil Biology and Biochemistry* **12**: 337–342.

Seastedt, T. R., and D. A. Crossley, Jr. 1981a. Microarthropod response following cable logging and clear-cutting in the southern Appalachians. *Ecology* **62**: 126–135.

Seastedt, T. R., and D. A. Crossley, Jr. 1981b. Sodium dynamics in forest ecosystems and the animal starvation hypothesis. *American Naturalist* **117**: 1029–1034.

Seastedt, T. R., and D. A. Crossley, Jr. 1983. Nutrients in forest litter treated with naphthalene and simulated throughfall: A field microcosm study. *Soil Biology and Biochemistry* **15**: 159–165.

Seastedt, T. R., and D. A. Crossley, Jr. 1984. The influence of arthropods on ecosystems. *BioScience* **34**: 157–161.

Seastedt, T. R., and C. M. Tate. 1981. Decomposition rates and nutrient contents of arthropod remains in forest litter. *Ecology* **62**: 13–19.

Seastedt, T. R., D. A. Crossley, Jr., and W. W. Hargrove. 1983. The effects of low-level consumption by canopy arthropods on the growth and nutrient dynamics of black locust and red maple trees in the southern Appalachians. *Ecology* **64**: 1040–1048.

Seastedt, T. R., R. A. Ramundo, and D. C. Hayes. 1988. Maximization of densities of soil animals by foliage herbivory: Empirical evidence, graphical and conceptual models. *Oikos* **51**: 243–248.

Seastedt, T. R., M. V. Reddy, and S. P. Cline. 1989. Microarthropods in decomposing wood from temperate coniferous and deciduous forests. *Pedobiologia* **33**: 69–78.

Setälä, H., and V. Huhta. 1991. Soil fauna increase *Betula pendula* growth: Laboratory experiments with coniferous forest floor. *Ecology* **72**: 665–671.

Setälä, H., V. G. Marshall, and J. A. Trofymow. 1996. Influence of body size of soil fauna on litter decomposition and ^{15}N uptake by poplar in a pot trial. *Soil Biology and Biochemistry* **28**: 1661–1675.

Shaw, C. G., III, and B. B. Eav. 1993. Modeling interactions. In *Beetle–pathogen Interactions in Conifer Forests* (T. D. Schowalter and G. M. Filip, Eds.), pp. 199–208. Academic Press, London.

Shaw, D. C. 1998. Distribution of larval colonies of *Lophocampa argentata* Packard, the silver spotted tiger moth (Lepidoptera: Arctiidae), in an old growth Douglas-fir/western hemlock forest canopy, Cascade Mountains, Washington State, USA. *Canadian Field Naturalist* **112**: 250–253.

Shaw, P. B., D. B. Richman, J. C. Owens, and E. W.

Huddleston. 1987. Ecology of the range caterpillar, *Hemileuca oliviae* Cockerell. In *Integrated Pest Management on Rangeland: a Shortgrass Prairie Perspective* (J. L. Capinera, Ed.), pp. 234–247. Westview Press, Boulder, CO.

Shea, S. R., M. McCormick, and C. C. Portlock. 1979. The effect of fires on regeneration of leguminous species in the northern jarrah (*Eucalyptus marginata* Sm) forest of Western Australia. *Australian Journal of Ecology* 4: 195–205.

Shelford, V. E. 1907. Preliminary note on the distribution of the tiger beetles (Cicindela) and its relation to plant succession. *Biological Bulletin* 14: 9–14.

Sheppard, P. M., J. R. G. Turner, K. S. Brown, W. W. Benson, and M. C. Singer. 1985. Genetics and the evolution of Muellerian mimicry in *Heliconius* butterflies. *Philosophical Transactions of the Royal Society of London B Biological Sciences* 308: 433–613.

Sherratt, T. N., and P. C. Jepson. 1993. A metapopulation approach to modelling the long-term impact of pesticides on invertebrates. *Journal of Applied Ecology* 30: 696–705.

Shettleworth, S. J. 1984. Learning and behavioural ecology. In *Behavioural Ecology: An Evolutionary Approach* (J. R. Krebs and N. B. Davies, Eds.), pp. 170–194. Blackwell Scientific, Oxford, UK.

Shugart, H. H., D. C. West, and W. R. Emanuel. 1981. Patterns and dynamics of forests: An application of simulation models. In *Forest Succession: Concepts and Application* (D. C. West, H. H. Shugart, and D. B. Botkin, Eds.), pp. 74–94. Springer-Verlag, New York.

Shure, D. J., and D. L. Phillips. 1991. Patch size of forest openings and arthropod populations. *Oecologia* 86: 325–334.

Shure, D. J., and L. A. Wilson. 1993. Patch-size effects on plant phenolics in successional openings of the southern Appalachians. *Ecology* 74: 55–67.

Siepel, H., and E. M. de Ruiter-Dijkman. 1993. Feeding guilds of oribatid mites based on their carbohydrase activities. *Soil Biology and Biochemistry* 25: 1491–1497.

Silva, S. I., W. P. MacKay, and W. G. Whitford. 1985. The relative contributions of termites and microarthropods to fluff grass litter disappearance in the Chihuahuan Desert. *Oecologia* 67: 31–34.

Simberloff, D. S. 1969. Experimental zoology of islands: A model for insular colonization. *Ecology* 50: 296–314.

Simberloff, D. S. 1974. Equilibrium theory of island biogeography and ecology. *Annual Review of Ecology and Systematics* 5: 161–182.

Simberloff, D. S. 1978. Colonization of islands by insects: Immigration, extinction, and diversity. In *Diversity of Insect Faunas* (L. A. Mound and N. Waloff, Eds.), pp. 139–153. Symposium of the Royal Entomological Society, London No. 9., Royal Entomological Society, London, UK.

Simberloff, D. S., and T. Dayan. 1991. The guild concept and the structure of ecological communities. *Annual Review of Ecology and Systematics* 22: 115–143.

Simberloff, D. S.. and E. O. Wilson. 1969. Experimental zoogeography of islands: The colonization of empty islands. *Ecology* 50: 278–295.

Simons, P. 1981. The potato bites back. *New Scientist* 20: 470–474.

Sims, P. L., and J. S. Singh. 1978. The structure and function of ten western North American grasslands. III. Net primary production, turnover and efficiencies of energy capture and water use. *Journal of Ecology* 66: 573–597.

Sinclair, A. R. E. 1975. The resource limitation of trophic levels in tropical grassland ecosystems. *Journal of Animal Ecology* 44: 497–520.

Skarmoutsos, G., and C. Millar. 1982. *Adelges* aphids and fungi causing premature defoliation of larch. *European Journal of Forest Pathology* 12: 73–78.

Skellam, J. G. 1951. Random dispersal in theoretical populations. *Biometrika* 38: 196–218.

Slansky, F., Jr. 1978. Utilization of energy and nitrogen by larvae of the imported cabbageworm, *Pieris rapae*, as affected by parasitism by *Apanteles glomeratus*. *Environmental Entomology* 7: 179–185.

Smalley, A. E. 1960. Energy flow of a salt marsh grasshopper population. *Ecology* 41: 672–677.

Smedley, S. R., and T. Eisner. 1995. Sodium uptake by puddling in a moth. *Science* 270: 1816–1818.

Smith, J. M. 1964. Group selection and kin selection. *Nature* 201: 1145–1147.

Smith, J. M. B. 1989. An example of ant-assisted plant invasion. *Australian Journal of Ecology* 14: 247–250.

Smith, W. H. 1981. *Air Pollution and Forests: Interactions Between Air Contaminants and Forest Ecosystems*. Springer-Verlag, New York.

Sollins, P., S. P. Cline, R. Verhoeven, D. Sachs, and G. Spycher. 1987. Patterns of log decay in old-growth Douglas-fir forests. *Canadian Journal of Forest Research* 17: 1585–1595.

Solomon, A. M., D. C. West and J. A. Solomon. 1981. Simulating the role of climate change and species immigration in forest succession. In *Forest Succession: Concepts and Application* (D. C. West, H. H. Shugart, and D. B. Botkin, Eds.), pp. 154–177. Springer-Verlag, New York.

Somlyódy, L., and G. van Straten (Eds.). 1986. *Modeling and Managing Shallow Lake Eutrophication, with Application to Lake Balaton*. Springer-Verlag, Berlin, Germany.

Soulé, M. E., and D. Simberloff. 1986. What do genetics and ecology tell us about the design of nature refuges? *Biological Conservation* 35: 19–40.

Sousa, W. P. 1979. Disturbance in marine intertidal

boulder fields: The nonequilibrium maintenance of species diversity. *Ecology* 60: 1225–1239.

Sousa, W. P. 1985. Disturbance and patch dynamics on rocky intertidal shores. In *Ecology of Natural Disturbance and Patch Dynamics* (S. T. A. Pickett and P. S. White, Eds.), pp. 101–124. Academic Press, New York.

Southwood, T. R. E. 1975. The dynamics of insect populations. In *Insects, Science, and Society* (D. Pimentel, Ed.), pp. 151–199. Academic Press, San Diego, CA.

Southwood, T. R. E. 1977. The relevance of population dynamics theory to pest status. In *Origins of Pest, Parasite, Disease and Weed Problems* (J. M. Cherrett and G. R. Sagar, Eds.), pp. 35–54. Symposium of the British Ecological Society 18, British Ecological Society, London, UK.

Southwood, T. R. E. 1978. *Ecological Methods with Particular Reference to the Study of Insect Populations.* Methuen, Inc., London, UK.

Spain, A. V., and R. P. Le Feuvre. 1987. Breakdown of four litters of contrasting quality in a tropical Australian rain forest. *Journal of Applied Ecology* 24: 279–288.

Spain, A. V., and R. P. Le Feuvre. 1997. Stable C and N isotope values of selected components of a tropical Australian sugarcane ecosystem. *Biology and Fertility of Soils* 24: 118–222.

Spencer, H. J., and G. R. Port. 1988. Effects of roadside conditions on plants and insects. II. Soil conditions. *Journal of Applied Ecology* 25: 709–715.

Spencer, H. J., N. E. Scott, G. R. Port, and A. W. Davison. 1988. Effects of roadside conditions on plants and insects. I. Atmospheric conditions. *Journal of Applied Ecology* 25: 699–707.

Springett, B. P. 1968. Aspects of the relationship between burying beetles, *Necrophorus* spp., and the mite, *Poecilochirus necrophori* Vitz. *Journal of Animal Ecology* 37: 417–424.

Stachurski, A., and J. R. Zimka. 1984. The budget of nitrogen dissolved in rainfall during its passing through the crown canopy in forest ecosystems. *Ekologia Polska* 32: 191–218.

Stadler, B., and T. Müller. 1996. Aphid honeydew and its effect on the phyllosphere microflora of *Picea abies* (L.) Karst. *Oecologia* 108: 771–776.

Stadler, B., B. Michalzik, and T. Müller. 1998. Linking aphid ecology with nutrient fluxes in a coniferous forest. *Ecology* 79: 1514–1525.

Städler, E. 1984. Perceptual mechanisms. In *Chemical Ecology of Insects* (W. J. Bell and R. T. Cardé, Eds.), pp. 3–35. Chapman and Hall, London, UK.

Stamp, N. E. 1992. Relative susceptibility to predation of two species of caterpillars on plantain. *Oecologia* 92: 124–129.

Stamp, N. E., and M. D. Bowers. 1990. Variation in food quality and temperature constrain foraging of gregarious caterpillars. *Ecology* 71: 1031–1039.

Stamp, N. E., Y. Yang, and T. L. Osier. 1997. Response of an insect predator to prey fed multiple allelochemicals under representative thermal regimes. *Ecology* 78: 203–214.

Stanton, M. L. 1983. Spatial patterns in the plant community and their effects upon insect search. In *Herbivorous Insects: Host-seeking Behavior and Mechanisms* (S. Ahmad, Ed.), pp. 125–157. Academic Press, New York.

Stanton, N. 1975. Herbivore pressure on 2 types of forests. *Biotropica* 7: 8–11.

Starzyk, J. R., and Z. Witkowski. 1981. Changes in the parameters describing the cambio- and xylophagous insect communities during the secondary succession of the oak–hornbeam association in the Niepołomice Forest near Kraków. *Zeitschrift für angewandte Entomologie* 91: 525–533.

Steffan-Dewenter, I., and T. Tscharntke. 1997. Bee diversity and seed set in fragmented habitats. In *Pollination: From Theory to Practise* (K. W. Richards, Ed.), pp. 231–234. Proceedings of the 7th International Symposium on Pollination. Acta Horticulturae #437.

Stephen, F. M., C. W. Berisford, D. L. Dahlsten, P. Fenn, and J. C. Moser. 1993. Invertebrate and microbial associates. In *Beetle-pathogen Interaction in Conifer Forests* (T. D. Schowalter and G. M. Filip, Eds.), pp. 129–153. Academic Press, London, UK.

Stephens, D. W., and J. R. Krebs. 1986. *Foraging Theory.* Princeton Univ. Press, Princeton, NJ.

Stermitz, F. R., J. N. Tawara, M. Boeckl, M. Pomeroy, T. A. Foderaro, and F. G. Todd. 1994. Piperidine alkaloid content of *Picea* (spruce) and *Pinus* (pine). *Phytochemistry* 35: 951- 953.

Sticher, L., B. Mauch-Mani, and M. P. Métraux. 1997. Systematic acquired resistance. *Annual Review of Phytopathology* 35: 235–270.

Stiles, J. H., and R. H. Jones. 1998. Distribution of the red imported fire ant, *Solenopsis invicta*, in road and powerline habitats. *Landscape Ecology* 13: 335–346.

Stiling, P. D. 1996. *Ecology: Theories and Applications.* Prentice-Hall, Upper Saddle River, NJ.

Stiling, P. D., D. Simberloff, and B. V. Brodbeck. 1991. Variation in rates of leaf abscission between plants may affect the distribution patterns of sessile insects. *Oecologia* 88: 367–370.

Stork, N. E. 1987. Guild structure of arthropods from Bornean rain forest trees. *Ecological Entomology* 12: 69–80.

Stout, J., and J. Vandermeer. 1975. Comparison of species richness for stream-inhabiting insects in tropical and mid-latitude streams. *American Naturalist* 109: 263–280.

Streams, F. A. 1994. Effect of prey size on attack components of the functional response by *Notonecta undulata. Oecologia* 98: 57–63.

Strom, B. L., L. M. Roton, R. A. Goyer, and J. R. Meeker. 1999. Visual and semiochemical disruption of host finding in the southern pine beetle. *Ecological Applications* 9: 1028–1038.

Strong, D. R. 1992. Are trophic cascades all wet? Differentiation and donor-control in speciose ecosystems. *Ecology* 73: 747–754.

Strong, D. R., J. H. Lawton, and T. R. E. Southwood. 1984. *Insects on Plants: Community Patterns and Mechanisms*. Harvard Univ. Press, Cambridge, MA.

Strong, W. B., B. A. Croft, and D. H. Slone. 1997. Spatial aggregation and refugia of the mites *Tetranychus urticae* and *Neoseiulus fallacis* (Acari: Tetranychidae, Phytoseiidae) on hop. *Environmental Entomology* 26: 859–865.

Sturgeon, K. B., and J. B. Mitton. 1986. Allozyme and morphological differentiation of mountain pine beetles *Dendroctonus ponderosae* Hopkins (Coleoptera: Scolytidae) associated with host trees. *Evolution* 40: 290–302.

Swank, W. T., J. B. Waide, D. A. Crossley, Jr., and R. L. Todd. 1981. Insect defoliation enhances nitrate export from forest ecosystems. *Oecologia* 51: 297–299.

Swetnam, T. W., and A. M. Lynch. 1989. A tree-ring reconstruction of western spruce budworm history in the southern Rocky Mountains. *Forest Science* 35: 962–986.

Swift, M. J. 1977. The ecology of wood decomposition. *Science Progress (Oxford)* 64: 175–199.

Swift, M. J., O. W. Heal, and J. M. Anderson. 1979. *Decomposition in Terrestrial Ecosystems*. Blackwell Scientific, Oxford, UK.

Sword, M. A. 1998. Seasonal development of loblolly pine lateral roots in response to stand density and fertilization. *Plant and Soil* 200: 21–25.

Szujko-Lacza J. 1982. *The Flora of the Hortobágy National Park*. Akadémiai Kiadó, Budapest, Hungary.

Szujko-Lacza J., and D. Kovacs (Eds.). 1993. *The Flora of the Kiskunság National Park*. Akadémiai Kiadó, Budapest, Hungary.

Tabashnik, B. E. 1994. Evolution of resistance to *Bacillus thuringiensis*. *Annual Review of Entomology* 39: 47–79.

Tabashnik, B. E., N. Finson, F. R. Groeters, W. J. Moaar, M. W. Johnson, K. Luo, and M. J. Adang. 1994. Reversal of resistance to *Bacillus thuringiensis* in *Plutella xylostella*. *Proceedings of the National Academy of Sciences USA* 91: 4120–4124.

Tabashnik, B. E., F. R. Groeters, N. Finson, Y. B. Liu, M. W. Johnson, D. G. Heckel, K. Luo, and M. L. Adang. 1996. Resistance to *Bacillus thuringiensis* in *Plutella xylostella*: The moth heard round the world. In *Molecular Genetics and Evolution of Pesticide Resistance* pp. 130–140. American Chemical Society Symposium Series 645, Washington, DC.

Tabashnik, B. E., Y. B. Liu, N. Finson, L. Masson, and D. G. Heckel. 1997. One gene in diamondback moth confers resistance to four *Bacillus thuringiensis* toxins. *Proceedings of the National Academy of Sciences USA* 94: 1640–1644.

Tahvanainen, J. O., and R. B. Root. 1972. The influence of vegetational diversity on the population ecology of a specialized herbivore, *Phyllotreta cruciferae* (Coleoptera: Chrysomelidae). *Oecologia* 10: 321–346.

Tallamy, D. W., and F. T. Halaweish. 1993. Effects of age, reproductive activity, sex, and prior exposure on sensitivity to cucurbitacins in southern corn rootworm (Coleoptera: Chrysomelidae). *Environmental Entomology* 22: 922–925.

Tallamy, D. W., J. Stull, N. P. Ehresman, P. M. Gorski, and C. E. Mason. 1997. Cucurbitacins as feeding and oviposition deterrents to insects. *Environmental Entomology* 26: 678–683.

Tallamy, D. W., D. P. Whittington, F. Defurio, D. A. Fontaine, P. M. Gorski, and P. W. Gothro. 1998. Sequestered cucurbitacins and pathogenicity of *Metarhizium anisopliae* (Moniliales: Moniliaceae) on spotted cucumber beetle eggs and larvae (Coleoptera: Chrysomelidae). *Environmental Entomology* 27: 366–372.

Tanada, Y., and H. Kaya. 1993. *Insect Pathology*. Academic Press, San Diego. CA.

Tansley, A. G. 1935. The use and abuse of vegetational concepts and terms. *Ecology* 16: 284–307.

Tantawi, T. I., E. M. El-Kady, B. Greenberg, and H. A. El-Ghaffar. 1996. Arthropod succession on exposed rabbit carrion in Alexandria, Egypt. *Journal of Medical Entomology* 33: 566–580.

Teal, J. M. 1957. Community metabolism in a temperate cold spring. *Ecological Monographs* 27: 283–302.

Teal, J. M. 1962. Energy flow in the salt marsh ecosystem of Georgia. *Ecology* 43: 614–624.

Temple, S. A. 1977. Plant–animal mutualism: Coevolution with dodo leads to near extinction of plant. *Science* 197: 885–886.

Terborgh, J. 1973. On the notion of favorableness in plant ecology. *American Naturalist* 107: 481–501.

Terborgh, J. 1985. The vertical component of plant species diversity in temperate and tropical forests. *American Naturalist* 126: 760–776.

Thomas, C. D., and I. Hanski. 1997. Butterfly populations. In *Metapopulation Biology: Ecology, Genetics and Evolution* (I. A. Hanski and M. E. Gilpin, Eds.), pp. 359–386. Academic Press, San Diego, CA.

Thomas, M. D., S. D. Wratten, and N. W. Sotherton. 1992. Creation of "island" habitats in farmland to manipulate populations of beneficial arthropods: Predator densities and species composition. *Journal of Applied Ecology* 29: 524–531.

Thompson, D. C., and K. T. Gardner. 1996. Importance of grasshopper defoliation period on southwestern blue grama-dominated rangeland. *Journal of Range Management* 49: 494–498.

Thornhill, R. 1976. Sexual selection and nuptial feeding behavior in *Bittacus apicalis* (Insecta: Mecoptera). *American Naturalist* 110: 529–548.

Tian, G., L. Brussaard, and B. T. Kang. 1995. Breakdown of plant residues with contrasting chemical compositions under humid tropical conditions: Effects of earthworms and millipedes. *Soil Biology and Biochemistry* 27: 277–280.

Tilman, D. 1978. Cherries, ants, and tent caterpillars: Timing of nectar production in relation to susceptibility of caterpillars to ant predation. *Ecology* 59: 686–692.

Tilman, D., and J. A. Downing. 1994. Biodiversity and stability in grasslands. *Nature* 367: 363–365.

Tilman, D., and S. Pacala. 1993. The maintenance of species richness in plant communities. In *Species Diversity in Ecological Communities: Historical and Geographic Perspectives* (R. E. Ricklefs and D. Schluter, Eds.), pp. 13–25. Univ. of Chicago Press, Chicago, IL.

Tilman, D., J. Knops, D. Wedin, P. Reich, M. Ritchie, and E. Siemann. 1997. The influence of functional diversity and composition on ecosystem processes. *Science* 277: 1300–1302.

Tinbergen, L. 1960. The natural control of insects in pinewoods. I. Factors influencing the intensity of predation by songbirds. *Archives Neerlandaises de Zoologie* 13: 265–343.

Tisdale, R. A., and M. R. Wagner. 1990. Effects of photoperiod, temperature, and humidity on oviposition and egg development of *Neodiprion fulviceps* (Hymenoptera: Diprionidae) on cut branches of ponderosa pine. *Environmental Entomology* 19: 456–458.

Torres, J. A. 1988. Tropical cyclone effects on insect colonization and abundance in Puerto Rico. *Acta Cientifica* 2: 40–44.

Torres, J. A. 1992. Lepidoptera outbreaks in response to successional changes after the passage of Hurricane Hugo in Puerto Rico. *Journal of Tropical Ecology* 8: 285–298.

Townsend, C. R., and R. N. Hughes. 1981. Maximizing net energy returns from foraging. In *Physiological Ecology: An Evolutionary Approach to Resource Use* (C. R. Townsend and P. Calow, Eds.), pp. 86–108. Blackwell Scientific, Oxford, UK.

Traugott, M. S., and N. E. Stamp. 1996. Effects of chlorogenic acid- and tomatine-fed prey on behavior of an insect predator. *Journal of Insect Behavior* 9: 461–476.

Trivers, R. L. 1971. The evolution of reciprocal altruism. *Quarterly Review of Biology* 46: 35–57.

Trlica, M. J. and L. R. Rittenhouse. 1993. Grazing and plant performance. *Ecological Applications* 3: 21–23.

Trumble, J. T., D. M. Kolodny-Hirsch, and I. P. Ting. 1993. Plant compensation for arthropod herbivory. *Annual Review of Entomology* 38: 93–119.

Tullis, K., and M. L. Goff. 1987. Arthropod succession in exposed carrion in a tropical rainforest on O'ahu Island, Hawai'i. *Journal of Medical Entomology* 24: 332–339.

Tumlinson, J. H., and P. E. A. Teal. 1987. Relationship of structure and function to biochemistry in insect pheromone systems. In *Pheromone Biochemistry* (G. D. Prestwich and G. J. Blomquist, Eds.), pp. 3–26. Academic Press, Orlando, FL.

Tuomi, J., P. Niemela, E. Haukioja, S. Siren, and S. Neuvonen. 1984. Nutrient stress: An explanation for plant anti-herbivore responses to defoliation. *Oecologia* 61: 208–210.

Turchin, P. 1988. The effect of host–plant density on the numbers of Mexican bean beetles, *Epilachna varivestis*. *American Midland Naturalist* 119: 15–20.

Turchin, P. 1990. Rarity of density dependence or population regulation with lags? *Nature* 344: 660–663.

Turchin, P. 1998. *Quantitative Analysis of Movement*. Sinauer Associates, Sunderland, MA.

Turgeon, J. J., A. Roques, and P. de Groot. 1994. Insect fauna of coniferous seed cones: diversity, host plant interactions, and management. *Annual Review of Entomology* 39: 179–212.

Turlings, T. C. J., J. H. Tumlinson, and W. J. Lewis. 1990. Exploitation of herbivore-induced plant odors by host-seeking parasitic wasps. *Science* 250: 1251–1253.

Turnbow, R. H., R. N. Coulson, L. Hu, and R. F. Billings. 1982. *Procedural Guide for Using the Interactive Version of the TAMBEETLE Model of Southern Pine Beetle Population and Spot Dynamics*. Texas Agricultural Experiment Station, Miscellaneous Publication MP-1518, Texas A&M Univ. College Station, TX.

Turner, G. B. 1970. The ecological efficiency of consumer populations. *Ecology* 51: 741–742.

Turner, M. G. 1989. Landscape ecology: The effect of pattern on process. *Annual Review of Ecology and Systematics* 20: 171–197.

Tyler, C. M. 1995. Factors contributing to postfire seedling establishment in chaparral: Direct and indirect effects of fire. *Journal of Ecology* 83: 1009–1020.

Ulanowicz, R. E. 1995. Utricularia's secret: The advantage of positive feedback in oligotrophic environments. *Ecological Modelling* 79: 49–57.

Van Cleve, K., and S. Martin. 1991. *Long-term Ecological Research in the United States* (6th ed.). Univ. of Washington, Seattle, WA.

Van Cleve, K., and L. A. Viereck. 1981. Forest succession in relation to nutrient cycling in the boreal forest of Alaska. In *Forest Succession: Concepts and*

Application (D. C. West, H. H. Shugart, and D. B. Botkin, Eds.), pp. 185–211. Springer-Verlag, New York.

van den Bosch, R., P. S. Messenger, and A. P. Gutierrez. 1982. *An Introduction to Biological Control.* Plenum Press, New York.

van der Maarel, E., and A. Titlyanova. 1989. Aboveground and below-ground biomass relations in steppes under different grazing conditions. *Oikos* 56: 364–370.

van Driesche, R. G., and T. Bellows. 1996. *Biological Control.* Chapman & Hall, New York.

van Hook, R. I., Jr., M. G. Nielsen, and H. H. Shugart. 1980. Energy and nitrogen relations for a *Macrosiphum liriodendri* (Homoptera: Aphididae) population in an east Tennessee *Liriodendron tulipifera* stand. *Ecology* 61: 960–975.

Vanni, M. J., and G. D. Layne. 1997. Nutrient recycling and herbivory as mechanisms in the "top-down" effect of fish on algae in lakes. *Ecology* 78: 21–40.

Vannote, R. L., G. W. Minshaw, K. W. Cummins, J. R. Sedell, and C. E. Cushing. 1980. The river continuum concept. *Canadian Journal of Fisheries and Aquatic Science* 37: 130–137.

Varley, G. C., and G. R. Gradwell. 1970. Recent advances in insect population dynamics. *Annual Review of Entomology* 15: 1–24.

Varley, G. C., G. R. Gradwell, and M. P. Hassell. 1973. *Insect Population Ecology: An Analytical Approach.* Blackwell Scientific, Oxford, UK.

Veblen, T. T., K. S. Hadley, E. M. Nel, T. Kitzberger, M. Reid, and R. Villalba. 1994. Disturbance regime and disturbance interactions in a Rocky Mountain subalpine forest. *Journal of Ecology* 82: 125–135.

Via, S. 1990. Ecological genetics and host adaptation in herbivorous insects: The experimental study of evolution in natural and agricultural systems. *Annual Review of Entomology* 35: 421–446.

Via, S. 1991a. The genetic structure of host plant adaptation in a spatial patchwork: Demographic variability among reciprocally transplanted pea aphid clones. *Evolution* 45: 827–852.

Via, S. 1991b. Specialized host plant performance of pea aphid clones is not altered by experience. *Ecology* 72: 1420–1427.

Vinson, M. R., and C. P. Hawkins. 1998. Biodiversity of stream insects: Variation at local, basin, and regional scales. *Annual Review of Entomology* 43: 271–293.

Visser, J. H. 1986. Host odor perception in phytophagous insects. *Annual Review of Entomology* 31: 121–144.

Vitousek, P. 1982. Nutrient cycling and nutrient use efficiency. *American Naturalist* 119: 553–572.

Vitousek, P. M., and D. U. Hooper. 1993. Biological diversity and terrestrial ecosystem biogeochemistry. In

Biodiversity and Ecosystem Function (E. D. Schulze and H. A. Mooney, Eds.), pp. 3–14. Springer-Verlag, Berlin, Germany.

Vitousek, P. M., H. A. Mooney, J. Lubchenco, and J. M. Melillo. 1997. Human domination of Earth's ecosystems. *Science* 277: 494–499.

Volney, W. J. A., J. E. Milstead, and V. R. Lewis. 1983. Effect of food quality, larval density and photoperiod on the feeding rate of the California oakworm (Lepidoptera: Diopidae). *Environmental Entomology* 12: 792–798.

Volterra, V. 1926. Fluctuations in the abundance of a species considered mathematically. *Nature* 118: 558–560.

Vossbrinck, C. R., D. C. Coleman, and T. A. Woolley. 1979. Abiotic and biotic factors in litter decomposition in a semiarid grassland. *Ecology* 60: 265–271.

Wagner, D. 1997. The influence of ant nests on *Acacia* seed production, herbivory and soil nutrients. *Journal of Ecology* 85: 83–93.

Wagner, D., M. J. F. Brown, and D. M. Gordon. 1997. Harvest ant nests, soil biota and soil chemistry. *Oecologia* 112: 232–236.

Wagner, R. 1991. The influence of the diel activity pattern of the larvae of *Sericostoma personatum* (Kirby & Spence) (Trichoptera) on organic matter distribution in stream sediments: A laboratory study. *Hydrobiologia* 224: 65–70.

Wagner, T. L., R. M. Feldman, J. A. Gagne, J. D. Cover, R. N. Coulson, and R. M. Schoolfield. 1981. Factors affecting gallery construction, oviposition, and reemergence of *Dendroctonus frontalis* in the laboratory. *Annals of the Entomological Society of America* 74: 255–273.

Waide, R. B., M. R. Willig, G. Mittelbach, C. Steiner, L. Gough, S. I. Dodson, G. P. Judy, and R. Parmenter. 1999. The relationship between primary productivity and species richness. *Annual Review of Ecology and Systematics* (in press).

Wallace, A. R. 1876. *The Geographical Distribution of Animals, with a Study of the Relations of Living and Extinct Faunas as Elucidating the Past Changes of the Earth's Surface.* Macmillan, London, UK.

Wallace, A. R. 1911. *Island Life, or the Phenomena and Causes of Insular Faunas and Floras Including a Revision and Attempted Solution of the Problem of Geological Climates.* Macmillan, London, UK.

Wallace, J. B., and J. J. Hutchens, Jr. 2000. Effects of invertebrates in lotic ecosystem processes. In *Invertebrates as Webmasters in Ecosystems* (D. C. Coleman and P. Hendrix, Eds.), CAB International, Wallingford, Oxon, UK (in press).

Wallace, J. B., and J. O'Hop. 1985. Life on a fast pad: Waterlily leaf beetle impact on water lilies. *Ecology* 66: 1534–1544.

Wallace, J. B., and J. R. Webster. 1996. The role of

macroinvertebrates in stream ecosystem function. *Annual Review of Entomology* **41**: 115–139.

Wallace, J. B., T. F. Cuffney, J. R. Webster, G. J. Lugthart, K. Chung, and G. S. Goldwitz. 1991. Export of fine organic particles from headwater streams: Effects of season, extreme discharges, and invertebrate manipulation. *Limnology and Oceanography* **36**: 670–682.

Wallace, J. B., J. R. Webster, and R. L. Lowe. 1992. High-gradient streams of the Appalachians. In *Biodiversity of Southeastern United States: Aquatic Communities* (C. T. Hackney, S. M. Adams, and W. A. Martin, Eds.), pp. 133–191. John Wiley, New York.

Wallace, J. B., S. L. Eggert, J. L. Meyer, and J. B. Webster. 1997. Multiple trophic levels of a forest stream linked to terrestrial litter inputs. *Science* **277**: 102–104.

Wallner, W. E. 1996. Invasive pests ("biological pollutants") and US forests: Whose problem, who pays? *EPPO Bulletin* **26**: 167–180.

Waloff, N., and P. Thompson. 1980. Census data of populations of some leafhoppers (Auchenorrhyncha, Homoptera) of acid grassland. *Journal of Animal Ecology* **49**: 395–416.

Walter, D. E., and D. J. O'Dowd. 1995. Life on the forest phyllophane: hairs, little houses, and myriad mites. In *Forest Canopies* (M. D. Lowman and N. M. Nadkarni, Eds.), pp. 325–351. Academic Press, San Diego, CA.

Ward, J. V., and J. A. Stanford. 1982. Thermal responses in the evolutionary ecology of aquatic insects. *Annual Review of Entomology* 97–117.

Waring, G. L., and N. S. Cobb. 1992. The impact of plant stress on herbivore population dynamics. In *Plant–Insect Interactions* (Vol. 4) (E. A. Bernays, Ed.), pp. 167–226. CRC Press, Boca Raton, FL.

Waring, G. L., and P. W. Price. 1990. Plant water stress and gall formation (Cecidomyiidae: *Asphondylia* spp.) on creosote bush (*Larrea tridentata*). *Ecological Entomology* **15**: 87–95.

Waring, R. H., and G. B. Pitman. 1983. Physiological stress in lodgepole pine as a precursor for mountain pine beetle attack. *Zeitschrift für angewandte Entomologie* **96**: 265–270.

Waring, R. H., and S. W. Running. 1998. *Forest Ecosystems: Analysis at Multiple Scales*. Academic Press, San Diego, CA.

Watson, M. A., and H. L. Nixon. 1953. Studies on the feeding of *Myzus persicae* (Sulz.) on radioactive plants. *Annals of Applied Biology* **40**: 537–545.

Watt, A. D., J. B. Whittaker, M. Docherty, G. Brooks, E. Lindsay, and D. T. Salt. 1995. The impact of elevated atmospheric CO_2 on insect herbivores. In *Insects in a Changing Environment* (R. Harrington and N. E. Stork, Eds.), pp. 197–217. Academic Press, London, UK.

Webb, W. L. 1978. Effects of defoliation and tree energetics. In *The Douglas-fir Tussock Moth: A Synthesis* (M. H. Brookes, R. W. Stark, and R. W. Campbell, Eds.), pp. 77–81. USDA Forest Service Tech. Bull. 1585, USDA Forest Service, Washington, DC.

Webb, W. L., W. K. Lauenroth, S. R. Szarek, and R. S. Kinerson. 1983. Primary production and abiotic controls in forests, grasslands, and desert ecosystems in the United States. *Ecology* **64**: 134–151.

Webster, J. R., J. B. Waide, and B. C. Patten. 1975. Nutrient recycling and the stability of ecosystems. In *Mineral Cycling in Southeastern Ecosystems* (F. G. Howell, J. B. Gentry, and M. H. Smith, Eds.), pp. 1–27. CONF-740513, USDOE Energy Research and Development Administration, Washington, DC.

Wedin, D. A., L. L. Tieszen, B. Dewey, and J. Pastor. 1995. Carbon isotope dynamics during grass decomposition and soil organic matter formation. *Ecology* **76**: 1383–1392.

Wegener, A. L. 1924. *Entstehung der Kontinente und Ozeane* (English translation, 3rd ed.). Methuen, London, UK.

Wellington, W. G. 1980. Dispersal and population change. In *Dispersal of Forest Insects: Evaluation, Theory and Management* (A. A. Berryman and L. Safranyik, Eds.), pp. 11–24. Proc. International Union of Forest Research Organizations Conference, Washington State Univ. Cooperative Extension Service, Pullman, WA.

Wellington, W. G., P. J. Cameron, W. A. Thompson, I. B. Vertinsky, and A. S. Landsberg. 1975. A stochastic model for assessing the effets of external and internal heterogeneity of an insect population. *Researches in Population Ecology* **17**: 1–28.

Wells, J. D., and B. Greenberg. 1994. Effect of the red imported fire ant (Hymenoptera: Formicidae) and carcass type on the daily occurrence of postfeeding carrion-fly larvae (Diptera: Calliphoridae, Sarcophagidae). *Journal of Medical Entomology* **31**: 171–174.

West, D. C., H. H. Shugart, and D. B. Botkin (Eds.). 1981. *Forest Succession: Concepts and Application*. Springer-Verlag, New York.

Westoby, M., K. French, L. Hughes, B. Rice, and L. Rodgerson. 1991. Why do more plant species use ants for dispersal on infertile compared with fertile soils? *Australian Journal of Ecology* **16**: 445–455.

Weygoldt, P. 1969. *The Biology of Pseudoscorpions*. Harvard Univ. Press, Cambridge, MA.

Wheeler, G. S., M. Tokoro, R. H. Scheffrahn, and N. Y. Su. 1996. Comparative respiration and methane production rates in Nearctic termites. *Journal of Insect Physiology* **42**: 799–806.

White, P. S., and S. T. A. Pickett. 1985. Natural disturbance and patch dynamics: An introduction. In

Ecology of Natural Disturbance and Patch Dynamics (S. T. A. Pickett and P. S. White, Eds.), pp. 3–13. Academic Press, New York.

White, T. C. R. 1969. An index to measure weather-induced stress of trees associated with outbreaks of psyllids in Australia. *Ecology* 50: 905–909.

White, T. C. R. 1976. Weather, food and plagues of locusts. *Oecologia* 22: 119–134.

White, T. C. R. 1984. The abundance of invertebrate herbivores in relation to the availability of nitrogen in stressed food plants. *Oecologia* 63: 90–105.

Whitford, W. G. 1978. Foraging by seed-harvesting ants. In *Production Ecology of Ants and Termites* (M. V. Brian, Ed.), pp. 107–110. Cambridge Univ. Press, Cambridge, UK.

Whitford, W. G. 1986. Decomposition and nutrient cycling in deserts. In *Pattern and Process in Desert Ecosystems* (W. G. Whitford, Ed.), pp. 93–117. Univ. of New Mexico Press, Albuquerque, NM.

Whitford, W. G. 1992. Effects of climate change on soil biotic communities and soil processes. In *Global Warming and Biological Diversity* (R. L. Peters and T. E. Lovejoy, Eds.), pp. 124–136. Yale Univ. Press, New Haven, CT.

Whitford, W. G. 2000. Arthropods as keystone webmasters in desert ecosystems. In *Invertebrates as Webmasters in Ecosystems* (D.C. Coleman and P. Hendrix, Eds.), CAB International, Wallingford, Oxon, UK (in press).

Whitford, W. G., P. Johnson, and J. Ramirez. 1976. Comparative ecology of the harvester ants *Pogonomyrmex barbatus* (F. Smith) and *Pogonomyrmex rugosus* (Emery). *Insectes Sociaux* 23: 117–132.

Whitford, W. G., V. Meentemeyer, T. R. Seastedt, K. Cromack, Jr., D. A. Crossley, Jr., P. Santos, R. L. Todd, and J. B. Waide. 1981. Exceptions to the AET model: Deserts and clear-cut forest. *Ecology* 62: 275–277.

Whitford, W. G., Y. Steinberger, and G. Ettershank. 1982. Contributions of subterranean termites to the "economy" of Chihuahuan Desert ecosystems. *Oecologia* 55: 298–302.

Whitham, T. G. 1983. Host manipulation of parasites: Within-plant variation as a defense against rapidly evolving pests. In *Variable Plants and Herbivores in Natural and Managed Systems* (R. F. Denno and M. S. McClure, Eds.), pp. 15–41. Academic Press, New York.

Whittaker, R. H. 1953. A consideration of climax theory: The climax as a population and pattern. *Ecological Monographs* 23: 41–78.

Whittaker, R. H. 1970. *Communities and Ecosystems*. Macmillan, London, UK.

Wickler, W. 1968. *Mimicry in Plants and Animals* [English translation by R. D. Martin]. Weidenfeld and Nicolson, Ltd., London, UK.

Wickman, B. E. 1964. Attack habits of *Melanophila*

consputa on fire-killed pines. *Pan-Pacific Entomologist* 40: 183–186.

Wickman, B. E., 1980. Increased growth of white fir after a Douglas-fir tussock moth outbreak. *Journal of Forestry* 78: 31–33.

Wickman, B. E. 1992. *Forest Health in the Blue Mountains: The Influence of Insects and Diseases*. USDA Forest Serv. Gen. Tech. Rpt. PNW-GTR-295. USDA Forest Serv., Pacific Northwest Res. Stn., Portland, OR.

Wiegert, R. G. 1964. Population energetics of meadow spittlebugs (*Philaenus spumarius* L.) as affected by migration and habitat. *Ecological Monographs* 34: 217–241.

Wiegert, R. G. 1968. Thermodynamic considerations in animal nutrition. *American Zoologist* 8: 71–81.

Wiegert, R. G., and F. C. Evans. 1967. Investigations of secondary productivity in grasslands. In *Secondary Productivity of Terrestrial Ecosystems: Principles and Methods* (K. Petrusewicz, Ed.), pp. 499–518. Państwowe Wydawnictwo Naukowe, Warszawa, Poland.

Wiegert, R. G., and C. E. Petersen. 1983. Energy transfer in insects. *Annual Review of Entomology* 28: 455–486.

Williams, D. W., and A. M. Liebhold. 1995. Forest defoliators and climatic change: Potential changes in spatial distribution of outbreaks of western spruce budworm (Lepidoptera: Tortricidae) and gypsy moth (Lepidoptera: Lymantriidae). *Environmental Entomology* 24: 1–9.

Williams, J. E., and R. G. Wiegert. 1971. Effects of naphthalene application on a coastal plain broomsedge (*Andropogon*) community. *Pedobiologia* 11: 58–65.

Williams, K. S., and C. Simon. 1995. The ecology, behavior, and evolution of periodical cicadas. *Annual Review of Entomology* 40: 269–295.

Williamson, M. 1972. *The Analysis of Biological Populations*. Edward Arnold, London, UK.

Williamson, S. C., J. K. Detling, J. L. Dodd, and M. I. Dyer. 1989. Experimental evaluation of the grazing optimization hypothesis. *Journal of Range Management* 42: 149–152.

Willig, M. R., and G. R. Camilo. 1991. The effect of Hurricane Hugo on six invertebrate species in the Luquillo Experimental Forest of Puerto Rico. *Biotropica* 23: 455–461.

Willig, M. R., and M. A. McGinley. 1999. Animal responses to natural disturbance and roles as patch generating phenomena. In *Ecosystems of the World: Ecosystems of Disturbed Ground* (L. R. Walker, Ed.), pp. 667–689. Elsevier Science, Amsterdam, Netherlands (in press).

Willig, M. R., and S. K. Lyons. 1998. An analytical model of latitudinal gradient in species richness with an empirical test for marsupials and bats in the New World. *Oikos* 81: 93–98.

Willig, M. R., and L. R. Walker. 1999. Disturbance in terrestrial ecosystems: Salient themes, synthesis, and future directions. In *Ecosystems of the World: Ecosystems of Disturbed Ground* (L. R. Walker, Ed.). Elsevier Science, Amsterdam, Netherlands (in press).

Willmer, P. G., J. P. Hughes, J. A. T. Woodford, and S. C. Gordon. 1996. The effects of crop microclimate and associated physiological contraints on the seasonal and diurnal distribution patterns of raspberry beetle (*Byturus tomentosus*) on the host plant *Rubus idaeus*. *Ecological Entomology* 21: 87–97.

Wilson, D. S. 1976. Evolution on the level of communities. *Science* 192: 1358–1360.

Wilson, E. O. 1969. The species equilibrium. In *Diversity and Stability in Ecological Systems* (G. M. Woodwell and H. H. Smith, Eds.), pp. 38–47. Brookhaven Symposium in Biology 22, Brookhaven National Laboratory, Upton, NY.

Wilson, E. O. 1973. Group selection and its significance for ecology. *BioScience* 23: 631–638.

Wilson, E. O. 1975. *Sociobiology: the New Synthesis*. Belknap Press of Harvard Univ. Press, Cambridge, MA.

Wilson, E. O. 1992. *The Diversity of Life*. Harvard Univ. Press, Cambridge, MA.

Wilson, E. O., and T. Eisner. 1957. Quantitative studies of liquid food transmission in ants. *Insectes Sociaux* 4: 157–166.

Wilson, E. O., and D. S. Simberloff. 1969. Experimental zoogeography of islands: Defaunation and monitoring techniques. *Ecology* 50: 267–278.

Wilson, M. V., P. C. Hammond, and C. B. Schultz. 1997. The interdependence of native plants and Fender's blue butterfly. In *Conservation and Management of Native Plants and Fungi* (T. N. Kaye, A. Liston, R. M. Love, D. L. Luoma, R. J. Meinke, and M. V. Wilson, Eds.), pp. 83–87. Native Plant Society of Oregon, Corvallis, OR.

Winchester, N. N. 1997. Canopy arthropods of coastal Sitka spruce trees on Vancouver Island, British Columbia, Canada. In *Canopy Arthropods* (N. E. Stork, J. Adis, and R. K. Didham, Eds.), pp. 151–168. Chapman & Hall, London, UK.

Windsor, D. M. 1990. *Climate and Moisture Variability in a Tropical Forest: Long-term Records from Barro Colorado Island, Panamá*. Smithsonian Institution Press, Washington, DC.

Wint, G. R. W. 1983. Leaf damage in tropical rain forest canopies. In *Tropical Rain Forest: Ecology and Management* (S. L. Sutton, T. C. Whitmore, and A. C. Chadwick Eds.), pp. 229–240. Blackwell Scientific, Oxford, UK.

Winter, K., and J. A. C. Smith. 1996. An introduction to crassulacean acid metabolism: Biochemical prin-

ciples and ecological diversity. In *Crassulacean Acid Metabolism: Biochemistry, Ecophysiology, and Evolution* (K. Winter and J. A. C. Smith, Eds.), pp. 1–13. Springer-Verlag, New York.

Wisdom, C. S., C. S. Crawford, and E. F. Aldon. 1989. Influence of insect herbivory on photosynthetic area and reproduction in *Gutierrezia* species. *Journal of Ecology* 77: 685- 692.

Wise, D. H. 1975. Food limitation of the spider *Linyphia marginata*: Experimental field studies. *Ecology* 56: 637–646.

Wise, D. H., and M. Schaefer. 1994. Decomposition of leaf litter in a mull beech forest: comparison between canopy and herbaceous species. *Pedobiologia* 38: 269–288.

Witcosky, J. J., T. D. Schowalter, and E. M. Hansen. 1986. The influence of time of precommercial thinning on the colonization of Douglas-fir by three species of root-colonizing insects. *Canadian Journal of Forest Research* 16: 745–749.

Witkamp, M. 1971. Soils as components of ecosystems. *Annual Review of Ecology and Systematics* 2: 85–110.

Witkamp, M., and D. A. Crossley, Jr. 1966. The role of microarthropods and microflora in the breakdown of white oak litter. *Pedobiologia* 6: 293–303.

Witkamp, M., and J. S. Olson. 1963. Breakdown of confined and unconfined oak litter. *Oikos* 14: 138–147.

Wold, E. N., and R. J. Marquis. 1997. Induced defense in white oak: Effects on herbivores and consequences for the plant. *Ecology* 78: 1356–1369.

Wood, D. M., and M. C. Andersen. 1990. The effect of predispersal seed predators on colonization of *Aster ledophyllus* on Mount St. Helens, Washington. *American Midland Naturalist* 123: 193–201.

Woods, P. V., and R. J. Raison. 1982. An appraisal of techniques for the study of litter decomposition in eucalypt forests. *Australian Journal of Ecology* 7: 215–225.

Woods, P. V., and R. J. Raison. 1983. Decomposition of litter in sub-alpine forests of *Eucalyptus delegatensis, E. pauciflora* and *E. dives*. *Australian Journal of Ecology* 8: 287–299.

Woodwell, F. I. 1993. How many species are required for a functional ecosystem? In *Biodiversity and Ecosystem Function* (E. D. Schulze and H. A. Mooney, Eds.), pp. 271–291. Springer-Verlag, Berlin, Germany.

Woolhouse, H. W. 1981. Aspects of the carbon and energy requirements of photosynthesis considered in relation to environmental constraints. In *Physiological Ecology: An Evolutionary Approach to Resource Use* (C.R. Townsend and P. Calow, Eds.), pp. 51–85. Blackwell Scientific, Oxford, UK.

Wotton, R. S., B. Malmqvist, T. Muotka, and K. Larsson. 1998. Fecal pellets from a dense aggregation of suspension-feeders in a stream: an example of ecosystem engineering. *Limnology and Oceanography* **43**: 719–725.

Wynne-Edwards, V. C. 1963. Intergroup selection in the evolution of social systems. *Nature* **200**: 623–626.

Wynne-Edwards, V. C. 1965. Self-regulating systems in populations of animals. *Science* **147**: 1543–1548.

Yoder, J. A., G. C. Theriot, and D. B. Rivers. 1996. Venom from *Nasonia vitripennis* alters water loss from the flesh fly, *Sarcophaga bullata*. *Entomologia Experimentalis et Applicata* **81**: 235–238.

Yodzis, P. 1980. The connectance of real ecosystems. *Nature* **284**: 544–545.

Zera, A. J., and R. F. Denno. 1997. Physiology and ecology of dispersal polymorphism in insects. *Annual Review of Entomology* **42**: 207–230.

Zeringue, H. J., Jr. 1987. Changes in cotton leaf chemistry induced by volatile elicitors. *Phytochemistry* **26**: 1357–1360.

Zhong, H., and T. D. Schowalter. 1989. Conifer bole utilization by wood-boring beetles in western Oregon. *Canadian Journal of Forest Research* **19**: 943–947.

Zimmerman, J. K., M. R. Willig, L. R. Walker, and W. L. Silver. 1996. Introduction: disturbance and Caribbean ecosystems. *Biotropica* **28**: 414–423.

Zimmerman, P. R., J. P. Greenberg, S. O. Wandiga, and P. J. Crutzen. 1982. Termites: A potentially large source of atmospheric methane, carbon dioxide, and molecular hydrogen. *Science* **218**: 563–565.

Zlotin, R. I., and K. S. Khodashova. 1980. *The Role of Animals in Biological Cycling of Forest–steppe Ecosystems* [English translation by N. R. French]. Dowden, Hutchinson & Ross, Stroudsburg, PA.

AUTHOR INDEX

TAXONOMIC INDEX

SUBJECT INDEX